IGNEOUS PETROGENESIS

Marjorie Wilson

Department of Earth Sciences, University of Leeds

KLUWER ACADEMIC PUBLISHERS

DORDRECHT / BOSTON / LONDON

Published by Chapman & Hall, 2-6 Boundary Row, London SE1 8HN, UK

Published by Kluwer Academic Publishers,
P.O. Box 17, 3300 AA Dordrecht, The Netherlands.

Sold and distributed in North, Central and South America
by Kluwer Academic Publishers,
101 Philip Drive, Norwell, MA 02061, U.S.A.

In all other countries, sold and distributed
by Kluwer Academic Publishers,
P.O. Box 322, 3300 AH Dordrecht, The Netherlands.

Originally published by Chapman & Hall

First edition 1989
Reprinted 1991, 1993, 1994, 1995, 1996, 1997, 2000

© 1989 Marjorie Wilson

Typeset in 10/12 Plantin light by The Design Team of Ascot
Printed in the Netherlands

ISBN 0 412 53310 3 (PB)

A Catalogue record for this book is available from the British Library

Library of Congress Cataloging-in-Publication Data available

Printed on acid-free paper

This work is dedicated to my mother, Blanche Pearce, for a lifetime of love and understanding, and to Mary and Andrew

PREFACE

My main objective in writing this book has been to review the processes involved in present-day magma generation and their relationship to global tectonic processes. Clearly, these are fundamental to our understanding of the petrogenesis of ancient volcanic and plutonic sequences, the original tectonic setting of which may have been obscured by subsequent deformation and metamorphism.

Until fairly recently, undergraduate courses in igneous petrology tended to follow rather classical lines, based on the classification of igneous rocks, descriptive petrography, volcanic landforms, types of igneous intrusions and regional petrology. However, the geologist of the late 1980s requires, in addition, an insight into the fundamental processes which operate within the Earth, and are responsible for the diversity of igneous rocks presently exposed. Unfortunately, attempts to make petrology

courses more petrogenesis-orientated arc immediately confronted with a basic problem; the average student does not have a strong enough background in geochemistry to understand the finer points of most of the relevant publications in scientific journals. It is virtually impossible to find suitable reading material for such students, as most authors of igneous petrology textbooks have deliberately steered clear of potentially controversial petrogenetic models. Even the most recent texts place very little emphasis on the geochemistry of magmas erupted in different tectonic settings, despite extensive discussions of the processes responsible for the chemical diversity of magmas. Perhaps even more surprisingly, few texts have attempted to relate igneous activity to global tectonic processes, except at the most elementary levels.

This book has arisen out of ten years of teaching igneous petrology to undergraduate and graduate students in the Department of Earth Sciences, University of Leeds, UK, and the resultant frustration at having no appropriate text which discusses the geochemical characteristics of magmas in relationship to their tectonic setting. It should appeal to the more advanced undergraduate and beginning postgraduate students, and their teachers, who already have a basic foundation in petrological principles, including elementary geochemistry, phase diagrams, mineralogy, regional geology and global tectonics. The text has been deliberately put together in such a way that certain sections can be omitted at first reading if the background of the student is not appropriate. Thus, for example, a beginning student could omit the more detailed sections on the trace element and isotope geochemistry of magmas generated in a particular tectonic setting. As far as possible, continuity in the presentation of data for individual provinces has been maintained for ease of inter-province comparisions. Thus, in each chapter, sections will be found covering, for example, major element, trace element and radiogenic isotope geochemistry.

Clearly, a text of this length cannot hope to cover all examples of a particular type of igneous activity. Consequently, I have tended to focus on volcanic rocks from those recently active provinces for which we have the most comprehensive geochemical data sets. Obviously, new data are appearing all the time and therefore the shape of many of the fields depicted in the geochemical variation diagrams may change with time. Nevertheless, I believe that we have reached a stage in our understanding of the petrogenesis of igneous rocks at which most of the fundamental principles are well established. Consequently, the time is right for a text of this nature. I have deliberately not devoted much space to the discussion of topics such as phase diagrams and the fluid properties of magmas, as these have been extensively covered in several recent texts.

I would like to thank the following for their help and encouragement throughout the completion of this project: K. G. Cox for his much valued comments on the original manuscript; P.J. Wyllie and J.B. Gill for useful discussion of the format of the book and the early chapters; my colleagues in the Department of Earth Sciences, G.R. Davies, P.H. Nixon, E.H. Francis and R.A. Cliff, for their enthusiastic approach to all aspects of igneous petrology which has led to many fruitful discussions; S. Caunt for word processing the large number of references, and for much needed moral support during the final stages.

I have been fortunate to have had a particularly lively group of research students over the past ten years, including I. Luff, Z. Palacz, H. Downes, J.P. Davidson, I. Wilson, D. Chaffey, S. Amini and S. Caunt, who have kept me on my toes. Many people, including the above, have contributed to my ideas on igneous petrogenesis over the years, in particular R. Powell, J.D. Bell, S.W. Richardson, K.G. Cox, M.J. O'Hara, P.E. Baker, C.J. Hawkesworth, M.J. Norry, the late P.J. Betton, I.S.E. Carmichael, A. Ewart, W. Hildreth, B.D. Marsh, C. Bacon, H. Helgeson, G.M. Brown and C.H. Emeleus.

Thanks must go finally to my family and friends for putting up with my reclusive habits over the past few months, especially to my mother, my sister Mary and her husband Andrew, my step-children Jane and Gary (and step son-in-law David), the latest addition to the family, my step-grandson Anthony, and to Anne and Brian Frost. Thanks also to Roger Jones of Unwin Hyman for being such an understanding editor.

Marjorie Wilson
Leeds

ACKNOWLEDGEMENTS

We are grateful to the following individuals and organizations who have kindly given permission for the reproduction of copyright material (figure numbers in parentheses):
K. G. Cox (1.2, 3.23, 4.8, 4.9); G. Faure and by permission of John Wiley & Sons (2.13); J. H. Pearce and Elsevier (2.14); Elsevier (2.15, 2.16, 3.12, 5.26, 6.19, 8.7, 12.5); D. A. Wood and Elsevier (2.17); P. J. Wyllie (3.5, 3.14, 3.19, 6.10); A. A. Finnerty and by permission of John Wiley & Sons Ltd (3.9); B. Harte (3.10); Lunar and Planetary Institute (3.13, 5.31, 6.15, 9.10); A. Zindler and by permission from *Nature* **298**, Copyright © 1982 McMillan Magazines Ltd (3.16, 3.17); G. C. Brown (3.18, 5.1, 5.3, 5.4, 5.8); H. S. Yoder Jr and National Academy of Sciences (3.22); D. S. Barker (3.20); D. Green and by permission of the Director, Bureau of Mineral Resources, Geology and Geophysics, Canberra; Springer-Verlag (3.26, 3.27, 3.29); Figure 3.28 Copyright Mineralogical Society of America; R. N. Thompson and Carnegie Institution of Washington (4.2); Z. A. Palacz and Elsevier (4.4, 4.6); Figures 5.5 and 5.13 adapted from *Earth* (F. Press and R. Siever) by permission of W. H. Freeman and Company, © 1974, 1978, 1982 W. H. Freeman and Company; M. G. Best (5.7, 5.24); R. D. Ballard and Elsevier (5.9, 5.12); R. Hekinian and Elsevier (5.10, 5.21, 5.30); R. S. White and by permission of the Geological Society (5.14); J. M. Sinton and Elsevier (5.16); S. Le Douaran and Elsevier (5.17, 5.46); P. Thy and Springer-Verlag (5.18); J.-G. Schilling and by permission from *Nature* **313**, Copyright © 1985 McMillan Magazines Ltd (5.19, 5.49); J. M. Edmond and Copyright © 1983 Scientific American, Inc. (5.23); Figure 5.25 adapted by permission from Ito *et al.*, *Acta* **47**, Copyright © 1983 Pergamon Journals Ltd; J. R. Cann and Elsevier (5.27); E. G. Nisbet and Royal Astronomical Society (5.28); J. F. G. Wilkinson (5.33); H. Staudigel and Elsevier (5.43, 5.44, 5.45); A. Zindler and Elsevier (5.47); J. C. Allegre and Elsevier (5.48); P. J. Wyllie and by permission from *Geological Society of America Bulletin* **93** (6.3, 6.11); J. B. Gill (6.4, 6.5, 6.9, 6.12, 7.6); H. K. Acharya (6.6); R. N. Anderson and *The Journal of Geology* **88**, Copyright © 1980, by permission of the University of Chicago Press (6.7, 6.8); V. Renard and Elsevier (6.11); A. Ewart and by permission of John Wiley & Sons Ltd (6.28, 6.30, 6.31); C. J. Hughes and Elsevier (6.39); R. J. Arculus (6.40); C. J. Allegre and by permission from *Nature* **299**, Copyright © 1982 McMillan Magazines Ltd (6.46); Figures 6.47 and 6.48 reproduced with permission from *Annual Review of Earth and Planetary Sciences* **9**, © 1981 Annual Reviews Inc; R. H. Pilger and by permission from *Geological Society of America Bulletin* **93** (7.3); C. Condie and Copyright © 1982 Pergamon Books Ltd (7.8); P. J. Wyllie and The Royal Society (7.9, 7.10, 7.11); W. S. Pitcher and Blackie & Son (7.12); R. D. Beckinsale and Blackie & Son (7.13); W. S. Pitcher (7.14); R. S. Harmon and by permission of the Geological Society (7.19,

7.31); J. A. Pearce (7.25, 7.26); R. J. Stern and by permission from *Geological Society of America Bulletin* **93** (8.1); A. J. Crawford and Elsevier (8.2); Figure 8.11 adapted with permission from Woodhead and Fraser, *Acta* **49** Copyright © 1985 Pergamon Journals Ltd; K. C. Burke (9.1, 9.2, 9.4); E. M. Parmentier (9.6); A. B. Watts and by permission from *Nature* **315**, Copyright © 1985 McMillan Magazines Ltd (9.7); M. P. Ryan (9.8); Figure 9.19 adapted with permission from Humphris *et al*, *Acta* **49**, Copyright © 1985 Pergamon Journals Ltd; S. R. Hart and by permission from *Nature* **309**, Copyright © 1984 McMillan Magazines Ltd (9.27); Figure 9.31 adapted with permission from Kurz *et al.*, *Nature* **297**, Copyright © 1982 McMillan Magazines Ltd; K. G. Cox and by permission from *Nature* **272**, Copyright © 1978 McMillan Magazines Ltd (10.1); M. A. Menzies and The Royal Society (10.3); K. G. Cox and Oxford University Press (10.4, 10.13); R. B. Smith and National Academy Press (10.6); F. Barberi (11.2); C. E. Keen (11.4); J. D. Fairhead and Elsevier (11.5); R. E. Long (11.7a); R. W. Girdler and Elsevier (11.7b); Y. A. Sinno (11.8); W. Bosworth and by permission from *Nature* **316**, Copyright © 1985 McMillan Magazines Ltd (11.11); Figure 11.33 adapted with permission from McDonough *et al.*, *Acta* **49**, Copyright © Pergamon Journals Ltd; Plenum Press (12.3); C. B. Smith and the Geological Society of South Africa (12.14); S. C. Bergman and by permission of the Geological Society (12.15); R. N. Thompson and Springer-Verlag (12.18); A. LeRoex and by permission from *Nature* **324**, Copyright © 1986 McMillan Magazines Ltd (12.19, 12.22); Figure 12.21 adapted with permission from Nelson *et al.*, *Acta* **50**, Copyright © Pergamon Journals Ltd; G. Ferrara and Springer-Verlag (12.27); D. K. Bailey and the Geological Society of South Africa (12.28, 12.29).

CONTENTS

LIST OF TABLES

SOME COMMON ABBREVIATIONS USED IN THE TEXT

A	alkalic	**Elements**	
AFM	the variation diagram of F (FeO + Fe_2O_3) – M (MgO) – A (Na_2O + K_2O)		
BABI	*Basaltic Achondrite Best Initial*; the best estimate of the initial $^{87}Sr/^{86}Sr$ ratio of basaltic achondrite meteorites, assumed to be similar to the primordial isotopic composition of Sr in the Earth	Al	aluminium
		Ar	argon
		Ba	barium
		Be	beryllium
CA	calc-alkaline	Ca	calcium
CFB	continental flood basalt	C	carbon
CHUR	chondritic uniform reservoir	Ce	cerium
CMAS	the system $CaO-MgO-Al_2O_3-SiO_2$	Cl	chlorine
D	the distribution or partition coefficient for the partitioning of a trace element between a mineral and a silicate melt (see Appendix)	Cr	chromium
		Co	cobalt
		Cu	copper
DM	depleted mantle	Dy	dysprosium
DSDP	Deep Sea Drilling Project	Er	erbium
EM	enriched mantle	Eu	europium
EPR	East Pacific Rise	F	fluorine
fO_2	fugacity of oxygen	Gd	gadolinium
Ga	thousand million years	Au	gold
IAB	island-arc basalt	Hf	hafnium
Ma	million years	He	helium
MAR	Mid-Atlantic Ridge	Ho	holmium
MORB	mid-ocean ridge basalt	H	hydrogen
OIB	ocean-island basalt	Fe	iron
P	pressure (kbar)	La	lanthanum
REE	rare earth element (LREE = light REE, HREE = heavy REE)	Pb	lead
		Li	lithium
SMOW	standard mean ocean water	Lu	lutetium
T	Temperature (°C)	Mg	magnesium
TH	tholeiitic		
V_P	seismic velocity of P waves		
V_S	seismic velocity of S waves		
ρ	density (g cm^{-3})		

Mn	manganese
Nd	neodymium
Ni	nickel
Nb	niobium
N	nitrogen
O	oxygen
P	phosphorus
K	potassium
Pr	praesodymium
Rb	rubidium
Sm	samarium
Sc	scandium
Si	silicon
Ag	silver
Na	sodium
Sr	strontium
S	sulphur
Ta	tantalum
Tb	terbium
Th	thorium
Tm	thulium
Ti	titanium
U	uranium
V	vanadium
Yb	ytterbium
Y	yttrium
Zn	zinc
Zr	zirconium

PART ONE

Magmatism and global tectonic processes

CHAPTER ONE

Relation of present-day magmatism to global tectonic processes

1.1 Introduction

Petrogenetic studies of igneous rocks involve characterization of the source regions of the magmas, the conditions of partial melting (Ch. 3), and the extent of subsequent modification of primary mantle derived magmas during transport and storage in high-level magma chambers (Ch. 4). Such studies must be based on sound field observations, involving careful mapping and sampling of the range of rock types exposed at a particular locality, and on a comprehensive knowledge of the petrography, major, minor and trace element and radiogenic and stable isotope geochemistry of the samples. Additionally, if the igneous activity is not recent, its age must be constrained, ideally by isotopic dating techniques.

The main emphasis in this text is on the petrogenesis of recent volcanic rocks, the origins of which can be related directly to their global tectonic setting. In general, associated plutonic rocks are not considered in any detail, as these are much harder to investigate geochemically and in active volcanic provinces are rarely exposed. Clearly, without a detailed understanding of present-day magma generation processes there is little chance of interpreting the formation conditions of older volcanic and plutonic rocks within the geological record. However, we must be cautious in assuming that the present is necessarily always the key to the past. While this may be essentially correct for the Phanerozoic (< 600 Ma), Precambrian magma generation processes may have been significantly different from those of today. For example, during the Archaean geothermal gradients would have been much steeper, causing greater degrees of

4 PRESENT-DAY MAGMATISM AND GLOBAL TECTONIC PROCESSES

partial melting at shallower depths, and the scale of plate tectonics may have been much smaller, with large numbers of microplates moving at higher velocities. Obviously, if we can correlate particular geochemical characteristics of modern volcanic rocks with specific tectonic settings we can use these to characterize the palaeotectonic setting of older volcanic sequences, in which deformation and metamorphism may have obscured the original tectonic context. However, as we shall see in Chapter 2, such an approach is unlikely to give an unambiguous determination of tectonic setting by itself.

The theory of plate tectonics provides an excellent framework for the discussion of the different styles and geochemical characteristics of present-day igneous activity. For those readers unfamiliar with the more detailed concepts involved, Cox & Hart (1986) provides an excellent introduction. The relatively thin lithosphere of the Earth (Ch. 3), encompassing part of the uppermost mantle and the crust, is divided into about a dozen separate plates (Fig. 1.1). These have thicknesses ranging from only a few kilometres close to the axes of the mid-oceanic ridges to greater than 200 km beneath some stable continental regions, and horizontal dimensions of thousands of kilometres. They move as a consequence of plastic flow or convection within the underlying asthenosphere (Ch. 3), although the relationship is by no means simple. More than 90% of all recent igneous activity is located at or near the boundaries of these lithospheric plates.

On the basis of tectonic setting we can define four distinct environments in which magmas may be generated:

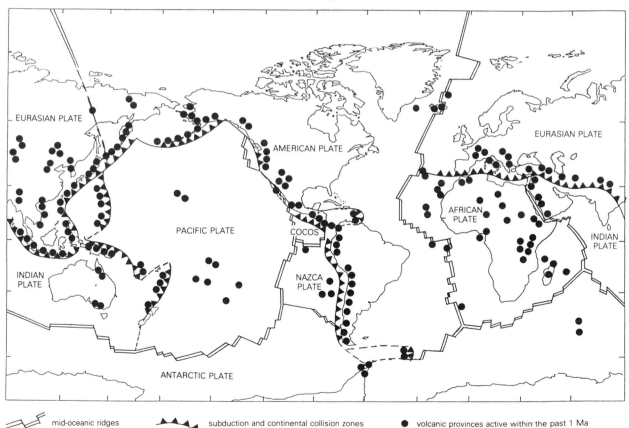

Figure 1.1 Global tectonic map showing the distribution of present-day volcanic activity. Plate boundaries after Condie (1982). Locations of volcanic provinces active within the past 1 Ma after Best (1982).

(1) *Constructive plate margins:* these divergent plate boundaries include the system of mid-oceanic ridges (Ch. 5) and back-arc spreading centres (Ch. 8).

(2) *Destructive plate margins:* these convergent plate boundaries include island arcs (Ch. 6) and active continental margins (Ch. 7).

(3) *Oceanic intra-plate settings:* oceanic islands (Ch. 9).

(4) *Continental intra-plate settings:* including continental flood basalt provinces (Ch. 10), continental rift zones (Ch. 11), and occurrences of potassic and ultrapotassic magmatism (including kimberlites) not related to zones of rifting (Ch. 12).

Estimates of magma production rates (in $km^3 yr^{-1}$) in these different environments are compared in Table 1.1.

Volumetrically, the most important sites of magma generation at the present time are the divergent plate boundaries or mid-oceanic ridges. Here, diapiric upwelling of upper mantle material induces partial melting by adiabatic decompression (Ch. 3) to form basaltic magma (MORB – *Mid-Ocean Ridge Basalt*) which is both erupted and intruded to form the oceanic crustal layer. Newly formed crust continuously pushes the older crust away from the ridge axis and thus the age of the oceanic crust increases symmetrically away from the ridge, as shown by the system of magnetic stripes on the ocean floor (Ch. 5). The oceanic crust

is metamorphosed to an uncertain extent by interaction with the hydrosphere, particularly as a consequence of deep convective penetration of sea water into the hot basaltic crust close to the axes of the mid-oceanic ridges. This involves both hydration and chemical exchange of some mobile elements between the basalt and sea water.

As the oceanic plate moves away from the axis of the mid-oceanic ridge it cools and thickens, and may occasionally be injected with basaltic magma, generated in localized mantle upwellings or hot spots, which may result in the formation of oceanic islands (Ch. 9). The migration of a relatively fast-moving oceanic plate over such a hot spot may produce linear chains of islands such as the Hawaiian islands.

As they increase in age, oceanic plates eventually become so dense, as a consequence of cooling, that they sink back into the mantle at convergent plate boundaries or subduction zones (Part 3). As indicated in Table 1.1, such boundaries are the second greatest magma producers at the present time, and most of the world's sub-aerially active volcanoes and earthquakes, including nearly all those with intermediate and deep foci, are associated with them. Upon subduction, basaltic rocks of the oceanic crustal layer, and any sediments which resist scraping off at the trench, are progressively heated as they are transported deeper into the mantle, and undergo a complex series of dehydration reactions which reverse the effects of previous ocean-floor metamorphism. Ultimately, they may become hot enough to partially melt. Hydrous fluids, released during dehydration, or silica-rich partial melts of the oceanic crustal layer, ascend into the mantle wedge above the subducted slab where they may induce partial melting. The resultant hydrous basaltic magmas then rise up into the crust where they differentiate in subvolcanic magma chambers to form a range of more silica-rich magma types. Volcanism in this environment may be highly explosive due to the high volatile (H_2O) contents of the magmas.

Plate convergence is clearly polar and produces asymmetric patterns of magmatism, tectonism and metamorphism in the overriding plate. If this is an oceanic plate (Ch. 6) linear chains of volcanic

Table 1.1 Global rates of Cenozoic magmatism (after McBirney 1984).

Location	Rate ($km^3 yr^{-1}$)	
	Volcanic rocks	Plutonic rocks
constructive plate boundaries	3	18
destructive plate boundaries	0.4−0.6	2.5−8.0
continental intra-plate	0.03−0.1	0.1−1.5
oceanic intra-plate	0.3−0.4	1.5−2.0
global total	3.7−4.1	22.1−29.5

islands develop forming an island arc, whereas if it is a continental plate the volcanism creates an active continental margin (Ch. 7). Magma generation in these two environments is broadly similar, but in the latter the geochemical characteristics of the magmas may be modified by continental crustal contamination. Behind some volcanic arcs secondary seafloor spreading occurs, resulting in the development of back-arc or marginal basins (Ch. 8). Processes here are similar to those operating at mid-oceanic ridges, although the resultant basaltic magmas may be more complex geochemically, having characteristics of both destructive and constructive margin tectonic settings.

The process of subduction recycles material which has been in equilibrium with the continental crust (sea water or pelagic sediments formed in equilibrium with sea water) or has been derived from it (oceanic terrigenous sediments) back into the mantle. This has fundamental implications for the long-term chemical evolution of the mantle. Additionally, some authors (Ch. 9) have suggested that recycled oceanic lithosphere may provide a source component for some oceanic-island basalts. Subduction-related magmatism appears to be the dominant mechanism for crustal growth at the present, and probably throughout the Phanerozoic.

In terms of Table 1.1 we can clearly see that much of present-day volcanic activity is concentrated within or adjacent to zones of plate divergence or convergence. Nevertheless, within both oceanic and continental plates active volcanoes occur, often at considerable distances from plate boundaries, forming intra(within)-plate volcanic provinces (Part 4). Magmatism within continental plates is volumetrically insignificant at the present time, being primarily associated with intra-plate rift systems such as the East African rift (Ch. 11). However, this may have been a much more important magma generation environment in the past, for example during the formation of large-scale continental flood basalt provinces (Ch. 10) which appear to predate continental fragmentation. Similarly, kimberlite magmatism (Ch. 12) is not a present-day phenomenon.

In general, the preservation potential of igneous rocks generated in oceanic-island, island-arc or mid-oceanic ridge tectonic settings is low, as these form part of the oceanic lithosphere which is recycled back into the mantle on a timescale of the order of 100 Ma. Only rarely are these rocks preserved as obducted slices in ophiolite complexes (Ch. 5). As a consequence intracontinental plate and active continental margin igneous rocks should, in theory, be more common within the geological record.

It is generally accepted that partial melting of mantle material produces primary magmas of basic or ultrabasic composition in most tectonic settings (Ch. 3), and that subsequent differentiation processes, including fractional crystallization, magma mixing and crustal contamination (Ch. 4), are responsible for the generation of the wide compositional spectrum of terrestrial igneous rocks. The geochemical characteristics of these primary magmas depend upon parameters such as the source composition and mineralogy and the depth and degree of partial melting; factors which may vary from one tectonic setting to another. Primary magmas appear to be generated within a very restricted depth range within the upper $100-200$ km of the mantle, although in detail their precise depths of origin are poorly constrained. Diamond-bearing kimberlites (Ch. 12) are probably the deepest terrestrial magmas, originating from depths greater than $200-250$ km.

A fundamental aim of petrogenetic studies of igneous rocks is to distinguish source characteristics that are inherited by the primary partial melts at their depth of segregation from those arising from subsequent processes. Variations in the isotopic compositions of Sr, Nd and Pb in oceanic basalts (Ch. 5 & 9) have provided important constraints on the structure and compositional heterogeneity of the upper mantle. In particular, they have provided support for a two-layer convection model (Ch. 3) with the boundary between the two layers probably located at the 670 km seismic discontinuity. The upper layer is depleted in incompatible elements and is the source of MORB. The formation of the continental crust throughout geological time is generally held responsible for the depletion of this layer. In contrast, the lower layer is considered to be less depleted and must have been isolated from the upper layer for periods in excess of 1 Ga (Ch. 9). It may contain near-primordial mantle material unmodified by partial melting processes since the formation of the Earth. Additionally, it may be internally heterogeneous due to the migration of partial melts and fluids. Although normal MORB (Ch. 5) are fairly uniform in terms of their trace element and isotopic characteristics, oceanic-island basalts (OIB; Ch. 9) are not. OIB sources may involve mixtures of the depleted and enriched mantle layers in addition to components derived from recycled oceanic crustal material and subcontinental lithosphere.

Within the ocean basins the relatively refractory mantle part of the oceanic lithosphere is not generally considered to play a dominant role in magma generation processes. Nevertheless, some oceanic-island basalts do appear to contain a significant lithospheric component (Ch. 9). In contrast, the subcontinental lithosphere may be the dominant source of some intracontinental plate volcanic rocks (Ch. 10−12). Unlike the oceanic lithosphere, the stable mantle keel beneath some cratonic nuclei may have been attached to the overlying crust for periods of 1−2 Ga and, during this time, may have developed significant geochemical heterogeneities due to the generation and migration of magmas and fluids. The subcontinental lithosphere may potentially be regarded as a reservoir of enriched mantle material.

There is clearly a close correlation between the processes of magma generation in some tectonic settings and the formation of economic ore deposits. Hutchison (1983) has presented an extensive review of this subject, and the reader is referred to this work for further details. Hydrothermal fluids play a fundamental role in the formation of ore deposits and are characteristically associated with certain tectonic settings. For example, detailed studies of the ocean floors have revealed that oceanic sediments may become enriched in metals while they are accumulating. The agent responsible is hot brine, which circulates through the unconsolidated sediments and the underlying basaltic oceanic crust. The circulation is driven by localized high heat fluxes associated with spreading axes. Examples include small-scale spreading centres in

the Red Sea and Salton Sea, in addition to many areas along the crests of active mid-oceanic ridges (Ch. 5), where hydrothermal fluids have been observed to vent onto the ocean floor, producing so-called 'black smokers'.

Volcanic activity in island arcs is also likely to drive circulation cells of sea water within the submarine volcanic pile. These may emanate on to the ocean floor to form submarine hydrothermal ore deposits within sequences of tuffaceous sediments derived from the arc volcanism, known as Kuroko deposits. Porphyry copper type ore deposits may occur when the arc volcanoes are situated on continental or thickened island-arc crust. These are large, low-grade deposits of disseminated copper and iron sulphides, often associated with Au, Mo and Ag, occuring in and around the roof zones of batholiths, which represent the plutonic root zones of former stratovolcanoes (Ch. 7). In both Kuroko and porphyry copper type deposits, the hydrothermal fluids may originally be of magmatic origin but quickly become diluted by meteoric water. As the circulation cells enlarge the hydrothermal fluids may leach metals from older volcanic and volcaniclastic rocks, which then become concentrated in the ore body.

1.2 Characteristic magma series associated with specific tectonic settings

A Harker variation diagram (Ch. 2) of wt. % Na_2O + K_2O versus wt.% SiO_2 (Fig. 1.2a) provides a useful way of displaying the wide compositional range of terrestrial volcanic rocks and their nomenclature. This diagram is based on that of Cox et al. (1979), and is consistent with that proposed by the IUGS Subcommission on the Systematics of Igneous Rocks (Le Bas et al. 1986). A simple diagram like this is preferable in the classification of igneous rocks as it makes direct use of their major element chemical composition, expressed in terms of wt. % constituent oxides. The reader may be surprised (or perhaps relieved) to find no mention of norms in this text. They have little intrinsic value in the description of the geochemical characteristics

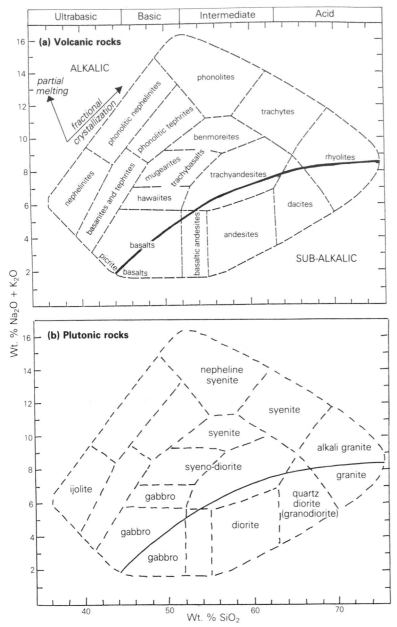

Figure 1.2 Nomenclature of normal (i.e. non-potassic) igneous rocks)(a) after Cox *et al.* 1979). The dividing line between alkalic and sub-alkalic magma series is from Miyashiro (1978).

of igneous rocks within a modern petrogenetic framework and consequently, with due regard to the memory of Cross, Iddings, Pirsson and Washington (1903), have been laid to rest. Figure 1.2b shows the nomenclature which we will use to describe the plutonic equivalents of the volcanic rocks shown in Fig 1.2a. This is consistent with the QAPF classification, based on modal proportions of constituent minerals (Streckeisen 1976). Figure 1.2a is only applicable to the classification of the commonly occurring non-potassic volcanic rocks, and the nomenclature of their rarer potassic equivalents is given in Table 1.2. Clearly, Figure 1.2a should only be used to classify fresh unmetamorphosed volcanic rocks, as alkalis are mobile during weathering and metamorphism.

Table 1.2 Approximate equivalents of some K-rich and normal rock types (after Cox *et al.* 1979).

Potassic	Normal
leucitophyre	phonolite
K-trachyte	trachyte
K-rhyolite	rhyolite
tristanite	benmoreite
latite	trachyandesite
leucitite	nephelinite
leucite basanite	basanite
leucite tephrite	tephrite
absarokite shoshonite }	basalt

In terms of Figure 1.2a, volcanic rocks may be subdivided into members of two major magma series, alkalic and sub-alkalic, separated by the solid line in this diagram. Such a division was first proposed by Iddings in 1892 and elaborated upon by Harker (1909), Macdonald & Katsura (1964), Macdonald (1968), Irvine & Baragar (1971) and Miyashiro (1978). Each of these magma series contains rocks ranging from basic to acid in composition, and although the boundary between them is marked as a solid line it is actually gradational. Broadly speaking, the compositional range of volcanic rocks displayed in this diagram may be regarded as a consequence of two fundamental processes (indicated by vector arrows), partial melting and fractional crystallization. As we shall see in Chapters 5—12, these are not the only processes responsible for the compositional diversity of magmas but they are almost certainly the dominant ones.

Diagrams of wt. % K_2O and Na_2O versus wt. % SiO_2 (Fig. 1.3) may also be used to differentiate between basaltic members of the alkalic and sub-alkalic series (Middlemost 1975). Occasionally, samples may plot in the alkalic field on one diagram and in the sub-alkalic field on the other, and these are termed *transitional basalts*. In terms of Figure 1.3a, sub-alkalic basalts can be further subdivided into normal and low-K types.

In general, the sub-alkalic magma series can be subdivided into a high alumina or calc-alkali series and a low-K tholeiitic series. The basaltic end-members of these series plot respectively within the

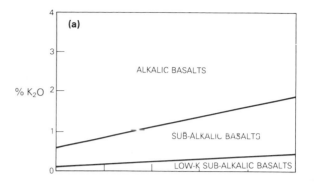

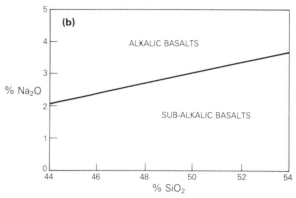

Figure 1.3 Classification of alkalic and sub-alkalic basalts in terms of (a) wt. % K_2O versus wt. % SiO_2 and (b) wt. % Na_2O versus wt. % SiO_2 (after Middlemost 1975).

sub-alkalic and low-K sub-alkalic fields in Figure 1.3a. The two series can be differentiated in terms of their trends on an AFM diagram (Fig. 1.4), in which tholeiitic suites commonly show a strong trend of iron enrichment in the early stages of differentiation, whereas calc-alkaline suites trend directly across the diagram, due to the suppression of iron enrichment by the early crystallization of Fe-Ti oxides. Additionally, a plot of Alkali Index (A.I.) versus wt. % Al_2O_3 (Fig. 1.5) usefully distinguishes between tholeiitic and calc-alkali basalts. A very small number of low-K basalts may plot in the high-alumina field on this diagram but this is usually readily explained in terms of accumulation of plagioclase crystals. The most prominent chemical difference between the more basic end-members of typical tholeiitic and calc-alkaline series is in their Al_2O_3 content: calc-alkali basalts and andesites contain 16—20% whereas

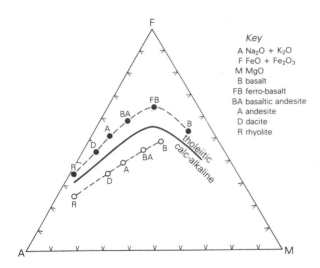

Figure 1.4 AFM diagram showing typical tholeiitic and calc-alkaline differentiation trends.

their tholeiitic counterparts contain only 12–16% Al_2O_3. Calc-alkali basalts may be further subdivided into low-, medium- and high-K types on the basis of a K_2O versus SiO_2 diagram (Fig. 6.15).

Rocks of the alkalic magma series may be subdivided into sodic, potassic and high-K types in terms of a plot of K_2O versus Na_2O (Fig. 1.6). Members of the high-K series tend to be silica poor and have been described by a variety of names, including absarokite, lamproite, leucite basalt, leucite basanite and leucitite (Ch. 12). These may differentiate to produce more silica rich high-K magmas in some cases.

Table 1.3 shows the characteristic magma series based on the above classification associated with each of the tectonic environments discussed in Section 1.1. Sub-alkalic basalts are the most common type of volcanic rock found within both the continents and ocean basins. Low-K sub-alkalic basalts, or tholeiitic basalts, are the dominant magma type produced at mid-oceanic ridges (MORB, Ch. 5) and within many continental flood basalt provinces (Ch. 10). These basalts are normally strongly depleted in K and other large cations (Rb, Ba, U, Th, Pb, Zr and light REE) compared with other types of basalt.

Analysis of volcanic rocks from the ocean floor (Ch. 5) has revealed considerable compositional heterogeneity. Although tholeiitic basalts predominate, transitional and alkalic varieties occur in some areas, particularly in slow-spreading oceans such as the Atlantic. The chemical characteristics of

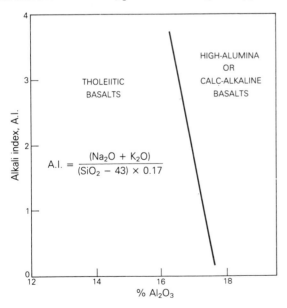

Figure 1.5 Diagram of Alkali Index (A.I) versus wt. % Al_2O_3 for the classification of tholeiitic and high-alumina (calc-alkaline) basalts (after Middlemost 1975).

$$A.I. = \frac{(Na_2O + K_2O)}{(SiO_2 - 43) \times 0.17}$$

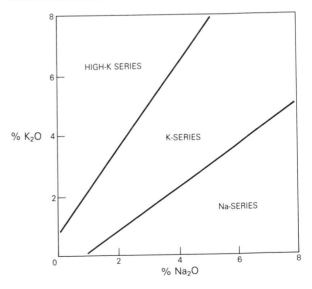

Figure 1.6 K_2O versus Na_2O (wt. %) diagram, showing the subdivision of the alkalic magma series into high-K, K and Na sub-series (after Middlemost 1975).

Table 1.3 Characteristic magma series associated with specific tectonic settings.

Tectonic setting	Plate margin		Within plate	
	Convergent (destructive)	Divergent (constructive)	Intra-oceanic	Intra-continental
volcanic feature	island arcs, active continental margins	mid-oceanic ridges back-arc spreading centres	oceanic islands	continental rift zones continental flood-basalt provinces
characteristic magma series	tholeiitic calc-alkaline alkaline	tholeiitic — —	tholeiitic — alkaline	tholeiitic — alkaline
SiO$_2$ range	basalts and differentiates	basalts	basalts and differentiates	basalts and differentiates

MORB appear to vary as a function of spreading rate and ridge crest elevation. Seafloor spreading may also occur in back-arc basins associated with subduction-related volcanic arcs (Ch. 8). In general, the erupted basalts are similar to MORB in terms of their major element characteristics, although they may have different trace element compositions.

At the present time, magmas of the calc-alkaline series are totally restricted in their occurrence to subduction-related tectonic settings. Consequently, the recognition of calc-alkaline characteristics in the geochemistry of ancient volcanic sequences may be an important petrogenetic indicator. The products of volcanism in island arcs vary with the stage of evolution of the arc, with vertical distance above the Benioff zone and, in some cases, laterally along the arc (Ch. 6). Volcanic arc rocks may be divided into tholeiitic, calc-alkaline and alkaline types, which are gradational. Tholeiitic magmas typically occur in young immature arcs, or closest to the trench in more mature arcs, whereas calc-alkaline magmas are typical of more mature arcs and active continental margins. The chemical characteristics of volcanic arc rocks are clearly much more variable than MORB. The proportions of more evolved (more SiO$_2$-rich) lavas are greater, particularly in the calc-alkaline series where andesites predominate.

Alkali basalts and their differentiates are commonly found in intra-plate tectonic settings such as oceanic islands (Ch. 9) and intracontinental plate rifts (Ch. 11). They also occur rather more rarely in some subduction-related tectonic settings (Ch. 6, 7 & 12). Oceanic-island basalts (OIB) exhibit considerable compositional diversity, ranging from tholeiitic (Hawaii, Iceland and Galapagos) through sodic-alkalic (Canary Islands and St Helena) to potassic-alkalic (Tristan da Cunha and Gough). More evolved magma types than basalt are quite common, often with a bimodal basalt−trachyte/phonolite distribution.

Continental basalts are very restricted at the present time, and are dominantly alkalic in the early stages of continental rifting (Ch. 11). However, in regions in which large amounts of crustal extension have occurred, transitional and tholeiitic types may be common. Tholeiitic continental flood basalt provinces (Ch. 10) appear to have been much more important in the past, associated with major phases of successful continental rifting and the generation of new ocean basins. Kimberlites and ultrapotassic magmas form an extremely diverse group of continental alkalic magmas, generated in a wide variety of tectonic settings (Ch. 12).

Further Reading

Cox, K. G., J. D. Bell & R. J. Pankhurst 1979. *The interpretation of igneous rocks*. London: Allen and Unwin.

Hutchison, C. S. 1983. *Economic deposits and their tectonic setting*. London: Macmillan.

Irvine, T. N. & W. R. A. Baragar 1971. A guide to the chemical classification of the common volcanic rocks. *Can. J. Earth Sci.* 8, 523–48.

Le Bas, M. J., R. W. Le Maitre, A. Streckeisen & B. Zanettin (1986). A chemical classification of volcanic rocks based on the total alkali-silica diagram. *J. Petrology* 27, 745–50.

Middlemost, E. A. K. 1975. The Basalt Clan. *Earth Sci. Rev.* 11, 337–64.

Middlemost, E. A. K. 1980. A contribution to the nomenclature and classification of volcanic rocks. *Geol Mag.* 117, 51–7.

CHAPTER TWO

Geochemical characteristics of igneous rocks as petrogenetic indicators

2.1 Introduction

In Chapter 1 we noted that magmas with distinctive major element characteristics are associated with specific tectonic settings. For example, calc-alkaline series magmas are apparently associated uniquely with subduction, while low-K tholeiitic basalts are the typical products of magma generation at constructive plate margins. However, in general, the major element characteristics of primary mantle derived magmas are· not particularly sensitive indicators of tectonic setting. Thus tholeiitic basalts are generated at mid-oceanic ridges but also in back-arc basins, oceanic islands, island arcs, active continental margins and continental flood basalt provinces. Fortunately, it is now well established that distinctive trace element and Sr–Nd–Pb isotopic signatures are associated with

different magma generation environments, although their petrogenetic interpretation in some instances remains ambiguous (Chs. 5–12).

On the basis of the major element chemical variation diagrams discussed in Chapter 1, we established the existence of three dominant magma series, tholeiitic, calc-alkaline and alkaline. Within each of these series there is a continuous spectrum of rock types ranging from basic to acid, which appear to be genetically related. In terms of Figure 1.2 we can consider the existence of a primary magma spectrum, generated by partial melting processes within the upper mantle (Ch. 3), and a range of more differentiated (more SiO_2-rich) magmas related to the primary magmas by processes of fractional cystallization, magma mixing and crustal contamination (Ch. 4), which often occur in high-level magma chambers.

In investigating the geochemical characteristics of suites of cogenetic igneous rocks we thus have two fundamental objectives:

(a) to understand the processes involved in the petrogenesis of the primary magma spectrum (Ch. 3); and
(b) to understand the processes involved in the differentiation of the primary magma spectrum (Ch. 4).

In this chapter we shall focus our attention on ways of looking at major and trace element data for volcanic rocks graphically, in order to place constraints on petrogenetic processes. Detailed discussion of the processes responsible for the diverse compositions of terrestrial magmas is deferred until Chapters 3 and 4. Additionally, we shall consider those aspects of the radiogenic and stable isotope geochemistry of igneous rocks which may be of petrogenetic significance.

2.2   Chemical analysis of igneous rocks

A wide variety of instrumental techniques are commonly used for silicate rock analysis, permitting the determination of an extensive range of major and trace elements on a routine basis. Potts (1987) presents a comprehensive review of the various methods used, and the reader is referred to this source for further information.

X-ray fluorescence (XRF) is one of the most widely used instrumental methods for analysing rock samples for major elements (Na, Mg, Al, Si, P, K, Ca, Ti, Mn, Fe) and selected trace elements (Rb, Sr, Y, Nb, Zr, Cr, Ni, Cu, Zn, Ga, Ba, Pb, Th, U± La, Ce, Nd, Sm). Instrumental neutron activation analysis (INAA) is used for the analysis of specific trace elements (Sc, Co, Cr, Cs, Hf, Ta, Th, U) down to detection limits in the ppm and ppb range, and is especially useful for the analysis of the rare earth elements (La, Ce, Nd, Sm, Eu, Tb, Yb, Lu). Alternatively, the REE may be determined by isotope dilution mass spectrometry. Inductively coupled plasma (ICP) techniques are now being applied to the analysis of geological

materials but, as yet, the number of samples in the literature analysed by this method is limited.

None of the commonly used techniques can provide analyses of H_2O and CO_2 or the ratio of Fe^{2+}/Fe^{3+} in igneous rocks. Consequently, these must be determined independently by other methods.

In the tables of analyses of igneous rocks from different tectonic settings, presented in Chapters 5 − 12, the analytical techniques employed have not been specified. In all cases major elements (expressed as wt. % constituent oxides) have been determined by XRF and trace elements (expressed as parts per million, or ppm) by a combination of XRF, INNA and isotope dilution. In general, analyses of H_2O, CO_2 and the Fe^{2+} / Fe^{3+} ratio are not listed, and the Fe content is expressed as total FeO or Fe_2O_3. In some XRF analyses the total volatile content is approximately expressed as the loss on ignition (LOI). When data for H_2O are given, H_2O^+ represents water present in a combined state within the rock in hydrous minerals (e.g. amphibole or biotite), whereas H_2O^- represents pore water, or that present in low-temperature alteration products (e.g. zeolites).

2.3   Chemical variation diagrams

A variation diagram is a simple display of the chemical differences and trends shown by a related suite of rocks (lavas) in which the compositional variation is a consequence of crystal−liquid fractionation processes, either partial melting or fractional crystallization. They may be plotted in terms of major elements or trace elements, or combinations of both. Variation diagrams provide a useful way of synthesizing a large volume of analytical data, which is clearly difficult to compare in table form. Additionally, they can provide the basis for the derivation of models to explain the petrogenesis of a particular suite.

A fundamental assumption in plotting variation diagrams for cogenetic suites of volcanic rocks is that they illustrate the course of chemical evolution of magmatic liquids. However, only analyses of phenocryst-poor or aphyric volcanic rocks can be regarded as representative of actual magma compositions. Consequently, much of the scatter in variation diagrams is a result of background 'noise' introduced by using analyses of porphyritic samples. This is a particular problem for subduction-related suites (Ch. 6) which, in general, are highly porphyritic.

2.3.1 Major elements

One of the most commonly used types of variation diagram in igneous petrology is the Harker diagram (Harker 1909), in which the weight percent of a constituent oxide is plotted against wt.% SiO_2 as abscissa. This is one of a family of binary plots in which one element or oxide is plotted against another. Figure 1.1, the plot of $Na_2O + K_2O$ versus SiO_2 used to classify the members of the alkalic and sub-alkalic magma series, is a typical example. SiO_2 is chosen as abscissa as an index of differentiation, but any other element or oxide which displays a wide range of variation within the suite can be used (e.g. MgO or Zr).

In general, for suites of cogenetic igneous rocks, pairs of oxides are strongly correlated (Fig. 2.1), either positively or negatively. Such correlations or trends may be generated as a consequence of partial melting, fractional crystallization, magma mixing or crustal contamination, either individually or in combination (Chs 3 and 4). Chayes (1964) has argued that at least some negative correlations with SiO_2 are to be expected as a consequence of the constant sum effect, i.e. as SiO_2 ranges between 40 and 75% in suites of igneous rocks, the sum of all other oxides must fall from 60 to 25% as SiO_2 increases. However, Cox et al. (1979) argue that this effect is not significant and does not negate the usefulness of Harker diagrams as a basis for petrogenetic modelling.

Coherent trends on Harker diagrams are generally considered to represent the course of chemical evolution of magmas and are referred to as liquid lines of descent. In reality, individual members of a series are erupted in a random time sequence, suggesting that such trends are actually the average of the evolutionary trends of numerous batches of parental magma of similar composition. Additional-

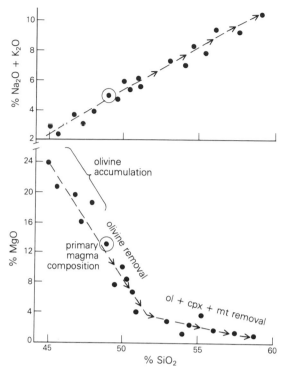

Figure 2.1 Harker variation diagrams of wt. % $Na_2O + K_2O$ and wt. % MgO versus wt. % SiO_2 for a suite of cogenetic volcanic rocks related by fractional crystallization of olivine, clinopyroxene and magnetite. The highly magnesian basalts (MgO >12%) may have accumulated olivine by crystal settling. This should be evident in their petrography, i.e. the samples should be highly olivine phyric.

ly, in a natural system, it is highly unlikely that each batch of parental magma will be identical in composition and, consequently, the differentiation process may not be exactly the same in each case. This accounts for some of the scatter in addition to that introduced by sample heterogeneity and analytical error. In some magmatic provinces, a range of primary magma compositions may have evolved by broadly similar low-pressure fractional crystallization processes, producing a series of subparallel liquid lines of descent.

Perhaps the single most important property of Harker-type variation diagrams is the applicability of the Lever Rule for mass balance (Cox et al. 1979). Thus it is possible to calculate graphically the way in which a liquid composition changes as a particular mineral is removed from it or a particular contaminant is added to it. This will be considered in detail in Chapter 4. Thus if we have a suite of volcanic rocks, related by processes of fractional crystallization, which display coherent trends in several different variation diagrams, we can potentially constrain the nature of the fractionating mineral assemblage.

Variation diagrams may display strongly segmented trends (Fig. 2.2), which provides powerful evidence for the operation of crystal−liquid separation during magmatic evolution (Ch. 4). In general, the inflections in the trends are interpreted to mark the onset of crystallization of a new mineral (or group of minerals). Sometimes marked inflections are present in one two-element plot, while being undetectable in another. This is because the magnitude of the inflection is dependent on the relative positions in the diagram of the two extracts, one with the new mineral and one without it. Thus, for example, the onset of apatite crystallization may be strongly marked in a plot of P_2O_5 versus SiO_2 but not in Na_2O versus SiO_2.

Another type of variation diagram employed for igneous rocks is the triangular AFM diagram (A = $Na_2O + K_2O$, F = $FeO + Fe_2O_3$, M = MgO; Fig. 1.4). This is very useful for distinguishing between tholeiitic and calc-alkaline differentiation trends in the sub-alkalic magma series.

2.3.2. Trace elements

The behaviour of trace elements during the evolution of magmas may be considered in terms of their partitioning between crystalline and liquid phases, expressed as the partition coefficient, D:

$$D_{xal}^{liq} = \frac{\text{concentration in mineral}}{\text{concentration in liquid}} \quad \text{for any trace element}$$

Elements that have values of $D \ll 1$ are termed *incompatible* and are preferentially concentrated in the liquid phase during melting and crystallization. Those elements which are incompatible with respect to normal mantle minerals (olivine, pyroxene, spinel and garnet; Ch. 3) are termed *lithophile* or *large-ion lithophile* (LIL), e.g. K, Rb, Sr, Ba, Zr, Th and light REE. In contrast, those with $D > 1$ (e.g. Ni, Cr) are termed *compatible*, and these are preferentially retained in the residual solids during partial melting and extracted in the crystallizing solids during fractional crystallization. Trace element partition coefficients between the major rock-forming minerals and magmatic liquids vary widely (see the Appendix). As a consequence, some elements or groups of elements may be used to identify those minerals involved in magmatic differentiation processes; these are summarized in Table 2.1.

Harker-type variation diagrams may be plotted using trace elements instead of major element oxides and may be interpreted in a similar way. Highly incompatible trace elements such as Zr may be useful as an index of differentiation if SiO_2 or MgO are inappropriate. In this section we shall concentrate on other types of trace element varia-

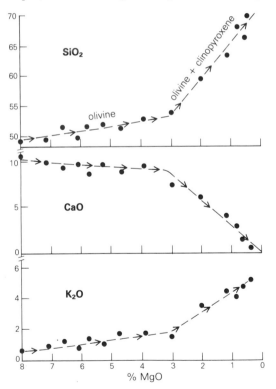

Figure 2.2 Harker-type variation diagrams, with wt.% MgO as abscissa, for a cogenetic suite of volcanic rocks related by fractional crystallization of olivine and clinopyroxene.

Table 2.1 Summary of key trace element parameters useful in evaluating petrogenetic models (after Green 1980).

Element	Interpretation
Ni, Co, Cr	High values (e.g. Ni = 250–300 ppm, Cr = 500–600 ppm) for these elements are good indicators of derivation of parental magmas from a peridotite mantle source. Decrease of Ni (and to a lesser extent Co) through a rock series suggests olivine fractionation. Decrease in Cr suggests spinel or clinopyroxene fractionation.
V, Ti	These elements show parallel behaviour in melting and crystallization processes. They are useful pointers to the fractionation of Fe–Ti oxides (ilmenite or titanomagnetite). When V and Ti show divergent behaviour, Ti substitution into some accessory phase such as sphene or rutile may be indicated.
Zr, Hf	These are classic incompatible elements, not readily substituting in major mantle phases. However, they may substitute for Ti in accessory phases such as sphene and rutile.
Ba	Substitutes for K in K-feldspar, hornblende and biotite. Changes in Ba content or K/Ba ratio may indicate the role of one of these phases.
Rb	Substitutes for K in K-feldspar, hornblende and biotite. K/Rb ratios provide possible indicators of the role of these phases in petrogenesis.
Sr	Substitutes readily for Ca in plagioclase and for K in K-feldspar. Sr or Ca/Sr ratio is a useful indicator of plagioclase involvement at shallow levels. Sr behaves more as an incompatible element under mantle conditions.
REE	Garnet and possibly hornblende readily accommodate heavy REE and so strongly fractionate light REE. Sphene has the opposite effect accommodating the light REE. Clinopyroxene fractionates the REE but only slightly. Eu is strongly fractionated into feldspars and Eu anomalies may reflect feldspar involvement.
Y	Generally behaves as an incompatible element resembling the heavy REE. It is readily accommodated in garnet and amphibole, less so in pyroxene. The presence of accessory phases such as sphene or apatite could have a major effect on the abundance of Y, since these phases readily concentrate it.

tion diagram which provide constraints for the petrogenesis of different magma types.

Rare earth elements (REE)

The rare earth elements are a group of 15 elements (La, Ce, Pr, Nd, Pm, Sm, Eu, Gd, Tb, Dy, Ho, Er, Tm, Yb, Lu) with atomic numbers ranging from 57 (La) to 71 (Lu), 14 of which occur naturally. Those with lower atomic numbers are generally referred to as light REE, those with higher atomic numbers as heavy REE, and those with intermediate atomic numbers as middle REE. They are particularly useful in petrogenetic studies of igneous rocks because all the REE are geochemically similar. All, except for Eu and Ce, are trivalent under most geological conditions. Eu is both trivalent and divalent in igneous systems, the ratio Eu^{2+}/Eu^{3+} depending upon oxygen fugacity (fO_2). Eu^{2+} is geochemically very similar to Sr. Ce may be tetravalent under highly oxidizing conditions.

In order to compare REE abundances for different rocks graphically, it is necessary to eliminate the Oddo-Harkins effect, which is the existence of higher concentrations of those elements with even atomic numbers as compared to those with odd atomic numbers. This is achieved by normalizing the concentrations of individual REE in a rock to their abundances in chondritic meteorites (Nakamura 1974). This smooths out the concentration variations from element to element. Chondrites are used in the normalization procedure because they are primitive solar system material which may have been parental to the Earth. Fig. 2.3 shows a range of chondrite-normalized REE patterns for basaltic rocks.

Particular minerals will have a characteristic effect upon the shape of the REE pattern of the melt during partial melting and fractional crystallization,

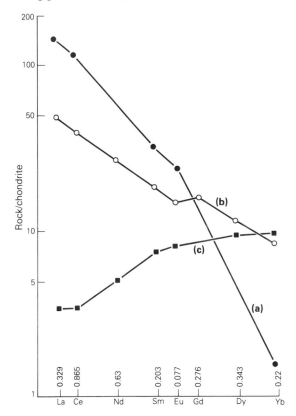

Figure 2.3 Schematic chondrite-normalized REE patterns for basaltic rocks. Normalization constants from Nakamura (1974) are indicated on the diagram. (a) Strongly light-REE enriched basalt with very low concentrations of heavy REE suggests the presence of residual garnet in the source: extremely high concentrations of light REE suggest very small degrees of partial melting or a light-REE enriched source. (b) Basalt with a slight negative Eu anomaly which may have fractionally crystallized plagioclase or may have been in equilibrium with a plagioclase-bearing mantle source: heavy-REE concentrations of 10 × chondritic suggest that garnet is absent from the source. (c) Basalt showing strong light-REE depletion, suggesting derivation from a light-REE depleted garnet-free source.

depending upon their D values (see the Appendix), which allows identification of their role in magmatic differentiation processes. The magnitude of the effect produced by a particular mineral will depend upon its relative abundance and on the magnitude of its D value for a particular element. Feldspars have low D's for all of the REE except Eu. Consequently, feldspar has a minor effect on the REE pattern of the melt except for the production of a negative Eu anomaly as a consequence of the high D value for Eu. The size of the Eu anomaly decreases with increasing fO_2 and temperature, but is probably significant in all magmatic systems (Hanson 1980). Garnet has very low D values for the light REE and increasingly larger D's for the heavy REE. Thus its presence in equilibrium with a magma leads to a depletion of heavy REE. Orthopyroxene and Ca-rich clinopyroxene generally have $D < 1$, with values for the light REE that are slightly lower than for the heavier, which may lead to light REE enrichment in the melt. Olivine D values for REE are all less than 0.1, and thus its presence leads to essentially equivalent enrichment for all the REE. D's for hornblende show a strong dependence on composition and may be greater than 10 for the middle REE in silica-rich systems. Thus the fractional crystallization of hornblende from intermediate-composition magmas may lead to the relative depletion of the middle REE. In contrast, biotite has generally low D's for the REE and its presence should have relatively little effect on the REE pattern of the melt. In evolved liquid compositions, zircon, apatite and sphene may strongly fractionate the REE (Le Marchand *et al.* 1987).

If the Earth originally had chondritic abundances of the REE and then differentiated to form the early mantle and the core, then the mantle would have retained the REE because they are lithophile. This would result in a primordial mantle with greater abundances of the REE (perhaps 2 to 3 times chondritic), but with a parallel pattern relative to chondrites. Partial melting of the mantle to form basalt results in a relative depletion of the light REE in the residuum, and therefore the mantle should have become progressively depleted in light REE over the course of geological time. During partial melting the presence of residual garnet in the source leads to less enrichment of the heavy REE in the magma relative to melts derived from garnet-free mantle sources. However, even in garnet-bearing sources, once the degree of partial melting is sufficiently high to eliminate garnet from the residue (Ch. 3) fractionation of the light and heavy REE is suppressed and the REE pattern of the melt becomes subparallel to that of the source. Thus, for relatively large degrees of partial melting, we may consider that the ratios of light REE in a magma should essentially be similar to those in the source. For common mantle mineralogies (olivine, orthopyroxene, clinopyroxene, garnet and spinel) there is no phase which preferentially concentrates light REE relative to the heavy. Thus we may deduce that magmas such as the basalt 'c' in Fig. 2.3, whch display light-REE depleted patterns, must be derived from mantle sources which are themselves depleted in light REE.

In terms of Fig. 2.3, the degree of enrichment for a particular REE relative to chondritic abundances is a function of the initial concentration of that element in the source and the degree of partial melting (Ch. 3), and subsequent fractional crystallization (Ch. 4).

Spiderdiagrams

In order to understand the pattern of trace element abundances in basalts (or indeed any igneous rock) it is useful to have a reference frame to which the elemental abundances in a particular rock can be compared. We have already used this approach in the previous section by plotting chondrite-normalized REE patterns. Developing this idea further leads us to the concept of spiderdiagrams (Wood *et al.* 1979, Sun 1980, Thompson 1982, Thompson *et al.* 1984), in which the abundances of a range of incompatible trace elements are normalized to estimates of their abundances in the primordial Earth.

The elements Ba, Sr, U, Th, Zr, Nb, Ti and REE are believed to have condensed at high temperatures from a gas of solar composition during planetary formation. Archaean high-MgO basalts (komatiites) have the same ratios of these elements (e.g. Ba/Sr, U/Th) as chondritic

meteorites, suggesting that their ratios in meteorites may approximate those in the primordial mantle before significant crust formation. However, while the absolute abundances of these elements in the bulk Earth may approximate to chondritic values, those in the primordial mantle may be greater due to the concentration effects of core formation.

Several variants of the spiderdiagram plot have been used in the literature, in which the order of the elements plotted varies slightly, and different normalization constants have been adopted. For example, Wood *et al.* (1979) normalize to a hypothetical primordial mantle composition, whereas Thompson *et al.* (1984) and Sun (1980) normalize their data to chondritic abundances, with the exception of K and Rb, which may be volatile during planetary formation, and P, which may be partly contained in the core. In subsequent chapters both Sun (1980) and Thompson *et al.* (1984) spiderdiagram plots have been used to discuss the trace element geochemistry of basaltic magmas formed in different tectonic settings. These diagrams are essentially identical to each other (Fig. 2.4) except for the order of some of the elements. This is somewhat arbitrary, being designed to give a smooth pattern for average MORB (Sun 1980). It is probably one of increasing incompatibility of the elements from right to left in a four-phase lherzolite (Ch. 3) undergoing partial fusion. The elements plotted all behave incompatibly ($D < 1$) during most partial melting and fractional crystallization processes. The main exceptions to this are Sr, which may be compatible with plagioclase, Y and Yb with garnet, and Ti with magnetite. Following Thompson *et al.* (1984), we shall refer to the individual patterns as spiderdiagrams.

Analytical error may be the source of many apparently spurious inflections in spiderdiagram patterns. However, once these effects have been taken into consideration, the peaks, troughs, slopes and curvature of the patterns may provide invaluable petrogenetic information concerning crystal−liquid equilibria. For example, troughs at Sr probably result from the fractional crystallization of plagioclase from many basalts. In contrast, a trough at Th and Rb combined with one at Nb−Ta may suggest contamination of a magma by lower

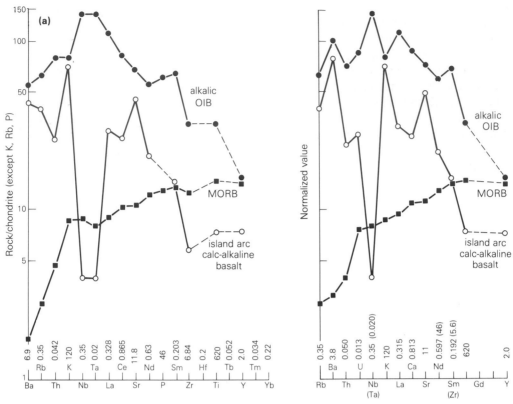

Figure 2.4   Typical spiderdiagram patterns for mid-ocean ridge (MORB), oceanic-island (OIB) and island-arc, basalts normalized according to (a) Thompson *et al.* (1984) and (b) Sun (1980). Values of the normalization constants used are given at the foot of each diagram.

continental crustal rocks (Ch. 4). The marked trough in the spiderdiagrams of some ultrapotassic intracontinental plate lavas (Ch. 12) may reflect residual phlogopite in the source.

Spiderdiagram patterns for typical mid-ocean ridge, island-arc and oceanic-island basalts are shown in Figure 2.4. MORB (Ch. 5) are considered to be the products of relatively large degrees of partial melting, and consequently their trace element patterns should reflect those of their mantle source. In general, the less incompatible elements on the right-hand side of the spiderdiagram pattern should be less enriched during partial melting, tilting the curve up to the left. Additionally, fractional crystallization subsequent to magma segregation should tilt the patterns even further up to the left. There is no known mechanism for tilting the curve down to the left, as shown, during the type of partial melting capable of creating the very large volumes of magma erupted at mid-oceanic ridges (Norry & Fitton 1983). Thus the source of MORB must clearly be depleted in the more incompatible elements such as K and Rb and, to a lesser extent, Nb and La. It has been argued that this depletion has been accomplished by the formation of the continental crust throughout geological time.

In contrast to the MORB spiderdiagram pattern, that for an alkalic oceanic-island basalt (OIB) shows extreme degrees of incompatible element enrichment with a peak at Nb−Ta, suggesting that the source may also be enriched in incompatible elements (Ch. 9). In general, the spiderdiagrams of all magnesian OIB (both tholeiitic and alkalic) approximate to smooth convex curves (Thompson *et al.* 1984).

Compared to the relatively smooth shapes of the MORB and OIB spiderdiagrams, that of the subduction-related basalt is strongly spiked. As we shall see in Chapter 6, the positive spikes are mostly a consequence of components added to the mantle source of the basalts by subduction-zone fluids. The most persistent feature of the spiderdiagrams of volcanic-arc basalts is the marked Nb–Ta trough, which has been explained in terms of retention of these elements in the source during partial melting. However, a similar trough is also a typical feature of the spiderdiagram patterns of basalts which have experienced crustal contamination (Ch. 4), and therefore an element of caution

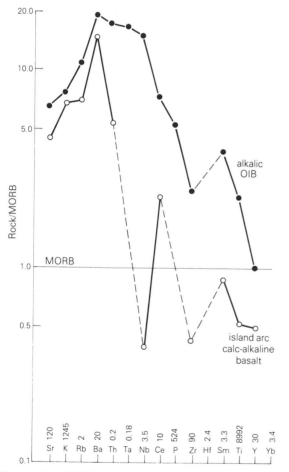

Figure 2.5 MORB-normalized trace element variation diagram for typical island-arc and oceanic-island basalts. The order of the elements and values of the normalizing constants are from Pearce (1983).

must be exercised when interpreting these sorts of patterns.

A modified version of the spiderdiagram is most useful for comparing the trace element characteristics of different types of basalts. For example, if we wish to know how continental tholeiites differ from their oceanic counterparts we could normalize their compositions to those of typical MORB or oceanic-island tholeiites. The sequence of elements plotted may be the same as that in a conventional spiderdiagram, or different (e.g. Pearce 1983). Figure 2.5 shows the island-arc and oceanic-island basalts from Figure 2.4 normalized to values in an average MORB. In this type of diagram (after Pearce op. cit.) the elements are divided into two groups based on their relative mobility in aqueous fluids. Sr, K, Rb and Ba are mobile and plot at the left of the pattern, while the remaining elements are immobile. The elements are arranged such that the incompatibility of the mobile and immobile elements increases from the outside to the centre of the pattern. Pearce considers that the shape of these patterns is not likely to be greatly changed by fractional crystallization or variable degrees of partial melting, and that they may consequently be used to discuss source characteristics. Such MORB-normalized trace element variation diagrams have been used to constrain the nature of the mantle source of subduction-related basalts (Chs 6 & 7) and continental flood basalts (Ch. 10), and the reader is referred to the appropriate sections of these chapters for a more detailed discussion of their interpretation (Sections 6.11.2 & 7.7.3).

2.4 Geochemical characteristics of primary magmas

Primary magmas are those formed by partial melting of the upper mantle, the compositions of which have not been modified subsequently by differentiation processes (fractional crystallization, crustal contamination, magma mixing, liquid immiscibility and volatile loss; see Ch. 4). As we have seen in Chapter 1, such magmas encompass a variety of types including tholeiitic, calc-alkaline and alkaline basalts. Clearly, it is of great petrogenetic significance to be able to recognize primary magma compositions as these are parental, giving rise through differentiation processes to more evolved (more silica-rich) magma types. Unfortunately, there are no generally accepted criteria which can be used to distinguish them unequivocally from those basaltic magmas with compositions that have been modified subsequent to segregation.

O'Hara (1968) considers that most of the basaltic lavas erupted at the Earth's surface are not primary and have evolved by fractional crystallization of olivine (Ch. 4) from more MgO-rich picritic precursors. The rarity of erupted high-MgO liquids has been used to argue against this, although various authors have suggested that such liquids would be difficult to erupt by virtue of their high density. As a consequence, we generally consider that the most primitive basalts of a particular suite are those with the highest MgO contents. These may actually be parental to the suite of lavas but need not necessarily be primary. Clearly, it is much easier to demonstrate the parental nature of a magma than to confirm its primary nature.

As olivine and orthopyroxene are the most abundant minerals in the upper mantle source rocks (Ch. 3), primary magmas must be in equilibrium with these phases at their depth of segregation, and should therefore crystallize them at the liquidus in melting experiments at the appropriate pressure (Yoder 1976, Basaltic Volcanism Study Project 1981; and see Ch. 3). Additionally, the Fe^{2+}/Mg ratios of primary magmas should lie within the range expected for liquids in equilibrium with typical upper mantle olivine compositions (Fo_{86-90}). Roeder & Emslie (1970) showed that the distribution of Fe and Mg between olivine and coexisting melt, as defined by the relation:

$$K_D = (Fe^{2+}/Mg)_{olivine}/(Fe^{2+}/Mg)_{melt}$$

is relatively insensitive to temperature, melt composition and oxygen fugacity, with a K_D value of 0.3 ± 0.03. Consequently, the Mg' values (Mg/(Mg + Fe^{2+})) of basalts in equilibrium with mantle olivine compositions can be calculated, and lie in the range $0.68-0.75$. In general, the Mg' value is insensitive to the degree of partial melting but is

highly sensitive to the amount of subsequent fractional crystallization, particularly of olivine.

The partitioning of the trace element Ni between olivine and melt has similarly been used to constrain the compositions of primary mantle derived magmas (Ch. 3). Estimated Ni contents are relatively insensitive to the degree of partial melting and range between 400 and 500 ppm, depending upon the assumptions made about mantle mineralogy and Ni content and on the choice of values for Ni partition coefficients. Similar estimates may be made of the Cr contents of primary magmas.

In general, primary magmas in equilibrium with typical upper mantle mineralogies (olivine + orthopyroxene + clinopyroxene ± garnet ± spinel) should have high Mg' values (>0.7), high Ni (>400−500 ppm), high Cr (>1000 ppm) and SiO_2 not exceeding 50%. However, if they are not derived from normal mantle but from metasomatized source regions (Chs 3, 6, 7, 11 & 12) these criteria may no longer be applicable. The metasomatism of the source may be so extreme that harzburgite (olivine + orthopyroxene) is no longer the residue from partial melting (Ch. 3), and thus these phases do not buffer the Mg' values, Ni and Cr contents of the partial melts. In the absence of the above criteria the presence of high-pressure mantle-derived ultramafic xenoliths may be used to infer relatively primary characteristics for the host magma, as these would be expected to settle out if significant fractional crystallization had occurred. However, even this is not completely diagnostic, as high-pressure olivine fractionation could have occurred before incorporation of the xenoliths into the magma.

2.5 Isotopes as petrogenetic indicators

Isotope geochemical studies are now regarded as a fundamental part of the petrogenetic interpretation of igneous rocks. They are based on two groups of isotopes, radiogenic and stable. In the former, isotopic variations are caused by the radioactive decay of elements, whereas in the latter they are the consequence of mass fractionation in chemical reactions. In general, mass fractionation effects are comparatively small, except for the lighter elements, O, H, C and S.

The naturally occurring long lived radioactive decay schemes of K, Rb, Sm, Th and U are critically important in establishing the chronology of magmatic events. The isochron method for dating co-magmatic suites of igneous rocks, which may be applied to all of these decay schemes, depends upon the fact that in a given sample the rate of accumulation of the radiogenic daughter isotope, relative to a non-radiogenic isotope of the same element, is related to the concentration ratio of the parent and daughter elements, and to the decay constant of the parent element. The dating of igneous rocks is a complex subject and the reader is referred to Faure (1986) for a detailed discussion. In the context of this chapter we shall focus our attention on the radiogenic daughters of Rb, Sm, U and Th as petrogenetic tracers in evaluating the evolution of magmas.

The importance of radiogenic isotopic variations is that they frequently survive the chemical fractionation events which accompany the formation and evolution of magmas, as isotopes of the heavier elements are not separated from each other through crystal–liquid equilibria. Thus, during partial melting, a magma will inherit the isotopic composition of its source, and this will remain constant during subsequent fractional crystallization processes, provided that the magma does not become contaminated by interaction with isotopically distinct wall rocks or other batches of magma (Ch. 4). As a consequence, estimates of the present-day isotopic characteristics of the mantle source region of basaltic magmas may be obtained from studies of young oceanic volcanic rocks (MORB and OIB), which have not been significantly contaminated by crustal rocks en route to the surface.

In general, rocks of the continental crust have very different radiogenic and stable isotope compositions from those of the mantle, and thus isotopic studies can provide important constraints on the extent of crustal contamination in lavas erupted in continental-plate tectonic settings (Ch. 4). Despite the fact that stable isotopes may become fractionated by crystal–liquid differentiation processes, the magnitude of these effects is small compared to the difference in isotopic composition between crustal and mantle reservoirs. As a consequence, oxygen isotope studies have been extensively used to assess the importance of crustal contamination effects in continental volcanic suites.

Isotopes of the rare gases, particularly He, have proved to be useful tracers of the role of primordial mantle components in the petrogenesis of oceanic-island basalts (Ch. 9). Additionally, cosmogenic radionuclides such as [10]Be are important as potential tracers of the role of subducted sediment in island-arc and active continental margin magmatism (Ch. 6 & 7).

In many of the provinces discussed in Chapters 5–12, the volcanic rocks are so young that their present-day radiogenic isotope ratios are identical to those at the time of their formation. Clearly, for older rocks, the isotopic compositions must always be age-corrected, as it is only the initial isotopic ratios which are directly of petrogenetic significance.

2.5.1 Radiogenic isotopes

Rb–Sr

Rb has two naturally occurring isotopes, [85]Rb and [87]Rb, of which [87]Rb is radioactive and decays to stable [87]Sr by beta emission. Sr has four naturally occurring isotopes, [88]Sr, [87]Sr, [86]Sr and [84]Sr. The precise isotopic composition of Sr in a rock or mineral that contains Rb depends upon its age and Rb/Sr ratio. This forms the basis of the Rb–Sr dating technique, using the following equation:

$$\frac{^{87}Sr}{^{86}Sr} = \left(\frac{^{87}Sr}{^{86}Sr}\right)_{initial} + \frac{^{87}Rb}{^{86}Sr}(e^{\lambda t} - 1)$$

where λ is the decay constant of [87]Rb($= 1.42 \times 10^{-11}yr^{-1}$) and t is the age in years.

During fractional crystallization of basaltic magma, Sr tends to be concentrated in plagioclase whereas Rb remains in the residual magma. Consequently, the Rb/Sr ratio of the magma increases gradually during the course of progressive crystallization. Thus a suite of cogenetic igneous rocks, related by processes of fractional crystallization to a parent magma, will tend to have increasing Rb/Sr ratios with increasing degree of differention. However, provided that the system has remained closed, all members of the suite should have identical initial ratios, although they will have different present-day ratios depending upon their age and Rb/Sr ratio.

It is generally agreed upon that the matter which accreted from the solar nebula to form the proto-planet Earth had a relatively uniform $^{87}Sr/^{86}Sr$ ratio, which we refer to as the primordial value. Since accretion, the isotopic composition of Sr in the Earth has become increasingly heterogeneous as a consequence of the geochemical differentiation of the planet. Unfortunately, it is not possible to measure the isotopic composition of the Sr that was incorporated into the Earth at the time of its formation, as rocks formed at this time are either inaccessible, being located deep within the mantle, or have been destroyed by subsequent geological processes. As a consequence, we must rely on the study of meteorites and lunar samples to determine the primordial $^{87}Sr/^{86}Sr$. The value of the initial $^{87}Sr/^{86}Sr$ ratio of basaltic achondrite meteorites (BABI) (*Basaltic Achondrite Best Initial*), 0.69897 ± 0.00003, is consequently taken to be that of the solar nebula at a very early stage in the formation of the planet. Thus we can consider that the Earth formed 4.5 ± 0.1 Ga ago with an $^{87}Sr/^{86}Sr$ ratio of approximately 0.699.

Crustal rocks of granitic composition, enriched in Rb relative to Sr, began to form at a very early stage in the history of the Earth and have played a dominant role in the isotopic evolution of Sr. Such materials, evolving with high Rb/Sr ratios, have developed much higher present-day $^{87}Sr/^{86}Sr$ ratios than the upper mantle, which has evolved with a very much lower Rb/Sr (Fig. 2.6). This difference is fundamental to petrogenetic studies of igneous rocks, as it enables us to trace the effects of continental crustal contamination of mantle-derived magmas. However, to do this we must know the range of $^{87}Sr/^{86}Sr$ ratios that characterize crustal materials, in addition to understanding the isotopic evolution of Sr in the mantle throughout the course of geological time.

The present-day isotopic composition of Sr in the

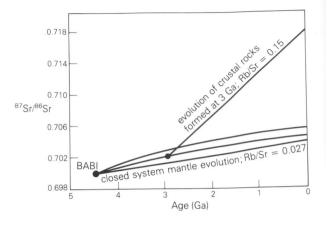

Figure 2.6 The isotopic evolution of terrestrial Sr. The curved lines represent the hypothetical evolution of Sr in the mantle from the primordial value (BABI) at 4.5 Ga to the present day. The curvature of these lines implies a time-dependent decrease in the bulk Rb/Sr ratio of the mantle. The straight line represents the closed system evolution of mantle material with an Rb/Sr ratio of 0.027.

mantle can be deduced from studies of young oceanic basalts from mid-oceanic ridges and oceanic islands, which clearly could not have been contaminated by continental crustal rocks en route to the surface. Such magmas could have assimilated oceanic crustal rocks, but as these are composed of similar materials, perhaps modified by seawater interaction, the effects will not in general be great. Figure 2.7a clearly demonstrates the considerable heterogeneity of $^{87}Sr/^{86}Sr$ in oceanic basalts, which suggests that the Rb/Sr ratio of the mantle is variable. If the Earth had evolved as a closed system with a constant Rb/Sr ratio of 0.027 to the present day, it would have an $^{87}Sr/^{86}Sr$ ratio of only 0.7040. In those regions of the mantle which have been depleted by the extraction of partial melts the Rb/Sr ratio will have been decreased, whereas in others it may have been increased as a consequence of metasomatic enrichment processes, involving the migration of incompatible element enriched fluids or partial melts. As a consequence of the differentiation of the mantle to form the rocks of the continental crust, its bulk Rb/Sr may have decreased with time.

The isotopic composition of Sr in the continental crust is markedly heterogeneous, as shown in

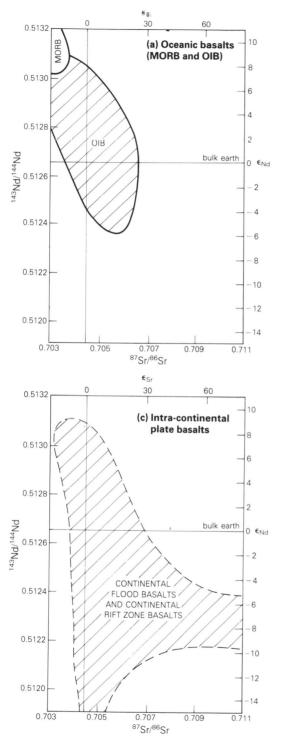

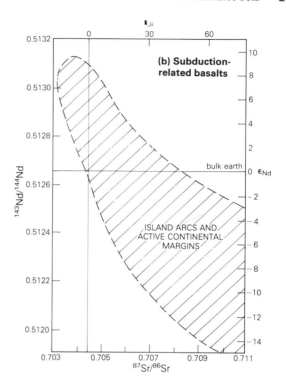

Figure 2.7 The variation of ^{143}Nd/^{144}Nd versus ^{87}Sr/^{86}Sr for volcanic rocks from different tectonic settings.

Figure 2.8, as a consequence of the wide age range and Rb/Sr ratios of crustal rocks. In general, the ^{87}Sr/^{86}Sr ratios are much higher than those of the upper mantle, as exemplified by the oceanic basalt array (Fig. 2.7a). Clearly, basaltic magmas erupted through the continental crust may interact with ^{87}Sr/^{86}Sr-rich basement rocks and, as a consequence, their isotopic compositions cannot be unambiguously attributed to source mantle characteristics.

The initial isotopic composition of Sr in volcanic rocks at the time of their formation can provide important information about the mantle sources from which the magmas originate, and the processes by which their chemical and isotopic compositions may be subsequently modified during ascent to the surface. In terms of Figure 2.7, we can clearly see that there are systematic differences in the ^{87}Sr/^{86}Sr ratios of volcanic rocks formed in different tectonic settings, implying that the upper mantle is isotopically heterogeneous on a large scale. The lowest ^{87}Sr/^{86}Sr ratios occur in mid-oceanic ridge basalts (MORB), while those of oceanic-island basalts (OIB) are in general signifi-

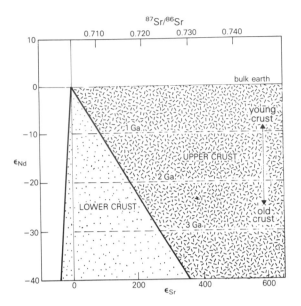

Figure 2.8   The Nd and Sr isotopic composition of continental crustal rocks at the present day; contours show the relative age of the crustal rocks in Ga (after DePaolo & Wasserburg 1979).

cantly higher. If we consider basalts generated in subduction-related and intracontinental plate-tectonic settings (Fig. 2.7b,c) an even greater range of $^{87}Sr/^{86}Sr$ ratios exists, some of which may be a consequence of continental crustal contamination. The high $^{87}Sr/^{86}Sr$ ratios of island-arc volcanic rocks (Ch. 6) have been variously attributed to the involvement of subducted sediments or sea water or, in some cases, to contamination of the magmas by interaction with terrigenous clastic sediments in the base of the island-arc crust. In Chapter 12 it is suggested that some intracontinental plate magmas have high $^{87}Sr/^{86}Sr$ ratios as a consequence of their derivation from Rb/Sr-enriched mantle sources within the subcontinental lithosphere.

The difference in $^{87}Sr/^{86}Sr$ between MORB and OIB suggests that MORB are produced by partial melting of mantle with a significantly lower Rb/Sr ratio than OIB source mantle. This difference must have existed for at least 1−2 Ga to be reflected in the isotopic compositions. It has been suggested (Ch. 5) that MORB are derived by partial melting of a depleted upper mantle layer from which the

materials of the continental crust have been progressively extracted throughout the past 4 Ga or so. In contrast, OIB appear to be derived from more enriched mantle sources which may involve primordial mantle components and recycled oceanic and continental lithosphere components (Ch. 9).

Sm−Nd

Samarium and neodymium are light REE, the concentrations of which in igneous rocks increase with increasing degree of differentiation, as they are incompatible. However, the Sm/Nd ratio decreases as Nd is concentrated in the liquid relative to Sm during the course of fractional crystallization. Sm and Nd are joined in a parent−daughter relationship by the alpha decay of ^{147}Sm to stable ^{143}Nd, with a half-life of 106×10^9 years. The decay of ^{147}Sm is described by the equation:

$$\frac{^{143}Nd}{^{144}Nd} = \left(\frac{^{143}Nd}{^{144}Nd}\right)_{initial} + \frac{^{147}Sm}{^{144}Nd}(e^{\lambda t} - 1)$$

where stable ^{144}Nd is used as a reference isotope.

The abundance of radiogenic ^{143}Nd, and hence the $^{143}Nd/^{144}Nd$ ratio of the Earth, has increased with time because of the decay of ^{147}Sm to ^{143}Nd. This can be described by a model based on the age and Sm/Nd ratio of the Earth and its primordial $^{143}Nd/^{144}Nd$ ratio (Fig. 2.9). The latter two parameters are assumed to equal the ratios in chondritic meteorites (referred to as a *Chondritic Uniform Reservoir* − CHUR; DePaolo & Wasserburg 1976). Partial melting of a chondritic uniform reservoir increases the Sm/Nd ratio of the residuum, which therefore evolves higher $^{143}Nd/^{144}Nd$ than CHUR. Those parts of the mantle which have not been involved in partial melting events should contain Nd, the isotopic composition of which evolves along the CHUR line in Figure 2.9.

There is still some uncertainty concerning the value of the primordial $^{143}Nd/^{144}Nd$ ratio, caused by differences in analytical procedures between different laboratories (Wasserburg *et al.* 1981). Mea-

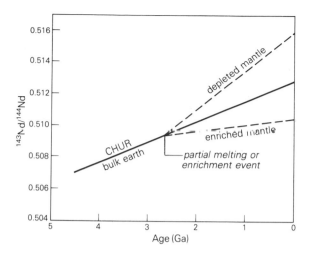

Figure 2.9 Isotopic evolution of Nd in a chondritic uniform reservoir (CHUR). Depleted and enriched mantle reservoirs evolve to higher and lower $^{143}Nd/^{144}Nd$ ratios than CHUR respectively.

sured $^{143}Nd/^{144}Nd$ ratios have been normalized in different ways (Faure 1986). For data corrected to $^{146}Nd/^{142}Nd = 0.636151$, CHUR has $^{143}Nd/^{144}Nd = 0.511847$, whereas for data corrected to $^{146}Nd/^{144}Nd = 0.7219$, the value for CHUR is 0.512638. All Nd isotopic data discussed in Chapters 5–12 have been corrected to the latter value.

To compare differences between the $^{143}Nd/^{144}Nd$ ratios of igneous rocks and CHUR, DePaolo & Wasserburg (1976) introduced the epsilon parameter:

$$\varepsilon_{Nd} = \left[\frac{(^{143}Nd/^{144}Nd)}{I^t_{CHUR}} \text{initial} - 1 \right] \times 10^4$$

where I^t_{CHUR} is the $^{143}Nd/^{144}Nd$ ratio of CHUR at the time of formation of the rock, t. Calculation of this parameter also circumvents the problem of different inter-laboratory normalization procedures. A positive ϵ value implies that the magmas were formed from depleted mantle, whereas a negative value indicates that they were derived from enriched mantle sources that had a lower Sm/Nd than CHUR.

Combined Nd–Sr

As shown in Figure 2.7a, the $^{143}Nd/^{144}Nd$ and $^{87}Sr/^{86}Sr$ ratios of recent oceanic basalts (MORB and OIB) form a strongly correlated array. This correlation has been used to define an $^{87}Sr/^{86}Sr$ ratio for the bulk Earth complementary to the Nd data. Thus, for a $^{143}Nd/^{144}Nd$ ratio of 0.512638 for CHUR (bulk Earth) at the present time, $^{87}Sr/^{86}Sr$ lies between 0.7045 and 0.7055. In all the Nd–Sr isotope diagrams in Chapters 5–12 a value of 0.7045 has been assumed. The $^{87}Sr/^{86}Sr$ ratio of the bulk Earth derived in this way can be used to define an epsilon parameter analogous to that defined for Nd:

$$\varepsilon_{Sr} = \left[\frac{(^{87}Sr/^{86}Sr)_{initial}}{(^{87}Sr/^{86}Sr)^t_{UR}} - 1 \right] \times 10^4$$

where UR represents 'uniform reservoir', equivalent to bulk Earth. However, in view of the increasing spread of Nd–Sr isotopic compositions being recorded in young volcanic rocks, the bulk earth parameter for the Rb/Sr system should be viewed with caution (White & Hofmann 1982) and most recent articles no longer use the ε_{Sr} notation.

The Nd and Sr isotopic compositions of MORB indicate that they originate from sources that have higher Sm/Nd and lower Rb/Sr than the chondritic reservoir (bulk Earth). Such source regions are said to be depleted, because they appear to have lost Rb and other LIL elements. However, the mantle beneath the ocean basins also appears to contain magma sources which are enriched and have higher Rb/Sr and lower Sm/Nd ratios than bulk Earth.

Typical continental crustal rocks have lower Sm/Nd and therefore lower $^{143}Nd/^{144}Nd$ ratios (negative εNd values) than those derived from the upper mantle (Fig. 2.8). As a consequence, combined Nd–Sr isotopic studies potentially provide a powerful tracer for contamination of magmas by continental crustal rocks.

U–Th–Pb

Uranium has three naturally occurring radioactive isotopes, ^{238}U, ^{235}U and ^{234}U, while Th exists primarily as one radioactive isotope, ^{232}Th. However, in addition five radioactive isotopes of Th occur in nature as short-lived intermediate daughters of ^{238}U, ^{235}U and ^{232}Th. Pb has four naturally occurring isotopes, ^{208}Pb, ^{207}Pb, ^{206}Pb and ^{204}Pb. Of these, only ^{204}Pb is not radiogenic and is therefore used as a stable reference isotope. ^{238}U, ^{235}U and ^{232}Th are each the parent of a chain of radioactive daughters ending with a stable isotope of Pb:

$$^{238}U \rightarrow {}^{234}U \rightarrow {}^{206}Pb$$

$$^{235}U \rightarrow {}^{207}Pb$$

$$^{232}Th \rightarrow {}^{208}Pb$$

The isotopic composition of Pb in rocks containing U and Th is given by the following equations:

$$\frac{^{206}Pb}{^{204}Pb} = \left(\frac{^{206}Pb}{^{204}Pb}\right)_{initial} + \frac{^{238}U}{^{204}Pb}(e^{\lambda_1 t} - 1)$$

$$\frac{^{207}Pb}{^{204}Pb} = \left(\frac{^{207}Pb}{^{204}Pb}\right)_{initial} + \frac{^{235}U}{^{204}Pb}(e^{\lambda_2 t} - 1)$$

$$\frac{^{208}Pb}{^{204}Pb} = \left(\frac{^{208}Pb}{^{204}Pb}\right)_{initial} + \frac{^{232}Th}{^{204}Pb}(e^{\lambda_3 t} - 1)$$

The ratio $^{238}U/^{204}Pb$ is known as μ.

Following the same logic that we have adopted for the Nd–Sm and Rb–Sr systems, the isotopic composition of Pb at the time of formation of the Earth should be the same as that in meteorites. Chen & Wasserburg (1983) have measured the following isotopic compositions of Pb in troilite (FeS) from the Canyon Diablo iron meteorite, which may be taken to represent the primordial values:

$$\frac{^{206}Pb}{^{204}Pb} = 9.3066, \quad \frac{^{207}Pb}{^{204}Pb} = 10.293, \quad \frac{^{208}Pb}{^{204}Pb} = 29.475$$

Figure 2.10 shows three growth curves for Pb isotopes for typical μ values of 8,9 and 10, assuming that the age of the Earth is 4.55 Ga. These growth curves fan out from the point representing primordial Pb. The straight lines in this diagram are isochrons for ages 0, 1, 2 and 3 Ga. All single-stage leads that were removed from their sources at time t must lie on these isochrons, even if they grew in different source regions (Faure 1986). The line for $t = 0$ is called the geochron, and all modern single-stage leads in the Earth and in meteorites must lie on it.

Igneous rocks contain Pb, the isotopic composition of which reflects multistage histories, having evolved in systems with varying U/Pb and Th/Pb ratios for varying lengths of time. Nevertheless, the isotope ratios of cogenetic suites of igneous rocks should define straight lines (isochrons) in plots of $^{206}Pb/^{204}Pb$ versus $^{207}Pb/^{204}Pb$ or $^{208}Pb/^{204}Pb$, similar to those of Figure 2.10, provided that all had the same initial Pb isotopic composition and evolved subsequently with different U/Pb and Th/Pb ratios. However, not all linear arrays on Pb–Pb diagrams need have age significance, as they may also result from mixing of leads of different isotopic composition.

Figure 2.11 shows the variation of $^{207}Pb/^{204}Pb$ versus $^{206}Pb/^{204}Pb$ for oceanic basalts, subduction-related basalts and intracontinental plate basalts, analogous to Figure 2.7 for Nd–Sr isotopes. The Pb isotopic compositions of MORB and OIB lie to the right of the geochron, defining marked linear arrays, the significance of which remains controversial (Chs 5 & 9). Clearly, the sub-oceanic upper mantle is extremely heterogeneous in terms of its Pb isotopic composition. The Pb isotopic compositions of subduction-related basalts are displaced to the high $^{207}Pb/^{204}Pb$ side of the MORB–OIB array (Chs 6 & 7) which, at least for oceanic-island arcs, has been attributed to mixing between mantle Pb and Pb from subducted oceanic sediments. Basalts from intracontinental plate-tectonic settings. (Chs 10–12) display a wide range of $^{207}Pb/^{204}Pb$ and $^{206}Pb/^{204}Pb$ ratios, which can often be interpreted in terms of crustal contamination of mantle-derived magmas.

U and Th are both preferentially concentrated in

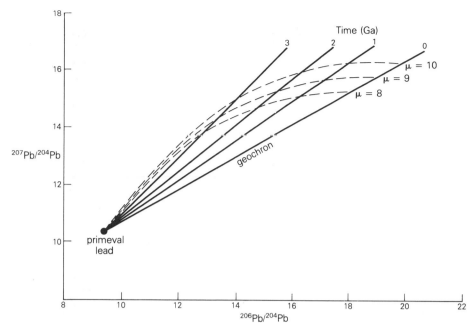

Figure 2.10   Growth curves showing the isotopic evolution of Pb in the Earth. The dashed curved lines are lead growth curves for U−Pb systems having present-day μ values of 8, 9 and 10. The straight solid lines are isochrons for ages 0, 1, 2 and 3 Ga.

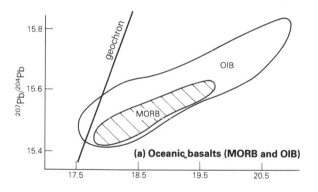

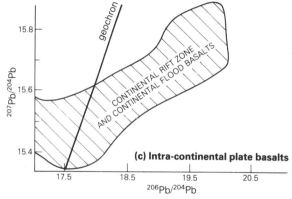

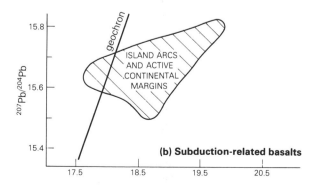

Figure 2.11   Variation of $^{207}Pb/^{204}Pb$ versus $^{206}Pb/^{204}Pb$ for volcanic rocks from different tectonic settings.

silicate melts compared to Pb and, consequently, the U/Pb and Th/Pb ratios of crustal rocks are higher than those of the mantle. Additionally, U and Th are preferentially concentrated in upper crustal rocks and, consequently, the upper and lower crust may have distinctly different Pb isotope signatures (Fig. 2.12). Thus Pb isotopes may provide powerful constraints for the nature of crustal contaminants in continental volcanic suites.

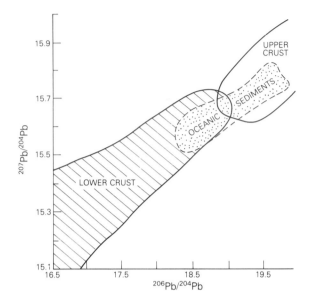

Figure 2.12 Variation of $^{207}Pb/^{204}Pb$ versus $^{206}Pb/^{204}Pb$ for rocks from the upper and lower continental crust. Shown for comparison is the field of oceanic sediments (after Zartman & Doe 1981).

U-series disequilibrium

The decay series arising from ^{238}U and ^{235}U contain radioactive isotopes of many different elements. These daughters of U may be separated from their parents and from each other during partial melting and subsequent fractional crystallization, because of their different geochemical properties. The resulting radioactive disequilibrium may be used for dating over time periods ranging from a few tens of years to one million years (Faure 1986). Additionally, for young volcanic rocks, it may provide useful information concerning the nature of partial melting processes (Ch. 6), and the time elapsed between the initial melting and subsequent extrusion of the lava (Allègre & Condomines 1982). For practical reasons, the abundances of the daughters of U are measured in terms of their activities by means of sensitive radiation detectors.

^{230}Th is a radioactive isotope of Th that is produced in the decay series of ^{238}U. Its immediate parent is ^{234}U and its daughter is ^{226}Ra. The activity of ^{230}Th in young volcanic rocks is described by the following equation (Faure 1986):

$$\left(\frac{^{230}Th}{^{232}Th}\right)_A = \left(\frac{^{230}Th}{^{232}Th}\right)_{Ax} e^{-\lambda_{230}t}$$

$$+ \left(\frac{^{238}U}{^{232}Th}\right)_A (1 - e^{-\lambda_{230}t})$$

where ^{232}Th is used as a reference isotope because of its long half-life (1.4×10^{10} years). The first term describes the decay of unsupported ^{230}Th, while the second term represents the growth of ^{230}Th that is supported by ^{238}U. This is the equation of a straight line in coordinates of $(^{230}Th/^{232}Th)_A$ versus $(^{238}U/^{232}Th)_A$ when t is constant (Fig. 2.13).

At the time of crystallization ($t = 0$), suites of cogenetic volcanic rocks define an isochron with slope equal to zero. Subsequently, the isochron rotates about the equipoint and its slope increases and approaches unity. The isochron with a slope of one is called the *equiline*, and is the locus of points for which $(^{230}Th/^{238}U)_A = 1$, as required by secular equilibrium. The slope of the isochron begins to deviate detectably from zero about 10^3 after crystallization and approaches unity about 10^6 years later. Thus $10^3 - 10^6$ years is the useful range of this geochronometer.

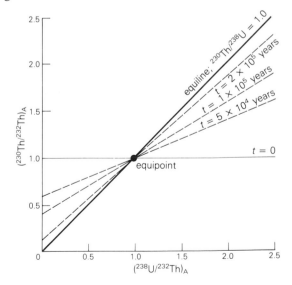

Figure 2.13 $^{230}Th/^{238}U$ isochron diagram (after Faure 1986, Fig. 21.7, p. 379). It should be noted that the $(^{230}Th/^{232}Th)_A$ ratio at $t = 0$ is not constrained to equal unity.

He isotopes

The variation of ^{3}He/^{4}He in volcanic rocks is due to the balance between primordial He and radiogenic He generated by radioactive decay of ^{232}Th, ^{235}U and ^{238}U. The ratio (R) is generally normalized to the atmospheric value of ^{3}He/^{4}He (R_a). R/R_a values for OIB range from 1 to 32, whereas MORB define a more restricted range from 5 to 15. Basalts from the oceanic islands of Hawaii and Iceland (Chs 5 & 9) have very high ^{3}He/^{4}He ratios, which has been considered to reflect the involvement of relatively undegassed (?primordial) mantle components in their petrogenesis (Allègre et al. 1986, Zindler & Hart 1986). In contrast, basalts from the oceanic islands of Tristan da Cunha and Gough are characterized by low ^{3}He/^{4}He ratios of around 5, which may suggest the involvement of recycled source components (Ch. 9).

2.5.2. Cosmogenic radionuclides:^{10}Be

Cosmogenic radionuclides are produced by the interaction of cosmic rays with atoms in the atmosphere and on the Earth's surface. Be has seven isotopes, the mass numbers of which range from 6 to 12, of which only ^{9}Be is stable. Unstable ^{7}Be and ^{10}Be occur in nature because they are produced by nuclear reactions caused by cosmic rays. ^{7}Be decays to ^{7}Li with a half-life of 53 days, whereas ^{10}Be has a half-life of 1.5×10^6 years and decays by beta emission to stable ^{10}B.

The cosmogenic Be isotopes are rapidly removed from the atmosphere by precipitation, and are transferred to the sediment at the bottom of the oceans and to the continental ice sheets of Greenland and Antarctica. Their production rates vary with latitude, altitude in the atmosphere and with time. The residence time ($\sim$ 16 years) of ^{10}Be in the surface layer of the oceans is, however, long enough to permit horizontal mixing, which reduces the effect of latitudinal variations in its production rate. Clearly, the abundance of ^{10}Be in a terrestrial reservoir depends upon the production rate, the sedimentation rate and the time elapsed since deposition. High sedimentation rates tend to dilute the concentration of the radionuclide.

The identification of ^{10}Be in some recent subduction-related lavas (Chs 6 & 7) has been used to argue for the role of subducted oceanic sediments in their petrogenesis. However, its absence does not argue against because of its relatively short half-life.

2.5.3   Stable isotopes

Oxygen

Oxygen has three stable isotopes, ^{16}O, ^{17}O and ^{18}O. Its isotopic composition in a sample is generally reported in terms of a parameter δ^{18}O, which is the difference between the ^{18}O/^{16}O ratio of the sample and that of a standard called SMOW (Standard Mean Ocean Water):

$$\delta^{18}O = \left[\frac{(^{18}O/^{16}O)_{sample} - (^{18}O/^{16}O)_{SMOW}}{(^{18}O/^{16}O)_{SMOW}} \right] \times 10^3$$

Crystal−liquid fractionation processes during partial melting and subsequent fractional crystallization may potentially cause variations in the oxygen isotope composition of magmas (Kyser et al. 1982). However, at magmatic temperatures the mass fractionation of oxygen isotopes is much less pronounced than at atmospheric temperatures and, consequently, the effects are only small (Faure 1986). Thus, broadly speaking, primitive basaltic magmas should have oxygen isotopic compositions which directly reflect those of their mantle source.

Conventionally mantle-derived magmas were considered to have a very narrow range of δ^{18}O values (from +5.5 to +6‰; James 1981) compared to crustal rock types (Fig.6.47) which have δ^{18}O > 6. However, Kyser et al. (1982) have demonstrated that oceanic basalts actually display a much wider range of δ^{18}O values (from +4.9 to +8.3‰), which they attribute to oxygen isotope heterogeneity in the mantle source. Nevertheless, oxygen isotopes remain useful indicators of crustal contamination processes because of the contrasting isotopic compositions of continental crustal rocks, which have equilibrated with the hydrosphere, and mantle-derived magmas.

Interaction between crustal material and basic magmas may occur through direct contamination of the magmas as they rise through the continental crust (Chs 7, 10 & 11) or by recycling of crustal materials (sediments) back into the mantle source region of basalts in subduction zones (Chs 6 & 7). Such contamination may involve bulk assimilation of crustal rocks, with resulting elemental mixing of the two reservoirs, or selective elemental or isotopic exchange. In subduction zones the release of fluids or partial melts, from sedimentary rocks and seawater-altered basalt in the subducted ocean crust, into the overlying mantle wedge may be a mechanism for transferring a continental crustal isotopic and trace element signature to the mantle source of the arc basalts.

The isotopic composition of oxygen in young volcanic rocks has been used in conjunction with radiogenic isotopes of Sr, Nd and Pb to detect the contamination of basaltic magmas by crustal rocks. Compared to mantle-derived magmas, the latter are enriched in ^{18}O and radiogenic ^{87}Sr, but depleted in radiogenic ^{143}Nd. As a consequence, addition of O, Sr and Nd from ancient granitic crustal components to basaltic magmas can cause positive correlations between $\delta^{18}O$ and $^{87}Sr/^{86}Sr$ and negative correlations between $\delta^{18}O$ and $^{143}Nd/^{144}Nd$ (e.g. see Ch. 6). However, the shape of the mixing curve may differ significantly depending upon the actual mechanism of contamination (Fig. 6.48), which may be useful in petrogenetic modelling. James (1981), Graham & Harmon (1983) and Faure (1986) have presented useful discussions of the use of stable isotopes in studying the roles of crustal contamination in magmas, and the reader is referred to these works for further details.

2.6   Geochemical criteria for the identification of the palaeotectonic setting of ancient volcanic sequences

Clearly, if we can correlate particular geochemical characteristics of modern volcanic rocks with their specific tectonic setting, we can use these data to identify the tectonic setting of ancient volcanic sequences. However, this approach in itself is unlikely to give an unambiguous determination of the tectonic setting. Instead, we need to consider the overall geological setting of the magmatism (see, e.g., Cas & Wright 1987) including, for example, the nature of the basement and the percentage of pyroclastic rocks within the sequence. The compositions of basaltic magmas are dependent upon their source composition and mineralogy, the depth and degree of partial melting, the mechanism of partial melting and the various fractionation and contamination processes they may have undergone en route to the surface. Only when some or all of these are unique to a particular tectonic setting will we be able to use basalt geochemistry as a diagnostic indicator of tectonic setting.

During the past 15 years, a number of papers have appeared in which the major, minor and trace element compositions of young basaltic rocks have been related to the tectonic environment in which the basalts were generated. These have led to the development of 'tectonomagmatic discrimination diagrams' which may be used to elucidate the tectonic setting of ancient volcanic suites (Pearce & Cann 1973, Floyd & Winchester 1975, Pearce et al. 1975, 1977, Wood et al. 1979, Shervais 1982, Pearce 1982, Mullen 1983, Meschede 1986). Such discriminant diagrams may be divided into three types (Duncan 1987):

(1)   Those which utilize relatively immobile trace and minor elements, the concentration of which can be measured by XRF. The most commonly used are the ternary diagrams $Ti/100-Zr-Y.3$ (Fig. 2.14; Pearce & Cann 1973) and $2Nb-Zr/4-Y$ (Fig. 2.15; (Meschede 1986).

(2)   Those which use major and minor elements, e.g. $TiO_2-K_2O-P_2O_5$ and $MgO-FeO-Al_2O_3$ (Pearce et al. 1975, 1977) and $TiO_2-MnO-P_2O_5$ (Fig. 2.16; Mullen 1983).

(3)   Those which use immobile trace elements, the concentrations of which can be determined by INAA. The most commonly used diagrams are the $Th-Hf/3-Ta$ ternary (Fig. 2.17; Wood et al. 1979) and Th/Yb versus Ta/Yb (Pearce 1982; e.g. Fig. 7.26).

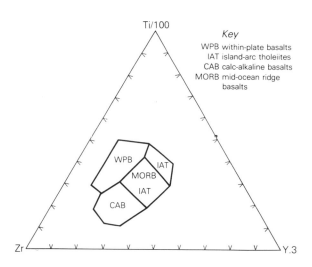

Figure 2.14 Ti/100–Zr–Y.3 tectonomagmatic discrimination diagram for basaltic rocks (after Pearce & Cann 1973, Fig. 3, p. 295).

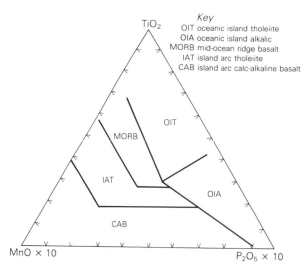

Figure 2.16 TiO₂–MnO–P₂O₅ tectonomagmatic discrimination diagram for oceanic basaltic rocks (after Mullen 1983, Fig. 1, p.54).

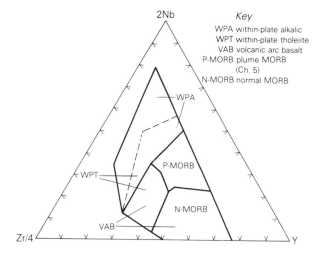

Figure 2.15 2Nb–Zr/4–Y tectonomagmatic discrimination diagram for basaltic rocks (after Meschede 1986, Fig. 1, p. 211).

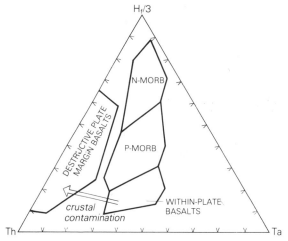

Figure 2.17 Hf/3–Th–Ta tectonomagmatic discrimination diagram for basalts and more differentiated rocks (after Wood *et al.* 1979a, Fig. 3, p. 332.

In order to use these diagrams (except that of Wood *et al.* shown in Fig. 2.17), it is necessary to know the degree of fractionation of the volcanic rock being classified, as they are only strictly applicable to basic volcanic rocks. However, this is not always apparent from the major element chemistry after the rock has been weathered or metamorphosed.

On the whole, major element geochemistry is not particularly useful for tectonomagmatic discrimination. Nevertheless, as we can see from Table 1.3, specific magma types are associated with particular tectonic settings, for example the association of calc-alkaline magmas with subduction zones. It is uncertain whether these discriminant diagrams are valid for very old rocks, as magma genesis in the Precambrian may have been very different from that of today. In particular, the Precambrian mantle would have experienced fewer melting episodes and would therefore be richer in incompatible elements.

Using the above types of discriminant diagrams it is found that, in general, the correct identification of tectonic setting is highest for magmas not erupted in within-continental-plate environments. Unfortunately, this is precisely the environment in which volcanic rocks are most easily preserved. In general, relatively few continental flood basalt data sets were used to establish the discriminant boundaries, and several recent papers (Holm 1982, Prestvik & Goles 1985, Duncan 1987, Marsh 1987) have noted that some continental flood basalts (CFBs) do not plot in the within-plate fields. However, this does not necessarily invalidate the diagrams for, as we shall see in Chapter 10, CFBs may actually be generated in a variety of tectonic settings. Many clearly have mixed geochemical characteristics; for example, features of both intraplate and subduction-related tectonic settings.

For a combination of elements to be useful in characterizing magma types from different tectonic settings they must ideally have a much greater variation in concentration between samples from different environments than between samples from the same environment. Also, the trace element classification employed should distinguish as many different environments as possible. Clearly, to be of any value in determining the palaeotectonic setting of altered volcanic rocks the selected trace elements must be immobile, i.e. they must not be transported significantly in the fluid phase during weathering and metamorphism. Elements such as Na, K, Ca, Ba, Rb and Sr, and possibly the light REE, are mobile and therefore are not useful for tectonomagmatic discrimination purposes. Thus we cannot in general use the alkali-silica diagram (Fig. 1.1) to reliably classify the volcanic rocks of ancient sequences. In contrast, elements such as Fe, Ti, Ni, Cr, V, Zr, Nb, Ta and Hf may be relatively immobile. The establishment of which element to use is clearly an empirical process, and problems frequently arise because elements which are immobile during weathering may become mobile during the lower grades of metamorphism.

In addition to these simple discriminant diagrams, it is possible to compare a much wider range of trace elements in basalts from known and unknown tectonic settings using normalized trace element variation diagrams or spiderdiagrams (Section 2.3.2). A similar approach has been used in Chapter 10 in which, instead of normalizing to chondritic or primordial mantle abundances, we normalize to the composition of a basalt type generated in a known tectonic setting, e.g. MORB or oceanic-island tholeiite. The problem then becomes one of choosing an average composition to define the normalizing constants which may be considered typical of a particular magma generation environment. Pearce (1987) has developed a computerized system for the identification of the eruptive setting of ancient volcanic rocks, based on this approach, which integrates geological, petrological, mineralogical and geochemical data.

On the basis of the geochemical data presented in Chapters 5−12 it is clear that particular trace element abundance patterns (spiderdiagrams) and Sr−Nd−Pb isotopic signatures are associated with different magma generation environments. However, in many provinces there is still some controversy concerning the detailed petrogenetic interpretation of these data. Additionally, there are clearly some geochemical signatures which are not necessarily unique to a single tectonic environment (Arculus 1987, Duncan 1987). For example, the trace element characteristics of subduction-related basalts are rather similar to those of intracontinental plate basalts which have become contaminated by the continental crust.

Further Reading

Cox, K. G., J. D. Bell & R. J. Pankhurst 1979. *The interpretation of igneous rocks*, London: Allen and Unwin; Ch. 2.

Faure, G. 1986. *Principles of isotope geology*, 2nd edn New York: John Wiley

Graham, C. M. & R. S. Harmon 1983. Stable isotope evidence on the nature of crust-mantle interactions. In *Continental basalts and mantle xenoliths*, C. J. Hawkesworth & M. J. Norry (eds), 20–45. Nantwich: Shiva.

Hanson, G. N. 1980. Rare earth elements in petrogenetic studies of igneous systems. *Ann. Rev. Earth Planet. Sci.* **8**, 371–406.

James, D. E. 1981. The combined use of oxygen and radiogenic isotopes as indicators of crustal contamination. *Ann. Rev. Earth Planet. Sci.* **9**, 311–44.

Potts, P. J. 1987. *A handbook of silicate rock analysis*. London: Blackie.

Thompson, R. N., M. A. Morrison, G. L. Hendry & S. J. Parry 1984. An assessment of the relative roles of crust and mantle in magma genesis. *Phil Trans R. Soc. Lond.* A310, 549–90.

CHAPTER THREE

Partial melting processes in the Earth's upper mantle

3.1 Introduction

The mantle of the Earth is an essentially solid shell separating the metallic and partially molten core from the cooler rocks of the crust. Extending to a depth of some 2900 km, it accounts for 83% of the Earth's volume and 67% of its mass. It can be subdivided into two main seismic regions that are broadly concentric with the surface; the upper mantle and lower mantle, separated by the 670 km seismic discontinuity (Fig. 3.1). In this chapter we shall focus our attention on the physical state, chemical composition, mineralogy and partial melting behaviour of the upper mantle. Everything that happens at the surface of the Earth − the building of mountain ranges, the formation of ocean basins, volcanism and even changes in sedimentation patterns − is a response to events taking place within this part of the mantle. In addition, this is also the zone in which the driving forces that move the lithospheric plates about the surface of the Earth originate.

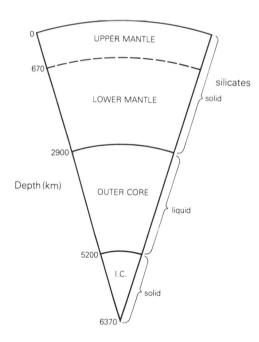

Figure 3.1 A sector through the Earth, showing its layered structure. I. C. = inner core.

During the 4.6 Ga since the formation of the Earth, melting of the more fusible constituents of the upper mantle has produced magmas which rise to the surface and solidify, adding new rocks to the crust and giving off water vapour and other gases which make up the oceans and atmosphere. The upper mantle has thus undergone an apparently irreversible differentiation process throughout geological time. However, as we shall see in Chapter 6, subduction of oceanic lithosphere provides a mechanism for recycling material back into the mantle to depths at least as far as the 670 km seismic discontinuity.

Unfortunately, our knowledge of the structure, physical properties and chemical composition of even the uppermost parts of the mantle is limited because of the impossibility of direct access. The velocity of seismic waves imposes certain constraints on the density and other physical properties of mantle materials. Evidence for the chemistry and mineralogy of the upper mantle is provided by studies of the major and trace element and isotope geochemistry of oceanic basalts (Chs 5 & 9), studies of small in-thrust slices of possible upper mantle material within some fold belts and ophiolite complexes, and, most importantly, studies of xenoliths in continental and oceanic alkali basalts and kimberlites.

3.2 The physical state of the upper mantle

In this section we shall discuss the thermal structure of the Earth and the nature of convective motions within the mantle, in addition to aspects of the structure and physical properties of the mantle as evidenced by seismic studies. For a more detailed review of the physics of the Earth, the reader is referred to Brown & Mussett (1981).

3.2.1 Seismic data

The main body of data about the physics of the Earth comes from the study of earthquake waves. The release of energy at the focus of an earthquake produces several types of wave, the energy of which is transmitted by different phenomena. The primary P waves are compressional, while the slower secondary, or S, waves are shear waves. Both types of waves pass through the interior of the Earth. P waves can be transmitted by both solids and liquids, whereas S waves can only be transmitted through solid media capable of supporting shear stresses.

Earthquake waves can be both reflected and refracted within the Earth. Reflection occurs at levels where there is a distinct change in physical properties between layers, and allows identification of the major boundaries shown in Figure 3.1. Three major first-order seismic discontinuities occur, of which the largest, at 2900 km, is the core−mantle interface. At 10−12 km beneath the oceans and 30−50 km beneath the continents is the Mohorovičić discontinuity, or Moho, separating the crust from the mantle, and at about 5200 km is the inner core − outer core interface. These major discon-

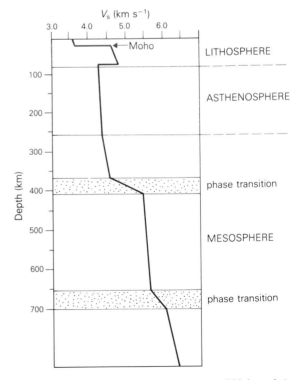

Figure 3.2 The structure of the outermost 700 km of the Earth, showing the variation of S-wave velocity (V_s) with depth. Stippled bands represent major velocity changes associated with high-pressure phase transitions.

tinuities are mainly compositional boundaries, although in the case of the inner − outer core boundary there is also a change of state from solid to liquid. Additional seismic discontinuities occur at 400 and 670 km, corresponding to major changes in the structure of silicate minerals (Fig. 3.2).

The major regions of the upper 670 km of the Earth can be summarized thus (Fig. 3.3):

(a) The *crust* consists of the region above the Moho and ranges in thickness from about 3 km in some oceanic regions to about 80 km in some continental areas.

(b) The *upper mantle* extends from the Moho down to about 670 km, and includes the lower part of the *lithosphere* and the *asthenosphere*. The *lithosphere* (50 to 150−200 km thick) is the strong outer layer of the Earth, including the crust, that reacts to stresses as a brittle solid. It forms the plates which move about the surface of the Earth according to plate tectonic theory. The lithosphere is characterized by high velocity and efficient propagation of seismic waves. Its lower boundary is marked by an abrupt decrease in shear wave velocity and may approximate an isothermal boundary layer ($T \sim 1200°C$) beneath the oceans. In general, the base of the continental lithosphere is much harder to define seismically than the base of the oceanic lithosphere. Nevertheless, most workers agree that it is much thicker than the oceanic lithosphere, particularly beneath ancient cratonic nuclei. The *asthenosphere*, extending from the base of the lithosphere to about 250 km, is by comparison a weak layer that readily deforms by creep. This is the zone in the upper mantle in which major convective motions are most likely to occur. The asthenosphere is a low-velocity zone in which seismic waves are attenuated strongly, possibly indicating the presence of a partial melt phase.

Earthquake waves provide not only a static picture of a concentrically layered Earth but also a dynamic picture of plate tectonics. Since earthquakes only occur in rocks which are rigid enough to fracture, the distribution of earthquake foci may be used to delineate the boundaries of the stable lithospheric plates. The depths of foci along constructive plate margins (mid-oceanic ridges) are <100 km, indicating the presence of anomalously hot mantle at shallow depths. At convergent plate boundaries, where one lithospheric plate sinks beneath another, foci can occur at depths up to 700 km, and can be used to delineate the upper surface of the subducting plate, the Benioff zone.

Seismology can clearly provide a great deal of information about the layered structure of the Earth but relatively little about the physical and chemical properties of the layers. The body wave seismic velocities, V_P and V_S, depend upon the density (ρ) of the medium through which they are passing and on its elastic moduli, μ and κ:

$$V_P = \sqrt{\frac{\kappa + 4/3\mu}{\rho}}, \qquad V_S = \sqrt{\frac{\mu}{\rho}}$$

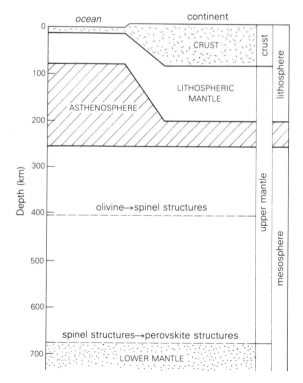

Figure 3.3 The major regions of the upper 700 km of the Earth.

where κ is the bulk or compressibility modulus, and μ is the rigidity or shear modulus. If we had a third relationship between these parameters the body wave data could be solved directly for ρ, μ and κ, providing valuable information about the physical properties of the Earth's interior. Unfortunately, only empirical relationships can be formulated, but nevertheless these are sufficient to allow fairly precise calculation of the variation of density with depth within the Earth (Fig. 3.4). Such physical constraints can then be used to determine possible rock types that may exist at depth (Section 3.3).

Given a knowledge of the density of mantle rocks we can then calculate the pressure at a given depth, using the relationship:

$$P = h\rho g$$

where P is the pressure, h the depth, ρ the density of the overlying column of rock, and g the acceleration due to gravity. If, for simplicity, we assume an average density of 3.3 g cm^{-3} in the upper few hundred kilometres of the Earth (Fig. 3.4), then:

$$P = 3.3 \times 10^7 \; h \quad Nm^{-2} \quad \text{or} \quad P = 0.33h \; kbar$$

where h is in km.

3.2.2 Temperature variation with depth within the Earth

Considerable uncertainty exists regarding the detailed variation of temperature with depth within the Earth. It is dependent on such features as:

(a) the initial temperature distribution in the newly formed planet;
(b) the amount of heat generated subsequently as a function of depth and time;
(c) the process of core formation.

Thermal gradients in near-surface rocks have been measured in deep boreholes and usually range between 20 and 40°C km^{-1}. Such gradients obviously cannot be extrapolated throughout the entire 2900 km of the mantle, as they would predict unrealistically high temperatures for the Earth's core. Consequently, we must rely on indirect modelling techniques to determine the variation of temperature with depth. Most models are based on combinations of the following:

(a) modelling the Earth's thermal evolution, including various models for core formation;
(b) constraints imposed by variations in seismic wave velocity, electrical conductivity, thermal conductivity and other physical properties with depth;
(c) models for the redistribution of radioactive heat sources in the Earth by partial melting and convective processes.

Generally, all such models converge on an estimate for the temperature of the core/mantle boundary of 3000 ± 500°C.

Any thermal model must be able to account for the total surface heat flow which varies over the

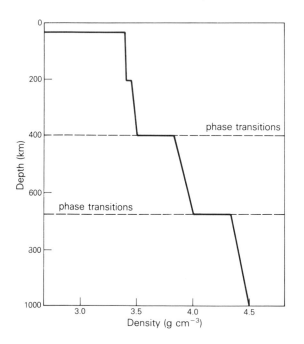

Figure 3.4 Density distribution in the outer 1000 km of the Earth (after Dziewonski & Anderson 1981).

Table 3.1 Selected heat flow data for the Earth (Brown & Mussett 1981, Table 8.2a, p.146).

Heat flow	Continental average (W m^{-2})	Oceanic average (W m^{-2})	Total (W)
through surface of the Earth	53×10^{-3}	62×10^{-3}	3×10^{13}
through Moho	28×10^{-3}	57×10^{-3}	2.4×10^{13}
through core/mantle boundary			$0.4 - 1.6 \times 10^{13}$

surface of the Earth, and which is chiefly attributed to radioactive heat production in the crust and mantle, with a smaller contribution from the core (Table 3.1.) The input of heat from the core to the mantle is inferred because of the need for an energy source to drive the geomagnetic dynamo. This could be due to radioactive decay of ^{40}K or to growth of the inner core.

The major source of the heat supplied to the base of the crust, the Moho, is considered to be derived from the radioactive decay of U, Th and K within the mantle. Theory suggests that the major radioisotopes of these elements should have been concentrated near the surface of the Earth as a consequence of the irreversible differentiation of the mantle throughout geological time. The mantle is considered to be highly depleted in radioactive elements, by a factor of about 100 compared to the continental crust (Table 3.2). However, because of its greater total volume the heat productivity of the mantle is probably similar to that of the crust.

As the continental crust is both richer in radioactive elements (Table 3.2) and thicker than the

oceanic crust, its total heat production must be higher. Thus, if the amount of heat supplied to the Moho from the mantle and core is constant then the heat flow through the continental crust should be significantly higher than that through the oceanic crust. However, as shown in Table 3.1, average oceanic heat flow is, if anything, slightly greater than the continental average. This must mean that heat flow through the sub-oceanic and subcontinental Mohos must differ, that through the oceanic Moho being about twice that through the continental Moho. This may mean that the subcontinental mantle is highly depleted in radioactive heat-producing elements.

The variation of temperature with depth within the Earth is known as the *geothermal gradient*, or *geotherm*, and this varies with tectonic setting. Figure 3.5 shows typical geotherms for a stable continental shelf area, an oceanic plate well away from the ridge and a mid-oceanic ridge. These geotherms are significantly different within the upper 200 km of the mantle but ultimately converge towards an adiabatic gradient of 0.3°C km^{-1}.

Table 3.2 Estimated radioactive contents and heat productivities (Brown & Mussett 1981, Table 8.2b, p.146).

	U (ppm)	Th (ppm)	K (%)	Total heat production (μW m^{-3})
average continental crust	1.6	5.8	1.7-3.0	1.0-1.1
average oceanic crust	0.9	2.7	0.4	0.5
undepleted mantle	0.015	0.08	0.1	0.02

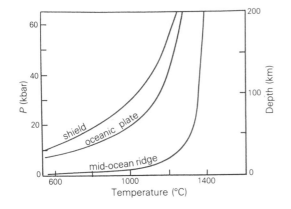

Figure 3.5 Variations in geothermal gradient within the upper 200 km of the Earth (after Wyllie 1981, Fig 1).

Geothermal gradients must differ between continental and oceanic regions because of the near-equivalence of oceanic and continental heat flow. As continental rocks have much higher heat productivity than oceanic rocks, temperatures in the sub-oceanic mantle must be higher than in the subcontinental mantle, as shown in Figure 3.5.

There are two dominant mechanisms by which heat is transported in the Earth, conduction and advection by convective flow. Although advection is the primary mode of outward heat transfer on a global scale, vertical conduction becomes important wherever the convective flow is mainly horizontal rather than vertical (Jeanloz & Morris 1986). Within the lithosphere (both continental and oceanic) the vertical transfer of heat must be predominantly by conduction, and it is in this region of the Earth that temperature gradients are most marked. Within the asthenospheric upper mantle, where convective heat transport dominates, the average temperature changes little with depth. The sharp bend in the geotherms shown in Figure 3.5 reflects the transition from relatively inefficient conductive heat transfer near the surface to relatively more efficient advective heat transfer at depth. This clearly accounts for the very rapid increase in temperature with depth associated with the constructive plate margin (mid-oceanic ridge) in Figure 3.5.

Knowledge of the geothermal gradient is essential if we are to understand partial melting processes within the upper mantle (Section 3.4). Unfortunately, because of the imprecise nature of the modelling techniques involved, geotherms can only be defined within rather broad limits for specific tectonic settings.

3.2.3   Convection within the mantle

Plate-tectonic theory requires some system of horizontal forces that can cause plates to collide and combine or to break up. The only possible mechanism seems to be *convection*. This is essentially motion induced by buoyancy, with lighter material rising and denser material sinking. In the mantle the convection is predominantly *thermal convection* in which the density variations are a result of temperature variations.

During the first half of the 20th century, geophysicists widely believed that convection could not occur in a solid, rigid mantle and this was one of the reasons why the theory of continental drift took so long to gain general acceptance. However, it was eventually shown that such convection is possible because of the solid-state deformation, or creep, of mantle minerals. Details of the movements and the scale of convection remain uncertain, although there is no doubt that the rates are so slow that the mantle is effectively motionless within a human framework of time. Schemes have been proposed involving large convection cells extending through the entire mantle (Loper 1985) or independently convecting upper and lower mantle layers (Richter & McKenzie 1981, Kenyon & Turcotte 1983, O'Nions 1987; Fig. 3.6).

The vertical movement of mantle material clearly must cause changes in the temperature distribution; at a given depth the temperature is increased where hotter material is rising and decreased where cooler material is sinking. This movement changes the shape of the geotherm from place to place and time to time (Fig. 3.5), and in many environments is fundamental to the magma generation process (Section 3.4.8).

Early plate-tectonic models considered that the lithospheric plates moved as passive passengers on large-scale mantle convection cells. In such a passive model the location of ridges and subduction zones is determined by the location and size of

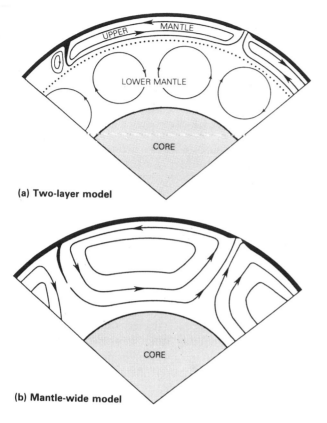

(a) Two-layer model

(b) Mantle-wide model

Figure 3.6 Models of convection in the mantle (after Basaltic Volcanism Study Project 1981).

deep-mantle convection cells, with ridges along the upwelling limbs and subduction zones along the descending limbs. However, it is now more generally accepted that the plates themselves are an active part of the convection process (Cox & Hart 1986). Subduction occurs because, as the oceanic lithosphere ages, it becomes cold and dense and eventually sinks back into the mantle. In this model the ridges are simply cracks between the diverging plates filled from local sources of magma within the asthenosphere.

It is possible that there may be two scales of convection in the mantle, a large-scale process in which the plates are active elements and a secondary smaller-scale process of convection confined to the asthenosphere. The latter delivers heat to the bottom of plates, but does not constitute a significant driving force for plate tectonics. This mode of

convection may be the source of the so-called 'hot spots' which generate intra-plate volcanism (Ch. 1).

One of the major sources of debate concerning mantle convection surrounds its geometrical relationship, if any, to the density structure of the mantle (Fig. 3.4) responsible for the known seismic discontinuities (Fig. 3.2; O'Nions 1987). The change in density around 670 km is important, as it may be sufficient to separate upper from lower mantle convection (Richter & McKenzie 1981). Whether or not this density change corresponds to a compositional change remains uncertain (e.g. Knittle *et al.* 1986).

Evidence for penetration or non-penetration of the 670 km seismic discontinuity by subducted oceanic lithosphere plates should provide evidence concerning the nature of any convective communication between seismically defined upper and lower mantles. Unfortunately, available data are conflicting; for example, Giardini & Woodhouse (1984) argue for penetration while Creager & Jordan (1984) argue against.

Geochemical studies of oceanic basalts (MORB and OIB, Chs 5 & 9) also have implications concerning mantle convection. The source of MORB is strongly depleted in incompatible elements, probably as a consequence of the extraction of continental crustal materials throughout the course of geological time. In contrast, the source of OIB is variably enriched in incompatible elements. Geochemical mass balance calculations can provide estimates for the relative masses of depleted (0.3−0.5) and non-depleted (0.7−0.5) portions of the mantle but do not specify their geometrical arrangement. In general, most models identify the depleted MORB-source mantle with the upper mantle and the undepleted or enriched portion with the lower mantle, on the assumption that chemical and convective layering will parallel the density structure of the mantle (O'Nions 1987). Additional arguments for the existence of a less depleted and chemically isolated part of the mantle arise from studies of the isotopic composition of terrestrial rare gases (Ch. 2), particularly He, in oceanic basalts. The major oceanic intra-plate hot spots of Hawaii and Iceland have [3]He anomalies compared to MORB, suggesting that the magmas contain a component derived from plumes of relatively un-degassed (?primordial) material rising from the lower mantle.

O'Nions (1987) suggests that an essentially layered structure, both chemically and convective-ly, has existed in some form for most of the Earth's history. The principal changes that have occurred from 4.4 Ga to the present are associated with the development of the continental crust.

3.3   Chemical composition and mineralogy of the upper mantle

As we cannot sample the Earth's upper mantle directly we must rely on indirect lines of evidence to elucidate its chemical composition and mineralo-gy. There are two approaches which are useful in this respect:

(1)   The study of terrestrial basaltic magmas be-lieved to be generated by partial melting of the upper mantle.
(2)   The study of ultramafic rocks of presumed mantle origin exposed at the Earth's surface.

Seismic data and chemical data from extraterrestrial bodies provide additional constraints.

Basaltic magma has been erupted in great volumes throughout geological time and over most of the Earth's surface. Broadly speaking, ancient and present-day basalts have similar chemistry and therefore we can make the following observations:

(a)   the mantle material must be capable of pro-ducing basaltic magma by partial melting;
(b)   the mantle composition and the nature of partial melting processes cannot have changed significantly over geological time.

Ultramafic rocks of potential mantle origin occur as tectonically emplaced massifs, including ophiolite complexes, and as xenoliths within highly SiO_2-undersaturated basaltic and kimberlitic mag-mas from both intracontinental and intra-oceanic plate tectonic settings. The freshest samples, least modified chemically and structurally during or after their ascent to the surface, are found amongst the ultramafic xenoliths, and only this group will be considered in detail here. The reader is referred to Chapter 5 for a more detailed discussion of ophiolite complexes.

3.3.1   Evidence from xenoliths in kimberlites and alkali basalts

Ultramafic xenoliths of presumed mantle derivation are extremely abundant in most kimberlite occur-rences (Dawson 1980, Nixon 1987). They are less common, though still of frequent occurrence, in basaltic rocks of the alkali basalt−basanite−nephelinite suite from both intra-oceanic and intra-continental plate-tectonic settings (Menzies 1983, Nixon op.cit.). The alkali basalt−nephelinite suite lavas are believed to be derived from somewhat shallower depths in the mantle than kimberlites and therefore their entrained xenoliths should reflect sampling of a more restricted (shallower) depth range. These xenolith suites represent the widest range of potential samples of upper mantle mate-rial.

Xenoliths in kimberlite

Kimberlite is the host rock of diamond and there-fore diamondiferous kimberlites must originate from depths of at least 150 km (i.e. within the di-amond stability field; Ch. 12). This is deeper than the estimated depths of generation of most other igneous rock types, and thus the kimberlite mag-mas have had the potential to sample a large cross section of the mantle on their ascent to the surface.

The xenoliths average 10−30 cm in maximum diameter but can be greater than 1 m. The com-monest type is a rock containing greater than 40% modal olivine called garnet lherzolite (Fig. 3.7). In addition, harzburgite, dunite, pyroxenite and glim-merite (mica-rich rocks) also occur, along with a megacryst suite of high-Ti garnet, pyroxenes, oli-vine and ilmenite, thought to be high-pressure phenocrysts crystallizing from the kimberlite magma (Ch. 12).

Garnet lherzolite consists of four major minerals:

OLIVINE−ORTHOPYROXENE−
CLINOPYROXENE−GARNET

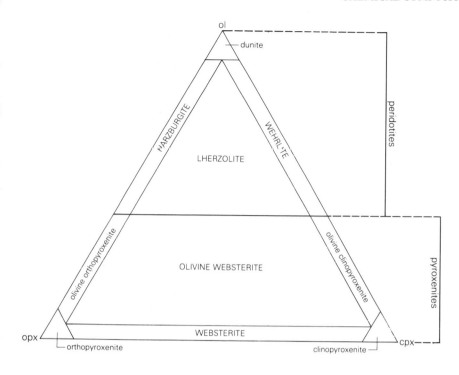

Figure 3.7 The modal classification of ultramafic rocks. Within the peridotite field the addition of spinel or garnet to the assemblage yields the following additional rock types; garnet lherzolite — garnet harzburgite, and spinel lherzolite — spinel harzburgite (after Harte 1983).

in order of decreasing abundance. In this assemblage garnet is the major Al-bearing mineral, and is referred to as the aluminous phase. Phlogopite and amphibole occur occasionally as primary accessory phases, attesting to the presence of volatiles in the upper mantle, although more commonly they are of a secondary or metasomatic origin. In some instances the introduction of serpentine, phlogopite and carbonates can be related to fluids associated with the host kimberlite magma. Average chemical compositions of the major mineral phases are given in Table 3.3.

The lherzolite xenoliths show a remarkable uniformity in mineralogy and chemistry when compared to other xenolith types with which they are associated. However, texturally, they are highly variable and have metamorphic fabrics reflecting varying degrees of deformation and recrystallization. Most commonly, they are coarse-grained with equant crystals, but they can exhibit strong mineral banding and deformation fabrics (Fig. 3.8). Such fabrics are akin to those found in crustal cataclastites and mylonites and are attributed to superplastic flow within the upper mantle, which may or

may not be associated with the kimberlite magmatic event.

Despite their rather uniform major element chemistry (Table 3.4), the xenoliths show significant variation in their contents of more fusible chemical components and thus in their magma-yielding potential. On this basis they can be classified as *fertile* (enriched) or *barren* (depleted). In terms of the major element chemistry, depletion is reflected in terms of decreased Al, Ca, Ti, Na, K and increased $Mg/(Mg + Fe)$ and $Cr/(Cr + Al)$ (Nixon *et al.* 1981).

The sequence of xenolith types observed in kimberlites from garnet lherzolite to garnet harzburgite to harzburgite to dunite can thus be regarded as a sequential series, with garnet lherzolite the most fertile and dunite the most barren. Dunite is thus the residue after complete extraction of basaltic magma from a lherzolite source (Section 3.4). Study of their major element chemistry reveals a spectrum of garnet lherzolite types which have had variable degrees of partial melt extracted from them (Gurney & Harte 1980).

A variety of experimentally calibrated geo-

Table 3.3 Mineral chemistry of the major phases in typical spinel and garnet lherzolites (data from Basaltic Volcanism Study Project 1981, Table 1.2.11.2, p. 286).

Spinel lherzolite

	Olivine	Orthopyroxene	Clinopyroxene	Spinel
SiO_2	38.7	53.7	50.8	0.08
TiO_2	—	0.19	0.78	0.31
Al_2O_3	0.03	3.9	6.0	53.5
Cr_2O_3	0.02	0.28	0.58	10.0
ΣFeO	14.4	8.9	4.5	16.3
MnO	0.22	0.20	0.07	0.12
MgO	46.4	31.40	15.2	13.6
NiO	0.13	0.02	0.02	0.30
CaO	0.07	0.73	21.2	—
Na_2O	—	0.07	0.89	—
K_2O	—	—	—	—
	99.97	99.39	100.06	99.21

Garnet lherzolite

	Olivine	Orthopyroxene	Clinopyroxene	Garnet
SiO_2	40.6	57.6	55.9	41.8
TiO_2	0.02	0.03	0.07	0.08
Al_2O_3	—	0.75	2.23	21.6
Cr_2O_3	0.03	0.28	1.72	4.0
ΣFeO	7.5	4.6	2.55	7.4
MnO	0.08	0.07	0.07	0.36
MgO	50.3	35.0	17.1	20.8
NiO	—	—	—	—
CaO	0.04	0.56	19.7	5.1
Na_2O	—	0.11	2.1	—
K_2O	—	—	—	—
	98.57	99.00	101.44	101.14

thermometers and geobarometers can be applied to garnet lherzolite mineral assemblages. Finnerty & Boyd (1987) have presented a useful critique of the various methods currently employed, and the reader is referred to this work for more detailed information. Unfortunately, the techniques of thermobarometry remain limited by inadequate means of applying experimental calibrations in simple analogue systems to complex natural mineral compositions. Consequently, there is considerable disagreement between the results of the various methods and such data must be interpreted with caution.

If a range of compositions are available for a suite of garnet lherzolite xenoliths from the same locality, then the results of geothermometry−barometry calculations may be used to define a 'fossil geotherm' (Fig. 3.9). Suites of garnet peridotites from kimberlites in many continents fall into two groups that differ in their ranges of equilibration temperature, a low-temperature group and a high-temperature group of deeper origin. These may occur together,

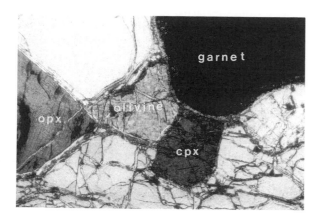

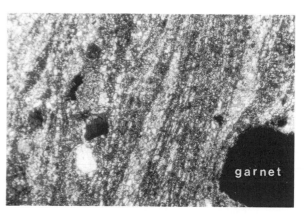

Figure 3.8 (a) Granular garnet lherzolite from the Wesselton Mine, Kimberley, South Africa (×40 magnification, crossed polars). (b) Sheared garnet lherzolite from Kenilworth Floors, South Africa (×40 magnification, crossed polars). Note the isotropic garnet in both (a) and (b) photomicrographs.

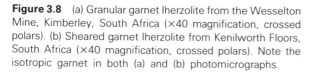

but there are localities at which one has been erupted in the absence of the other. Equilibration conditions for many low-temperature suites appear to approximate to the ambient geotherm at the time and place of eruption, whereas the high-temperature suites commonly deviate to higher temperatures. A gap in $P-T$ space commonly separates the low- and high-T suites, and may approximate to the lithosphere/asthenosphere boundary (Finnerty & Boyd 1987). In terms of Figure 3.9 we can see that high-temperature peridotites from mobile belts surrounding the Kapvaal craton in southern Africa have originated from depths of 130–160 km, compared with 170–220 km within the craton. This suggest that the craton has a deeper lithospheric root than the mobile belts. Sparse garnet lherzolite xenoliths from oceanic areas have estimated depths of origin that are substantially shallower than the continental geotherm shown (Fig. 3.9c). A major problem with thermobarometry calculations involves the interpretation of what event the calculated $P-T$'s actually represent. It is unlikely that the ultramafic xenoliths are the solid

Table 3.4 Major element compositional ranges for spinel and garnet lherzolites (Maaløe & Aoki 1977).

	Composition range for garnet lherzolite	Composition range for spinel lherzolite	Average for garnet lherzolite	Average for spinel lherzolite
SiO_2	43.8–46.6	42.3–45.3	45.89	44.2
TiO_2	0.07–0.18	0.05–0.18	0.09	0.13
Al_2O_3	0.82–3.09	0.43–3.23	1.57	2.05
Cr_2O_3	0.22–0.44	0.23–0.45	0.32	0.44
FeO	6.44–8.66	6.52–8.90	6.91	8.29
MnO	0.11–0.14	0.09–0.14	0.11	0.13
NiO	0.23–0.38	0.18–0.42	0.29	0.28
MgO	39.4–44.5	39.5–48.3	43.46	42.21
CaO	0.82–3.06	0.44–2.70	1.16	1.92
Na_2O	0.10–0.24	0.08–0.35	0.16	0.27
K_2O	0.03–0.14	0.01–0.17	0.12	0.06
P_2O_5	0.00–0.08	0.01–0.06	0.04	0.03
$100Mg/(Mg + Fe)$	89–92.95	89.1–92.6		
$100Cr/(Cr + Al)$	7.4–18.6	7.0–31.7		

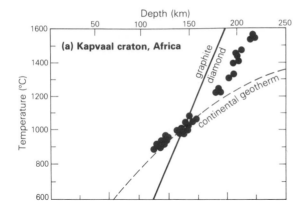

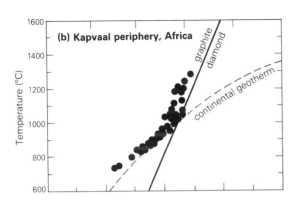

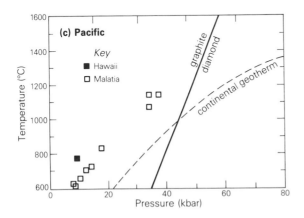

Figure 3.9 Calculated equilibration pressures and temperatures for garnet lherzolite xenoliths from kimberlites (after Finnerty & Boyd 1987, Figs 197,198 & 200). The continental geotherm is from Pollack & Chapman (1977)

residue from the partial fusion event generating the magma which contains them. Instead, most are accidental xenoliths, detached from the conduit wall by the kimberlite magma ascending from greater depth.

Harte (1983) has proposed a subdivision of the garnet peridotite specimens found in kimberlite pipes into coarse, modally metasomatized and deformed types (Fig. 3.10). Of the various subgroups shown, the coarse Mg-rich types form the dominant mantle xenoliths at many kimberlite localities. These usually contain less than 10−20% combined clinopyroxene and garnet and commonly yield temperature and depth estimates of 850−1100°C and 90−140 km, suggesting that these represent fragments of the subcontinental lithosphere. These lherzolites are generally depleted in basaltic constituents.

The binary grouping that appears in $P-T$ plots such as Figure 3.9 correlates with the modal, textural and compositional characteristics of the xenoliths. Those in the low-temperature group have olivines more magnesian than $Fo_{91.5}$, exhibit intergranular REE enrichment and contain diopsides with enriched $Nd-Sr$ isotopic signatures (Section 3.3.6). In contrast, the high-temperature peridotites contain less magnesian olivines (< $Fo_{91.5}$), have chondritic REE patterns and depleted $Nd-Sr$ isotopic signatures. Compared to the coarse-grained low-T peridotites, the majority of the high-T types are deformed, and these textures are believed to have originated during the early stages of kimberlite eruption. Most of the low-T peridotites are relatively depleted in basaltic constituents, especially Fe and Ti. In addition, they commonly contain coarse phlogopite and chromite and sparse diamond or graphite. These phases are not usually found in the high-temperature types.

The incompatible trace element geochemistry of lherzolite xenoliths is frequently difficult to study because of the late-stage introduction of metasomatic minerals such as amphibole and biotite (Hawkesworth et al. 1984, Richardson et al. 1985, Harte 1987). In many cases, major element and incompatible element data provide conflicting evidence as to the history of the xenolith. For example, there are xenoliths with apparently depleted major

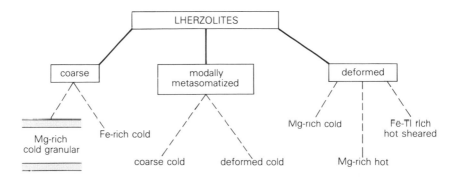

Figure 3.10 Major subdivision of the lherzolite group of xenoliths found in kimberlites (after Harte 1983). The subdivisions in solid lines are based on petrographic observations; and subdivisions in dashed lines on mineral composition data.

Modally metasomatized rocks are those showing evidence of an infiltration metasomatic event, involving the formation of phlogopite, K-richterite, ilmenite, oxides and sulphides. The term *modal* is used to emphasize a distinction with other rocks showing addition of elements (e.g. Fe and Ti) prior to kimberlite entrainment but lacking additional phases. The subdivision into *Mg-rich* and *Fe-rich* types is based on whether $Mg/(Mg + Fe)$ of olivine is greater or less than 0.91. In a simplistic way this is equivalent to the subdivision between depleted and fertile lherzolites used to indicate poverty or richness of the rocks in basaltic components, particularly CaO, Al_2O_3, FeO, TiO_2, and Na_2O. The subdivision into *cold* or *hot* types is based on geothermometric estimates derived from mineral compositions.

Coarse lherzolites have a grain size of 2−10 mm, equant grains with irregular boundaries and slight internal strain.

Deformed lherzolites have two texturally distinct sets of grains: (1) porphyroclasts of mainly pyroxene and garnet, 2−10 mm in diameter, showing variable strain; (2) neoblasts − recrystallized grains, dominantly of olivine, <0.3 mm in diameter, showing mosaic or granuloblastic grain shapes.

element characteristics which nevertheless have high incompatible element contents and display marked light-REE enrichment, apparently inconsistent with a magma extraction event. Such specimens have been interpreted as reflecting a later metasomatic event superimposed upon an earlier depletion event, which is not necessarily revealed by the presence of hydrous phases due to subsequent annealing. Much of the subcontinental lithosphere may be of this nature (Harte 1983).

Xenoliths in alkali basalts

The xenolith suites in continental and oceanic alkali basalts are mineralogically and chemically heterogeneous, showing a range of compositions from dunite to lherzolite comparable to those in kimberlites (Menzies 1983). They are most commonly contained in the more vesicular and tuffaceous members of the alkali basalt−basanite−nephelinite suite, indicating the volatile rich nature of the host magmas. Lherzolite is the most common xenolith

type, but unlike the kimberlite occurrences the major aluminous phase is spinel rather than garnet (Fig. 3.11). The xenoliths show the same range of textures displayed by the garnet lherzolites but, in contrast, show an apparent predominance of deformed types.

The spinels within these spinel lherzolites show a significant range of compositional variation in the quaternary system $MgAl_2O_4-MgCr_2O_4-FeAl_2O_4-FeCr_2O_4$, which can be used as a basis for classification. Carswell (1980) has proposed a threefold subdivision into Al-spinel lherzolites, Cr-spinel lherzolites and chromite lherzolites, based on the $Cr/(Cr + Al)$ ratio in the spinel (Fig. 3.12). Comparison of whole-rock analyses for these different types (Table 3.5) shows that with increasing $Cr/(Cr + Al)$ ratio in the spinel the bulk rock becomes progressively more depleted.

Calculation of the pressure−temperature conditions of equilibration of the spinel lherzolite xenoliths is much more uncertain than for the garnet

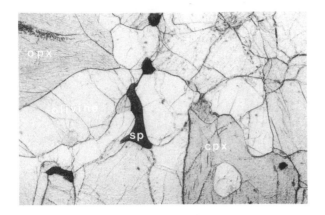

Figure 3.11 Granular spinel lherzolite xenolith from New South Wales, Australia (×40 magnification, ordinary light). Note the dark spinel grains in the centre of the field of view.

lherzolite xenoliths (Carswell 1980). Available temperature estimates mostly lie in the range 900−1150°C, while from petrographic considerations of the depth of origin of the alkali basalt−nephelinite magmas it seems likely that the depth of their incorporation in the magma must be less than 70−80 km.

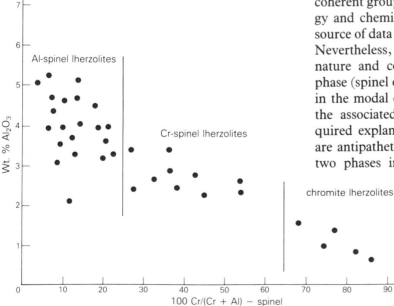

Spinel lherzolites and garnet lherzolites compared

Maaløe & Aoki (1977) made a detailed comparison of garnet and spinel lherzolite suites from kimberlites and alkali basalts and deduced, as shown in Table 3.4, that there are considerable overlaps in their compositional ranges. They suggest the following average modes for the two types:

Spinel lherzolite (%)		Garnet lherzolite (%)
66	olivine	63
24	orthopyroxene	30
8	clinopyroxene	2
2	spinel	—
—	garnet	5

These must be taken merely as an approximation, as lherzolite xenoliths are mineralogically highly variable.

Lherzolite xenoliths represent an abundant and coherent group of comparatively uniform mineralogy and chemistry, and have proved an invaluable source of data about the nature of the upper mantle. Nevertheless, there are important variations in the nature and composition of the stable aluminous phase (spinel or garnet), complemented by changes in the modal content and chemical composition of the associated pyroxenes (Table 3.3), which required explanation. In general, spinel and garnet are antipathetic but some xenoliths do contain the two phases in apparent primary textural equili-

Figure 3.12 Plot of $100Cr/(Cr + Al)$ ratio in primary spinels against wt.% Al_2O_3 in coexisting orthopyroxene from spinel lherzolite xenoliths (after Carswell 1980, Fig. 1).

Table 3.5 Comparison of bulk rock compositions of spinel lherzolite xenoliths (Carswell 1980, Table 6).

	Al–spinel lherzolite	Chrome spinel lherzolite	Chromite lherzolite
SiO_2	44.48	42.30	45.31
TiO_2	0.18	tr	0.11
Al_2O_3	1.80	1.21	0.43
Cr_2O_3	0.42	0.35	0.25
FeO	8.90	6.89	6.52
MnO	0.14	0.17	0.09
NiO	—	0.18	0.34
MgO	41.77	48.26	46.03
CaO	2.34	0.44	0.56
Na_2O	0.13	0.10	0.13
K_2O	0.04	0.04	0.17
P_2O_5	< 0.01	0.06	0.04
$100Mg/(Mg + Fe)$	89.3	92.6	92.6
$100Cr/(Cr + Al)$	13.5	16.2	28.1

brium. These should not be confused with those in which secondary spinel is developed in garnet lherzolite xenoliths as a result of olivine–garnet breakdown reactions. Clearly, the spinel–garnet stability relationships in these rocks are complex, determined by both chemical and physical (P–T) controls (Section 3.3.4).

Figure 3.13 shows the typical REE abundance patterns for spinel and garnet lherzolites. Chondritic abundances (i.e. enrichment factors of 1) are common, but there is a wide range from light REE depleted to light-REE enriched types, with a maximum light-REE enrichment of 10× chondrite. Both spinel and garnet lherzolites show similar ranges of patterns. Such data are essential for quantitative modelling of partial melting processes.

From the spectrum of xenoliths observed we can conclude that the upper mantle is mineralogically complex and extremely heterogeneous. Nevertheless, its bulk chemical composition can be closely represented by five major element oxides: SiO_2, 44%; Al_2O_3, 2%; FeO, 8%; CaO, 2%; MgO, 42%; rest, 2%. MgO and SiO_2 together account for 86% by weight of the mantle, thus explaining the predominance of olivine $(Mg,Fe)_2SiO_4$ and orthopyroxene $(Mg,Fe)SiO_3$ as major mantle phases.

The common occurrence of garnet lherzolite xenoliths in kimberlites and spinel lherzolite xeno-

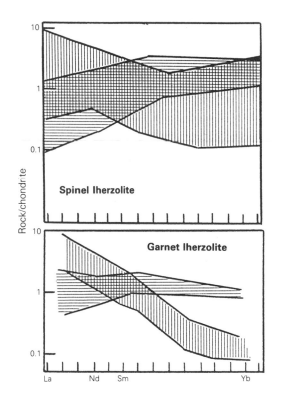

Figure 3.13 REE abundance patterns, normalized to chondritic abundances, for spinel and garnet lherzolites (after Basaltic Volcanism Study Project 1981, pp. 292–7).

liths in alkali basalts must be of major significance with respect to the overall structure of the mantle (Nixon & Davies 1987). As the alkali basalt—nephelinite suite magmas originate at rather shallower depths in the mantle than kimberlite, it seems logical to assume that the spinel lherzolites are derived from a shallower depth zone than the garnet lherzolites. Given that both lherzolite types have broadly similar bulk rock compositions, the change in the aluminous phase from spinel to garnet may simply reflect some pressure-dependent metamorphic transition. This will be considered further in Section 3.3.4.

In constructing mantle models from lherzolite xenolith data, it is important to consider the extent to which the xenoliths are representative of the whole length of the mantle section traversed by the ascending magmas. Comparison of xenolith suites at closely spaced kimberlite localities reveal systematic differences suggestive of localized sampling and localized mantle heterogeneities (Richardson *et al.* 1985). In addition, it is also prudent to consider whether the calculated equilibration conditions of the minerals in the xenoliths represent past rather than present $P-T$ conditions, and to what extent they represent modification of the conduit walls by the ascending magmas.

3.3.2 Evidence from meteorites

To obtain an estimate of the composition of the Earth as a whole, and hence of the core and mantle, it is necessary to rely on evidence for the chemistry of extraterrestrial bodies, including the stars, the Sun and meteorites, assuming that such bodies have an essentially similar bulk composition to the Earth. The Solar System formed approximately 4.6 Ga ago through the gravitational collapse of matter previously dispersed in interstellar space. Since the Sun accounts for 99.6% of the mass of the Solar System, its composition, as determined by spectroscopic methods, can be considered to be that of the system as a whole.

Meteorites travel through the Solar System in elliptical orbits that occasionally intersect the Earth. They vary widely in chemistry, mineralogy and structure and can be subdivided into two major groups, iron meteorites and chondrites. The chondritic meteorites are composed mainly of silicate minerals, together with varying proportions of metallic alloy and iron sulphide. Their relative abundances of non-volatile elements (Mg, Si, Al, Ca, Fe) are similar to those of the Sun, and therefore it is argued that their chemical composition can be used to estimate the overall abundances of elements in the Earth.

A specific group of chondrites, the carbonaceous chondrites, richest in C, H_2O and volatile trace elements, have radiometric ages of 4.6 Ga, and are presumed to reflect the generalized chemical composition of the primordial solar nebula. Assuming that this also represents the composition of the material which condensed to form the Earth, then calculations can be made to estimate the bulk composition of the mantle after segregation of the core.

From such calculations it is possible to make the following deductions about the primordial mantle composition:

(1) Greater than 90% by weight of the mantle is composed of SiO_2, MgO and FeO. No other oxide exceeds 4%.
(2) Al_2O_3, CaO and Na_2O total 5–8%.
(3) Greater than 98% of the mantle can be represented by these six oxides, with no other oxide reaching a concentration greater than 0.6% (Wyllie 1981).

Obviously, the present-day mantle will have a slightly different bulk composition due to its irreversible differention throughout geological time to form the Earth's crust. However, such differences are insignificant as far as major elements are concerned, as the crust accounts for <1% of the Earth's mass, compared to 68% for the mantle.

Of all the possible samples of mantle material found within the Earth's crust, only lherzolites correspond to the above estimate of the bulk chemical composition of the mantle.

3.3.3 Pyrolite

Pyrolite is a synthetic model mantle composition

devised by A. E. Ringwood, which has frequently been used as the starting material in experimental investigations of phase equilibria in the upper mantle (Ringwood 1975). Its composition is calculated by combining a Hawaiian basalt composition with that of a sterile ultramafic rock assumed to represent the solid residue from partial fusion. The proportions of these two components are adjusted so that the ratios of non-volatile major components in the pyrolite match those in chondritic meteorites, which are assumed to represent the primordial material from which the Earth condensed.

3.3.4 High $P-T$ experimental studies

High pressure−temperature experimental studies have provided important evidence for the relationship between spinel and garnet lherzolite mineralogies in rocks of the same bulk chemical composition, in addition to providing constraints on the conditions necessary for the onset of partial melting (Section 3.4; Wyllie 1981, 1984). Figure 3.14a shows that, for lherzolite bulk compositions, the subsolidus mineral assemblage changes from plagioclase to spinel to garnet lherzolite with increasing pressure (depth). At still greater depths the garnet lherzolite assemblage transforms to a combination of high-pressure mineral phases (Ringwood 1975).

The subsolidus boundaries between the different lherzolite mineral assemblages are in fact metamorphic transitions which can be represented by the following end-member reactions.

(a) Plagioclase to spinel lherzolite:

$$\underset{\text{olivine}}{2Mg_2SiO_4} + \underset{\text{plagioclase}}{CaAl_2Si_2O_8} =$$

$$\underset{\text{opx}}{2MgSiO_3} + \underset{\text{cpx}}{CaMgSi_2O_6} + \underset{\text{spinel}}{MgAl_2O_4}$$

(b) Spinel to garnet lherzolite:

$$\underset{\text{spinel}}{MgAl_2O_4} + \underset{\text{opx}}{4MgSiO_3} =$$

$$\underset{\text{olivine}}{Mg_2SiO_4} + \underset{\text{garnet}}{Mg_3Al_2Si_3O_{12}}$$

Clearly, the position of such boundaries must be dependent upon the bulk chemical composition of the lherzolite sample. For example, it has been shown that the transition pressure increases with increasing $Cr/(Cr + Al)$ ratio (Carswell 1980), and thus it is possible that certain chromite lherzolites might be stable under the same pressure conditions as some garnet lherzolites. However, as many lherzolite xenoliths have a rather restricted range of $Cr/(Cr + Al)$ ratios, the boundaries shown in Figure 3.14 may be taken as generally applicable.

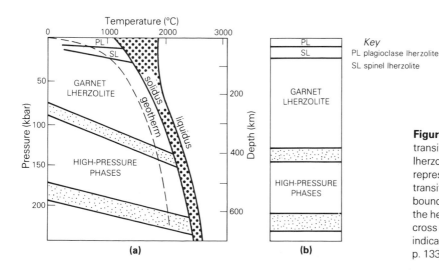

Figure 3.14 (a) Subsolidus phase transitions in anhydrous mantle lherzolite. The lightly stippled bands represent high-pressure phase transitions. The region of partial melting, bounded by the solidus and liquidus, is the heavily stippled band. (b) Mantle cross section along the geotherm indicated in (a). (After Wyllie 1981, Fig 2, p. 133).

Key
PL plagioclase lherzolite
SL spinel lherzolite

The sequence of lherzolite mineral assemblages observed in a vertical section through the mantle will clearly be dependent upon the geothermal gradient, as shown in Figure 3.14b. Plagioclase lherzolite xenoliths will rarely be found due to their limited stability range (generally <30 km) and in any case are likely to be overlooked due to their gabbroic mineralogy.

3.3.5   Volatile contents and redox state

Studies of mantle-derived lherzolite samples suggest that total volatile contents are low, such that for most practical purposes the mantle can be considered to be essentially anhydrous. Quite obviously, this cannot be correct as the oceans and atmosphere have been derived by degassing of the mantle throughout geological time, and most volcanic eruptions are accompanied by volatile emissions, some of which can be shown to be of primordial origin.

The presence of the hydrous minerals phlogopite and amphibole in some lherzolite xenoliths confirms the presence of H_2O in the mantle, although it probably amounts to less than 0.1%. Additionally, olivine and pyroxene crystals from mantle samples often contain CO_2-rich fluid inclusions, providing evidence for the presence of CO_2.

Volatiles and the oxidation−reduction environment exerted by gas species in the system C−O−H−S are critical to petrogenetic models for the generation of magmas within the upper mantle, influencing both the degree of partial melting and the composition of the resultant partial melt (Sections 3.4 & 3.5; Wyllie 1981, Olafsson & Eggler 1983).

An inevitable consequence of the progressive degassing of the mantle throughout geological time is that the redox state of the mantle must also have changed. A major question is whether it has become more reduced or more oxidized with time, or whether the condition is steady state, i.e. whether volatile loss through volcanism is offset by subduction of hydrous minerals (Ch. 6). Haggerty & Tompkins (1983) suggest that the early mantle may well have been reduced, with oxygen fugacities close to those defined by the synthetic oxygen

buffer magnetite−wustite (MW), and that some xenolith suites brought to the surface in kimberlites reflect this primitive condition. They consider that, with time, the mantle has become progressively more oxidized, resulting largely from the preferential loss of hydrogen and carbon.

If the mantle is geochemically layered, with a depleted lithosphere overlaying a more fertile asthenosphere, then it is likely that the redox state of each zone will differ. Figure 3.15 shows the Haggerty & Tompkins model for the variation of the redox state of the mantle, with the lithosphere being more reduced than the upper part of the asthenosphere. At present, this must be considered to be somewhat conjectural (Arculus & Delano 1987).

3.3.6   Isotope geochemistry

Evidence for the isotopic composition of the upper mantle has been mainly based on studies of young volcanic rocks believed to be derived by direct partial melting of mantle lherzolite (e.g. Zindler & Hart 1986), supplemented by a limited body of data on the isotope geochemistry of lherzolite xenoliths (Menzies & Hawkesworth 1987). Attention has been focused on oceanic volcanics because of the lesser possibility of crustal contamination (Ch. 4). In this section only Sr, Nd and Pb isotopic variations will be considered, as these are the isotopes which are of major petrogenetic significance.

The growth of the sialic continental crust over the past 4 Ga has progressively depleted the upper mantle in incompatible large-ion lithophile (LIL) elements. Selective extraction from mantle lherzolite of LIL elements such as Rb during partial melting will be recorded in isotopic tracers such as ^{87}Sr, measured within mantle samples. Since ^{87}Rb decays radioactively to ^{87}Sr, removal of any Rb in a partial melt from a mantle source region will cause the $^{87}Sr/^{86}Sr$ ratio to be less at some future time than a region of the mantle not previously melted. Any liquids formed during a second episode of partial melting of the Rb-depleted source rock will have a low $^{87}Sr/^{86}Sr$ signature relative to that of the original primordial system. Following the same

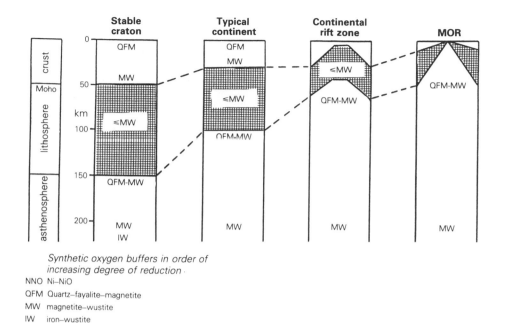

Figure 3.15 Redox state of the crust and mantle (after Haggerty & Tompkins 1983).

logic, parts of the mantle enriched by LIL-rich metasomatic fluids will evolve with higher Rb/Sr ratios and thus, if subsequently melted, would yield magmas with higher $^{87}Sr/^{86}Sr$ ratios than the primitive mantle. Thus the $^{87}Sr/^{86}Sr$ ratio of mantle-derived partial melts serves as a petrogenetic tracer for the relative fractionation of Rb from Sr by the extraction and migration of partial melts and metasomatic fluids. We thus have an isotopic analogue for the depleted, fertile and metasomatized lherzolites considered in Section 3.3.1.

The ratio $^{143}Nd/^{144}Nd$ also serves as a petrogenetic tracer in much the same way as $^{87}Sr/^{86}Sr$, because ^{143}Nd is a decay product of ^{147}Sm and the Sm/Nd ratio is also fractionated by partial melting processes. Nd^{3+} is slightly more incompatible than Sm^{3+} in peridotitic systems because of the relative differences in their ionic radii (1.20 Å and 1.17 Å respectively). Thus partial melts have a lower Sm/Nd ratio, but a greater Rb/Sr ratio, than their mantle source.

$^{143}Nd/^{144}Nd$ and $^{87}Sr/^{86}Sr$ ratios determined for a wide range of mantle-derived volcanic rocks from the ocean basins display an inverse correlation (Fig. 3.16), which has been termed the 'mantle array' (e.g. DePaolo & Wasserburg 1977, Zindler et al. 1982; and see Ch. 2). As chondritic meteorites are a good first approximation to the bulk chemical composition of the Earth, their present-day isotopic ratios can be used as a standard of reference to decide whether magmas are derived from relatively depleted or enriched sources. All magmas with $^{87}Sr/^{86}Sr$ ratios less than the bulk Earth value are derived from depleted sources, with MORB being derived from the most depleted source regions. Similarly, magmas with $^{87}Sr/^{86}Sr$ ratios greater than bulk Earth must be derived from enriched source regions. Clearly, most oceanic basalts are derived from variably depleted source regions. However, basalts from some oceanic islands deviate significantly from the mantle array (e.g. White & Hofmann 1982, White 1985, Zindler & Hart 1986; and see Ch. 9), and as more Nd−Sr isotope data become available for oceanic basalts the array is becoming increasingly blurred. Spinel and garnet lherzolite xenoliths from the subcontinental lithosphere show considerable overlap with the oceanic basalt array (Fig. 3.16b), but extend to consider-

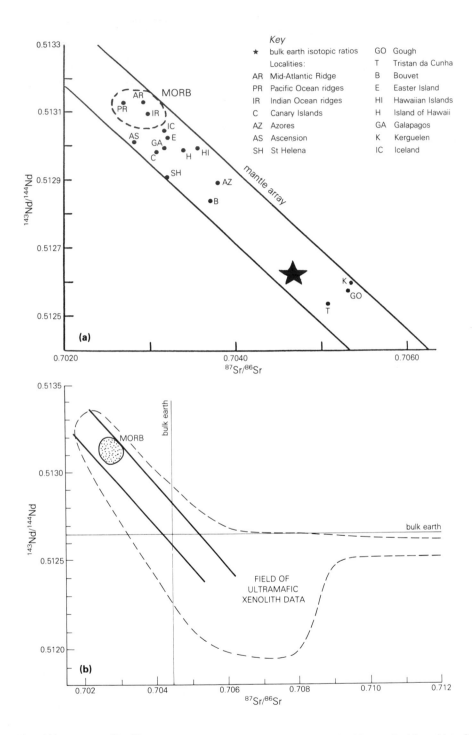

Figure 3.16 ^{143}Nd/^{144}Nd versus ^{87}Sr/^{86}Sr for (a) basalts from oceanic islands and mid-oceanic ridges (data from Zindler *et al.* 1982), and (b) ultramafic xenoliths from the subcontinental lithosphere (data from Menzies 1983).

ably more radiogenic compositions. This suggests that the subcontinental lithosphere may be considered as an enriched mantle reservoir.

Figure 3.17 shows a three-dimensional plot of average $^{206}Pb/^{204}Pb$ versus $^{143}Nd/^{144}Nd$ versus $^{87}Sr/^{86}Sr$ for the oceanic basalts used to define the Nd–Sr mantle array in Figure 3.16. The data form a planar array which could be described in terms of mixing of three chemically independent mantle components (Zindler et al. 1982). However, it is important to note that this 'mantle plane' is defined by average data and that many of the localities plotted display intra-island variations that do not lie in the plane (e.g. Stille et al. 1983, Staudigel et al. 1984). There are also several localities that, even when averaged, do not lie on this plane (e.g. Richardson et al. 1982, White & Hofmann 1982, White 1985). Thus, while the concept of a Nd–Sr–Pb mantle plane remains a useful way of considering the source components involved in the petrogenesis of oceanic basalts, it is clearly an oversimplification. Indeed, Allègre & Turcotte (1985) and White (1985) have proposed that there are five types of sources from which oceanic basalts may be produced by variable mixing relationships. Zindler & Hart (1986) require at least four end-member mantle components, plus two additional components, in their global geodynamic model.

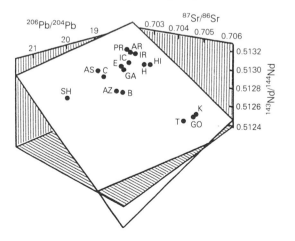

Figure 3.17 Three-dimensional plot of average $^{206}Pb/^{204}Pb$, $^{143}Nd/^{144}Nd$ and $^{87}Sr/^{86}Sr$ for basalts from oceanic islands and mid-ocean ridges, showing the best-fit plane to the data; localities as in Figure 3.16 (after Zindler et al. 1982, Fig. 2).

3.3.7 Mantle heterogeneity

From the discussions in the previous sections, we can conclude that although the mantle is mineralogically and chemically complex, we can consider it to be composed predominantly of four-phase lherzolites in which the aluminous phase changes from plagioclase to spinel to garnet with increasing pressure. Metasomatic introduction of minerals such as amphibole, phlogopite, apatite, sphene, perovskite and carbonate appear to be common, and it is possible that mantle previously depleted by a major magma extraction event could become subsequently re-enriched by the addition of incompatible element enriched fluids or partial melts.

Figure 3.18 shows a schematic cross section through the upper mantle beneath both continental and oceanic regions, which attempts to summarize all the conclusions reached in Sections 3.3.1–6. The continental lithosphere may consist of great thicknesses of depleted material, the depletion corresponding to major crust-forming events early in the history of the Earth. However, much of it may have been subsequently re-enriched (Menzies & Hawkesworth 1987), providing a now-fertile source for intracontinental plate alkaline magmatism (Chs 11 & 12). In contrast, the depletion of the oceanic lithosphere is related to a much more recent process, the extraction of basaltic magma at mid-oceanic ridges (Ch. 5). The sublithospheric upper mantle (the source of MORB) is also depleted (relative to the primordial mantle) as a consequence of the formation of the continental crust throughout geological time. The lower mantle may be considered to represent a reservoir of relatively enriched mantle components. Figure 3.18 does not include the complexities introduced by the recycling of subducted oceanic lithosphere back into the source region of oceanic basalts: this will be considered further in Ch. 9.

Clearly, if partial melting processes have been operating throughout geological time then the upper mantle must be heterogeneous, comprising both fertile and depleted lherzolite and still more refractory harzburgite and dunite. In addition, there must be innumerable 'pods' of 'frozen-in' partial melts which failed to segregate and form

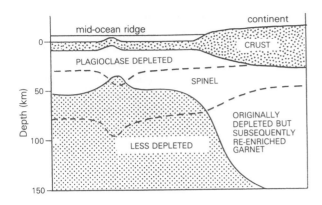

Figure 3.18 Schematic cross section through the top 150 km of mantle beneath oceanic and continental regions, ignoring the complications of a subduction zone (after Brown & Mussett 1981, Fig.7.10, p.122).

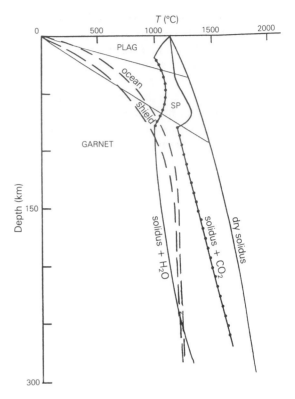

Figure 3.19 Minimum conditions necessary for partial melting of mantle lherzolite under anhydrous conditions and in the presence of small amounts of H_2O (<0.4%) and CO_2 (<5 wt.%). The dotted segments of the solidii are those along which a free vapor phase does not exist, all the H_2O and CO_2 being contained in amphibole and carbonate respectively. The intersection of the geothermal gradients ——— with the water present solidus may account for the low-velocity zone (after Wyllie 1981, Fig. 3).

magmas. The process of convection should effectively homogenize these differences in the sub-lithospheric upper mantle. In contrast, the rigid lithosphere should preserve all the heterogeneities imposed upon it since it was formed. This is particularly important with respect to the evolution of the continental lithosphere, which may have been mechanically coupled to the overlying crust for long periods of geological time (>1−2 Ga).

3.4   Partial melting processes in the mantle

3.4.1   The normal state of the mantle

Before considering the nature of magma generation processes in the Earth's upper mantle we must first consider its normal state, i.e. solid, liquid or partially molten. Figure 3.19 shows the dry lherzolite solidus, the pressure−temperature curve marking the minimum conditions necessary for the onset of partial melting. Also shown are typical oceanic and shield geotherms. Note how these differ at shallow depths, but converge towards the adiabatic gradient at depths greater than 150 km. From these geotherms we can see that under normal conditions the mantle beneath both continents and oceans must be totally solid, if it is completely anhydrous,

as the geotherms never intersect the solidus. However, the addition of small amounts of H_2O and CO_2 have the effect of lowering the solidus such that, for mantle with 0.4 wt. % H_2O, the geotherms intersect the solidus in the depth range 100−250 km. This could produce a small amount of partial melt (<1%) in the mantle within this depth range, which could explain the seismic characteristics of the low-velocity zone (LVZ) (Section 3.2.1).

From these data we can conclude that even if the mantle contains small amounts of volatiles, which from studies of ultramafic xenoliths it must do, its normal state must be essentially solid. How then do we achieve the degrees of partial melting necessary to generate the spectrum of magmas erupted at the Earth's surface, and what causes the link between plate-tectonic processes and magma generation?

Partial melting of a multicomponent source such as the lherzolite of the upper mantle must obviously be a complex phenomenon which inherently does not lend itself readily to direct study. However, we can approach the problem by looking at partial melting processes in simple binary, ternary and quaternary systems and then extrapolating these data to the mantle.

3.4.2 The different types of partial melting

Partial fusion, anatexis and partial melting are all general terms for any process that produces a melt of a system in some proportion less than the whole. There are two main ideal or end-member models for the process by which this occurs:

(a) Equilibrium or batch melting: the partial melt formed continually reacts and equilibrates with the crystalline residue until the moment of segregation. Up to this point the bulk composition of the system remains constant.
(b) Fractional or Rayleigh melting: the partial melt is continuously removed from the system as soon as it is formed, so that no reaction with the crystalline residue is possible. For this type of partial melting the bulk composition of the system is continuously changing.

In nature, the critical parameter controlling the nature of the partial melting process will be the ability of the newly formed magma to segregate from the residual crystals (Maaløe 1985). This in turn will depend upon the permeability threshold of the partially molten mantle lherzolite. Batch melting involves a degree of partial melting up to a permeability threshold, followed by magma segregation and accumulation. Fractional melting, with magma removal as soon as it is formed, could obviously only occur if the mantle becomes permeable at very low degrees of partial melting (<1%). Maaløe (1982) has suggested that the actual partial

melting process in the mantle is of a type intermediate between the batch and fractional models, which he terms *critical* melting. In this case the mantle becomes permeable after a certain degree of batch partial melting, and thereafter the magma is squeezed out continuously from the residuum and accumulated. The permeability threshold of the mantle is unknown and may differ between different tectonic settings; for example, in regions of sheared flow it may be substantially decreased. Experimental determinations show that unsheared lherzolite can become permeable at about 2−3% partial melting (Section 3.6).

A prime objective in attempting to understand the nature of partial melting processes in the mantle is to explain the diversity of apparently primary magma compositions (Section 3.5.1) erupted at the Earth's surface. Experimental determination of the compositions of partial melts of spinel or garnet lherzolite under controlled $P-T$ conditions would ideally provide the necessary data. However, such experiments are difficult to perform and because of the nature of the techniques involved only batch partial melting can be investigated. Thus an alternative approach is required to assess the influence of partial melting mechanism upon the chemical composition of the resultant liquids. This is provided by phase equilibrium studies of simplified ternary and quaternary systems, some of which may be considered to be mantle analogues.

3.4.3 Partial melting in a model ternary system

Figure 3.20 shows a model ternary system, ABC, with three solid phases a, b and c, the compositions of which are represented by the end-members of the system. Composition X (40% A, 30% B, 30% C) represents the bulk composition of the material in whose partial melting behaviour we are interested.

Regardless of the melting mechanism (batch or fractional) the first liquid will be formed at the ternary eutectic E_1. As further heat is supplied to the system more melt of the eutectic composition E_1 will be generated until one of the solid phases is completely consumed into the melt. In the case of bulk composition X this will be phase b, deduced from the projection of the line connecting E_1 and X

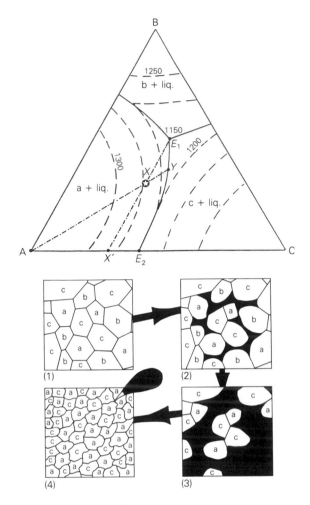

Figure 3.20 Partial fusion in a model ternary system A−B−C, in which the solid phases a, b and c are pure A, B and C respectively (after Barker 1983, Fig. 7.1, p.125): (1) Appearance of bulk composition X at a temperature below that of the ternary eutectic E_1 i.e. the subsolidus assemblage. (2) Partial fusion has begun to produce liquid (black) with a composition E_1 at those points where the three phases a, b and c are in mutual contact. (3) More advanced melting, still at E_1, at the point at which all of the phase b is used up. (4) Liquid escapes as b is consumed, leaving a crystalline residue of a and c of bulk composition X'. This residue will only be able to melt further if the temperature is raised to that of the binary eutectic E_2.

to the A−C sideline.

The maximum amount of the eutectic liquid E_1 which can be generated for either batch or fractional melting of X is given by the Lever Rule:

$$\text{maximum \% of liquid } E_1 = \frac{\text{distance } X - X'}{\text{distance } E_1 - X'} \times 100$$

Only with greater degrees of melting than this do batch and fractional melting differ. For batch melting, once all of phase b is used up further melts are generated along the cotectic curve $E_1 - Y$ until, at Y, A, X and Y are colinear and therefore all of phase c has been consumed. Further melting occurs along $Y - X$, until at X the system is completely molten.

For fractional melting the bulk composition of the system is constantly changing due to the removal of increments of melt as soon as they are formed. In this case once all of phase b is used up the bulk composition of the system is X', and no further melting can occur until the temperature of the system is raised to that of the binary eutectic E_2. The difference between the melting curves for batch and fractional melting of X is shown schematically in Figure 3.21.

One of the interesting features of ternary systems such as this is that, regardless of the partial melting mechanism, the initial partial melt will always have the same eutectic composition E_1. In addition, the closer the bulk composition of the system is to E_1 the greater the volume of partial melt E_1 which can be generated. A further important point to note is that, regardless of the bulk composition of the system, the initial partial melt will always have the eutectic composition E_1.

Figure 3.20 also shows the changing appearance of the system with progressive partial melting, starting with a crystalline solid X with the granular texture of a metamorphic rock. The initial liquid E_1 forms only at points where the three phases a, b and c are in direct contact. Other intersections, such as a−a−b, b−b−c etc, do not provide the components necessary for partial melting. With progressive partial melting the residual crystals effectively become disaggregated, until a point is reached at which the liquid can percolate upwards, rapidly driven by the density difference between it and the residual crystalline phases. Following melt extraction the residual solid may become annealed,

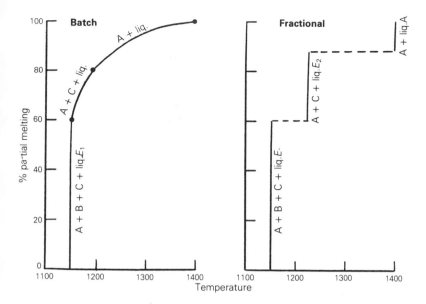

Figure 3.21 Comparison of batch and fractional melting curves for the system shown in Figure 3.20. Note that for fractional melting the percentages shown here are those of the original system.

leaving no trace of the partial melting process it has undergone.

3.4.4 The system forsterite−diopside−enstatite as a mantle analogue

Figure 3.22 shows a schematic ternary phase diagram of the anhydrous system forsterite−diopside−enstatite at a pressure of about 20 kbar. This system can be used as a fairly realistic mantle analogue, although obviously it does not include an aluminous phase (spinel or garnet). Additionally, in nature there is extensive solid solution along the diopside−enstatite join, which cannot be easily represented.

In this system a lherzolite with a bulk composition X would initially partially melt to produce a liquid of the eutectic composition Y. If this liquid could continually escape from the system in some fractional melting type process, the bulk composition of the residual solid would migrate along the line $X−Z$. Once the residue has reached Z and all of the clinopyroxene (diopside) has been consumed, then melting at the eutectic Y must cease, leaving a residue of solid harzburgite (Z). If the temperature continued to rise further, a second melting event might occur at the binary eutectic a, driving the composition of the residue towards

dunite (forsterite).

This simple phase diagram allows us to represent, albeit schematically, the relationships between the lherzolite, harzburgite and dunite xenoliths observed in kimberlites and alkali basalts

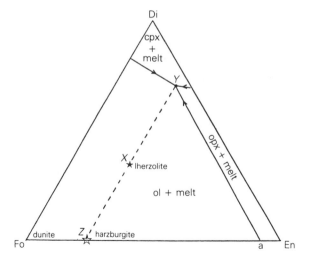

Figure 3.22 Schematic phase diagram of the anhydrous system forsterite−diopside−enstatite at a pressure of about 20 kbar as a potential mantle analogue (after Yoder 1976, Fig. 2. 14). Note that mineral phases in natural peridotites are solid solutions rather than the simple end-members shown here, and therefore the actual phase relations must be more complicated.

(Section 3.3.1). Additionally, we could construct melting curves similar to those shown in Figure 3.21 if we had the necessary data on the positions of the isotherms. Partial melting of lherzolite X would yeild something like 40% of the eutectic liquid Y by either batch or fractional melting.

If this system can indeed be considered as a fairly realistic mantle analogue then it has fundamental implications for magma genesis in the mantle. For example, primary partial melts of the mantle are predominantly basaltic (Section 3.5.1). Could basalt be a eutectic partial melt in some n−dimensional phase diagram? We shall consider this further in subsequent sections.

3.4.5   Partial melting in the quaternary system CMAS

O'Hara (1968) formulated a system of projecting natural rock analyses into a modification of the quaternary system $CaO-MgO-Al_2O_3-SiO_2$ (CMAS) in order to synthesize the available experimental data on the polybaric partial melting behaviour of mantle lherzolite. The reader is referred to Cox, Bell & Pankhurst (1979; pp.244−56) for a detailed discussion of this system. Here we will consider only its more important implications.

Figure 3.23 is a polybaric, polythermal projec-tion from enstatite into the plane $M_2S-A_2S_3-C_2S_3$ in CMAS. This gives us a pseudoternary system in which to consider the partial melting behaviour of a four-phase mantle lherzolite. Partial melting of a typical mantle composition (indicated by a star in Fig. 3.23) at 1 atm (0 kbar) produces a eutectic basaltic liquid in equilibrium with olivine, ortho-pyroxene, clinopyroxene and plagioclase. At 15 kbar pressure the eutectic liquid has much less silica (see inset of CMAS for confirmation of this) and is in equilibrium with a spinel lherzolite mineralogy. At 30 kbar the eutectic liquid, though still basaltic, is even poorer in SiO_2 (S) and is in equilibrium with a garnet lherzolite mineral assemblage.

From this simplified projection we can predict that with increasing pressure initial partial melts of mantle lherzolite become more SiO_2-poor and more MgO-rich as the coexisting mineral assemblage changes from plagioclase to spinel to garnet lherzo-lite. This leads us to ask the following questions about magma generation in the mantle:

(1) Does the compositional spectrum of primary basaltic magmas erupted at the Earth's surface simply reflect a polybaric continuum of in-variant points in some n−dimensional phase diagram?

(2) Are all mantle derived partial melts n−dimensional eutectic compositions? If this

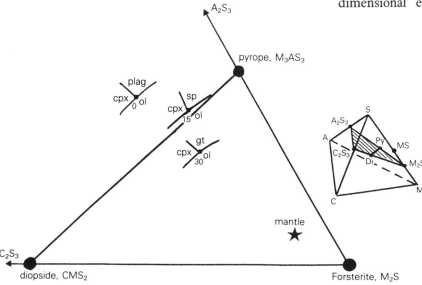

Figure 3.23 Schematic polythermal, polybaric projection from enstatite (MS) into the plane $M_2S-A_2S_3-C_2S_3$ in CMAS, showing the migration of the invariant melting composition with increasing pressure at 0, 15 and 30 kbar. Note how the nature of the coexisting aluminous phase changes from plagioclase to spinel to garnet with increasing pressure (after Cox *et al.* 1979, Fig. 9.11, p. 250).

ᴡᴇʀᴇ ᴄᴏʀʀᴇᴄᴛ we would obviously not need to worry about the nature of the partial melting process (i.e. batch or fractional or some other as yet unspecified type of partial melting).

3.4.6 Trace elements and partial melting processes

In the preceding sections we have shown that the major element composition of mantle-derived partial melts may be largely insensitive to the type of partial melting process. However, the trace element concentrations of partial melts may vary considerably during partial melting and, as such, may provide valuable information about the actual mechanisms of magma generation in the mantle.

Simple models have been developed to quantify the changes in trace element concentrations which occur during partial melting. The reader is referred to Wood & Fraser (1976; ch. 6), for derivation of the basic equations. It is important to remember that these are all based on highly idealized assumptions and, at best, can only be regarded as crude approximations to reality. They do, however, serve to illustrate the basic principles and the limiting effects of simple igneous processes.

For the case of batch partial melting, the concentration of a trace element in the liquid, C_L, is related to that in the original solid, C_0, by the expression:

$$\frac{C_L}{C_0} = \frac{1}{F + D - FD} \qquad (1)$$

where $D = \sum_\alpha X_\alpha D_\alpha$, and X_α is the weight fraction of phase α in the mineral assemblage and D_α its crystal–liquid partition coefficient. F is the weight fraction of melt formed, and D is the bulk distribution coefficient for the residual solids at the moment when melt is removed from the system. The original source mineralogy and the changes which it undergoes during melting are thus immaterial, except in so far as they determine respectively the value of C_0 and the mineralogical composition of the residue. Obviously, in order to define D it is necessary to calculate the proportions of residual minerals, and so the above expression is often formulated in terms of the initial proportions and the relative melting rates of the different solid phases. Clearly, with increasing degrees of partial melting, different minerals may be progressively consumed, causing discontinuous changes in the value of D. This is particularly important in the partial melting of possible mantle materials since minor phases such as garnet, amphibole, biotite and clinopyroxene would, in general, complete their melting significantly earlier than olivine and orthopyroxene.

If the proportions of phases entering the melt (p_α) are different from their initial proportions in the rock ($X^0\alpha$) then:

$$D = \frac{D_0 - FP}{1 - F}$$

where $D_0 = \sum_\alpha X^0_\alpha D_\alpha$ and $P = \sum_\alpha p_\alpha D_\alpha$. Substituting into (1) gives:

$$\frac{C_L}{C_0} = \frac{1}{D_0 + F(1 - P)} \qquad (2)$$

In this case some phases will enter the melt before $F = 1$. The fraction of melt, F, at which this occurs is given by

$$F = X^0_\alpha / p_\alpha$$

It is important to note that once a phase has completely melted the value of D_0 used in the equation must be changed.

This type of partial melting is referred to as *non-modal melting*. Similar equations can be derived for fractional melting:

modal fractional

$$\frac{C_L}{C_0} = \frac{1}{D}(1 - F)^{(1/D - 1)} \qquad (3)$$

non-modal fractional

$$\frac{C_L}{C_0} = \frac{1}{D_0}\left(1 - \frac{PF}{D_0}\right)^{(1/P - 1)} \qquad (4)$$

These equations give the trace element concentration in the infinitesimally small amount of liquid formed in equilibrium with the residue and then removed from the system after a fraction, F , has already melted.

Perfect fractional melting is obviously physically unrealistic as a model for magma generation. A more reasonable model would be one in which the melt increments were collected in a common reservoir in which they are perfectly mixed. For highly incompatible trace elements this would be practically indistinguishable from batch melting.

We can use the simple system forsterite–diopside–enstatite depicted in Figure 3.22 to investigate the numerical consequences of using these different partial melting equations. This system is shown schematically as an inset in Figure 3.24. The bulk composition of the source X can be expressed as 0.4Di, 0.2En, 0.4Fo (these are the X_α^0). The proportions (p_α) in which the various phases enter the melt are initially given by the proportions in the eutectic composition E, i.e. 0.7Di, 0.2En, 0.1Fo.

The fraction of melt (F) at which each phase is totally consumed is given by:

$$F = X_\alpha^0 / p_\alpha$$

Thus:

$$F_{cpx} = 0.4/0.7 = 0.57, \qquad F_{opx} = 0.2/0.2 = 1.0,$$

$$F_{ol} = 0.4/0.1 = 4.0$$

i.e. clinopyroxene (diopside) is the only phase to be completely consumed into the partial melt and this occurs at 57% partial melting. This should correspond to the fraction of melting at which clinopyroxene disappears, as determined by application of the Lever Rule to the phase diagram, i.e.

$$F = \frac{XY}{EY} \approx \frac{4}{7}$$

(which it does).

In Table 3.6 the results of the application of the various partial melting models to source composi-

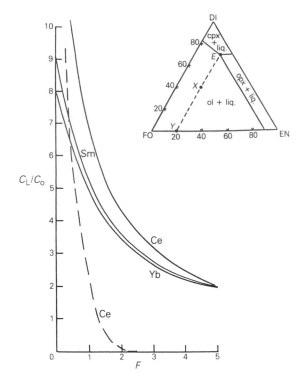

Figure 3.24 Trace element concentrations of the REE Ce, Sm and Yb as a function of the degree of partial melting of bulk composition X in the system diopside–forsterite–enstatite. Solid curves are for non-modal batch melting; the dashed curve is for non-modal fractional melting (data from Table 3.6).

tion X are compared for the elements Ce and Yb. Clearly, for highly incompatible elements like these (in non-garnet-bearing sources) the modal and non-modal equations give essentially identical results and thus, for simplicity, the modal equations can be used even though the actual melting process must be non-modal, as constrained by the phase diagram. However, as shown in Table 3.6 and in Figure 3.24, batch melting and fractional melting give significantly different results. For fractional melting, most of the incompatible trace elements are extracted in the early melt increments, such that subsequent melt fractions have very low concentrations of these elements. As far as mantle partial melting processes are concerned, perfect fractional melting is not a realistic model, and thus the batch melting equations are probably the most useful for

Table 3.3 Calculated values of C_L/C_O for modal and non-modal batch and fractional melting of bulk composition X in the system diopside—forsterite—enstatite for the highly incompatible elements Ce and Yb.

	F	Modal batch	Non-modal batch	Modal fractional	Non-modal fractional
Ce	0	24.39	24.39	24.39	24.39
	0.1	7.30	7.47	2.07	2.03
	0.2	4.30	4.11	0.13	0.094
	0.3	3.04	3.13	0.0058	0.0017
	0.4	2.36	2.42	0.00016	0.000005
	0.5	1.92	1.98	—	—
Yb	0	8.14	8.14		
	0.1	4.75	4.95		
	0.2	3.35	3.55		
	0.3	2.59	2.77		
	0.4	2.11	2.27		
	0.5	1.78	1.92		

$X_\alpha = 0.4Di, 0.2En, 0.4Fo$
$P_\alpha = 0.7Di, 0.2En, 0.1Fo$

Partition coefficients:

	Ce	Yb
ol	0.001	0.002
opx	0.003	0.05
cpx	0.10	0.28

calculation purposes.

This example highlights some of the problems associated with attempts to model quantitatively partial melting processes in the mantle:

(1) What is the modal mineralogy of the source? Given the modal variations amongst lherzolite xenoliths what do we choose?
(2) In what proportions do the various phases enter the melt, i.e. how do we determine the p_α? For the example shown in Figure 3.24 this was easy because we had a ternary phase diagram to constrain the melting behaviour of the system. However, for natural lherzolites we would need a multi-dimensional phase diagram to adequately portray the phase relationships.
(3) Which melting model should we choose?

Obviously, all of these are essentially unknown for the partial melting of mantle lherzolite. However, as we showed in the above example, the equations are fairly insensitive to some of the parameters involved and therefore they can still be used to provide important constraints on the magma generation processes.

3.4.7 More complex partial melting models and the role of accessory phases

The equilibrium batch partial melting equations described above have been successfully used to model the distribution of trace elements in igneous rocks with low total REE enrichment factors. However, the equations apparently have serious limitations when applied to strongly light-REE enriched samples such as kimberlites, nephelinites and alkali basalts. We have stated previously that mantle metasomatism, with the introduction of incompatible element enriched accessory phases such as amphibole, biotite, apatite and sphene, may be a necessary precursor to the generation of these magma types. Such accessory phases would most likely be the first to melt during subsequent anatexis of the metasomatized source, and would be entirely consumed by very small degrees of partial melting. Their melting behaviour would not necessarily be an equilibrium process, and this type of melting has been referred to as *disequilibrium melting*.

A range of equations for more complex partial melting models exist in the literature, but will not be considered here. Attempts to use equations more complex than those for non-modal batch and fractional melting are not justifiable at present because of our lack of understanding about the actual melting behaviour of the mantle.

3.4.8   What causes partial melting in the mantle?

In Section 3.4.1 we showed that the normal state of the mantle is essentially solid. How then do we achieve the degrees of partial melting necessary to generate large volumes of magma, and how is this related to tectonic processes? Figure 3.25 shows the melting interval of a typical lherzolite. At the solidus the first trace of partial melt appears, while at the liquidus the system is completely molten. The interval between the solidus and liquidus can be contoured for the degree of melting, as shown in the diagram. As we shall show in Section 3.5 the

degree of melting necessary to generate the spectrum of terrestrial basaltic magmas is between 5 and 30% on average. To achieve such degrees of melting the geotherm must intersect the 5–30% lherzolite melting contours. There are several ways in which this might occur:

(1)   Anomalous thermal perturbation of the geotherm.
(2)   Lowering of the mantle solidus (and liquidus) by the addition of volatiles to the system.
(3)   Adiabatic decompression (pressure-release melting) of mantle lherzolite.

All of these are possible, but (3) is probably the most likely mechanism for generating large volumes of magma. Pressure-release melting of ascending mantle material is probably responsible for generating a large proportion of the magmas reaching the surface of the Earth. For example, mid-oceanic ridges are the sites of upwelling of deeper hotter mantle material, and also sites of one of the most voluminous magma generation environ-

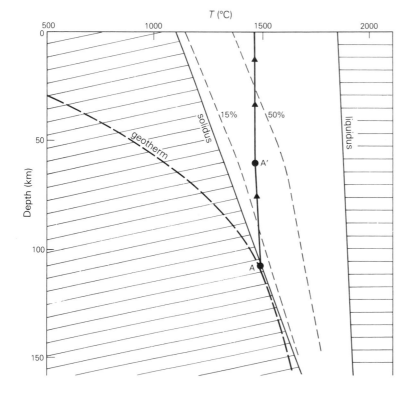

Figure 3.25 Partial melting of mantle lherzolite by adiabatic decompression. Magma segregates at A′ (20% partial melt) and ascends rapidly towards the surface. Diagonal shading marks the subsolidus field and horizontal shading the completely molten field (which can never be achieved under present geothermal gradients). The interval between the solidus and the liquidus is contoured for the degree of melting.

mantle on Earth. The process can be envisaged in terms of Figure 3.25 as follows. Consider a volume of mantle lherzolite at A, whose temperature is constrained by the geothermal gradient close to the solidus. Assume that mantle A is now caught up in the ascending limb of a convection cell and rises adiabatically towards the surface. The pressure on this volume of lherzolite is reduced dramatically, but its temperature remains essentially constant, as shown by the ascent path A–A′. As soon as the mantle A starts to rise we can see from Figure 3.25 that it moves progressively further and further into the lherzolite melting interval and thus will become partially molten. At A′ the ascending mantle diapir will consist of ~20% partial melt and 80% residual crystals. A 20% partial melt should be physically separable from its source (Section 3.6) and thus, at this depth, the magma could segregate from the residual crystals and ascend rapidly to the Earth's surface. If the magma did not segregate and the diapir continued to rise then an even greater degree of partial melting would ensue.

Pressure-release melting may be the major process responsible for the generation of magma at mid-oceanic ridges, back-arc spreading centres and also at many intra-plate volcanic centres. However, in the subduction-zone environment (Ch. 6), the main trigger for partial melting may be a lowering of the mantle solidus by the addition of volatiles derived from the subducted oceanic crust.

3.5 The basaltic magma spectrum in relation to partial melting processes

3.5.1 The primary magma spectrum

Extensive laboratory experiments during the past 20 years on lherzolite bulk compositions under mantle conditions leave no doubt as to the wide range of silicate liquids that can be generated by partial melting of the mantle. Variations in the bulk composition and mineralogy of the source and in the depth and degree of melting can combine to generate magmas ranging from kimberlite to alkali basalt to tholeiitic basalt, and perhaps even andesite (Wyllie 1981).

The task for the petrologist is a rather daunting one: to determine the relative roles of partial melting and crystal fractionation in producing the diversity of eruptive magma compositions. A major problem in considering a particular basaltic magma is whether it represents an unmodified primary mantle melt, or whether it is a derivative magma modified by one or more subsequent processes after segregation from its source rock (Ch. 4).

Primary magmas are infrequently represented in the spectrum of eruptive magma compositions at the Earth's surface because of the range of near-surface processes (fractional crystallization, crustal contamination, magma mixing etc.) which modify their compositions. Their chemical characteristics are primarily determined by:

(a) the phase relations of their mantle source; and
(b) the nature of the partial melting process, including the dynamic processes which lead to the accumulation of magma.

A variety of geochemical criteria can be used to assess whether a magma has primary characteristics. These are described in Chapter 2.

Partial fusion of a lherzolitic source, either at a eutectic-like invariant point or a low-temperature cusp on the mantle solidus, can produce large volumes of compositionally uniform magma (Section 3.4). Consequently, continental flood basalts (Ch. 10) and ocean-floor basalts (Ch. 5) have frequently been interpreted as the products of moderate amounts of mantle partial melting modified by subordinate crystal fractionation. Paradoxically, such basaltic magmas do not generally have primary geochemical characteristics.

A basic tenet of experimental petrology is that primary magmas, produced by partial fusion of lherzolite, should be saturated with the residual mantle phases at temperatures and pressures corresponding to the depth of segregation. High-pressure melting studies on natural peridotites or synthetic analogues (Mysen & Kushiro 1977, Jaques & Green 1980) suggest that three or more phases (ol + opx + cpx $\mp$ Al-rich phase) will be on the liquidus up to $20-25\%$ partial melting, two or three phases (ol + opx $\pm$ cpx) between 25 and 40%

and one or two phases (ol ± opx) above about 40% melting. The aluminous phase will be either plagioclase, spinel or garnet, depending on the pressure. Thus all primary basalts derived from an olivine-rich mantle should have olivine, together with the other appropriate phases, on their liquidi at pressures corresponding to their depth of segregation.

Considerable controversy has developed in the interpretation of high-pressure melting experiments on terrestrial basalts because of the common absence of olivine from the liquidus at pressures above 10−15 kbar, corresponding to depths of magma segregation of 30−40 km. Clearly, this may be appropriate for mid-ocean ridge volcanism, but it is obviously too shallow for most other magma generation environments. Similar problems are encountered with orthopyroxene, which should commonly accompany olivine as a major liquidus phase at the depth of segregation. Few of the basalts so far studied have orthopyroxene on the liquidus at any pressure. However, this may reflect reaction relationships between olivine, orthopyroxene and liquid rather than indicating non-primary characteristics.

The spectrum of terrestrial basaltic magma types may be divided into three broad groups, komatiites, picrites and basalts. An important question is whether these represent a range of primary magma types or whether basalts are the products of extensive differentiation of primary picritic or komatiitic magmas.

Melting experiments on both depleted and enriched lherzolite source compositions (Jaques & Green 1980; Figs. 3.26 & 27), have provided invaluable data which help to resolve this problem. The experiments show that tholeiitic basalt magmas can be produced by moderate degrees of partial melting (20−30%) of either source at pressures below 15−20 kbar. At higher pressures picritic liquids are generated at the same degrees of partial melting. Alkali basaltic magmas appear to be generated by smaller degrees of partial melting (<20%) of enriched sources at pressures greater than 10 kbar, while liquids akin to peridotitic komatiites can be generated by 40−50% partial melting of a fertile lherzolite, or 30−40% partial

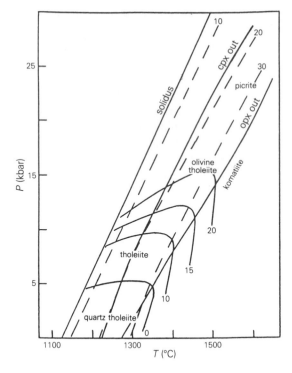

Figure 3.26 Experimentally determined partial melting characteristics of a depleted mantle lherzolite source. Dashed lines are partial melting contours for 10, 20 and 30% partial melting. Cpx-out and opx-out lines mark the degree of melting at which cpx and opx respectively are completely consumed into the melt. The strongly curved contours indicate the normative content of olivine in the melt (after Jaques & Green 1980).

melting of a depleted lherzolite.

A major problem in understanding the petrogenesis of komatiites is to account for the extremely high temperatures and extensive degrees of melting apparently required to produce them. It is particularly difficult to explain why the melt failed to segregate from the residue before the required 40−50% partial melting was achieved. Estimates of the ease of segregation of partial melts are varied, and the relative roles of permeability and shear unclear (Section 3.6). However, Spera (1980) has suggested that a 35% melt will segregate from the residue over 1000 times more effectively than a 5% melt. Combining this with the observation that alkali basalts, widely held to result from small (2−10%) degrees of melting, are common, attesting

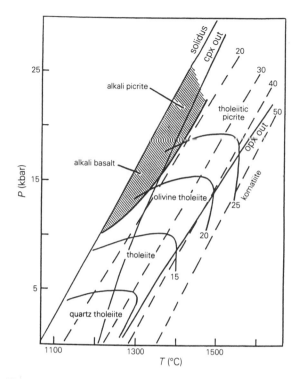

Figure 3.27 Experimentally determined partial melting characteristics of an enriched lherzolite source. Dashed lines are partial melting contours for 20, 30, 40 and 50% partial melting. The shaded area represents the conditions necessary for the generation of alkalic basaltic magmas. Other symbols as Figure 3.26 (after Jaques & Green 1980).

to their ability to segregate from the mantle, suggests that melt segregation should occur long before 40% partial melting. A more plausible explanation is that komatiites result from sequential melting processes in which the source material is the highly refractory residuum of a previous melting episode (incremental melting).

Picrites are crucial to the primary magma problem, because it is widely held that these are parental to the other basaltic magma types through extensive crystal fractionation. Unfortunately, there is an apparent rarity of picrites which might be considered to represent true liquid compositions. Explanations to account for this apparent rarity include the following:

(1) Picritic melts undergo extensive olivine frac-

tionation within the mantle to produce basaltic magmas.
(2) Picritic magmas mix with evolved magmas in shallow-crustal magma chambers to produce basaltic magmas.
(3) Dense picritic magmas are trapped at the base of or within less dense crust, where they differentiate until the density contrast between the crust and the magma is appropriate for ascent of the more differentiated magma.

Basalts may thus be both the direct products of mantle partial melting and the differentiates of more 'primitive' picritic partial melts. Melting experiments on lherzolites show that primitive basalts, encompassing both tholeiitic and alkaline varieties, represent a continuum of compositions produced by moderate amounts (<20%) of partial melting. More extensive melting leads to the production of picritic and komatiitic melts.

As considered in Section 3.3.5, the melting relations of mantle lherzolite may be considerably modified in the presence of volatile components such as H_2O and CO_2. At subsolidus temperatures amphibole may be generated by reaction of peridotites with H_2O-rich fluids at pressures less than 25 kbar, whereas in the presence of CO_2 the magnesian carbonate dolomite may be produced at pressures greater than 25 kbar (Wyllie 1981). H_2O derived from the subducted oceanic crust appears to be fundamental in the petrogenesis of subduction-zone magmas (Ch. 6) and hydrous partial melting of the subduction-modified mantle wedge may produce primary magmas with basaltic andesite or even andesite chemistries. In contrast, CO_2 appears to be important in the petrogenesis of kimberlites, carbonatites and highly alkalic silica-undersaturated magmas in intracontinental plate-tectonic settings (Ch. 12). Detailed discussion of the role of volatiles in the generation of these more unusual primary magma compositions is deferred until the appropriate chapters.

3.5.2 The eutectic nature of mantle partial melts?

An important problem in attempting to understand mantle partial melting processes is whether natural lherzolite melts in a eutectic-like fasion similar to the simple analogue systems considered in Section 3.4. Experimental studies can obviously provide vital information in this respect, but are restricted to batch partial melting processes.

Mysen & Kushiro (1977) investigated the melting behaviour of a natural fertile lherzolite under anhydrous conditions at 20 and 35 kbar. As shown in Figure 3.28, the melting curves have an identical form to the synthetic batch melting curves for the simple ternary system shown in Figure 3.21. At 35 kbar olivine−orthopyroxene−clinopyroxene and garnet coexist with liquid up to 30% partial melting. This could correspond to a eutectic-like partial melt in some multidimensional system, as the temperature of this segment of the curve is approximately constant. At greater degrees of partial melting phases disappear sequentially in the order garnet, clinopyroxene, orthopyroxene and their disappearance causes inflections in the melting curve.

Unfortunately, such melting behaviour is not a consistent feature of the experiments and the melting curves produced by Jaques & Green (1980), for the enriched and depleted lherzolite compositions shown in Figures 3.26 and 3.27 are smooth functions of temperature (Fig. 3.29).

However, regardless of the shape of the curves, equilibrium melting still results in a progressive elimination of phases with increasing degree of melting, in the order aluminous phase (spinel or garnet) clinopyroxene, orthopyroxene. In all of these melting experiments the residual phases change progressively to more refractory compositions with increasing degree of melting, i.e. their $Mg/(Mg + Fe)$ and $Cr/(Cr + Al)$ ratios increase.

The most important point to be made on the basis of these experiments is that mantle lherzolites do not undergo modal melting. Thus, in principle, non-modal melting equations should always be used in calculations.

3.6   Segregation and ascent of magma

During the initial stages of partial melting the melt will form an interconnected network within the crystalline matrix and may flow relative to the matrix as a consequence of its buoyancy. Such buoyancy-driven porous flow may be considered as the initial process leading to melt segregation (Stolper *et al.* 1981; McKenzie 1984a, 1985; Richter & McKenzie 1984; Ribe 1985; Scott & Stevenson 1986). Other transport processes will

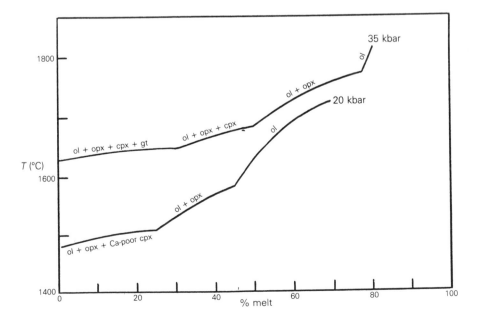

Figure 3.28   Batch partial melting behaviour of a natural fertile lherzolite at 20 and 35 kbar. The phases marked on each segment of the curve indicate the mineral assemblages coexisting with liquid. Note the segmented nature of the melting curve which might be considered analogous to that shown for the model ternary system in Figure 3.21 (after Mysen & Kushiro 1977, Fig. 3, p. 848).

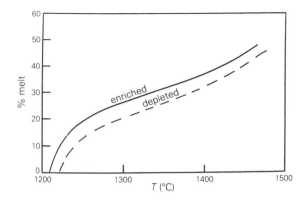

Figure 3.29 Batch partial melting behaviour of enriched and depleted lherzolites (Figs 3.26 & 7) at 10 kbar pressure. Note the smooth character of the melting curves in comparison to that shown in Fig. 3.28 (after Jaques & Green 1980, Fig. 2.).

clearly take place at higher levels. For example, transport through the brittle lithosphere probably occurs through cracks (Spera 1984, Spence & Turcotte 1985).

Almost every stage in the genesis of magma is influenced to some extent by the density contrast between silicate liquid and residual crystals ($\Delta\rho$). The rate of liquid segregation from partially molten source regions is proportional to $1/\Delta\rho^2$ (Stolper *et al.* 1981). As shown in Figure 3.30, the density contrast between basaltic liquid and the major mantle minerals decreases with increasing pressure. Thus the greater the depth of a partially molten source region the smaller the density contrast between the melt and the residue. As $\Delta\rho$ approaches zero melt segregation becomes increasingly difficult. This suggests that there could be a rough natural limit to the depth from which dry basaltic magmas can be derived. The critical depth range in which $\Delta\rho \rightarrow 0$ lies between 100–200 km (30–70 kbar), varying with the bulk composition of the partial melt. Melts generated at greater depths, in regions of small or even negative density contrast, could remain in their source regions, unsegregated from the residual crystals, indefinitely. This could explain how regions of the mantle apparently remain undepleted throughout geological time. Only when undepleted mantle rises above the critical depth, either in diapirs or as part of a

local or global convection pattern, will magma segregate.

During ascent by porous flow the matrix must deform and compact (Scott & Stevenson op. cit.). The compaction rate of the partially molten rock is likely to be rapid and melt-saturated porosities in excess of $\sim$3% are unlikely to persist (McKenzie 1984). Melt extraction is always controlled by the rate of deformation of the matrix, and may also be governed by its permeability and the viscosity of the melt. Low-viscosity magmas can separate rapidly whereas for high-viscosity melts separation is restricted (McKenzie 1985). The velocity with which a melt can move will govern its ability to escape from convecting regions of the upper mantle. McKenzie (1984) has shown that for melt fractions as large as 10% the velocity reaches 1 m yr^{-1}. In such cases the separation of melt and

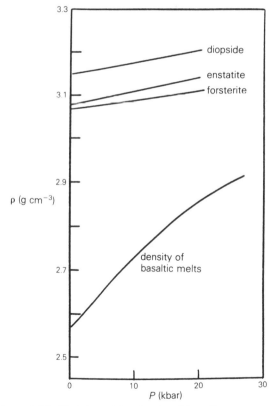

Figure 3.30 Variation of the density of basaltic magma and the major mantle minerals as a function of pressure (after Stopler *et al.* 1981, Fig. 3, p. 6265).

matrix is rapid, and consequently it is unlikely that such large melt fractions are ever present in the Earth.

If batch melting were the major process operating in the mantle then magma must be able to remain where it was formed for an extended period of time, despite the permeability of the source. This could only happen if the residuum were so rigid that it could resist deformation under its own weight. However, creep rates for olivine-rich materials are far too high, and the residuum would inevitably undergo compaction, squeezing the interstitial liquid out. The rheological behaviour of the partially molten mantle will thus eliminate batch melting and, instead, primary magmas may be formed by critical melting (Section 3.4.2). Fortunately, it has been suggested by Maaløe (1982) that the major and trace element compositions of primary magmas generated by critical melting will not differ much from those formed by batch melting.

Clearly, permeability is required for melt extraction and we must consider how this is achieved, and for what melt fraction. Nicolas (1986) has suggested that the stable melt fraction is ≪7%, and that above this level melt extraction starts. In the initial stages a network of connected melt veins is formed in the deforming peridotite, opened by fluid-assisted shear fracturing. Eventually, the melt pressure fractures the overlying peridotites, creating a dyke-like conduit for melt extraction, which must ultimately reach the Earth's surface. Such hydro-fracturing should be registered by earthquakes and tremors beneath active volcanoes. Nicolas (op. cit.) suggests that 50 km is about the maximum depth at which melt may be evacuated from the mantle through dykes.

The porous-flow stage of melt extraction may be particularly important with respect to the chemical characteristics of the magma, because the melt is in intimate contact with the matrix through which it is flowing and may therefore react chemically. Navon & Stolper (1987) have modelled such interactions in terms of ion-exchange processes similar to those in simple chromatographic columns, which have important analogies with zone refining (Ch. 4).

Further reading

Best, M.G. 1982. *Igneous and metamorphic petrology*. New York: W.H. Freeman.

Brown, G.C. & A.E. Mussett 1981. *The inaccessible Earth:* London: Allen and Unwin.

Cox, K.G., J.D. Bell & R.J. Pankhurst 1979. *The interpretation of igneous rocks*. London: Allen and Unwin.

Finnerty, A.A. & F.R. Boyd 1987. Thermobarometry for garnet peridotites; basis for the determination of thermal and compositional structure of the upper mantle. In *Mantle xenoliths*, P.H. Nixon (ed.) 381−402. Chichester: Wiley.

Jeanloz, R. & S. Morris 1986. Temperature distribution in the crust and mantle. *Ann Rev. Earth Planet. Sci.* 14, 377−415.

Maaløe, S. 1985. *Principles of igneous petrology*. Berlin: Springer-Verlag.

McKenzie, D. 1985. The extraction of magma from the crust and mantle. *Earth Planet. Sci. Lett.* 74, 81−91.

Nixon, P.H. 1987. *Mantle xenoliths*. Chichester: Wiley.

O'Nions, R.K. 1987. Relationships between chemical and convective layering in the Earth. *J.Geol. Soc. Lond.* 144, 259−74.

Ringwood, A.E. 1975. *Composition and petrology of the Earth's mantle*. New York: McGraw-Hill.

Yoder, H.S.Jr. 1976. *Generation of basaltic magma*. Washington D.C.: National Academy of Science.

CHAPTER FOUR

Processes which modify the composition of primary magmas

4.1   Introduction

In Chapter 3 we discussed the origins of the spectrum of primary basic and ultrabasic magmas generated by partial melting processes within the upper mantle. Once such magmas have segregated from their source region they may undergo a variety of complex fractionation, mixing and contamination processes en route to the surface, during transport and subsequent storage in high-level magma chambers. These processes are of fundamental importance in producing the diversity of igneous rocks presently exposed at the Earth's surface.

As primary magmas progress upwards from their depth of segregation, which may vary from greater than 100 km to less than 50 km in different tectonic settings (Chs 5−12), they will begin to cool and

ultimately to crystallize. When a magma solidifies it does so over a temperature range rather than at a specific temperature (Fig. 4.1). The temperature at which crystallization begins is called the *liquidus*, and that at which crystallization is complete the *solidus*. Both the liquidus and the solidus are pressure dependent and the two are subparallel in $P-T$ space. In general, the early stages of the ascent path are likely to be approximately adiabatic and therefore, given the $P-T$ slope of the liquidus (Fig. 4.1), basaltic magmas may not crystallize significantly until they reach crustal depths. However, more picritic primary magmas may crystallize significant amounts of olivine en route to the surface (O'Hara 1968).

Upon entering lower-density crustal rocks, rising basaltic magmas may reach a state of zero or negative buoyancy and pond, forming a magma

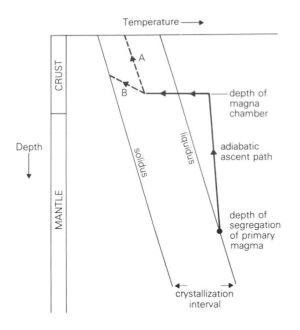

Figure 4.1 Schematic representation of the ascent path of a basaltic magma to the surface in relation to its crystallization interval.

chamber. For simplicity in Fig. 4.1, conditions within such a chamber are considered to be isobaric (i.e. constant pressure). However, within large systems of several kilometres vertical extent there must clearly be a small pressure gradient from the base to the top. Magma within the chamber cools by losing heat to the wall rocks, and as its temperature falls below the liquidus it begins to crystallize. Depending upon the degree of under-cooling below the liquidus the magma will contain varying amounts of suspended crystals. If the chamber is replenished subsequently by a new batch of primitive magma this may trigger an eruption of the porphyritic chamber magma (path A). Alternatively, the magma batch may solidify completely within the crust, forming a plutonic or hypabyssal igneous rock (path B).

Clearly, any magma which exists at a temperature between its liquidus and solidus must consist of a mixture of crystals and liquid. From petrographic and experimental studies it is apparent that the various minerals crystallizing from the melt do not usually form simultaneously. In general, a

single mineral crystallizes first and with cooling is joined by a second, then a third and so on. For basaltic systems, confining pressure can exert an important control on both the nature and sequences of minerals crystallizing. Figure 4.2 shows the crystallization interval of a basalt from the Snake River Plain (Ch. 10) between 0 and 35 kbar pressure (0-120 km depth), experimentally determined by Thompson (1972). At atmospheric pressure (0 kbar) with decreasing temperature the crystallization sequence is olivine < plagioclase < clinopyroxene < ilmenite. Such a sequence might be observed in the groundmass of a basalt erupted at the surface. With increasing pressure, plagioclase replaces olivine as the liquidus phase between 5 and 10 kbar and is in turn replaced by clinopyroxene between 10 and 32 kbar. Above 32 kbar garnet becomes the stable liquidus phase, i.e. the first crystalline phase (mineral) to precipitate from the crystal-free melt with falling temperature.

As the melt and the precipitating crystals are not usually of the same composition, the residual liquid will change composition during crystallization, following a liquid line of descent in some multi-component phase diagram. Crystallization may be equilibrium or fractional (Ch. 3; and see Cox *et al.* 1979, Maaløe 1985). During equilibrium crystallization, the crystals continuously re-equilibrate with the melt, and the bulk composition of the system remains constant, i.e. there is no opportunity for magmatic differentiation. Thus a magma of basaltic composition will crystallize to form a basalt, dolerite, gabbro or eclogite depending upon the depth. However, if a process operates within the magma body which effectively separates the crystal and liquid fractions, preventing equilibration, then magmatic differentiation can occur. This process is called *fractional crystallization*. Use of this term does not imply any particular mechanism for crystal—liquid separation, and a variety of alternatives have been proposed including gravitational crystal settling, flowage separation, filter pressing, incomplete reaction between crystals and melt where the two phases are not mechanically isolated and convective fractionation. These will be considered in more detail in subsequent sections.

The pseudoternary system $CaMgSi_2O_6$ (diop-

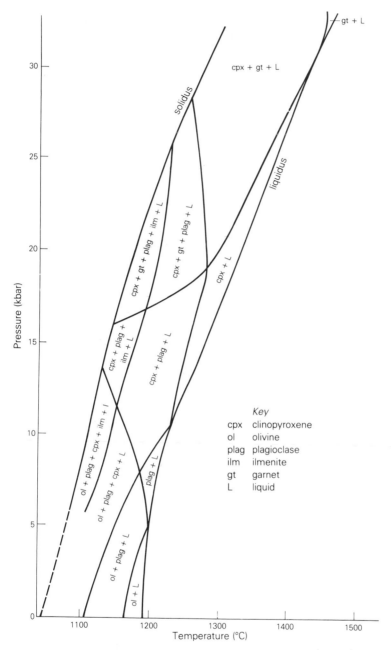

Figure 4.2 The crystallization interval of a Snake River Plain basalt between 0 and 35 kbar (after Thompson 1972, Fig. 1, p. 407).

side) − Mg_2SiO_4 (forsterite) − $CaAl_2Si_2O_8$ (anorthite) (Figure 4.3) can be used to compare processes of equilibrium versus fractional crystallization in basaltic magmas. A magma of composition A must crystallize to a mixture of diopside (clinopyroxene), forsterite (olivine) and anorthite (plagioclase), in the proportions fixed by the bulk composition A,

for equilibrium crystallization. However, for fractional crystallization the bulk composition of the system can change, depending upon the proportion of solid phases removed from equilibrium with the liquid. For both equilibrium and fractional crystallization the first mineral to crystallize from liquid A is olivine and the residual liquid is driven away

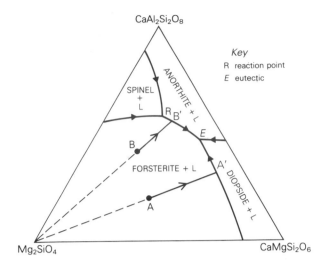

Figure 4.3 Liquidus projection for the system forsterite–diopside–anorthite at 1 atm pressure (Osborn & Tait 1952). Note that the spinel + liquid field is non-ternary. A and B are two hypothetical basalt magma compositions. Solid arrows denote directions of falling temperature along the cotectic curves.

from A along the line $A-A'$ until the forsterite–diopside cotectic curve is intersected. Olivine and pyroxene then crystallize together as the temperature falls until the eutectic E is reached, when plagioclase joins the assemblage. In the case of perfect fractional crystallization (Rayleigh fractionation), where magma and crystals are continuously separated, the residual liquid will ultimately attain the composition E. This is also the composition of the last dregs of liquid in the case of equilibrium crystallization. Thus by a fractional crystallization process we could clearly generate a continuous spectrum of liquid compositions lying along the lines $A-A'$ and $A'-E$, each of which could crystallize as an igneous rock. If, instead, the bulk composition of the original basaltic magma was B, the array of possible differentiated liquid compositions would now be $B-B'-E$, as the sequence of fractionating phases has changed. Path B may schematically represent the differentiation of a tholeiitic basalt and path A of an alkali basalt.

Clearly, such ternary phase diagrams do not enable us to predict quantitatively fractionation trends for actual magma compositions, as these are

multicomponent systems in which the phase relations of the minerals may differ significantly from those in the simple systems. However, as we have seen previously in Chapter 3, they are most useful in understanding the chemical consequences of fractionation processes. Nearly all the minerals of petrological importance in the evolution of basaltic magmas (olivine, pyroxene, plagioclase, amphibole and garnet) display wide ranges of solid solution, and hence the fractionation trends of the magmas cannot be constrained without taking into consideration the changing composition of the fractionating phases. This is a complex task, even in ternary systems with phases which display solid solution (Cox *et al.* 1979, Maaløe 1985).

The majority of basalts have major element compositions which appear to be controlled by low-pressure crystal-liquid equilibria (O'Hara 1968). These magmas must have evolved chemically in crustal magma chambers, in addition to any fractionation events which may have occurred at greater depths. Differentiation of basaltic magmas in such chambers, combined with variable amounts of crustal contamination (Section 4.4) in some provinces, appears to be the dominant mechanism for producing the spectrum of magma types more SiO_2-rich than basalt in both the alkalic and sub-alkalic magma series (Ch. 2).

One of the main objectives of petrogenetic studies of suites of cogenetic volcanic rocks (Chs 5–12) is to characterize the source region of the primary magmas generated in a particular tectonic setting. Clearly, in order to do this we must be able to see through the effects of both low- and high-pressure fractional crystallization processes and crustal contamination. Thus it is of fundamental importance to understand the chemical evolution of magmas stored for any length of time in high-level magma reservoirs, and to quantify the processes involved if possible.

Unfortunately, we cannot study the processes operating in magma chambers directly and therefore we must use indirect lines of evidence. A variety of geophysical techniques allow constraints to be placed on the location, shape and size of magma chambers beneath regions of Quaternary volcanism (Iyer 1984). Such studies are limited at present, but available data suggest that they may exist in many different shapes and sizes, located at depths ranging from a few tens of metres below the surface to below the crust—mantle interface (the Moho). Invaluable data are also provided by detailed studies of the deeply eroded plutonic root zones of former active volcanoes, including the group of basic and ultrabasic stratiform intrusions known as 'layered igneous rocks' (Section 4.3.2). These provide us with direct samples of the crystallization products of magmas stored in high-level reservoirs. In areas of strong vertical relief we can study both the three-dimensional geometry and the internal structure of such fossil magma chambers. For example, the immense Andean granite batholiths (Ch. 7) represent the plutonic root zones of formerly active andesitic stratovolcanoes. As such, their study has tremendous potential for understanding the differentiation of calc-alkaline magmas. In contrast, many large layered basic and ultrabasic intrusions appear to be associated with continental flood basalt provinces (Ch. 10), and have provided fundamental data concerning the evolution of continuously refluxed basaltic magma chambers. Additional constraints on magma chamber processes, particularly the nature of the fractionating phases during progressive crystallization, come from the inversion of geochemical data for suites of cogenetic volcanic rocks (Section 4.3.3).

Within the past decade, experimental modelling of fluid processes in saltwater systems has radically influenced our ideas concerning the mechanisms of magmatic differentiation (Sparks *et al.* 1984, Turner & Campbell 1986). There remains a general consensus of opinion that fractional crystallization is the dominant process involved in the diversification of the primary magma spectrum, generating the wide variety of igneous rock compositions observed. Until recently, the physical process involved was considered to be gravitational crystal settling (Section 4.3.1). However, Sparks *et al.* (op. cit.) consider that this is an inadequate and in some cases improbable mechanism, as convective motions in magma chambers (Section 4.2.1) are usually sufficiently vigorous to keep crystals in suspension. Instead, they propose that during crystallization, which predominantly takes place in

boundary layers at the margins of the magma chamber, the magma adjacent to the growing crystals develops a different density to the chamber magma and convects away. This efficiently separates crystals and liquid, just as gravitational crystal settling does. Sparks *et al.* (op.cit.) have proposed the term *convective fractionation* to describe this process.

It is highly unlikely that magma chambers beneath major volcanic structures can remain as closed systems. Instead, fresh batches of relatively primitive magma must be periodically injected into the base of the chamber from the underlying mantle source and this may trigger a contemporaneous volcanic eruption. The extent of mixing between such fluxes and the resident chamber magma depends on both the input rate and on the relative densities and viscosities of the two magmas (Turner & Campbell 1986; and see Section 4.2.2). Magma chambers in general must be regarded as open systems which are periodically replenished, periodically tapped and continuously fractionated (O'Hara & Mathews 1981). O'Hara (1977) considers that, in such a system, the compositions of the erupted magmas should approach a steady state if the parental magma composition and amount supplied in each cycle and the ratio of magma crystallization to magma escape were held constant. In general, there may be large differences in the concentrations of both major and trace elements between the steady-state magma and the parental magma (O'Hara & Mathews op. cit.).

Turner & Campbell (1986) have presented a comprehensive review of the wide range of fluid processes which are believed to be significant in magma chambers. Chamber geometry and variations in the density and viscosity of the magmas within it have been shown to play a major role in determining the dynamical behaviour of the system and the composition of both erupted and solidified products. Despite the existence of turbulent convection (Section 4.2.1) these authors consider that the magma in most high-level chambers is likely to be vertically stratified in terms of its density, composition and temperature. Gradients in these properties can be set up directly during filling or replenishment, or by various fractionation mechanisms driven by the interaction of the cooling and crystallizing magma with the boundaries of the chamber. Such stratified magma bodies may break down into a series of well-mixed horizontal layers, separated by thin diffusive interfaces. These are known as double-diffusive systems (Section 4.2.1).

At any stage during their ascent from source to surface, magmas may become variably contaminated by assimilation of their wall rocks (Section 4.4). Such assimilation requires heat, which may be provided by the latent heat of crystallization given off during simultaneous crystallization of the magma. The combination of assimilation and fractional crystallization provides a powerful mechanism for magmatic differentiation, known as AFC (DePaolo 1981; and see Section 4.4.1). Within the mantle such contamination may be considered as a variant of zone refining (Section 4.5) and may be difficult to prove on the basis of major element data alone. However, extreme enrichments in incompatible elements in primitive basaltic magmas may suggest that such interaction has occurred. Within the continental crust the wall rocks are likely to have significantly different major and trace element and Sr−Nd−Pb isotopic characteristics from those of the ascending basaltic magmas, and it is in this environment that crustal contamination effects should be most easily recognized. In contrast, in the oceanic environment, crustal magma chambers will in general have wall rocks of ocean-floor basalts (MORB and OIB) and contamination by such materials will be much harder to detect.

In general, it appears that fractional crystallization in combination with varying degrees of wall-rock assimilation, in both the crust and the mantle, is the dominant process involved in the differentiation of primary magmas. Locally liquid immiscibility (Section 4.6) and gaseous transfer (Section 4.7) processes may be important, but these cannot be considered as major agents of magmatic differentiation.

4.2   Convection and mixing in magma chambers

4.2.1   Convection

Prior to the late 1970s, almost all applications of convection theory to the study of fluid processes in magma chambers assumed that magmas could be described as simple one-component systems (e.g. Shaw 1965). However, in reality, they are multi-component systems and, as such, can exhibit a wide variety of convective phenomena not encountered in one-component fluids (McBirney & Noyes 1979, Sparks et al. 1984, Turner & Campbell 1986, Martin et al. 1987).

The form of convection in a filled magma chamber is controlled by the size and shape of the chamber, the density and viscosity of the magma and by processes which operate at the boundaries of the chamber, i.e. the walls, roof and floor (Turner & Campbell op. cit.). Unfortunately, detailed theoretical modelling of convective flow patterns and velocities, even in simply shaped bodies, is difficult because of the complex behaviour of multicomponent silicate melts.

Both thermally and compositionally induced density differences influence convective motions in magmas, and these can be expressed in terms of two dimensionless parameters, the thermal Rayleigh number (R_a) and the compositional Rayleigh number (R_s) (Martin et al. op. cit.). The effective Rayleigh number of the system, which can be used to predict its convective behaviour, is the sum of R_a and R_s. Turner & Campbell (op. cit.) consider that purely thermal convection in large basaltic magma chambers with vertical dimensions greater than 1 km will be highly turbulent ($R_a > 10^6$) in the early stages after the chamber is filled or refilled. However, such turbulence may diminish as crystallization proceeds, due to the combined effects of increasing viscosity, decreasing temperature and decreasing chamber size.

When crystallization occurs in a magma the fluid layer immediately adjacent to the growing crystals may become either enriched or depleted in heavy components, depending upon the minerals crystallizing. The density of this layer will therefore be different from that of the main chamber magma and, as a consequence, it will convect away from the crystal faces. Martin *et al.* (op. cit.) have shown that when crystallization is occurring R_s may be up to a factor of 10^6 greater than R_a, and thus that compositional convection, due to the fluid released during crystallization, often dominates over thermal convection. Sparks *et al.* (1984) call this process *convective fractionation*. They consider that it is capable of generating highly differentiated liquids at an early stage in the evolution of a magma chamber, without requiring large amounts of crystallization of the whole system. Clearly, in systems undergoing compositionally dominated convection, fluid motions may be even more turbulent than in the purely thermal case.

An important property of multicomponent fluids is that individual components (including heat) can have different diffusivities. As a consequence, such fluids may become vertically stratified with respect to density, composition and temperature. If opposing gradients of two components with different diffusivities are set up, the system may separate into a series of independently convecting layers, bounded by sharp diffusive interfaces, across which heat and chemical components are transported by molecular diffusion. This phenomenon is known as *double-diffusive convection*, and Sparks *et al.* (1984) consider that it will inevitably occur in silicate magmas.

It is the influence of gradients in components on the density of the system which is the controlling factor in the development of double-diffusive systems. This is shown schematically in Figure 4.4. The most important situation for magmatic systems is one in which the compositional gradients are stabilizing (i.e. causing density to decrease from the base to the top of the chamber), whereas the temperature gradients are destabilizing (causing density to increase from base to top) (Turner & Campbell 1986). The compositional gradients have a much larger effect on the density of the chamber magma than the temperature gradients and, as a consequence, the overall density distribution decreases from base to top.

A double-diffusive interface may form when a layer of hot dense magma is overlain by a layer of

cooler, less dense magma. Such a situation may occur when a new pulse of primitive magma is injected into the base of the chamber through vents in the floor. Heat is transported between the layers faster than chemical components, driving convection in both layers and maintaining a sharp interface between them (Turner & Campbell op. cit.). Eventually, crystallization of the lower layer will occur, causing a reduction in its density, and it may then mix with the overlying layer.

There is substantial evidence to suggest that many silicic magma chambers are both composi-

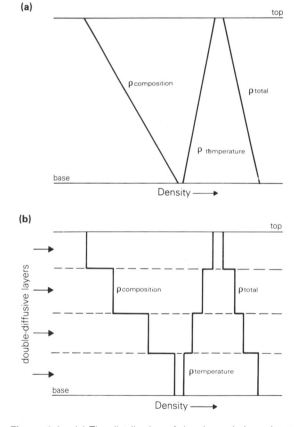

Figure 4.4 (a) The distribution of density variations due to compositional and temperature effects in a multi-component fluid layer before it breaks up into double-diffusive layers. (b) The breakdown of the system in (a) into double-diffusive layers. $\rho_{composition}$ is the variation of density in the system due to compositional gradients; $\rho_{temperature}$ is the variation of density in the system due to temperature gradients; ρ_{total} is the actual density of the system. (After Turner & Campbell 1986, Fig. 3. p. 265).

tionally and thermally zoned (Hildreth 1979, Smith 1979). Much of this is derived from detailed geochemical studies of voluminous ash flow eruptions, in which the earliest part of the eruptive sequence are derived from the cooler, more silicic, more volatile rich upper parts of the chamber, and successive products come from hotter, more mafic deeper levels. The occurrence of stable compositional gradients and unstable temperature gradients suggests that double-diffusive effects should be important in such systems. Magmas in such chambers can display large variations in trace element contents from top to bottom (Hildreth op. cit.) with comparatively small changes in major element chemistry. These are difficult to reconcile in terms of conventional differentiation mechanisms such as crystal fractionation and assimilation, and Sparks *et al.* (1984) consider that convective fractionation mechanisms may be important in their development.

4.2.2 Mixing

In recent years, magma mixing has played an increasingly important role in petrogenetic models for co-genetic suites of igneous rocks from a wide variety of tectonic settings. For example, many of the geochemical and petrographic characteristics of mid-ocean ridge basalts (Ch. 5) can be explained if the ridge axis magma chambers are periodically refluxed with new pulses of primitive magma, which mix with batches of fractionated magma in the chamber (Dungan & Rhodes 1978, Walker *et al.* 1979). Similarly detailed studies of cyclic layers (Section 4.3.2) in layered basic intrusions (Campbell 1977, Palacz 1985) suggest that when new pulses of magma enter the chamber they eventually mix with the residual chamber magma.

In general, it appears that basic magmas can mix fairly readily, and therefore we might predict that the eruptive products of frequently refluxed basaltic magma chambers may be predominantly well homogenized mixtures. The extent of mixing between the inflowing magma and the chamber magma depends on the flow rate and on the relative densities and viscosities of the two magmas (Turner & Campbell 1986). Slow dense inputs may mix

very little with a resident magma of comparable viscosity. However, a similar magma injected with a high upward momentum will form a turbulent fountain which is a very efficient mechanism for magma mixing (Campbell & Turner 1986). Laboratory experiments suggest that magma mixing is inhibited by large viscosity differences between the two magmas. There is ample field evidence for the incomplete mixing of basic and acid magmas to support this. For example, mixed intrusions are common in the British Tertiary Volcanic Province (Ch. 10), either as composite dykes or as pillows of basic rock enclosed in a matrix of felsite or granophyre.

When the magma in a chamber is compositionally stratified, crystallization can cause the density and composition of a layer to change, which may lead to mixing with an overlying layer. If hot primitive magma is injected into the base of such a zoned magma body, the thermal disturbance will be transmitted into the upper zone but without any immediate mixing of the old and new magmas. Eventually, however, the density of the lower layer

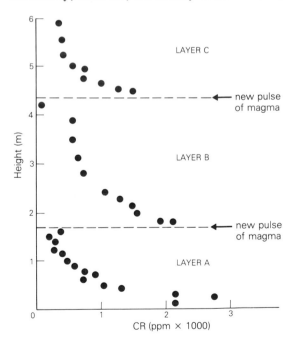

Figure 4.5 Variation in Cr content of magnetite with structural height in layered rocks of the Bushveld complex, South Africa (data from Cawthorn & McCarthy 1981).

will be reduced by crystallization, and convective overturn and mixing of the two layers will occur. In such a situation Turner & Campbell (op. cit.) consider that mixing would be limited to the lower parts of the chamber.

Overturning of a volatile rich mafic layer into a cooler silicic layer can cause sudden quenching, with the exsolution of a gas phase, which could trigger an explosive volcanic eruption. This may be particularly important in the evolution of calc-alkaline magmas (Huppert *et al.* 1982, Turner *et al.* 1983).

Some of the best evidence for the existence of double-diffusive layers and their overturning in natural systems comes from detailed studies of the distribution of compatible trace elements in rhythmically layered sequences in layered intrusions (Section 4.3.2; and Cawthorn & McCarthy 1981, Hiemstra 1985). In general, compatible trace element profiles in minerals such as magnetite and chromite display a sawtooth pattern as a function of stratigraphic height (Fig. 4.5) which may be related to the crystallization and mixing of double-diffusive layers. Palacz (1985) has observed similar trends in the $^{87}Sr/^{86}Sr$ ratio of cumulates from the Tertiary Rhum intrusion, NW Scotland (Fig. 4.6). These may be attributed to mixing of pulses of primitive mantle derived magma with crustally contaminated residual magma batches in the chamber (Section 4.4).

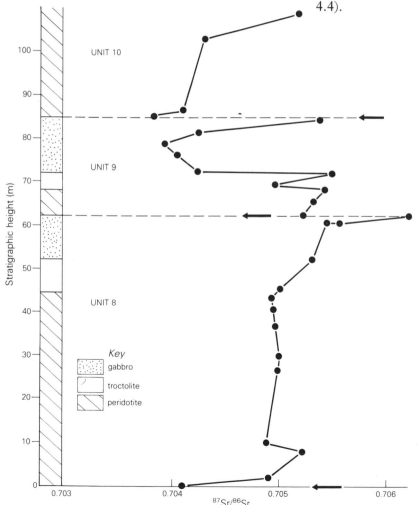

Figure 4.6 Variation of $^{87}Sr/^{86}Sr$ with stratigraphic height in macrorhythmic units 8, 9, and 10 of the Rhum intrusion, north-west Scotland. Arrows indicate the influx of batches of more primitive magma into the chamber (after Palacz 1985, Fig. 3, p. 38).

4.3   Fractional crystallization

In Section 4.1 we demonstrated that the separation of crystals from liquid in a magma body provides a powerful mechanism for magmatic differentiation. This process is generally referred to as *fractional crystallization*. The simplest way to achieve separation is by gravitational crystal settling. However, as proposed by Sparks *et al.* (1984), an important alternative involves convective separation of the residual liquid from the crystals (Section 4.2.1). In simple terms, we might conclude that the geochemical consequences of these two processes should be the same, as they are simply different physical methods for separating crystals from liquid. However, Sparks *et al.* (op. cit.) consider that convective fractionation and gravitational crystal settling may have fundamentally different geochemical consequences. Until more experimental data are available to constrain convective fractionation processes, this must remain a matter for speculation. If correct, this clearly has important implications with respect to theoretical modelling of the geochemical characteristics of suites of co-genetic volcanic rocks in terms of fractionation of the observed phenocryst phases (Section 4.3.2).

4.3.1   Gravitational crystal settling

Gravitative separation of crystals from liquid normally involves sinking of the crystals by virtue of their greater density. Occasionally, however, flotation may be the process if the density of the crystals is less than that of the liquid. Settling rates are a function of crystal size, the viscosity of the liquid and the density contrast between the liquid and the crystals. If we assume Newtonian fluid dynamics, then for spherical particles, the settling velocity, v, is given by the Stokes' Law equation:

$$v = \frac{2gr^2 \Delta \rho}{q\eta}$$

where g is acceleration due to gravity, r is the radius of the sphere, $\Delta \rho$ is the density contrast between crystals and liquid, and η is the viscosity of the liquid. Calculated settling velocities using this equation could permit effective crystal fractionation on the timescale on which large bodies of magma must cool ($10^4 - 10^6$ years). For example, the typical phenocrysts in basaltic magmas (plagioclase, olivine, augite and Fe−Ti oxides) of 1 mm radius should settle at rates from about 20 (plagioclase) to 500 (magnetite) metres per year (Cox *et al.* 1979). However, it is important to recognize that Stokes' Law only applies to Newtonian fluids, and that only basaltic magmas close to their liquidus are likely to behave as such. Partially crystallized basaltic magmas, and those with more SiO_2-rich compositions, have significant yield strengths which may be sufficient to inhibit crystal settling (McBirney & Noyes 1979).

Some of the best evidence for the occurrence of crystal settling comes from field descriptions of high-temperature mafic and ultramafic lava flows and thin sills, in which denser olivine crystals are clearly concentrated towards the base. Additionally, there is strong geochemical evidence that some highly porphyritic basic lavas have accumulated crystals of olivine, clinopyroxene and plagioclase (e.g. Wright 1971). However, in both these cases it is also possible that crystals could have been concentrated during flow.

Up to the late 1970s, the most widely cited evidence for crystal settling came from studies of layered igneous intrusions (Wager & Brown 1968, and see Section 4.3.2). These were thought to represent the piles of crystals (cumulates) settling on to the floors of large magma chambers under the influence of gravity. However, more recent studies of the same intrusions have been used to argue against the importance of crystal settling (Campbell 1978, McBirney & Noyes 1979, Irvine 1980a). The central theme of such arguments is the low probability of static conditions existing in most magma chambers. In general, convection will be highly turbulent (Sparks *et al.* 1984, Turner & Campbell 1986) and this will tend to maintain crystals in suspension. Sparks *et al.* (op. cit.) have shown that, in general, convective velocities are likely to be orders of magnitude greater than the settling velocities of individual crystals, calculated using the Stokes' Law equation.

4.3.2 Layered igneous rocks

Studies of layered igneous intrusions can clearly provide important constraints on the crystallization and chemical evolution of crustal magma chambers. Unlike the eruptive products of volcanoes, these accumulative rocks can preserve a continuous record of the differentiation of the chamber magma, involving influxes of new batches of primitive magma, mixing processes and wall-rock contamination effects.

Many of these layered intrusions are immense in size. For example, the Precambrian Bushveld intrusion in South Africa is exposed over an area of 65 000 km^2, with a maximum thickness of 7 km. In contrast, the Tertiary Skaergaard intrusion in Greenland is relatively small, with an area of 170 km^2 and an estimated volume of 500 km^3. Most of these intrustions appear to be funnel shaped, although the diameter of the funnel relative to its height varies greatly. Layering is, in general, discordant with the walls of the funnel. The reader is referred to Wager & Brown (1968), McBirney & Noyes (1979) and Irvine (1980a,b, 1982) for detailed descriptions of the field relations, mineralogy and internal structures of the classic layered intrusions.

There are many different varieties and scales of layering in mafic and ultramafic intrusions and, consequently, a variety of processes are likely to be involved in its production. It appears to mimic stratification in sedimentary rocks, with individual layers ranging from millimetres to hundreds of metres in thickness. In some of the largest intrusions, distinctive layers may be traceable laterally for several tens of kilometres. Layering is usually defined by variations in the relative proportions of the constituent minerals. Additionally, there may be gradational variations within a single layer, analogous to grading in sedimentary rocks. Crystals with a tabular habit, such as plagioclase feldspar, may sometimes show a marked preferred orientation within a layer, which is strongly suggestive of crystal settling. In some intrusions current bedding, channelling, grain-size grading and slumping provide unequivocal evidence for sedimentation from magmatic density currents (McBirney &

Noyes 1979, Irvine 1980a,b, 1982).

Within the lower parts of many of the larger intrusions there is commonly a regular repetition of sequences of cumulate layers (e.g. an olivine-rich layer, overlain by an olivine + clinopyroxene layer, overlain by an olivine + clinopyroxene + plagioclase layer; Fig. 4.6), which is termed macrorhythmic, or cyclic, layering. This occurs on a scale of metres to tens of metres, and has been attributed to the refluxing of the chamber with new batches of primitive magma. Detailed studies of these cyclic layers (e.g. Campbell 1977, Irvine 1980a, Palacz 1985) suggests that each unit records the influx of a new magma pulse and its subsequent mixing with the more differentiated chamber magma. During replenishment the hotter, denser, more primitive magma forms a layer at the base of the chamber, where it cools and crystallizes by exchanging heat across a double-diffusive interface with the more fractionated residual chamber magma. Once its density has been reduced by crystallization to that of the overlying layer, mixing can occur (Huppert & Sparks 1980, Sparks & Huppert 1984).

In contrast, McBirney & Noyes (1979) consider that much of the small scale (mm−cm) microrhythmic layering forms by periodic oscillations in the nucleation and growth of different minerals in boundary layers along the margins of the chamber. Carmichael et al. (1974) have suggested that the order of nucleation of minerals in basaltic magmas should be Fe−Ti oxides < olivine < clinopyroxene < plagioclase, which could account for the sequence of minerals commonly occurring in microrhythmic layers in stratiform intrusions. This sequence is the same as that which would be produced by gravity sorting of equal-sized crystals. If, during such in-situ crystallization, the fluid in contact with the growing crystals could convect away, then compositional gradients may develop in the magma chamber, as considered in Section 4.2.1.

Over the years there has been considerable debate as to whether the accumulation of crystals at the bottom of layered intrusions is due to crystal settling or in-situ growth. Irvine (1980b) and McBirney & Noyes (1979) have pointed out significant deficiencies in simplistic gravitational crystal settling models. For example, according to the original explanation of layering in the Skaergaard intrusion (Wager & Brown 1968), cyclic convection currents transported crystals growing near the roof zone down the walls and across the floor of the intrusion. Here, because of their greater density, they settled out of the melt and were sorted hydraulically according to their size and density. However, careful analysis of mineral densities and grain sizes in some layers of stratiform intrusions indicates that, in many cases, grains are not sorted according to hydraulic processes. This suggests that the settling of grains in magmatic systems may not be completely analogous to processes of clastic sedimentation.

In general, the majority of evidence now appears to favour *in-situ* growth (e.g. Campbell 1978, McBirney & Noyes 1979, Irvine 1980a, b, Sparks *et al.* 1984). Indeed, crystal settling is no longer believed by some authors to be the dominant process during crystallization in magma chambers (Sparks *et al.* op. cit.). Instead, crystals are thought to nucleate and grow *in situ* on the floor and walls of the chamber (Turner & Campbell 1986, McBirney *et al.* 1985, Nilson *et al.* 1985). If this is the case then fractional crystallization of the chamber magma is only possible if the depleted liquid is removed from contact with the growing crystals by a combination of diffusive and convective processes (Sparks *et al.* op. cit.). Such convective fractionation has been described in Section 4.2.1. Clearly, this does not rule out the possibility of crystal settling in the evolution of basaltic magmas, but this does not necessarily have to be the dominant process.

These new ideas on processes of crystallization in magma chambers obviously present problems for the petrogenetic modelling of suites of volcanic rocks in terms of simple fractional crystallization processes (Section 4.3.3). The compositions of lavas erupted from the top of a magma chamber may be related to those of the cumulates forming at the base, but only by complex and as yet poorly constrained processes.

4.3.3 Petrogenetic modelling of volcanic suites in terms of fractional crystallization processes

Suites of volcanic rocks which are closely associated in space and time often display coherent chemical variations which can be interpreted in terms of fractional crystallization processes. These are most simply illustrated in terms of Harker variation diagrams (Ch. 2), in which the wt. % of a constituent oxide, or ppm of a trace element, are plotted against an index of differentiation, usually wt. % SiO_2. The existence of marked inter-element correlations strongly supports the operation of fractional crystallization processes. In general, SiO_2 is the most useful index of differentiation in suites of rocks ranging from basic to acid in composition. However, if we are dealing with a suite of predominantly basic lavas then MgO is more useful. Highly incompatible trace elements such as Zr can also be used for this purpose.

Implicit in the study of such variation diagrams is that they illustrate the course of evolution of true liquids − a liquid line of descent. In general, only analyses of aphyric volcanic rocks should be regarded as representative of liquid compositions. However, as aphyric rocks are rather rare we must use data for the least porphyritic samples available. This presents particular problems for the petrogenetic modelling of calc-alkaline volcanic suites which are invariably highly porphyritic.

Generally, within the eruptive sequence of a volcano, there is no time progression with respect to the appearance of basic and acid magmas, and the two are frequently erupted contemporaneously. Thus the apparent liquid line of descent displayed on the Harker diagram does not represent the evolution of a single batch of magma, but of a series of broadly similar batches, evolving by similar processes in a high-level magma chamber. Rarely, as in the eruption of some zoned acidic ash flow tuffs, we do find evidence which may support the fractionation of a single batch of magma.

Linear trends on Harker variation diagrams may be continuous throughout the suite or may display marked inflections, corresponding to major changes in the fractionating mineral assemblage (Fig. 4.7; and see Ch. 2). With respect to major

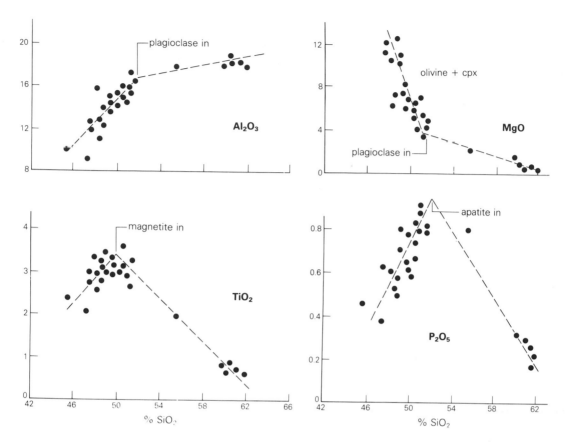

Figure 4.7 Harker variation diagrams for a suite of co-genetic volcanic rocks related by fractional crystallization of olivine, clinopyroxene, plagioclase, magnetite and apatite.

element diagrams, intermediate volcanic suites typically display inflection-free trends, despite wide variations in the crystallizing assemblage. Conversely, basaltic magmas, in which the number of phases crystallizing is normally low, frequently display strongly inflected trends. In modelling such trends we tend to use the phenocrysts present in the lavas themselves to constrain the composition and modal mineralogy of the fractionating assemblage. However, in terms of present ideas about crystallization processes in magma chambers (Sections 4.3.1 & 2) this may be substantially incorrect. Phenocrysts are conventionally regarded as crystals which have nucleated in a homogeneous magma. However, in terms of newly developed models for fluid processes in magma chambers (Sparks *et al.* 1984, Turner & Campbell 1986), they could also result from the redistribution of crystals nucleated and grown at the margins of the chamber, from crystallization consequent upon magma mixing or by nucleation at double-diffusive interfaces. The common observation that phenocrysts in volcanic rocks are variably zoned further suggests that they may represent crystals formed by a variety of processes. These may be quite different from those involved in the formation of contemporaneous cumulates on the floor of the chamber, which may actually constrain the fractionation process. Consequently, we should adopt a somewhat cautious approach to the theoretical modelling of fractional crystallization processes. In this respect it is important to recognize that fractional crystallization is not the only process capable of generating coherent trends on Harker variation diagrams. Partial melting, crustal contamination and magma mixing can have similar effects.

It is possible to constrain the dominant mineral assemblage which may be responsible for the development of a particular trend by using combinations of Harker variation diagrams (Cox *et al.* 1979). However, it is important to recognize at the outset that a given trend is in general made up of a series of samples which may vary quite widely in their SiO_2 content. It is clearly unrealistic, for example, to imagine that the separation of a constant-composition mineral assemblage could be responsible for the differentiation of a basalt to an andesite. From petrographic studies of volcanic rocks we know that the phenocryst mineralogy and mineral compositions change with progressive differentiation of the magma. Thus if we wish to model a particular trend in terms of fractional crystallization processes we must divide it into realistically small intervals, such that the differentiated liquid could be derived from the parent liquid in a single step.

The general approach is illustrated in Figure 4.8, in which we plot the two lava compositions (PM and DM) we wish to relate on a Harker diagram, along with the compositions of possible fractionating mineral phases. The latter are normally obtained from analyses of phenocrysts in the more primitive lava (PM), i.e. the parent magma. Clearly, there is an in-built assumption that minerals which do not occur in the phenocryst assemblage are not important. In some instances, as considered previously, this may be fundamentally incorrect. A useful example in this respect is the crystallization of amphibole from some calc-alkaline basaltic magmas. Amphibole is a stable phase at pressures appropriate to crystallization in subvolcanic magma chambers, and as a consequence amphibole-bearing cumulate xenoliths are common in many subduction-related basalts (Ch. 6). The latter are generally considered to be derived by fragmentation of cumulates in the magma chamber. However, at low pressures amphibole becomes unstable and is therefore rarely present in the phenocryst assemblage, other than as strongly resorbed crystals which may be xenocrysts. Thus, by using the phenocryst mineralogy to constrain the fractiona-

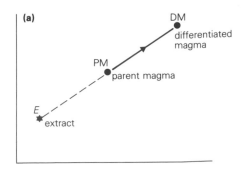

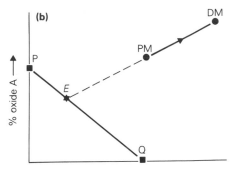

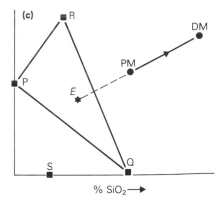

Figure 4.8 Schematic representation of the fractional crystallization of mineral assemblages involving one, two and three phases from a parental magma PM to produce a differentiated magma DM (after Cox *et al.* 1979, Fig. 6.2, p. 149).

tion assemblage, in this case we would overlook the importance of amphibole.

Figure 4.8 shows the consequences of fractionally crystallizing different numbers of minerals from the parent magma. In Figure 4.8a a monomineralic extract (E) is removed from PM, constraining the residual liquid to follow the trend PM−DM. The proportion of the extract which we need to remove to generate the differentiated magma DM can be calculated simply using the Lever Rule:

$$\% E \text{ removed} = \frac{\text{distance PM–DM}}{\text{distance E–DM}} \times 100$$

In Figure 4.8b, two minerals, P and Q, are removed in proportions such that the bulk composition of the extract plots at E. The liquid path again evolves away from E along the line PM−DM. In Figure 4.8c, three minerals, P, Q and R, are removed. The bulk composition of the extract in this case must lie within the triangle defined by the mineral compositions, and again the liquid evolves along PM−DM. Clearly, the trend PM−DM in Figure 4.8c could also be generated by fractionation of the bimineralic assemblages $P + Q$ and $R + Q$, in addition to the trimineralic assemblage $P + Q + R$. If we were to add a fourth mineral, S, to the crystallizing assemblage then the bulk composition of the extract must lie within the polygon $PQRS$.

In the general case of a system in which n minerals are fractionally crystallizing, the bulk composition of the extract must lie within the polygon defined by the mineral compositions and along the back projection of the line PM−DM (dashed line in Figure 4.8a−c). In such a case for a single variation diagram there is clearly no unique solution to the extract problem. However, if we have n independent variation diagrams we should be able to constrain the mineral proportions in the extract uniquely. The reader is referred to Cox et al. (1979) for a more detailed discussion of this approach. Geological systems are always over-constrained, in that we have a greater number of analysed oxides and trace elements than the number of minerals crystallizing. The mathematical consequences of this are considered by Maaløe (1985).

In the case of multiphase extracts where $n > 3$, graphical solutions to the problem become increasingly complex and we must instead use computer-based methods (e.g. Bryan et al. 1969,

Wright & Doherty 1970, Le Maitre 1980). The basis of such calculations involves computation, by least-squares methods, of the mixture of minerals (the analyses of which form part of the input data) which will produce the best fit to a regression line through the analysed lava compositions. Clearly, by increasing the number of minerals in the input data it will always be possible to find a solution. However, this does not necessarily mean that the calculated extract is a true reflection of the fractional crystallization processes actually operating in the magma chamber. Given the apparent complexity of crystallization phenomena in magma chambers (Sections 4.3.1 & 2) we should treat such calculations as at best semi-quantitative.

There have been several attempts in recent years to quantify certain aspects of mineral-melt equilibria involving natural silicate liquids in order to model fractional crystallization processes (e.g. Nathan & Van Kirk 1978, Hostetler & Drake 1980, Langmuir & Hanson 1981, Nielsen & Dungan 1983). All of these methods involve empirical or semi-empirical mathematical formulations of experimental phase equilibrium data. Ghiorso (1985) has developed an algorithm for describing chemical mass transfer in magmatic systems which may be used to model fractional crystallization and assimilation processes (Ghiorso & Carmichael 1985). This is much more complex than the earlier methods in that it incorporates thermodynamic models for the solid and liquid phases. However, its use is limited by the lack of thermodynamic data for many of the minerals of interest in magmatic systems.

Petrogenetic models which attempt to explain the geochemical characteristics of suites of volcanic rocks in terms of fractional crystallization tend to assume that the process is isobaric (i.e. constant pressure). However, O'Hara (1968) considers that many basalts may have evolved from more picritic primary magmas by polybaric olivine fractionation en route to the surface. This can be easily explained in terms of Figure 3.23, the enstatite projection into the plane $M_2S−A_2S_3−C_2S_3$ in the CMAS system (Ch. 3). A primary picritic magma generated at a depth of 100 km should have a composition which projects to the mantle side of the 30 kbar invariant pont, as garnet and clinopyroxene should be eliminated from the residue during the relatively high degrees of partial melting necessary to generate a melt of this composition. The marked expansion of the olivine phase field with falling pressure means that as the picritic magma rises towards the surface it will only be able to crystallize olivine (+ orthopyroxene) until it ponds in a high-level magma chamber. In general, the last major site of fractional crystallization will tend to blur if not erase completely the history of previous fractionation events at greater depths. However, in some volcanic suites, evidence for a precursor phase of high-pressure fractionation may remain imprinted in their geochemical characteristics, particularly in the trace element geochemistry.

A fundamental objective in geochemical studies of suites of volcanic rocks is to determine whether all the members of a particular suite are co-genetic, i.e. whether they are related by fractional crystallization to a common parent (or possibly by partial melting to a common source). The approach is normally to consider the major element variation

diagrams first, to constrain the possible fractionating mineral phases, and then to consider the trace element data to see if they are consistent. To prove a genetic relationship conclusively we also require Sr−Nd−Pb isotopic data for the lavas. Isotopic ratios of these elements should be identical in all members of the suite, provided that crustal contamination has not occurred, as they are not modified by crystal−liquid fractionation processes (Ch. 2).

If individual members of a suite of volcanic rocks are related by fractional crystallization processes then trace element contents and ratios should vary consistently throughout the series. For example, Figure 4.9 shows the variation of chondrite-normalized REE patterns and Cr, Ni, Ba and Sr contents in a suite of alkaline lavas ranging from basalt (A) through hawaiite and mugearite to trachyte (G and H), after Cox et al. (1979). The marked fall in the abundance of compatible trace elements such as Ni and Cr from A to C strongly suggests that fractional crystallization of olivine and clinopyroxene has been important in the early

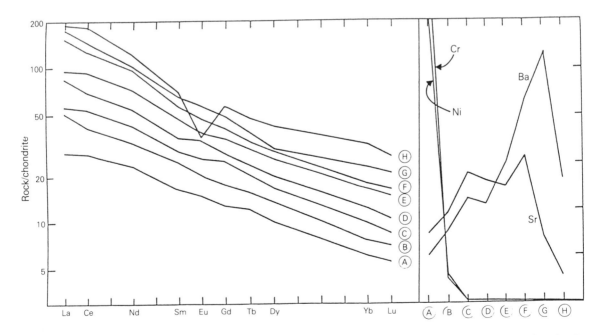

Figure 4.9 Chondrite-normalized rare earth patterns and abundances of Cr, Ni, Ba and Sr in a suite of co-genetic volcanic rocks ranging in composition from alkali basalt (A) through hawaiite and mugearite to trachyte (G & H) (after Cox *et al.* 1979, Fig. 14.2, p. 349).

evolution of the suite. The REE patterns are subparallel, with normalized abundances increasing with progressive differentiation (i.e. from basalt to trachyte). This is to be expected, as the REE are incompatible with respect to the major phases crystallizing (olivine, clinopyroxene, plagioclase, alkali feldspar and magnetite) and therefore they become increasingly concentrated in the more evolved liquids. The negative Eu anomaly and the fall in Ba and Sr contents in the trachytes suggests the fractionation of feldspar at this stage.

In many suites of volcanic rocks, despite an apparent coherence in major element trends, it is difficult to explain the trace element data in terms of simple models of crystal−liquid fractionation, either during crystallization or partial melting. In such cases we have to consider the possibility that the magmas are derived from a heterogeneous mantle source and that a range of similar primary magmas may exist, each of which can undergo subsequent fractional crystallization in a high-level magma chamber. Nd−Sr−Pb isotopic data should provide important constraints for such a model. Mantle heterogeneity has been shown to be important in the petrogenesis of many terrestrial basalt magmas, including MORB (Ch. 5), OIB (Ch. 9) and continental basalts generated in both intra-plate (Chs 10−12) and subduction-related (Ch. 7) tectonic settings.

4.3.4   Trace element modelling

The distribution of trace elements in a crystallizing magma can be predicted using the following equations.

If all the crystalline products remain in chemical equilibrium with the magma (equilibrium crystallization), the concentration of a trace element in the liquid (C_L) relative to that in the original liquid (C_L^0) before crystallization commenced is given by:

$$\frac{C_L}{C_L^0} = \frac{1}{F + D - FD}$$

where F is the fraction of liquid remaining and D is the bulk distribution coefficient $(D = \sum_\alpha X_\alpha D_\alpha$; see Ch. 3).

For the case of perfect fractional crystallization (Rayleigh fractionation):

$$\frac{C_L}{C_L^0} = F^{(D-1)}$$

Figure 4.10 shows that the enrichment in incompatible elements $(D< 1)$ for perfect fractional crystallization is comparable to that for equilibrium crystallization until greater than 75% of the magma has crystallized. However, for highly compatible trace elements $(D \gg 1)$ the two models differ significantly.

Given a set of trace element data for a suite of lavas and mineral-melt partition coefficient data $(D_\alpha$'s), it should be possible to invert the Rayleigh equation to calculate the proportions of phases crystallizing (X_α) for a given amount of fractional crystallization (e.g. Allègre et al. 1977, Allègre & Minister 1978). At present, such calculations are limited by the lack of high-quality mineral-melt partition coefficient data for many of the elements of interest.

4.4   Crustal contamination

In Section 4.1 we noted that ascending mantle derived magmas may become contaminated by assimilation of their wall rocks at any stage on their ascent path from source to surface. However, the importance of the process in relation to others, such as fractional crystallization, in producing the diversity of magma compositions remains controversial (McBirney 1979). Interaction between magmas and their wall rocks within the upper mantle is not conventionally referred to as contamination, and will not be considered further in this section. The reader is referred to Section 4.5 for additional information.

The effects of such contamination on the geochemical characteristics of magmas should be most marked when the major and trace element and Sr−Nd−Pb isotopic composition of the wall rocks contrasts strongly with that of the magma. As a consequence, in this section we shall focus our

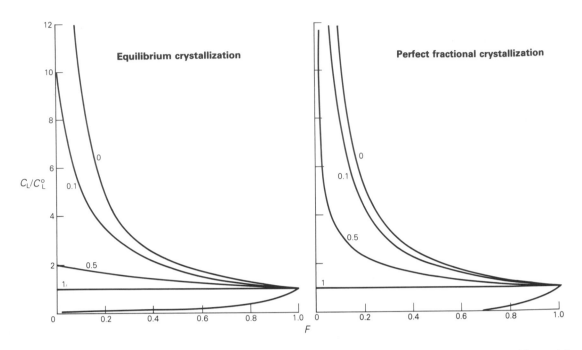

Figure 4.10 Comparison of trace element enrichment factors (C_L/C_L^0), as a function of the fraction of liquid (F) remaining for equilibrium and fractional crystallization. Curves are shown for D values of 0, 0.1, 0.5, 1 and 10.

attention on the contamination of mantle-derived magmas during their passage through the continental crust. Contamination processes may be equally important in the oceanic environment, but their effects are in general much harder to recognize. For example, in magma chambers beneath the axes of mid-oceanic ridges (Ch. 5) tholeiitic MORB may be continuously assimilating roof rocks of previously erupted seawater-altered basalt. This may also be a significant process in crustal magma chambers beneath oceanic-island (Ch. 9) and island-arc (Ch. 6) volcanoes.

Crustal contamination may explain some of the geochemical differences between oceanic tholeiites (MORB + OIB) and continental tholeiites (Ch. 10). The latter typically show greater enrichments in SiO_2 and incompatible elements and can have Nd–Sr–Pb isotopic compositions similar to those of continental crustal rocks (Campbell 1985). However, recent investigations have shown that enriched subcontinental upper mantle sources may be equally important in creating these differences (Ch. 10).

There are several ways in which mantle-derived magmas may interact with crustal rocks (Huppert & Sparks 1985):

(a) Partial melts from crust and mantle sources may mix in magma chambers or conduits.
(b) The walls, roof and floor of a magma chamber (or the walls of a conduit) may begin to melt, when their solidus temperature is exceeded and, in fluid dynamically favourable circumstances, this melt can contaminate the chamber magma by mixing.
(c) Blocks of wall rocks can be stoped into the magma and assimilated by complete melting (bulk assimilation).

In (b) and (c) thermal energy is required to melt the wall rocks, and this must come from the magma itself. As a consequence the magma must cool and may therefore begin to crystallize. Assimilation in magma chambers is commonly associated with concurrent fractional crystallization (DePaolo 1981). In general, the amount of assimilation of colder wall rocks is limited by the thermal energy of the magma and consequently is unlikely to be extensive, as magmas are rarely superheated. Maximum amounts will rarely exceed 10−20% on thermal arguments.

In addition, Patchett (1980) has suggested that hydrous fluids, released as the product of dehydration reactions in the heated wall rocks, may be absorbed into the magma, resulting in selective contamination by elements which are enriched in the fluid phase. Watson (1982) has suggested that selective contamination of alkali elements such as K may occur during wall-rock interaction because of their much greater chemical diffusivities.

The continental crust contains a complex variety of rock types, ranging from granite pegmatites with relatively low fusion temperatures to more refractory gabbros, granulites, amphibolites and ultramafic rocks. In general, contamination will tend to be selective towards rocks of low fusion temperature (Huppert & Sparks op. cit.) and therefore rock types with high proportions of alkali feldspar and sodic plagioclase should be preferentially assimilated. Clearly, it is highly unlikely that ascending basaltic magmas will become contaminated by material of a fixed chemical composition, and this has fundamental implications for quantitative modelling of contamination processes (Section 4.4.1).

The extent of thermal erosion of the wall rocks will be directly proportional to the difference between the magma temperature and the fusion temperature of the wall rocks. Thus more evolved (cooler) magmas are only likely to become contaminated by the most easily fusible crustal rock types, whereas high-temperature picrites may be able to assimilate a much wider range of rock types (Huppert & Sparks op cit.).

At any contact at which the melting point of the wall rocks is less than the temperature of the magma, partial melting will occur. If this melt can convect away from the boundary, by virtue of its density contrast with the chamber magma, the process will continue until the onset of crystallization at the contact (Turner & Campbell 1986). Once crystallization begins the wall rocks will be protected by a layer of crystalline rock which must be melted before further assimilation can occur.

This may only be possible if the system receives an additional heat flux from the injection of a new batch of more primitive magma. In general, assimilation of roof rocks in crustal magma chambers is not an efficient process for contamination, as the lower density crustal melts will pond against the roof, protecting it from further erosion.

When magma flows in dyke-like conduits the extent of wall-rock assimilation may be controlled by the nature of the flow (Huppert & Sparks op. cit.). If the flow is laminar the magma is likely to solidify against the walls, preventing significant contamination. However, if the flow is turbulent the products of partial melting of the wall rocks will be continuously swept away by the convective motion, allowing further assimilation to continue. Huppert & Sparks (op. cit.) and Campbell (1985) indicate that flows of high-temperature primitive magma will be fully turbulent if the dyke width exceeds 3 m. Extensive wall-rock interaction might be expected in such bodies.

Recent studies of continental basalt provinces (Chs 10 & 11) have revealed the occurrence of two contrasting styles of crustal contamination (Huppert & Sparks op. cit.). In one, contamination is accompanied by concurrent fractional crystallization, and this is widely considered to occur in magma chambers, where the heat released by crystallization allows fusion of the wall rocks. DePaolo (1981) has termed this process AFC (Assimilation and Fractional Crystallization). AFC has been widely used to interpret radiogenic isotope and trace element variations in suites of volcanic rocks. In such suites there are positive correlations between indices of differentiation, such as percent SiO_2 and Fe/(Fe + Mg), and incompatible trace element ratios or Sr−Nd−Pb isotopic ratios which reflect contamination.

The second style of crustal contamination apparently occurs in those provinces in which deep magma chambers develop at the base of the crust or within the upper mantle. Fractional crystallization processes may occur in such chambers without significant crustal contamination because of the refractory nature of the wall rocks. When magmas from these chambers subsequently ascend to higher levels in the crust they may become contaminated by more easily fusible upper crustal rocks. How-ever, unlike the AFC style of contamination, in this case only the hotter, more primitive magmas are likely to become significantly contaminated. Thus geochemical parameters which reflect contamina-tion should negatively correlate with those that reflect the degree of fractionation.

In general, radiogenic and stable isotope and trace element geochemical studies are invaluable in identifying the role of crustal contamination in the petrogenesis of volcanic suites. Unfortunately, they cannot always provide unequivocal evidence con-cerning the nature of the contamination process. The reader is referred to Chapters 7, 10 & 11 for more detailed case studies.

4.4.1   Theoretical modelling of AFC processes

DePaolo (1981) and Powell (1984) have developed equations describing both the isotopic and trace element consequences of concurrent assimilation and fractional crystallization.

For any trace element:

$$C_{L} = C_{L}^{0} f + \frac{r}{r - 1 + D} \cdot C_{*}(1 - f) \qquad (1)$$

where C_{L}^{0} is the concentration of the trace element in the original magma; C_{L} is the concentration of the trace element in the contaminated magma; C_{*} is the concentration of the trace element in the contaminant; r is the ratio of the rate of assimilation to the rate of fractional crystallization; D is the bulk distribution coefficient for the fractionating assemblage; $f = F^{-(r-1+D)/(r-1)}$; and F is the fraction of magma remaining. Note that the deriva-tion of this equation assumes that r and D are constants.

For any radiogenic isotope

$$\varepsilon_{L} = \varepsilon_{L}^{0} + (\varepsilon_{*} - \varepsilon_{L}^{0})\left(1 - \frac{C_{L}^{0}}{C_{L}}f\right) \qquad (2)$$

where ε_{L}, ε_{L}^{0} and ε_{*} are isotopic ratios (e.g. $^{87}Sr/^{86}Sr$, $^{206}Pb/^{204}Pb$, $^{143}Nd/^{144}Nd$ etc.) whose subscripts are defined above.

These equations are generally applicable to any magma and any contaminant, and thus they can also be used to model the interaction of rising basaltic magmas with their mantle wall rocks. When $r = 1$ this is analogous to zone refining (Section 4.5). Equations (1) and (2) may also be used to model the evolution of a body of magma which is being recharged with fresh magma at the same time as it is fractionally crystallizing. In this case C_{*} is the composition of the new batch of magma. The AFC equations can also be used to describe the assimilation of a partial melt of the wall rock in addition to the case of bulk assimilation, C_{*} being the concentration of the trace element in the partial melt.

For the case of perfect fractional crystallization without concurrent assimilation, $r = 0$ and equation (1) reduces to the Rayleigh fractionation equation (Section 4.3.4):

$$C_{L} = C_{L}^{0} F^{(D-1)}$$

In modelling the isotopic consequences of crustal contamination, many authors simply assume bulk assimilation of the contaminant without concurrent fractional crystallization. This is clearly a limiting case but one which is probably unrealistic in nature. Equation (3) below enables calculation of the isotopic ratio of an element in a magma contaminated by such a process.

For simple mixing, by mass balance:

$$\varepsilon_{L} = \frac{X_{L}^{0} C_{L}^{0} \varepsilon_{L}^{0} + (1 - X_{L}^{0}) C_{*} \varepsilon_{*}}{X_{L}^{0} C_{L}^{0} + (1 - X_{L}^{0}) C_{*}} \qquad (3)$$

where X_{L}^{0} is the weight fraction of the original magma in the mixture; and $(1 - X_{L}^{0})$ is the weight fraction of the contaminant in the mixture. By substituting the relation:

$$C_{L} = X_{L}^{0} C_{L}^{0} + (1 - X_{L}^{0}) C_{*}$$

we can eliminate X_{L}^{0} from equation (3) to generate the following equation (4), which is more readily comparable with the AFC equation (2) (Powell 1984):

$$\varepsilon_{\rm L} = \frac{1}{C_{\rm L}}\left[\frac{\varepsilon_* - \varepsilon_{\rm L}^0}{1/C_* - 1/C_{\rm L}^0}\right] - \left[\frac{\varepsilon_*/C_{\rm L}^0 - \varepsilon_{\rm L}^0/C_*}{1/C_* - 1/C_{\rm L}^0}\right] \quad (4)$$

Since $\varepsilon_{\rm L}^0$, ε_*, $C_{\rm L}^0$ and C_* are constants, then for simple mixing:

$$\varepsilon_{\rm L} \propto 1/C_{\rm L}$$

Thus, for a suite of lavas contaminated by bulk mixing without concurrent fractional crystallization, we should expect to see linear trends in diagrams of, for example, $^{87}\mathrm{Sr}/^{86}\mathrm{Sr}$ versus 1/Sr. In general, this will not be the case for a suite of lavas related by AFC, which will display a curved trend.

Figure 4.11 shows the trace element enrichments which may be generated by AFC processes, in comparison to those generated by perfect fractional crystallization, for a highly incompatible trace element ($D = 0.001$) and a highly compatible trace element ($D = 10$). In both cases r (the ratio of the rate of assimilation to the rate of fractional crystallization) is set at 0.5. Curves are drawn for different contaminants, the compositions of which are expressed in terms of their enrichment relative to the original magma ($C_*/C_{\rm L}^0$).

When $D \ll 1$ (Fig. 4.11a), incompatible elements became much more strongly enriched as the magma evolves by AFC, in comparison to simple fractional crystallization, the enrichment in the magma increasing as the concentration of the trace element in the wall rock increases (i.e. as $C_*/C_{\rm L}^0$ increases). In

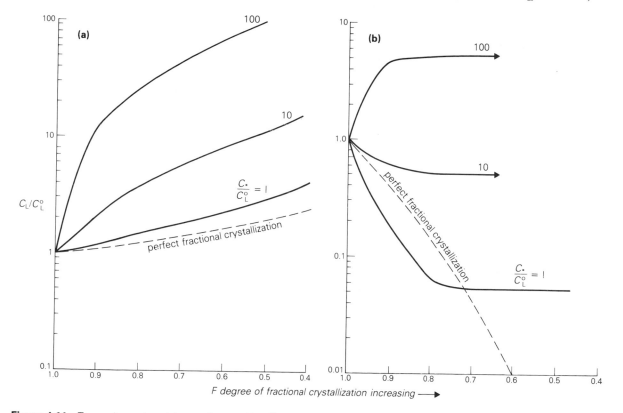

Figure 4.11 Trace element enrichment factors ($C_{\rm L}/C_{\rm L}^0$), as a function of the degree of fractional crystallization, for AFC (solid lines) compared to perfect fractional crystallization (dashed lines): (a) for a highly incompatible trace element for which $D = 0.001$; (b) for a highly compatible trace element for which $D = 10$. The ratio of the rate of assimilation to the rate of fractional crystallization, r, is set at a constant value of 0.5. Numbers on the AFC curves represent the composition of the contaminant in terms of the ratio $C_*/C_{\rm L}$.

the case where $D > 1$ the concentrations of compatible trace elements in the melt reach a steady state for AFC, given by the expression (DePaolo 1981):

$$C_L \approx \frac{rC_*}{r + D - 1}$$

This is shown in Figure 4.11b, in which the steady state is reached at between 10 and 20% fractional crystallization, depending upon the composition of the contaminant. As a consequence, we can clearly see that magmas with constant compatible trace element compositions could develop widely varying incompatible element enrichments as a consequence of AFC. However, in reality, neither r nor D may necessarily remain constant with progressive fractionation and therefore the steady state may never be reached. For example, during the early stages of fractional crystallization of basaltic magmas, D_{Sr} may be much less than 1, as olivine and clinopyroxene dominate the fractionating assemb-

lage. However, once plagioclase begins to crystallize Sr becomes compatible and D_{Sr} may become greater than 1.

4.4.2 Trace element characteristics of contaminated magmas

In Chapter 2 we considered the use of mantle-normalized trace element variation diagrams (spiderdiagrams) in comparing the trace element characteristics of different types of volcanic rocks. These can also be employed to illustrate the characteristic trace element signatures of contaminated magmas. Figure 4.12 shows the spiderdiagram pattern of a typical tholeiitic mid-ocean ridge basalt (MORB) compared to that of two typical continental crustal rocks, an upper crustal amphibolite facies gneiss (a) and a lower crustal granulite facies gneiss (b). Major and trace element analyses of the gneisses are given in Table 4.1, while analyses of typical MORB may be found in Chapter 5. MORB has been chosen to illustrate the trace element characteristics of a basaltic magma, derived by

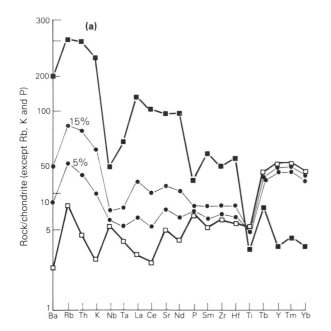

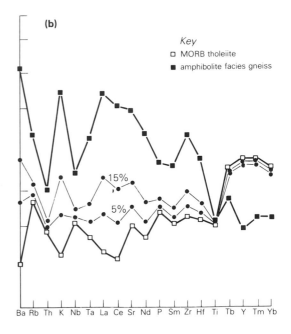

Figure 4.12 Spiderdiagram showing the effects of contamination of a MORB tholeiite by 5 and 15% bulk assimilation of (a) amphibole facies gneiss and (b) granulite facies gneiss. Normalization factors from Thompson (1982).

Table 4.1 Major and trace element analyses of some typical continental crustal rocks (data from Weaver & Tarney 1981).

	Average granulite facies gneiss	Average amphibolite facies gneiss
%		
SiO_2	61.2	66.7
TiO_2	0.54	0.34
Al_2O_3	15.6	16.0
Fe_2O_3	5.9	3.6
MnO	0.08	0.04
MgO	3.4	1.4
CaO	5.6	3.2
Na_2O	4.4	4.9
K_2O	1.0	2.1
P_2O_5	0.18	0.14
ppm		
Cr	88	32
Ni	58	20
Rb	11	74
Sr	569	580
Ba	757	713
Zr	202	193
Nb	5	6
Y	9	7
Pb	13	22
Th	0.42	8.4
Ta	0.56	0.45
Hf	3.6	3.8
La	22	36
Ce	44	69
Nd	18.5	30
Sm	3.3	4.4
Eu	1.18	1.09
Tb	0.43	0.41
Tm	0.19	0.14
Yb	1.2	0.76

partial melting of asthenospheric upper mantle, which cannot have interacted with continental crustal rocks en route to the surface. The continental crustal rocks are variably enriched in the whole range of incompatible elements from Ba to Hf relative to MORB and variably depleted in the elements Ti to Yb. Their spiderdiagram patterns are remarkably similar from K to Yb but differ significantly for the elements Rb and Th, which are strongly concentrated in the upper crust.

For the simplest possible case of contamination, that of bulk assimilation of the contaminant without concurrent fractional crystallization, Figure 4.12 shows calculated spiderdiagram patterns for contaminated basalts containing 5 and 15% of amphibolite and granulite facies crustal components (by weight). Figure 4.13 compares the patterns for the 15% contaminated basalts with that of the uncontaminated magma. This reveals some important features concerning crustal contamination effects in tholeiitic basalts. The elements Ti, Tb, Y, Tm and Yb appear essentially unmodified, even at significant degrees of contamination. These are thus important elements if we wish to see through the contamination process to the trace element characteristics of the primary mantle-derived magma. In contrast, the whole group of elements from Ba to Hf are variably enriched in the contaminated magmas. Those from Nb to Hf clearly do not discriminate between upper and lower crustal contaminants. The marked trough in the spiderdiagram pattern of both contaminated magmas at Nb−Ta is a distinctive signature for all magmas which have been contaminated by continental crustal rocks. As we shall see in Chapters 6 and 7, this is also a characteristic feature of subduction-related magmas, in which the signature may be inherited from subducted sedimentary rocks. The group of elements Ba, Rb, Th and K show very different patterns for upper and lower crustal contamination, as a consequence of the relative depletion of Rb and Th in lower crustal rocks. In Chapters 10 & 11 this has been used to infer the approximate depth of crustal contamination (i.e. upper versus lower crust) for many intracontinental plate basaltic magmas.

Clearly, as we have discussed in previous sections, crustal contamination is unlikely to take place without concomitant fractional crystallization. Thus, if we wish to model spiderdiagram patterns realistically, we should really use the AFC equation to do so (Section 4.4.1, Equation (1)). This involves additional assumptions concerning the rate of assimilation relative to the rate of fractional crystallization (r) and also the mineralogy of the fractionating assemblage in order to calculate D values.

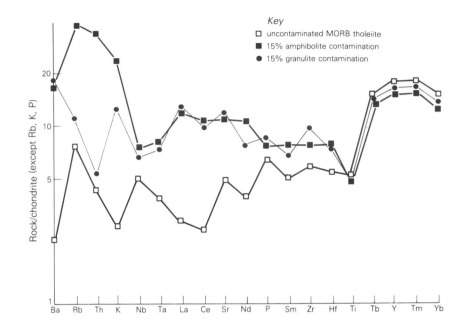

Figure 4.13 Spiderdiagrams for 15% contamination by granulite and amphibolite facies crustal rocks (from Fig. 4.12), compared to that of the uncontaminated magma.

4.5   Zone refining

Once primary magmas have segregated from their mantle source they must traverse considerable volumes of mantle rocks before reaching the crust. In general, it is difficult to imagine how they can do so without interacting with them in any way. The process of interaction of ascending magmas with mantle rocks is sometimes referred to as zone refining (Cox *et al.* 1979). It is simply a variant of the contamination processes we have already discussed in Section 4.4.

According to the original concepts of the process (Harris 1957), magmas may advance through the mantle by melting the wall rocks in front and depositing crystalline cumulates behind them. As we have considered in Section 4.4.1 this is simply a special case of AFC in which *r* (the ratio of the rate of assimilation to the rate of fractional crystallization) is equal to 1. Clearly, for a magma to advance in this way would require a considerable amount of superheat to assimilate the solid mantle material. As a consequence, its geological significance must be somewhat dubious. Alternative mechanisms for the ascent of magmas through the mantle are discussed in Chapter 3.

Some authors have suggested that the same effect could be achieved without melting as a result of diffusion of incompatible elements into the magma from the walls of the conduit through which it ascends (Green & Ringwood 1967). This could be important for elements such as K and Rb, as suggested by Watson (1982) for selective assimilation of crustal rocks. However, it is difficult to envisage sufficiently high diffusion rates for other incompatible elements such as the REE. Additionally, it could be argued (Cox *et al.* 1979) that in the immediate environment of the conduit the wall rocks would become rapidly depleted in incompatible elements, which would limit the process significantly.

The enrichment of incompatible elements in a magma ascending by a process of zone refining is given by the following equation (Cox *et al.* 1979):

$$\frac{C_L}{C_L^0} = \frac{1}{D} - \left[\frac{1}{D} - 1\right] e^{-nD}$$

where *n* is the number of equivalent volumes of mantle rock processed by the magma. This enrichment has a limiting value of $1/D$ where *n* is large.

4.6   Liquid immiscibility

The acceptance of liquid immiscibility as a viable petrogenetic process for the diversification of magmas has varied widely over the years (Roedder 1979). Experimental studies have demonstrated that it can occur in two types of silicate melt:

(a)   In highly Fe-rich basaltic melts, where one liquid is rich in Fe and P and poor in Si, and the other is rich in Si but poor in Fe and P.
(b)   In highly alkaline melts.

Criteria for the existence of immiscibility in natural systems are based on textural, mineralogical and chemical data. Clearly, two immiscible liquids must initially coexist as droplets of the minor phase suspended in the major. Thus, texturally, immiscibility should be recognized by the presence of rounded globules of one phase in a matrix of the other. If a density contrast exists between two immiscible liquids, the possibility of upward migration and concentration of the less dense phase exists. However, this is unlikely to be a major petrogenetic process.

Evidence for the existence of conjugate Fe-rich and Si-rich liquids can be found in the groundmass of many tholeiitic basalts, in which minute globules of dark brown glass are set in a matrix of paler glass (Philpotts 1979). Additionally, petrographic and experimental studies have established the immiscibility of sulphide and silicate melts, which may have implications for the genesis of some magmatic ore deposits (Hughes 1982). A substantial two-liquid field exists in simple experimental alkali silicate—carbonate systems (Edgar 1987), which may account for the origins of some carbonatite magmas (Le Bas 1987).

4.7   Gaseous transfer processes

Magmas contain variable amounts of dissolved volatiles, predominantly H_2O and CO_2, the solubility of which increases with increasing pressure. As pressure is decreased during their ascent to the surface, the solubility of volatile components decreases markedly and a gas phase may exsolve, in some cases triggering an explosive volcanic eruption. The loss of this gas phase from the system is clearly an important fractionation process, and thus the H_2O and CO_2 contents recorded in analyses of volcanic rocks may not necessarily be a true reflection of the volatile contents of the magmas which crystallized to form them.

High-temperature aqueous fluids are capable of dissolving considerable quantities of silicate components, particularly alkalis and silica. This can be demonstrated by the existence of intensely fenitized aureoles in the country rocks surrounding alkaline plutonic and subvolcanic complexes (Woolley 1982, Rubie & Gunter 1983, Le Bas 1987). Macdonald (1987) considers that equilibria involving such an alkali-rich fluid may be important in the geochemical evolution of peralkaline silicic magmas. Additionally, migrating hydrous fluids within the continental lithosphere (Ch. 12) and in the mantle above subduction zones (Chs 6 & 7) may be important agents of mantle metasomatism (Ch. 3).

Further reading

Best, M.G. 1982. *Igneous and metamorphic petrology*, New York: W.H. Freeman 630 pp.

Cox, K.G., J.D. Bell & R.J. Pankhurst 1979. *The interpretation of igneous rocks*. London: Allen and Unwin, 450 pp.

Maaløe, S. 1985. *Igneous petrology*. Berlin: Springer-Verlag, 374 pp.

McBirney, A.R. & R.M. Noyes 1979. Crystallisation and layering of the Skaergaard intrusion. *J. Petrol.* 20, 487−554.

Turner, J.S. & I.H. Campbell 1986. Convection and mixing in magma chambers. *Earth Sci. Rev.* 23, 255−352.

Wager, L.R. & G.M. Brown 1968. *Layered igneous rocks*. Edinburgh: Oliver and Boyd, 588 pp.

PART TWO

Magmatism at constructive plate margins

CHAPTER FIVE

Mid-ocean ridges

5.1 Introduction

A map of the ocean basins (Fig. 5.1) shows that their most conspicuous topographic feature is the system of mid-oceanic ridges, the crests of which rise on average 1000–3000 m above the adjacent ocean floor. Such ridges extend through all the major ocean basins, with a total length in excess of 60 000 km. With the exception of the East Pacific Rise, they occur in the middle part of the oceans and essentially form a submarine mountain range, which rises to its highest elevation at the ridge crest and slopes away symmetrically on either flank. Topographically, they vary throughout their length, the East Pacific Rise being much broader and less rugged than the other ridges (Section 5.4). A rare terrestrial expression of the mid-oceanic ridge system occurs in Iceland, whose central

graben is the extension of the Mid-Atlantic Ridge.

The ocean floor is cut by hundreds of fracture zones (Section 5.4.3) that, on an ocean-floor map, form a pattern of semi-parallel stripes cutting across and frequently offsetting the ridge axes. Such fracture zones are remarkably continuous features, normally extending for large distances across the flanks of the ridge, in some cases extending across the entire ocean floor to the continental margin.

Until the 1960s most geologists believed in the permanence of the continents and ocean basins. However, in 1962 H. H. Hess, in a classic paper, revolutionized thinking about the nature and origin of the oceanic crust and established the basic concepts of seafloor spreading, or plate tectonics, as we now know it. According to plate-tectonic theory, a mid-ocean ridge (or constructive plate margin) is a boundary between plates at which new

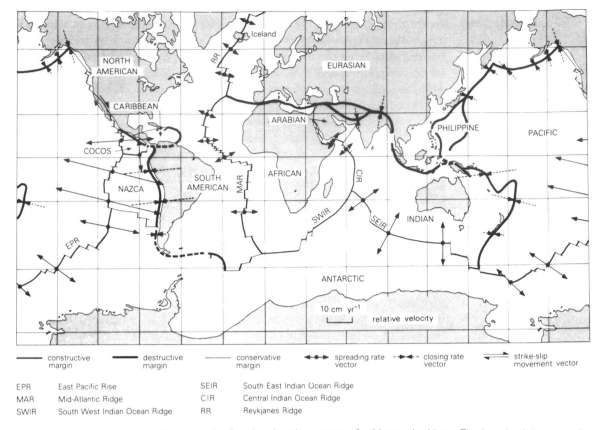

—— constructive margin	—— destructive margin	—— conservative margin	◄━●━► spreading rate vector ‑►◄‑ closing rate vector ⇐══⇒ strike-slip movement vector

EPR	East Pacific Rise	SEIR	South East Indian Ocean Ridge
MAR	Mid-Atlantic Ridge	CIR	Central Indian Ocean Ridge
SWIR	South West Indian Ocean Ridge	RR	Reykjanes Ridge

Figure 5.1 Tectonic map of the ocean basins showing the system of mid-oceanic ridges. The length of the spreading rate vector arrows is proportional to the spreading rate (after Brown & Mussett 1981, Fig. 8.16, p. 153).

oceanic lithosphere (crust + mantle) is generated, in response to partial melting of mantle lherzolite undergoing adiabatic decompression in a narrow zone of upwelling (Fig. 5.2). Partial melting results in the formation of basaltic magma, which is injected through tensional fissures into a narrow zone only a few kilometres wide at the ridge axis. Surface volcanism, sometimes in the form of pillow lava, occurs but most of the magma solidifies within dykes and layered intrusives at greater depths. The new rocks thus generated are then transported away from the ridge axis by the continuous process of seafloor spreading at half rates of $1-10$ cm yr.$^{-1}$

Since the size of the Earth is essentially constant, new lithosphere can only be created at mid-oceanic ridges if an equivalent amount of material is consumed elsewhere, at subduction zones (Part 3). Throughout geological time, a succession of ocean basins have been born and have grown, diminished

and closed again. The present episode of continental drift and seafloor spreading began about 200 Ma ago with the opening of the Atlantic and Indian oceans, which are still growing in size with respect to the Pacific, which is decreasing.

Hess' theory of seafloor spreading was confirmed by the discovery that periodic reversals of the Earth's magnetic field are recorded in the oceanic crust symmetrically about the ridge axis (Section 5.3.1), and the lateral migration of new crust away from mid-oceanic ridges is now well documented. Nevertheless, our understanding of the actual processes involved in the generation of new oceanic crust is still incomplete.

Essentially, the oceanic crust can be divided into two major domains:

(1)   the accreting plate boundary zone (mid-ocean ridge) at which new oceanic crust is created;

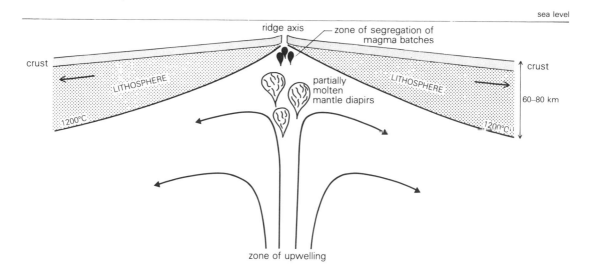

Figure 5.2 Schematic cross section of a constructive plate margin and its relation to the zone of upwelling of deep-mantle material. The oceanic lithosphere (crust + mantle) is generated at the ridge axis and increases progressively in thickness away from the ridge. The base of the lithosphere is represented as a thermal boundary layer (1200°C isotherm).

(2)   the passive crust, which after creation at the ridge axis has moved away.

In this chapter attention will be focused on the processes operating at the accreting plate boundary zone.

Information about the character of the oceanic crust has been obtained from a number of sources: dredging; deep-sea drilling; marine geophysical studies (especially seismic surveys); direct observation of the ocean floor from submersibles; and studies of rock sequences on land thought to represent uplifted segments of oceanic crust (ophiolite complexes). Until comparatively recently, seismic refraction studies and dredging provided most of the data.

Basalt was first dredged from the ocean floor at the turn of the century and since then extensive sampling programmes have shown that basaltic lavas with distinctive chemical characteristics are the major component of the oceanic crust (Section 5.10.1). Such basalts have been variously termed submarine basalts, ocean-floor basalts (OFB), abyssal basalts and mid-ocean ridge basalts (MORB). Initially, based on rather limited sampling, it was considered that ocean-floor basalts had rather restricted chemical compositions, with tholeiitic characteristics, constant SiO_2, low K and low incompatible element contents. However, since the mid-1970s, detailed sampling programmes involving drilling, dredging and the use of submersibles have revealed significant chemical and petrological diversity, often within a single site. In particular, basalts erupted along topographically 'normal' segments of ridges have different isotopic and trace element characteristics from those erupted along topographic highs or platforms associated with islands astride the ridge axis (e.g. Iceland, Azores, Galapagos, Bouvet and Reunion). In this respect the latter have more affinities with oceanic-island basalts but nevertheless are indistinguishable from normal MORB in terms of petrography, mineralogy and major element chemistry (Section 5.10).

In general, MORB are olivine tholeiites with a narrow range of major element compositions, indicating a relative constancy of sources and processes operating along most spreading ridges. They are the most voluminous eruptive rocks on Earth, and their generation has been a significant process in the differentiation of the upper mantle throughout geological time. The trace element variability of MORB is mainly attributable to

source heterogeneities and to shallow-level processes in open system steady-state magma chambers
(Section 5.8). Notable exceptions to the general
compositional uniformity include localized occurrences of Fe−Ti-rich basalt and rare silicic differentiates along the East Pacific Rise, Galapagos,
Juan de Fuca, SW Indian and SE Indian oceanic
ridges. In some instances these are associated with
propagating rifts (Section 5.4.4), but this is not true
for all ferrobasalt localities. Anomalous ridge segments such as Iceland, which have become emergent due to high magma production rates, exhibit
exceptionally high volumes of silicic differentiates
compared to normal ridge segments.

5.2   Simplified petrogenetic model

Basaltic magma generation at constructive plate
margins should theoretically represent the simplest
type of terrestrial magmatism. Nevertheless, as we
shall see in the following sections, the petrogenesis
of MORB is by no means simple. The apparent
regularity of the oceanic crustal layer over hundreds
of millions of square kilometres (Section 5.3) attests
to the continuity of magmatic processes over a
timespan of the order of 100 Ma (the average life of
an ocean basin). In contrast, detailed geochemical
studies of MORB (Section 5.10) reveal significant
source heterogeneities and a diversity of petrogenetic processes.

Figure 5.2 presents a summary of the processes
responsible for the generation of magma at mid-
oceanic ridges. They are the site of localized
upwellings of deep mantle material (Section 5.6)
which undergoes adiabatic decompression and, in
doing so, partially melts to produce basaltic magma. Initiation of such an upwelling seems a
necessary precursor to the fragmentation of a
continent (Ch. 10) and the subsequent generation
of a new ocean basin by seafloor spreading.

The oceanic lithosphere is generated at the ridge
and increases symmetrically in thickness away from
the axis, due to progressive cooling, to a maximum
of about 60−80 km. The base of the lithosphere,
marked by the 1200°C isotherm in Figure 5.2, is a
thermal boundary layer reflecting a marked change

in mechanical, but not necessarily chemical, properties from the underlying asthenosphere. The
lower parts of the lithosphere are probably composed of fairly fertile mantle lherzolite (Ch. 3),
indistinguishable from that of the asthenosphere,
whereas the upper parts must be variably depleted
due to magma extraction at the ridge axis.

The oceanic crust, comprising the upper 8−10
km of the lithosphere, also has its origins at the
ridge axis due to the extrusion and intrusion of
basaltic magma. It has a well layered layer structure
(deduced from seismic studies) comprising a surface layer of basalts underlain by a variety of
intrusive rocks, including dykes, cumulate gabbros
and ultramafics (Section 5.3). Many models have
been proposed to explain this layered structure,
mostly involving high-level magma chamber processes (Fig. 5.3; see also Section 5.8 and Cann
1970, 1974, Christensen & Salisbury 1975, Rosendahl et al. 1976, Rosendahl 1976, Nisbet & Fowler
1978).

The chemical composition of basalts generated at
mid-oceanic ridges must depend upon a variety of
factors, including the following:

(1)   The composition and mineralogy of the source
      mantle.
(2)   The degree of partial melting of the source and
      to a lesser extent the mechanism of partial
      melting (see Ch. 3).
(3)   The depth of magma segregation.
(4)   The extent of fractional crystallization and
      magma mixing processes during storage of
      magma in high-level sub-axial magma cham-
      bers.

With so many factors involved, the apparent
worldwide compositional uniformity of MORB
appears rather intriguing. Early workers used this
to argue that MORB compositions are those of
primary magmas unmodified by near-surface processes, but O'Hara (1968) showed fairly conclusively that most MORB are in fact highly fractionated.
This will be considered further in Section 5.10. If
MORB are considered in terms of their major
element chemistry alone (Section 5.10.2), they do
indeed represent an incredibly uniform magma

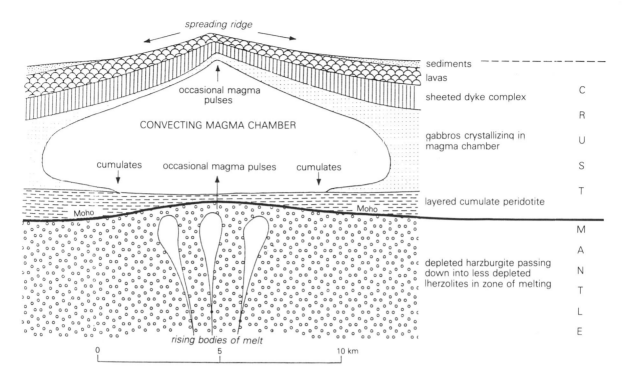

Figure 5.3 Hypothetical section through a mid-oceanic ridge, showing the development of the structure of the oceanic crust in response to magmatic processes at the ridge axis (after Brown & Mussett 1981, Fig. 7.8, p. 119).

type, the origins of which could be modelled in terms of rather simplistic processes. However, studies of their trace element and Sr, Nd and Pb isotopic composition (Sections 5.10.3 & 5) reveal the need for much more complex models.

Detailed discussion of the various parameters which control MORB chemistry is, of necessity, deferred until the appropriate sections, but the general conclusions to be drawn are presented here:

(a) *Source*. Generally depleted lherzolite in the spinel or possibly even plagioclase lherzolite facies. Some MORB associated with topographic highs along the ridge axis are apparently generated by partial melting of more enriched mantle sources (see Ch. 3 for explanation of the terms depleted and enriched). The trace element geochemistry of MORB shows no evidence for the existence of residual garnet in the source (Section 5.10.3).

(b) *Degree of partial melting*. As deduced from a variety of partial melting experiments on synthetic and natural lherzolites (Ch. 3 & Section 5.7) a minimum of about 20% partial melting appears to be required to generate the most primitive (MgO-rich) MORB compositions. If primary MORB are more picritic then the degree of melting would be somewhat greater (see Ch. 3).

(c) *Depth of partial melting and magma segregation*. Geophysical studies of the attenuation of P and S waves, although only available for a few ridge segments, suggest initiation of significant partial melting in rising mantle diapirs at depths of 60–80 km. Segregation of magma batches probably occurs at depths of about 20 km, feeding magma directly into high-level reservoirs (Hekinian 1982).

(d) *Fractional crystallization*. A major controversy in the study of MORB petrogenesis centres on the role of magma chamber processes in the control of their geochemistry and in the generation of the layered structure of the oceanic crust (Section 5.8). Large sub-axial magma chambers cannot persist on thermal grounds at slow-spreading (Atlantic-type) ridges and therefore high-level processes may be significantly different from those at fast-spreading (Pacific-type) ridges, where large magma chambers have been shown to exist.

5.3   Nature of the oceanic crust

5.3.1   Geophysical data

Palaeomagnetic studies in the early 1960s revealed the existence of magnetic stripes on the ocean floor and laid the foundation for modern theories of plate tectonics and seafloor spreading. Subsequently, detailed geophysical investigations (seismic, gravity and magnetic) of the world's ocean basins have provided important constraints on the thickness and structure of the oceanic crust. Seismic reflection methods have been mainly used to investigate the nature of the upper crust, particularly the veneer of oceanic sediments, whereas elucidation of the deep structure of the oceanic crust and upper mantle comes from seismic refraction work.

Paleomagnetic studies

The Earth's magnetic field is highly variable and has reversed its direction many times throughout geological time, i.e. north palaeopoles become south palaeopoles and vice versa. Figure 5.4 shows the palaeomagnetic record of reversals in the past 80 Ma. In the early 1960s it was discovered that the ocean floor off the west coast of North America exhibited a regular pattern of magnetic stripes of alternating normal and reversed polarity (Mason & Raff 1961). This led Vine & Matthews (1963) to propose a model for the generation of the oceanic crust in which molten magma injected at the axis of the mid-oceanic ridge becomes magnetized in the direction of the Earth's prevailing magnetic field as it cools. This newly cooled material is subsequently pushed away from the ridge by the injection of new magma, forming stripes of alternately normal and

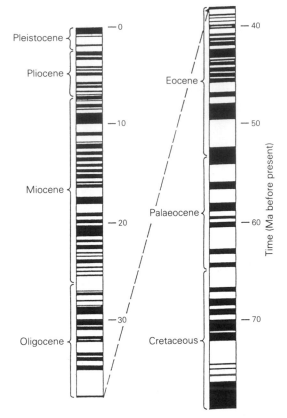

Figure 5.4 Palaeomagnetic record of reversals of the Earth's magnetic field over the past 80 Ma. The black intervals are normal (i.e. in the same sense as the present field) and the white intervals are reversed (after Brown & Mussett 1981, Fig. 6.4, p. 98).

reversed magnetism, depending on the polarity of the Earth's magnetic field when the magma solidified.

Magnetic anomaly patterns subsequently recorded in all the world's oceans (Fig. 5.5) reveal the same succession of magnetic stripes parallel to the mid-ocean ridge. The widths of individual stripes are different within each ocean due to the different spreading rates, but each ocean has the same sequence of reversals extending back to at least 76 Ma. Magnetic quiet zones occur beyond the oldest correlated anomalies, and in the Atlantic these lie near the margins of the continents. They are related to periods of infrequent polarity reversals in the late Mesozoic.

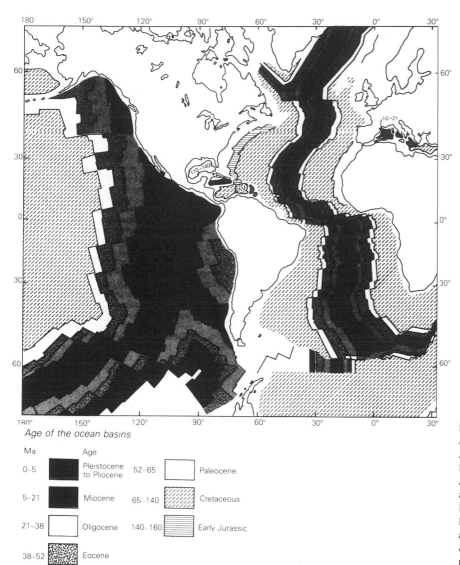

Age of the ocean basins

Ma	Age			
0–5	Pleistocene to Pliocene	52–65		Paleocene
5–21	Miocene	65–140		Cretaceous
21–38	Oligocene	140–160		Early Jurassic
38–52	Eocene			

Figure 5.5 The magnetic anomaly pattern in the Atlantic and Pacific oceans. Note that the pattern in the Atlantic is symmetrical about the Mid-Atlantic Ridge, whereas that in the Pacific is highly asymmetrical (after Press & Siever 1982, Fig. 18.21, p. 432).

The pattern of magnetic anomalies shows that spreading rates are not constant throughout the length of the mid-oceanic ridge system, but vary from region to region. The spreading rate is normally quoted as a half-rate of separation from a bilaterally symmetrical axis. Rates vary from a few mm yr^{-1} to 8 cm yr^{-1} along parts of the East Pacific Rise. Differences in spreading rate appear to influence MOR topography (Section 5.4).

Seismic refraction data

Seismic refraction studies have revealed that, although structurally complex on a local scale, the oceanic crust in general possesses a rather consistent downward seismic velocity gradient which can be modelled in terms of three major layers, 1, 2 and 3, which grade into one another. Table 5.1 and Figure 5.6 show this layered structure in greater detail (Christensen & Salisbury 1975, Houtz & Ewing 1976, Rosendahl *et al.* 1976, Kennett 1982).

Table 5.1   The layered structure of the oceanic crust.

	V_p (kms^{-1})[a]	Average thickness (km)[b]	Average density (g cm^{-3})[b]
water	1.5	4.5	1.0
layer 1 – sediment	1.7–2.0	0.5	2.3
layer 2 – basalt	2.0–5.6	1.75	2.7
layer 3 – gabbroic/ ultramafic cumulate layer	6.5–7.5	4.7	3.0
		Moho	
upper mantle	7.4–8.6		3.4

Data sources: [a] Basaltic Volcanism Study Project (1981), Table 1.2.5.1., p. 133; [b] Kennett (1982), Table 7–1, p.207.

Layer 1: comprises a thin veneer of oceanic sediments.

Layer 2: is composed of basalt and can be subdivided into two sub-layers. 2A is a layer the seismic velocities of which are less than those predicted by laboratory measurements on basalts. This has been interpreted as indicating high porosities and low densities caused by the presence of cavities and fractures in the upper crust (Kirkpatrick 1979). The layer thins away from the ridge crest and is often absent in crust older than 20–60 Ma (Houtz & Ewing 1976). This is probably due to cementation effects accompanying diagenesis as the crust ages. *2B* is a layer with higher seismic velocities than 2A, more appropriate to basalt and metabasalt.

Layer 3: the precise nature of this layer is less obvious on the basis of seismic velocity data alone. Petrological arguments combined with ophiolite studies suggest that it is composed of gabbros and cumulate ultramafic rocks formed in high-level magma chambers (Cann 1974, Dewey & Kidd 1977). As with layer 2 this layer can also be subdivided into an upper layer, *3A*, and a lower layer, *3B*. Layer 3B apparently increases in thickness away from the ridge axis (Christensen & Salisbury 1975), growing at the expense of underlying asthenosphere either by off-axis intrusions or by underplating with mantle material (Clague & Straley 1977). Layer 3A is well developed and is characterized by relatively uniform velocities and thicknesses away from the ridge axis, although its presence at the ridge axis is a subject of debate (Rosendahl 1976).

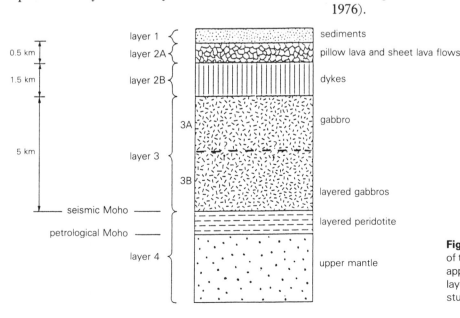

Figure 5.6   The layered structure of the oceanic crust, showing the approximate thickness of the layers as determined by seismic studies.

Mid-ocean ridges are elevated structures because they are composed of hotter and therefore less dense material than the surrounding plate. As the newly formed oceanic lithosphere moves away from the ridge axis it cools, becoming more dense, and subsides. A simple relation thus exists between the depth of the ocean and the age of the oceanic crust; mean ocean depth being proportional to the square root of the age. This is shown schematically in Figure 5.7a.

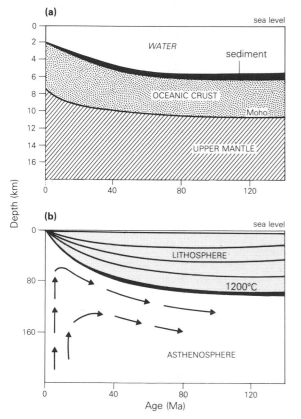

Figure 5.7 Structure of the oceanic lithosphere. (a) Schematic cross section through the upper lithosphere, showing the increasing thickness of the sediment cover with age away from the ridge crest, the deepening of the sea floor and the near-uniform thickness of the oceanic crust. (b) The part of the lithosphere beneath the oceanic crust is essentially a thickening lid of cooling ultramafic material overlying the hotter asthenosphere. The lines within the lithosphere are schematic isotherms and the base of the lithosphere is essentially an isothermal surface (~1200°C). Arrows within the asthenosphere indicate mantle flow lines away from the zone of upwelling at the ridge axis. (After Best 1982, Fig. 5.40, p. 183).

The thickness of the subcrustal lherzolitic lithosphere beneath layer 3 has been found to thicken exponentially with distance and therefore age away from the crest of mid-oceanic ridges (Fig. 5.7b; and see Leeds 1975). In the vicinity of the ridge crest it is virtually absent, but at distances from the axis where the crust exceeds 100 Ma in age its thickness becomes approximately 100 km. This thickening of the lithosphere as it ages is intimately related to its thermal evolution (Section 5.5.1).

5.3.2 Direct sampling of ocean-floor rocks

The first samples of ocean-floor rocks were obtained by dredging, and this technique is still in use largely because of its relatively low cost. Such samples obviously have no stratigraphic control and frequently come from anomalous areas such as fault scarps. Thus they can give us little information on spatial and temporal variations in MORB chemistry and mineralogy. The Deep Sea Drilling Project (DSDP), commencing in the late 1960s, gave a new dimension to the study of the oceanic crust, making it possible to define the vertical distribution of the various lithologies previously obtained by dredging. Unfortunately, it is still impossible to drill deep into the oceanic crust, maximum penetration being less than 1500 m. Additionally, most drill holes have to be sited on relatively old crust because a sediment cover of at least 80 m is required to hold the drilling units in place.

Most deep-crustal drilling has taken place in the North Atlantic, where it has been established that layer 2 is composed predominantly of pillow lavas with minor intercalated biogenic sediment to a depth of at least 600 m. However, until deeper penetration of the crust becomes possible the extact nature of the rocks at greater depths will remain unknown, and analogies with the lower parts of ophiolite complexes must be relied upon (Section 5.3.3). Deep-sea drilling has revealed a marked lack of lithologic and stratigraphic continuity in the crust, even between holes drilled only a few hundred metres apart. This indicates that seafloor eruptions are highly localized due to rapid chilling of the erupted magmas.

Technological advances since the 1970s have

made possible the direct observation and sampling of the mid-ocean ridges by the use of submersibles. The FAMOUS (French American Mid-Ocean Undersea Study) project initiated in 1971 involved a detailed study of the Mid-Atlantic Ridge near the Azores at 37°N, using submersibles and remote-controlled instruments to collect samples. New navigation techniques allowed the mapping of topographic features to a scale of a few tens of metres and, combined with the detailed sampling, provided the first comprehensive description of a section of an active mid-ocean ridge (Moore *et al.* 1974, Ballard & Van Andel 1977, Ballard *et al.* 1975).

The FAMOUS project was followed in the late 1970s by similar large-scale projects in the Pacific (Van Andel & Ballard 1979, Ballard *et al.* 1981), concentrating on the Galapagos spreading centre and the East Pacific Rise at 21°N. Detailed surveys employing submersibles and bottom photography allowed mapping of the ridge topography and the distribution of various different morphological types of lava. Additionally, these studies led to spectacular discoveries of hydrothermal fields venting onto the ocean floor (Section 5.5.2; see also Corliss *et al.* (1979).

5.3.3 Ophiolites

Plate-tectonic theory provides an explanation for the transience of the ocean floors, being derived from the mantle and returning to it within a timescale of 100 Ma. Occasionally, portions of oceanic plates escape destruction at subduction zones and during the final stages of ocean closure become obducted onto one of the colliding continental forelands to form an *ophiolite*. Many ophiolite sequences have now been documented in detail (Coleman 1977, Gass *et al.* 1984). Their study theoretically allows elucidation of the deep structure of the oceanic crust which cannot yet be sampled by direct drilling. However, not all ophiolites need necessarily represent sections of normal oceanic crust; some may be fragments of marginal basin crust (Ch. 8).

Figure 5.8 shows the theoretical structure of an ophiolite sequence and the correlation of the observed lithologies with the measured P-wave velocity profile of typical oceanic crust. Characteristically, seismic wave velocities are much lower in ophiolites than in normal oceanic crust. This may be related to the mineralogical changes attending their pervasive hydrothermal metamorphism. The upper portion of the model sequence comprises fine-grained Fe−Mn-rich mudstones, cherts, shales and limestones, i.e. typical deep-sea sediments. Both these sediments and the underlying pillow lava sequence contain Cu−Zn sulphides deposited by circulating hydrothermal solutions. The pillow lavas are the product of chilling of lava by sea water. In a typical succession they are apparently fed by a sequence of multiple dykes, the sheeted dyke complex, emanating from a basic magma chamber, now represented by coarse-grained gabbroic and ultramafic cumulates. A remarkable feature of such dykes is that they consistently exhibit one-way chilling, which is interpreted as the result of continued splitting in the axial zone.

The seismic distinction between ocean crust and upper mantle occurs where gabbros grade downwards into layered peridotites at the layer 3/layer 4 junction. However, the petrological Moho is in a different position, where layered peridotites, produced by crystal settling in magma chambers, change to massive peridotites that are part of the upper mantle.

Models for the formation of a typical ophiolite stratigraphy have had a profound influence upon models for magma chamber processes at mid-oceanic ridges (Cann 1970,1974, Robson & Cann 1982). These will be considered further in Section 5.8.

5.4 Structure of mid-ocean ridges

Mid-ocean ridges are made up of elevated volcanic mountains and valleys, in general rising about 2000 m from the adjacent ocean floor. Such ridges occupy some 33% of the total seafloor area and occur in all the major ocean basins. The system of ridges is essentially an oceanic phenomenon but in

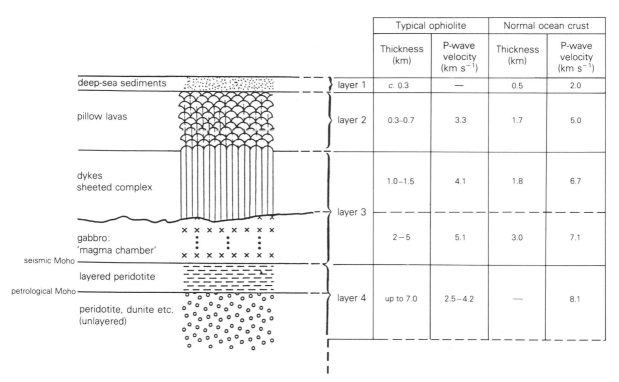

	Typical ophiolite		Normal ocean crust	
	Thickness (km)	P-wave velocity (km s^{-1})	Thickness (km)	P-wave velocity (km s^{-1})
layer 1	c. 0.3	—	0.5	2.0
layer 2	0.3–0.7	3.3	1.7	5.0
layer 3	1.0–1.5	4.1	1.8	6.7
	2–5	5.1	3.0	7.1
layer 4	up to 7.0	2.5–4.2	—	8.1

Labels: deep-sea sediments, pillow lavas, dykes sheeted complex, gabbro: 'magma chamber', seismic Moho, layered peridotite, petrological Moho, peridotite, dunite etc. (unlayered)

Figure 5.8 Petrological, seismic and thickness data for a typical ophiolite sequence compared with the layered structure of the oceanic crust, as deduced from seismic studies (after Brown & Mussett 1981, Fig. 7.4, p. 114).

places it passes laterally into continental rift zones (Ch. 11). For example, recent volcanism in the Afar and Ethiopian rift system is the continental continuation of spreading processes operating at the Mid-Indian Oceanic Ridge. Rarely, a mid-oceanic ridge segment may become emergent, due to volcanic overproduction, as in the case of Iceland.

Topographically, the mid-oceanic ridges are very variable and this has been shown to correlate well with spreading rate. Fast-spreading ridges have rather smooth profiles, whereas slow-spreading ridges have jagged profiles and an axial rift valley (Table 5.2).

The nature and extent of volcanic activity in the deep ocean environment has been recognized only recently due to photographic surveys and detailed sampling programmes made from manned submersibles (Section 5.3.2). Unfortunately, these are restricted to a few areas of the mid-ocean ridge system, none of which are necessarily typical ridge segments (e.g. 37°N – FAMOUS, Galapagos

Rise). The principal effects of the deep-sea environment are the almost complete elimination of explosive volcanism and a more rapid cooling rate due to quenching in water. Thus flows tend to be shorter and thicker than their subaerial counterparts. The major volcanic features are small volcanic cones, lava domes, flows and lava lakes.

5.4.1 Slow-spreading ridges

The Mid-Atlantic Ridge (MAR) is a typical example of a slow-spreading ridge (half-rate of 1–2 cm yr^{-1}; Hekinian 1982). Information about the structure of the ridge crest is limited to a few areas such as that between 36 and 37°N, known as the FAMOUS area (Section 5.3.2). The FAMOUS project commenced in 1971 using manned submersibles to document in detail the bathymetry, petrographic and geochemical variation of volcanic rocks from a small segment of the ridge crest. This segment is characterized by a broad 25–30 km

Table 5.2 Classification of mid-ocean ridges according to spreading rate.

		Latitude	Half-spreading rate (cm yr^{-1})	Source of data	
Fast	East Pacific Rise	21−23°N	3	[a]	No median valley; smooth topographic profile; basalts have higher FeO_T and TiO_2
		13°N	5.3		
		11°N	5.6	[a]	
		8−9°N	6	[a]	
		2°N	6.3	[a]	
		20−21°S	8	[a]	
		33°S	5.5	[a]	
		54°S	4	[a]	
		56°S	4.6	[a]	
Slow	Indian Ocean	south-west	1	[b]	Median valley; rugged topographic profile; basalts have lower FeO_T and TiO_2
		south-east	3−3.7	[c]	
		central	0.9	[c]	
	Mid-Atlantic Ridge	85°N	0.6	[c]	
		45°N	1−3	[a]	
		36°N	2.2	[a]	
		23°N	1.3	[c]	
		48°S	1.8	[a]	

Data sources: [a] Hekinian (1982); [b] Sclater *et al.* (1976); [c] Jackson & Reid (1983).

wide axial valley bounded by rift mountains (Fig. 5.9). Within this broad valley is a well defined narrow inner valley, 3−9 km wide, in which present volcanic activity appears to be concentrated. The flanks of this inner rift are fault controlled and have a veneer of thin lava flows originating on the valley walls and flowing inwards towards the rift axis. Small isolated volcanic hills, less than 300 m high, occur within the inner rift, defining the main focus of current activity. These are generally not split centrally as spreading continues but are carried as distinct units to one side of the valley (Ballard & Van Andel 1977), where they become dismembered by faulting. The fact that these volcanic hills are physically distinct features shows that volcanic activity is neither spatially nor temporally continuous.

The FAMOUS project documented marked compositional variations of basalts erupted within the inner rift zone. In general, the most primitive basalts (with high MgO, Ni and Cr contents) are associated with the central volcanic hills, whereas more evolved basalts are associated with eruptions at the margin of the rift (Fig. 5.10). Such compositional variations suggest the operation of low-pressure crystal fractionation processes and the

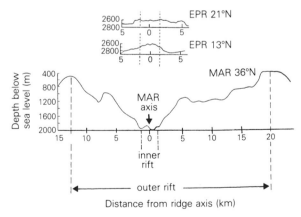

5.9 Schematic topographic profile across the central axis of the Mid-Atlantic Ridge in the FAMOUS area (36°N). The inset shows topographic profiles across the East Pacific Rise at 21°N and 13°N, drawn to the same vertical scale for comparison (after Francheteau & Ballard 1983, Fig.1).

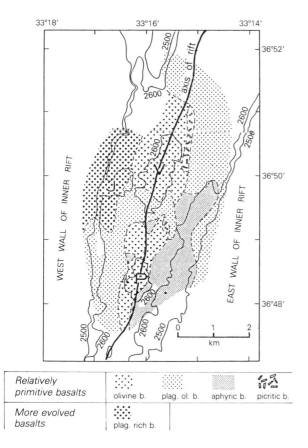

| Relatively primitive basalts | olivine b. | plag. ol. b. | aphyric b. | picritic b. |
| More evolved basalts | plag. rich b. | | | |

Figure 5.10 Distribution of the different types of basaltic rocks present in the inner rift valley of the Mid-Atlantic Ridge near 36° 50'N. Jupiter (J), Venus (V) and Pluto (P) are small volcanic hills defining the present focus of igneous activity along the ridge axis (after Hekinian 1982, Fig. 2.8, p. 73).

presence of some sort of high-level magma body beneath the axis of the inner rift. Two models for the nature of such magma bodies have been proposed (Hekinian 1982):

(a) A magma chamber the width of the inner floor (~ 3 km), in which case the central volcanic hills would simply be adventive cones.
(b) Each volcanic hill could have a small (<1.5 km wide) magma chamber beneath it, acting as an independent source for the volcanism.

The second model is most consistent with the geophysical data (Section 5.8), as a large magma chamber 3–4 km in diameter would attenuate S

waves and this is not observed in the FAMOUS area.

5.4.2 Fast-spreading ridges

Information on the nature of fast-spreading ridges comes mainly from the Pacific, specifically the East Pacific Rise at 21°N and the Galapagos Rift, whose half-spreading rates are 6-7 cm yr^{-1} (Crane & Ballard 1980, Spiess *et al.* 1980, Cyamex 1978, 1981, Francheteau & Ballard 1983). Unlike the Mid-Atlantic Ridge the ridge system of the Pacific is not symmetrical with respect to the bordering continents, most of the ridge being in southern latitudes and in the eastern part of the ocean basin (Fig. 5.1). Parts of the East Pacific Rise spread at half-rates of 8–9 cm yr^{-1}, greater than in any other part of the world.

Compared to slow-spreading ridges, the major difference in morphology is the lack of a well defined central rift valley (Fig. 5.11). In other respects there is a remarkable similarity in the dimensions and sequence of extrusive forms and the manner in which the ridge morphology changes from constructional volcanic along the axis to destructional tectonic along the margins. In both ridge types the central active extrusion zone is marked by a discontinuous line of small volcanic hills composed of sediment-free pillow lava (Fig. 5.12). In fast-spreading ridges these are flanked by plains of low relief made up of ponded lava lakes. These lava plains constitute the principal difference in volcanic morphology between fast- and slow-spreading ridges, and their formation is attributed to relatively high extrusion rates.

The total width of the extrusion zone along the East Pacific Rise is 2.5–3.0 km. Beyond this fissures and normal faults dominate, generating horst-like marginal highs. However, these faults have much less displacement than those bounding the inner rift in the FAMOUS area of the Mid-Atlantic Ridge. Active hydrothermal fields, 400–4000 m^2 in area, are associated with the ridge. These have only recently been discovered (Section 5.5.2) and may actually be a common feature of constructive plate margins, even slow-spreading ones, but have yet to be identified.

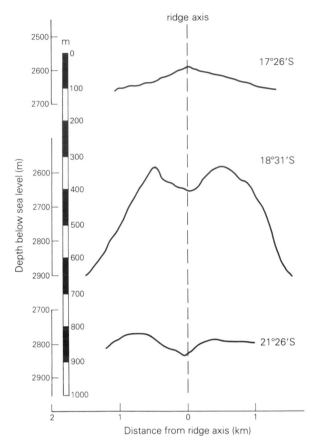

Figure 5.11 Bathymetric profiles across the fast-spreading East Pacific Rise at 17° 26′S, 18° 31′S and 21° 26′S, showing the essentially smooth topography of the ridge in comparison to the slow-spreading Mid-Atlantic Ridge (Fig. 5.9). Note the much expanded vertical scale in comparison to Figure 5.9. The vertical scale bar shows, for comparison, the relative depth of the inner rift valley (1000 m from floor to marginal highs) of the Mid-Atlantic Ridge in the FAMOUS area (after Renard *et al.* 1985, Fig. 3).

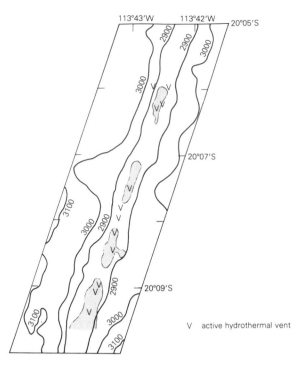

Figure 5.12 East Pacific Rise at 20°S, showing the bathymetry of the ridge axis (in metres). Shaded areas represent the youngest volcanic flows and these are clearly concentrated along the axial zone. Associated with these volcanics are active hydrothermal vents (after Francheteau & Ballard 1983, Fig. 13).

Petrographic and geochemical studies have shown the existence of a spectrum of primitive and more evolved basaltic magma types (e.g. Hekinian & Walker 1987). These are spatially distributed in a manner similar to that in the FAMOUS area, with the most primitive compositions erupted close to the ridge axis. However, fractionated basalts are the most common type sampled from the East Pacific Rise and this may be related to a fundamental difference in magma chamber processes beneath slow- and fast-spreading ridges. Certainly magma

production rates at the EPR must be greater to sustain the high spreading rate, and this may lead to the establishment of much larger magma reservoirs beneath the ridge axis in which fractional crystallization can occur. This will be considered further in Section 5.8.

5.4.3 Transform faults and fracture zones

The dominant grain of the ocean floor is produced by the magnetic lineations that reflect the location of the spreading centre at the time of formation of the oceanic crust. This magnetic anomaly pattern is frequently offset by transform faults (fracture zones), often forming arrays subparallel to the direction of spreading (Fig. 5.13). The fracture zones are remarkably continuous features, normally extending for large distances across the flanks of the ridge and in some cases extending across the entire

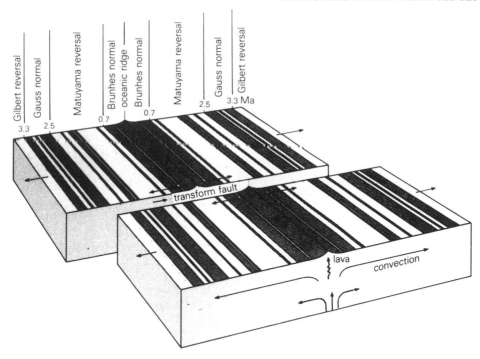

Figure 5.13 Offset of the magnetic anomaly pattern of the oceanic crust by a left lateral transform fault. Rocks of normal polarity are shown in black and of reversed polarity in white (after Press & Siever 1982, Fig. 18.17, p. 429).

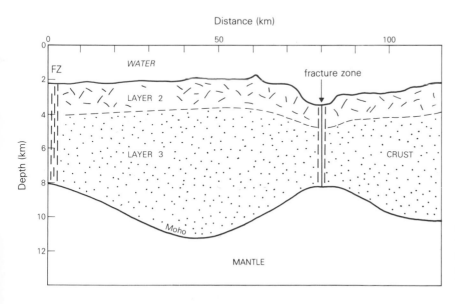

Figure 5.14 Schematic longitudinal section along the axis of a mid-oceanic ridge showing the variation in seafloor depth and crustal thickness in the vicinity of a fracture zone (FZ) (after White 1984, Fig. 4, p. 109).

ocean floor to the continental margin. This suggests that the locations of the active transform faults, from which the fracture zones originate, remain fixed with respect to the accreting plate boundary.

Fracture zones are usually marked by some irregularity in the topography of the ocean floor and are associated with shallow earthquakes generated by the lateral sliding of the adjacent segments of plate

(Fig. 5.14). Additionally, the oceanic crust thins adjacent to ridge–transform intersections, indicating a diminished rate of magma supply. Some fracture zones have a component of dip-slip movement and this may expose vertical sections through the oceanic crust. Dredge hauls from North Atlantic fracture zones have sampled a variety of lithologies including harzburgite, lherzolite, gabbro and amphibolite. A considerable amount of geophysical data from the North Atlantic ocean allows more detailed discussion of the nature of fracture zones (Johnson & Vogt 1973, Schouten & Klitgord 1982, Hekinian 1982). Here the average fracture zone spacing is ~50 km and many such fracture zones are traceable away from active transform faults at the ridge axis. Lateral offsets along the transforms vary from more than 20 km to effectively zero. However, even where the offset is small, the structural anomalies and discontinuities characteristic of the fracture zone persist.

A simplified model of the ocean floor (Fig. 5.15) suggests a pattern of strips of normal oceanic crust separated by narrow zones of anomalous crust. The normal oceanic crustal structure is formed in a string of individual spreading centre cells, whereas the anomalous crust is formed in the fracture zones. As more high-quality geophysical data become available, such a cellular pattern is becoming evident in all of the major oceanic ridge systems.

In some instances volcanism is associated with the transform, e.g. St Paul's Rocks and the Romanche Fracture Zone in the Atlantic and the Tamayo Fracture Zone at 23°N on the EPR. Basalts sampled from transform fault zones are characteristically more fractionated than average MORB. Bender et al. (1984) initiated a systematic study of the Tamayo zone to evaluate the effects of truncation of a normal spreading ridge by a large transform. For this particular section of the EPR, the normally broad swell morphology of the ridge changes to an axial rift valley morphology more characteristic of slow-spreading ridges within 20 km of the transform. Close to the transform a wider spectrum of magma compositions than normal MORB is observed, with a preponderance of differentiated ferrobasalts. Bender et al. (op. cit.) suggest that parental magmas close to the transform

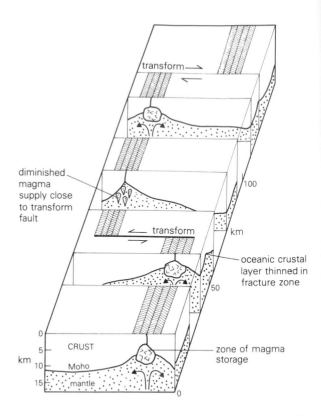

Figure 5.15 Simplified model of the ocean floor, showing the normal oceanic crustal structure developing in individual spreading centre cells separated by zones of anomalous crust generated in fracture zones (transform fault zones). Note the diminished rate of magma supply close to the fracture zone and the consequent thinning of the oceanic crustal layer. The zone of magma storage will only approximate to a large sub-axial magma chamber for fast-spreading ridge segments.

are derived from lower degrees of partial melting than those further away and, additionally, undergo more extensive fractional crystallization.

5.4.4 Propagating rifts

Propagating rifts are a special case of ridge–transform intersections in which one of the ridge segments begins to propagate into the older lithosphere adjacent to the transform, growing at the expense of the displaced ridge segment, which ceases to be active (Christie & Sinton 1981, Sinton et al. 1983). They are characterized by a diversity of erupted magma types extending from basalt through to rhyodacite, the latter representing the most differentiated magma type known from an oceanic spreading centre.

The process of propagation results in a failed rift that is offset from the propagating rift by a transform fault and a 'V'-shaped pair of magnetic anomaly offsets called pseudofaults, (Fig. 5.16), which strike obliquely to the spreading ridge and point in the direction of propagation. The pseudofaults mark the past positions of the propagating rift tip and separate crust formed at the propagating rift from the older crust through which the ridge propagated. In the vicinity of propagating rift—transform intersections, old and cold lithosphere bounds the active rift across the pseudo-

faults as well as across the transform. This unusual thermal regime may have a profound effect on the evolution of the propagating rift magmas.

The near uniformity of normal MORB compositions indicates that most ridges are characterized by steady-state magmatic processes involving balanced rates of supply and eruption. The association of highly differentiated lavas with propagating rifts suggests that this steady-state situation is not attained for some distance behind their tips. Propagating rift magmatic processes can be viewed as variations of those at normal ridge—transform intersections, in which the thermal environment is more conducive to high degrees of differentiation and thus the anomalous effects extend over an expanded length of ridge.

5.4.5   Anomalous ridge segments

Volcanics erupted along the strike of a mid-ocean ridge can show important compositional changes which appear to correlate with topographic and structural features (Le Douaran & Francheteau 1981, Francheteau & Ballard 1983, Schilling et al. 1983, Klein & Langmuir 1987). Topographic highs or volcanic platforms such as the Azores and the Galapagos have been termed 'hot spot' locations on their respective constructional plate boundaries. Such areas have crustal thicknesses intermediate between typical oceanic and continental values.

Figure 5.17 shows the variation of zero-age depth (i.e. depth to the floor of the inner rift) with latitude along the Mid-Atlantic Ridge (Le Douaran & Francheteau 1981). Sclater et al. (1975) demonstrated that these axial depth variations are not a young phenomenon, and have apparently persisted throughout the opening of the Atlantic. For example, the high between 26 and 31°N corresponds to the Canary Islands, and that between 13 and 22°N to the Cape Verde Islands. The major high between 34 and 42°N corresponds to the anomalous topography along the Azores—Gibraltar fracture zone, marking the triple junction between the European, North American and African plates. Superimposed upon the larger-scale anomaly pattern are relatively short-wavelength variations which are clearly associated with fracture zones. In general, an

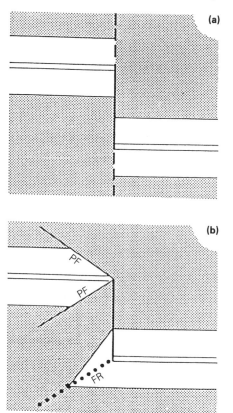

Figure 5.16 Schematic representation of normal ridge (a) and propagating ridge (b) transform intersections. Stippled areas denote all crust older than the axial anomaly which is unpatterned. Tectonic features specific to propagating rift systems include pseudofaults (PF) and a failed rift (FR) (after Sinton et al. 1983, Fig. 3).

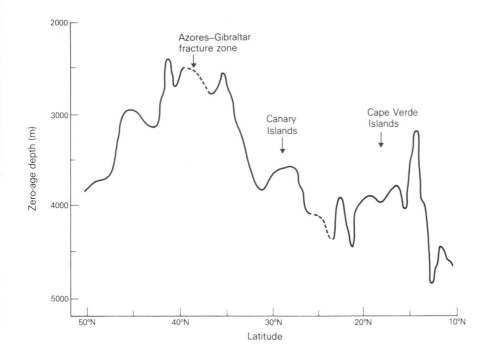

Figure 5.17 Zero-age depth variation with latitude along the axis of the Mid-Atlantic Ridge (after Le Douaran & Francheteau 1981, Fig. 3).

individual ridge segment displays a monotonic increase in depth towards the two adjoining transform faults from a single topographic high. The larger-scale depth anomalies correlate closely with geochemical anomalies in trace element and isotopic characteristics (Section 5.10) indicative of source heterogeneity. Such geochemical anomalies also seem to have been persistent throughout the opening of the Atlantic and testify to the relatively static nature of the mantle hot spots with respect to the accreting plate boundary. In comparison with the Atlantic, along-strike topographic variations along the East Pacific Rise are much smaller, of the order of 200–300 m, and the fracture zones are much more widely spaced.

The volcanism of Iceland (Imsland 1983, Gibson & Gibbs 1987), astride the Mid-Atlantic Ridge, can be taken as an extreme example of an anomalous ridge segment. This is the only large island that lies astride a spreading ridge and it is the most active volcanic region in the world. For these reasons Iceland has frequently been chosen as an ideal location for studying volcanic processes at constructional plate margins. However, its anomalous characteristics compared to the rest of the MAR in terms of topography, gravity, tectonic activity, crustal and mantle structure, seismicity and geochemistry must negate any generalizations.

The active plate boundary on Iceland is expressed as a series of tectonically and volcanically active rift zones, classified into axial and lateral types based on their tectonic structure and the chemical characteristics of the erupted magmas (Fig. 5.18). The axial rift zones are 30–60 km wide and erupt voluminous tholeiitic basalts from fissure swarms. These fissure swarms tend to be associated with central volcanic complexes in which volcanic activity is higher than normal, and subvolcanic magma chambers allow the fractionation of parental basalts to produce silicic differentiates. High-temperature geothermal fields are usually associated with the central complexes. In the south of Iceland two such axial zones run parallel to each other for a distance of approximately 100 km. In contrast, the lateral zones are characterized by the eruption of transitional and alkali basalts. Clearly, the constructive plate boundary on Iceland is considerably more complex than its submarine counterpart.

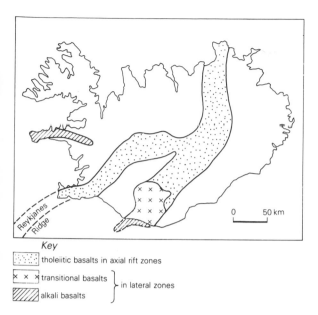

Key

[tholeiitic] tholeiitic basalts in axial rift zones

[x x x] transitional basalts
[////] alkali basalts
} in lateral zones

Figure 5.18 Outline of the active volcanic zones of Iceland, showing the distribution of the different basalt types (after Thy 1983, Fig. 1).

5.4.6 Aseismic ridges

Aseismic ridges are linear elevated volcanic structures rising 2000–4000 m above the surrounding ocean floor, varying from 250 to 400 km in width and 700 to 5000 km in length (Hekinian 1982). They comprise about 25% of the ocean floor, but are very poorly sampled and thus comparatively poorly understood features. They appear to represent chains of volcanic islands or seamounts that have subsided during their evolution.

The major aseismic ridges of the ocean basins are:

Atlantic   Iceland – Faeroe, Walvis Ridge – Rio
           Grande Rise
Pacific   Cocos, Carnegie
Indian   Ninety-East Ridge

Most are attached to a continental margin and they sometimes terminate in a volcanic island which generally forms a continuous structural feature with the ridge. For example, the Walvis Ridge in the

South Atlantic Ocean extends from the volcanic island of Tristan da Cunha on the flanks of the Mid-Atlantic Ridge to the continental margin of Africa (Fig. 5.19). The ridges lack seismic activity and yet many are fractured perpendicularly to their axis in a closely similar manner to mid-ocean ridge fracture zones which are seismically active. All of the aseismic ridges are quite old and represent features which were formed during the early history of opening of the present ocean basins. The bathymetry across the ridges and along their length is smooth in comparison with mid-ocean ridge topography, and they are essentially made up of basaltic and more differentiated volcanics.

Morgan (1971, 1972a,b, 1983) suggested that aseismic ridges are the surface expression of plates moving over fixed hot spots in the mantle, analogous to the model proposed for linear chains of islands and seamounts in the Pacific (Fig. 5.20; see also Ch. 9). If this is correct then the age of the volcanic rocks in the aseismic ridge should decrease in the direction of the adjacent mid-oceanic ridge. This appears to be broadly correct for the Ninety-East and Walvis ridges.

From both gravity and seismic studies it is inferred that the ridges have a thicker crustal structure (15–30 km) than young oceanic crust (7–10 km; Figure 5.21). They are, in general, in isostatic equilibrium and their structure may contain large volumes of mafic and ultramafic cumulates. Chemically and mineralogically the volcanic rocks differ from MORB, though still dominantly tholeiitic, more closely resembling oceanic-island and seamount eruptives.

5.5   Heat flow and metamorphism

5.5.1   Oceanic heat flow

Measurements of oceanic heat flow (Fig. 5.22; see also Parsons & Sclater 1977) show that heat flow through mid-oceanic ridges is several times greater than through average ocean floor. This is precisely what would be predicted from plate-tectonic theory, as the ridge crest is the focus for intrusion and extrusion of high-temperature basaltic magma.

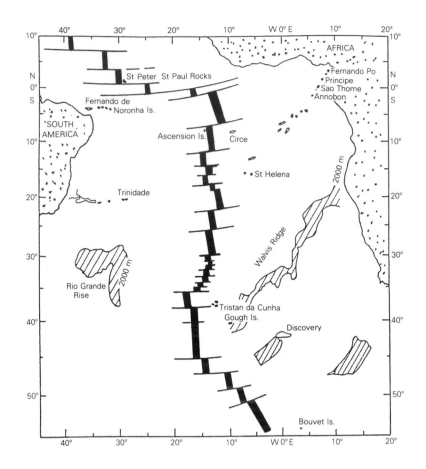

Figure 5.19 Paired aseismic ridges of the South Atlantic; Walvis Ridge – Rio Grande Rise (after Schilling *et al.* 1985, Fig. 1).

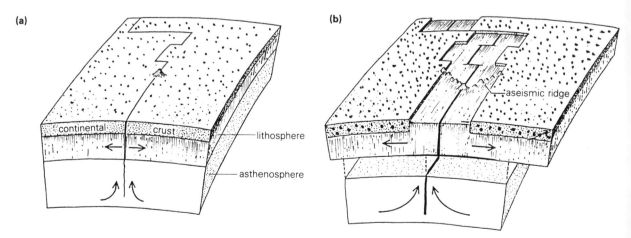

Figure 5.20 Hot spot model for the generation of paired aseismic ridges. (a) A hot spot exists at a continental rift zone, which becomes a site of successful rifting and the generation of a new ocean basin. (b) The hot spot remains a feature of the ridge axis in the growing ocean basin and is a focus of volcanic overproduction. Paired aseismic ridges develop, representing the trace of the hot spot in the newly formed oceanic crust.

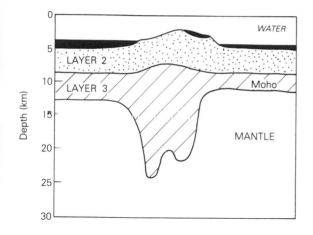

Figure 5.21 Crustal structure of the Walvis Ridge, showing the relative thickness of oceanic crustal layers 2 and 3 and the sedimentary layer (1), shaded black (after Hekinian 1982, Fig. 3.9, p. 163).

The oceanic plate is generated at the ridge and as heat flows out of its upper surface it cools and thickens (Fig. 5.7). Theoretical models of conductive heat flow give a simple equation to predict the heat flow through the plate (Sclater *et al.* 1980):

$$q = 11.3/\sqrt{t}$$

where q is the heat flow in HFUs (10^{-6} cal cm^{-2} s^{-1}) and t is the age of the crust in Ma. This equation is depicted as the solid curve in Figure 5.22 and clearly fits the measured heat flow data from the Pacific, Atlantic and Indian oceans quite well.

However, since the first oceanic heat flow measurements were made in the early 1950s it has become increasingly apparent that there is a very large scatter in the data obtained from mid-oceanic rift zones. Additionally, the mean of these data is consistently lower than the heat flow predicted by thermal models of seafloor spreading. This scatter

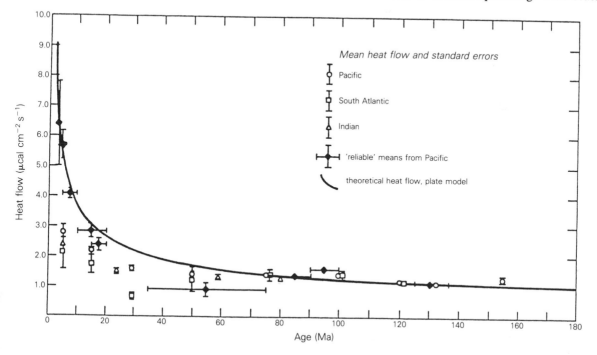

Figure 5.22 Changes in mean oceanic heat flow as a function of increasing crustal age away from the mid-oceanic ridge crest (after Parsons & Sclater 1977).

in the heat flow data can be attributed to permeation of sea water through fractures and cracks at the ridge crest, with the result that most of the heat loss is actually due to the advection of water through the crust. Evidence for this was provided during the 1970s by the discovery of extensive hydrothermal fields on the Galapagos ridge crest (Section 5.5.2). The scatter in the data decreases dramatically as the age of the crust increases and the heat flow values agree with the theoretical curves. This is due to the combined effects of the sealing of fractures due to ocean-floor metamorphism and the development of a sedimentary blanket over the crust.

5.5.2   Hydrothermal systems

The study of hydrothermal circulation in the oceanic crust is of primary importance in understanding its effect on the alteration of ocean-floor rocks. Exchange reactions between basaltic layer 2 and circulating hydrothermal solutions buffer the chemical and isotopic composition of seawater and can lead to the formation of metalliferous ore deposits near mid-ocean ridge crests. The amount of water incorporated into the crust by this process is a critical parameter in models for the generation of subduction-zone magmas (Ch. 6).

Water circulating through fissures and fractures in the oceanic crust near the ridge crest becomes heated, and ultimately re-emerges as hot springs on the ocean floor carrying various metals in solution. Initially, evidence for such hydrothermal activity came from studies of ophiolite complexes, particularly Troodos (Cyprus) and Oman (Section 5.3.3). In most ophiolites the pillow basalts are covered by several metres of metalliferous sediment called *umber*. These are often associated with lenticular iron sulphide ore bodies occupying depressions in the surface of the basalt, underlain by pipes of ore minerals. These pipes were clearly the conduits for ascending hydrothermal solutions.

Direct evidence for hydrothermal activity associated with mid-ocean ridge crests came in 1977 during a manned submersible study of the Galapagos spreading centre off the coast of Ecuador (Corliss *et al.* 1978). Subsequently, further north

on the crest of the East Pacific Rise a second hydrothermal field was identified in which 350°C fluids, blackened by sulphide precipitates, were blasting upward through chimney-like vents as much as 10 m tall − the so-called 'black smokers' (Fig. 5.23; see also Edmond & Von Damm 1984). These chimneys protrude in clusters from mounds of sulphide precipitates and provide direct evidence for the mechanism of generation of the ophiolite complex ore bodies described above.

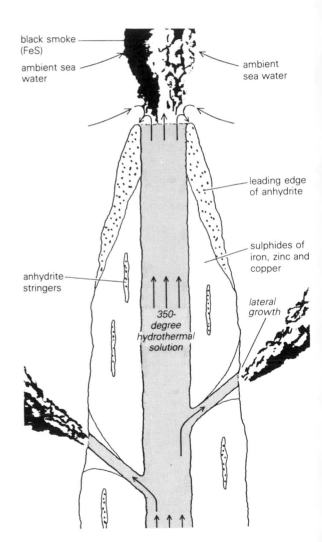

Figure 5.23 Schematic vertical section through a black smoker (after Edmond & Von Damm 1984).

5.5.3   Ocean-floor metamorphism

Studies of the oceanic crust in the 1960s and 1970s revealed that significant amounts of the crust are metamorphosed (Melson & Van Andel 1986, Cann 1969, Miyashiro *et al.* 1971, Hekinian & Aumento 1973). Dredge samples, especially those from the vicinity of fault scarps and transform faults, include greenstones, sepentinites and rare tectonized amphibolites. Such rocks have higher $^{87}Sr/^{86}Sr$ ratios than normal MORB, indicating that the water responsible for the growth of hydrous minerals was sea water. Much of this metamorphism has been ascribed to the circulation of hydrothermal brines through the oceanic crust in the vicinity of the ridge crest (Fig. 5.24).

Studies of the metamorphic mineral assemblages, field relations and tectonic fabrics of ophiolite complexes further support a model of dynamic hydrothermal metamorphism of the oceanic crust. Rapid increase in metamorphic grade stratigraphically downward into ophiolite complexes indicates geothermal gradients as much as several hundred degrees centigrade per kilometre, comparable with some of the highest heat flows measured at mid-oceanic ridges. However, the evidence for metamorphism within ophiolites must be viewed with some caution as, in many cases, it can be shown that they may have suffered several phases of metamorphism prior to and during obduction.

In marked contrast to the samples obtained by dredging, those from drill cores through the oceanic crust show rather limited alteration. This may in part be due to the restricted depth range of penetration of the crust (<600 m). Drill core samples in general show only zeolitic alteration along cracks and in vugs, with little evidence for major recrystallization.

The upper oceanic crust gains water by interaction with sea water through a series of metamorphic reactions involving the growth of chlorite, serpentine, smectite, illite and ultimately amphibole. The formation of such minerals is sequential and can be subdivided into three major stages (Staudigel *et al.* 1981):

I   formation of palagonite
II   formation of smectite
III   formation of carbonates

Stages I and II represent the stages of seawater–basalt interaction and involve large volumes of water, producing major chemical fluxes between the upper parts of layer 2 and the ocean reservoir. Both stages strongly deplete circulating solutions of alkalis and appear to end within a few Ma of formation of the crust. Stage III has a somewhat longer lifespan (< 10 Ma) and involves

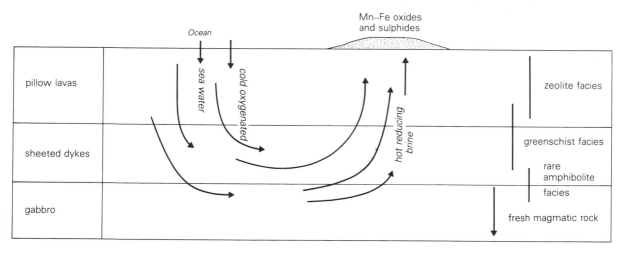

Figure 5.24   Convective circulation of sea water through the hot oceanic crust in the vicinity of a mid-ocean ridge crest (after Best 1982, Fig. 12.17, p. 430).

the release of Ca from basalt to the circulating solutions, culminating in carbonate precipitation to a depth of at least 500 m.

The basaltic layer 2 of young oceanic crust can be subdivided into an upper (low-velocity) layer 2A and a lower (high-velocity) layer 2B (Section 5.3.1). The depth of the boundary between 2A and 2B becomes shallower with increasing age, and layer 2A disappears completely at an age of approximately 70 Ma in the Atlantic. This disappearance is most readily explained by the filling of cracks and voids with secondary minerals.

Theoretically, the pervasiveness of ocean-floor metamorphism should decrease with depth, being directly related to the permeability of the oceanic crust. Generally, dredged samples are extensively altered because of biased sampling towards fault scarps and fracture zones. In contrast, cored samples show significantly less alteration, as considered previously. Figure 5.25 shows an estimate of the proportions of the various secondary minerals within the oceanic crust (Ito *et al.* 1983). This implies that perhaps only 15% of the oceanic crust is actually hydrothermally altered. This has profound implications for petrogenetic models for subduction-zone magmatism, which imply that subduction of altered oceanic crust recycles H_2O and Cl from the surface reservoir (atmosphere, hydrosphere and crust) back into the mantle (Ch. 6).

Ocean-floor metamorphism can produce significant compositional changes in the basaltic rocks of the oceanic crust. Some greenschist facies metabasalts preserve their original chemistry, apart from the addition of H_2O, but most show a marked decrease in CaO and significant variations in alkali and silica content. Thus caution must be exercised when considering the geochemistry of ocean-floor basalts, and fresh glassy material should be chosen whenever possible.

Albarede & Michard (1986) consider that the U content of the oceanic crust may be increased by 20% during hydrothermal alteration. In contrast, Pb is leached from the crust, resulting in subduction of oceanic crustal layers with high μ which may later contribute to the genesis of oceanic basalts with distinctive Pb isotopic compositions (Ch. 9).

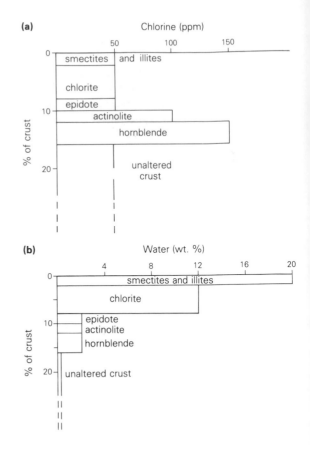

Figure 5.25 Proportions of various hydrous minerals in altered oceanic crust and the average concentration of (a) chlorine and (b) water in each mineral (after Ito *et al.* 1983, Figs. 2 & 3).

5.6 Convection systems at constructive plate margins

It is generally agreed that the generation of basaltic magma at constructive plate margins occurs in response to the upwelling of hot upper mantle material, which partially melts due to adiabatic decompression (Section 5.2). This upwelling is normally considered part of the large-scale convective motion of the asthenosphere. Convection is the principal form of heat transfer from the interior of the Earth, where heat is mainly generated by radioactive decay, and is linked via plate tectonic theory to the motions of the lithospheric plates (Ch.

3). However, it is still unclear as to how closely the shape, area and velocity of plates reflects the pattern of convection cells beneath. Additionally, the question of whether the convection extends in a single layer through the entire depth of the mantle (~3000 km) or is separated into two or more discrete layers is also undecided (Carrigan 1982, Houseman 1983a,b). One view is that convection in the upper mantle is separated from that in the lower mantle at a depth of about 650–700 km. This corresponds to the depth of significant phase changes in the upper mantle (marked by the 670 km seismic discontinuity) and also to the maximum depth to which subducted lithospheric plates can be traced. If the upper convecting layer is only 700 km deep the horizontal dimensions of the lithospheric plates imply large aspect ratios for upper mantle convection cells (Houseman 1983a).

Plastic flow in the upper mantle at the ridge crest must be driven by both large-scale phenomena related to plate drifting and by local phenomena related to partial fusion in the ascending mantle. The flow associated with the tectonic process is mainly driven by the cooling and sliding of the lithosphere and the buoyancy forces of the upwelling. However, superimposed upon this large-scale flow is a pattern of rapidly ascending partially molten mantle diapirs, which progressively expel their magma via dykes, lose their buoyancy and then flow away with the general circulation (Figure 5.26; see also Rabinowicz *et al.* 1984). In this model at shallow depths the upward flow is divided into two parts; one part closest to the axial plane is channelled into a narrow conduit with a 5–10 km

half-width, while the remaining flow is deflected away more or less horizontally. In between, matter circulates on closed trajectories and thus stays permanently close to the ridge. The major asthenospheric upwelling is thus channelled into a narrow zone, at most 20 km wide in cross section, beneath the ridge axis.

5.7 Partial melting processes

Partial melting beneath mid-oceanic ridges occurs in response to adiabatic decompression of ascending mantle lherzolite in the zone of upwelling (Ch. 3). In order to understand the processes involved, several important factors must be considered:

(a) The composition of primary mid-ocean ridge basalt magmas.
(b) The mineralogy of the source: plagioclase, spinel or garnet lherzolite.
(c) The degree of partial melting.
(d) The mechanism of partial melting (e.g. batch, fractional etc.).
(e) The depth of beginning of melting in the rising mantle material and the depth of segregation of the magma.

The enormous volume and apparent compositional uniformity of MORB (at least in terms of major element chemistry; see Section 5.10.2) led early workers to suppose that they were primary magmas, derived directly from the mantle without subsequent modification en route to the surface (Engel *et al.* 1965). However, O'Hara (1968) showed that this was unlikely and it is now accepted that the bulk of MORB are evolved magmas whose compositions have been modified by a variety of high-level processes including fractional crystallization, magma mixing and crustal contamination. It is therefore essential to characterize the compositions of primary MORB in order to provide constraints on the chemical and mineralogical composition of their source and on the extent of subsequent fractionation, contamination and mixing processes. Unfortunately, the nature of primary MORB compositions is still equivocal and detailed

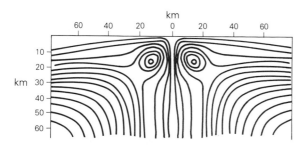

Figure 5.26 Theoretical flow pattern for mantle material rising beneath the axis of a fast-spreading oceanic ridge (after Rabinowicz *et al.* 1984, Fig. 5).

discussion will be deferred to Section 5.10, as many of the arguments are based on geochemical data. Essentially, the problem revolves around whether the most MgO-rich MORB compositions (10 wt.% MgO) are the primary magmas or whether the true parental basalts are in fact picrites.

Experimental melting studies provide some support for the picritic parental magma hypothesis, on the assumption that harzburgite (olivine + orthopyroxene) is the residue of the partial melting process (Ch. 3). If this is correct, then olivine and orthopyroxene should be on the MORB liquidus at pressures corresponding to the depth of segregation (O'Hara 1968). Green *et al.* (1979) and Stolper (1980) have shown that picritic liquids with 13–17% MgO are saturated with both olivine and orthopyroxene at 10–12 kbar pressure, whereas magmas with lower MgO contents do not have orthopyroxene on their liquidi at any pressure. Additionally, Presnall *et al.* (1979) predicted, by analogy with the CMAS system, that primitive MORB compositions should be saturated with olivine and orthopyroxene at about 9 kbar. These data, if directly applicable, would thus tend to suggest rather shallow depths of magma segregation of 30–40 km, within the stability field of spinel lherzolite. This is consistent with the trace element geochemistry of MORB (Section 5.10.3), which does not provide any evidence for the presence of residual garnet in the source.

Melting experiments on lherzolite bulk compositions can be used to deduce the degrees of partial melting necessary to generate typical MORB major element chemistry. Jaques & Green (1980; see also Ch. 3) have shown that tholeiitic basalt magmas can be produced by moderate degrees of batch partial melting (20–30%) of a lherzolite source at pressures below 15–20 kbar (50–60 km depth). At higher pressures, picritic liquids are generated at similar degrees of partial melting, while for greater degrees of melting (>35%) at all pressures the primary partial melts have komatiitic characteristics. It thus seems reasonable to assume degrees of partial melting somewhat in excess of 20% in the generation of MORB. However, this is apparently at variance with the much smaller degrees of partial melting predicted from studies of the trace element

geochemistry of MORB (Section 5.10.3).

Delineation of the extent of partial melting beneath mid-oceanic ridges is theoretically possible by the study of seismic wave velocities, as both P and S waves are attenuated by regions of partial melt. Unfortunately, such studies are very few in number. A zone of unusually high attenuation has been discovered beneath the crest of the East Pacific Rise, extending from 20 to 60–70 km depth. The upper bound probably reflects the depth of segregation of the primary basalts from their mantle source, while the lower bound marks the onset of significant degrees of partial melting (Hekinian 1982).

5.8 Magma storage and release

Models of oceanic crustal generation by seafloor spreading usually assume the presence of a magma chamber beneath the ridge axis (Greenbaum 1972, Cann 1974, Robson & Cann 1982). This is required to explain the chemical and mineralogical diversity of MORB (Sections 5.10 & 5.11) via fractional crystallization and magma mixing processes and also the structure of ophiolite complexes (Section 5.3.3). Considerable effort has been made in recent years to search for such magma reservoirs, mostly using seismological techniques (McClain *et al.* 1985, McClain & Lewis 1980, Lewis & Garmony 1982). Evidence for the presence of high-level magma bodies is provided by attenuation of S waves and a marked lack of seismic activity in areas in which brittle deformation should be taking place. Conversely, efficient propagation of S waves and the occurrence of earthquakes throughout a crustal volume can be taken to indicate that no substantial body of magma is present.

It seems inevitable that magma rising beneath the axis of a mid-oceanic ridge will form stationary pools in the crust prior to its eruption onto the ocean floor. However, very little is known about the size and shape of such pools or chambers, despite a variety of theoretical models. Inferences about the nature of sub-axial magma reservoirs are based on a limited amount of data from scattered ridge segments, which need not necessarily be typical of their respective ridge systems. For example, geophysical studies of segments of the East Pacific Rise have revealed the existence of a low-velocity zone beneath the ridge axis (Orcutt *et al.* 1975, Rosendahl *et al.* 1976, McClain *et al.* 1985), which could represent a magma reservoir. In contrast, similar studies of the much slower spreading FAMOUS area of the Mid-Atlantic Ridge have failed to reveal any substantial axial magma chamber (Fowler 1976).

Axial magma chambers may exist as closed systems or periodically replenished open systems (Ch. 4), of which the latter is the most likely. Their size and persistence is intimately related to the spreading rate. Thermal modelling (Sleep 1975, Kuznir 1980) indicates the potential existence of a magma chamber beneath ridge crests spreading faster than $0.5-0.9$ cm yr^{-1} (half-rate). Below this critical spreading rate, no permanent magma chamber can exist on thermal grounds. As the spreading rate increases above the critical value the chamber must expand in size until at a half-rate of 6 cm yr^{-1} it will underlie 10 km of ocean floor on either side of the ridge axis (Kuznir 1980). Thus fast-spreading ridges are predicted to have large continuous magma chambers, whereas slow-spreading ridges will have small discontinuous magma reservoirs.

The nature of the sub-axial magma reservoir will obviously control processes of fractional crystallization and magma mixing. In general, a large magma chamber might be expected to erupt compositionally uniform, somewhat differentiated magma due to the efficiency of mixing processes. Small magma reservoirs, on the other hand, may undergo rather extensive differentiation and will thus tend to erupt a much wider compositional range from primitive basalt to more evolved ferrobasalts. Comparative studies of basalts from the Mid-Atlantic Ridge and East Pacific Rise appear to support this idea.

From seismic, petrological and theoretical thermal modelling studies, several models have been proposed for magma storage and release beneath the axes of mid-oceanic ridges. Two limiting cases will be considered here, appropriate to fast- and slow-spreading ridges respectively.

5.8.1 Large magma chamber: fast-spreading ridge segments

Cann (1970, 1974) proposed the existence of a large magma chamber beneath the axes of mid-oceanic ridges, partly on theoretical grounds but also to explain the layered structure of the oceanic crust (Section 5.3.1) and the structure of ophiolite complexes (Section 5.3.3). Such a chamber is shown schematically in Figure 5.27. As the ridge spreads, increments of lava are erupted through dykes feeding from the roof and, concurrently, magma crystallizes at the walls of the chamber to form isotropic gabbros. Crystal fractionation processes within the chamber produce a sequence of layered basic and ultrabasic rocks at the floor. Cann termed this model the 'infinite onion', as it continuously peels off layers at the edges.

Robson & Cann (1982) have considered in detail the influence of such a large open system magma reservoir on the chemical composition of erupted MORB. Magma feeding the high-level chamber should be close in composition to the primary basalt generated by partial melting of the mantle at depth, assuming that high-pressure crystal fractionation processes are not particularly significant in the evolution of MORB chemistry (Section 5.11). Although magma production beneath the ridge may be essentially continuous, magma batches will only rise up buoyantly through the asthenosphere on reaching a certain critical size. Thus the supply of magma to any high-level reservoir will be periodic, the rate of supply being proportional to spreading rate. In periods between the addition of new magma batches to the chamber, the chamber magma will undergo continuous open-system Rayleigh fractionation, producing more evolved liquids enriched in incompatible elements. As a new magma batch enters the chamber it can either mix with the chamber contents or form a temporary pool on the chamber floor. Here it may crystallize olivine cumulates before becoming sufficiently fractionated (and therefore less dense) to mix with the magma chamber contents. The chamber magma will be evolved (incompatible element enriched), and mixing with a primitive (relatively incompatible element depleted) magma will result

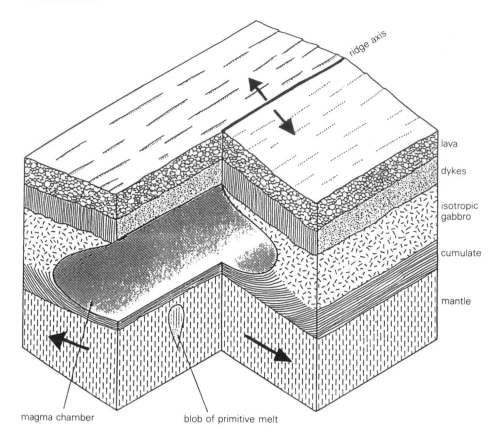

Figure 5.27 Schematic representation of the open-system magma chamber which may exist beneath all active mid-oceanic ridges (except those spreading extremely slowly). The chamber is periodically replenished by new magma batches, derived from partially molten mantle diapirs rising beneath the ridge axis, and is continually cooling and fractionating to produce new oceanic crust. The chamber size is directly proportional to the spreading rate; fast-spreading ridges have large continuous chambers while slow-spreading ridges have small discontinuous semi-permanent magma reservoirs (after Robson & Cann 1982, Fig. 1).

in its composition being reset to an intermediate composition more depleted in incompatible elements.

Lava will be erupted from the chamber when the magma pressure exceeds the lithostatic pressure and the strength of the chamber roof. This is most likely to be coincident, or nearly so, with the injection of a new batch of magma into the chamber. For small chambers this is likely to result in eruption of the fractionated chamber magma, followed closely by more primitive basalt. In a large magma chamber, appropriate to Cann's infinite onion model, magma mixing is likely to occur almost spontaneously and erupted magma will correspond to some fractionated 'perched state'.

A large axial magma chamber as proposed above should characteristically produce marked attenuation of S waves. Considerable support for such a model has come from seismic studies of the fast-spreading East Pacific Rise. However, the model cannot be considered generally applicable and, in particular, slow-spreading ridges appear to require a rather different model of magma storage and release (Nisbet & Fowler 1978).

5.8.2   Small ephemeral magma reservoir: slow-spreading ridge segments

Detailed studies of the FAMOUS area of the Mid-Atlantic Ridge in the late 1970s (Fowler 1976,

1978; Nisbet & Fowler 1978) cast serious doubts on the general applicability of Cann's infinite onion model to mid-oceanic ridge magmatic processes. In particular, there is no obvious attenuation of S waves beneath this segment of the MAR, suggesting that at present there can be no magma bodies larger than about 2 km across beneath the ridge axis. Additionally, thermal constraints (Sleep 1975) make the existence of a large axial chamber beneath such a slow-spreading ridge highly unlikely.

Nisbet & Fowler (1978) postulated an alternative model for crustal formation at slow-spreading ridges, assuming that no permanent sub-axial magma reservoir exists. This they termed the 'infinite leek', partly as a parody of Cann's 'infinite onion' model. In this model magma rises at high levels by a process of crack propagation through the brittle crust, allowing the existence of only very small storage reservoirs at very high levels. Figure 5.28 shows a comparison of the two models, emphasizing their capacity to produce a layered crustal structure. In general, the 'infinite leek' model will produce a poorly layered oceanic crust.

Closed-system fractional crystallization processes may dominate slow-spreading ridges if the magma is stored in very small ephemeral reservoirs. In contrast, open-system fractionation, combined with efficient magma mixing, must dominate the large axial reservoirs beneath fast-spreading ridges. This may partly account for the observed geochemical differences between MORB erupted at slow- and fast-spreading ridge segments (Section 5.10). The 'infinite leek' model may be regarded as a limiting case when the axial magma chamber is vanishingly small. As the spreading rate increases, small magma chambers may develop, eventually becoming large 'infinite onion' type chambers at half-rates in excess of a few centimetres per year.

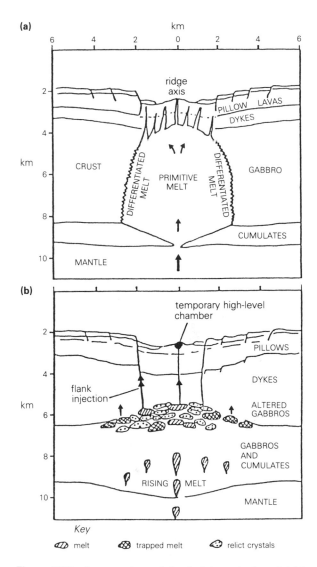

Figure 5.28 A comparison of the 'infinite onion' model (a) for magma chambers beneath fast-spreading ridge segments, and the 'infinite leek' model (b) for magma storage beneath slow-spreading ridge segments (after Nisbet & Fowler 1978, Fig. 6).

5.9 Petrography of mid-ocean ridge basalts

A detailed discussion of the petrography of the entire range of igneous rock types obtained from the oceanic crust by drilling and dredging is clearly beyond the scope of this work. For such detailed information the reader is referred to Hekinian (1982). Instead, attention is focused here on the petrography of mid-ocean ridge basalts (MORB), the most voluminous ocean-floor rocks sampled to date.

The petrographic characteristics of MORB reflect both the chemical composition of the magma and its cooling history. Fabrics reflect rapid cooling of near liquidus temperature magmas extruded into a cold submarine environment. Grain sizes are variable, from glassy to highly porphyritic types with 20–30% phenocrysts. Porphyritic basalts are common, and some highly phyric types are probably accumulative in origin.

The most commonly observed phenocryst assemblages are;

    olivine ± Mg–Cr spinel
    plagioclase + olivine ± Mg–Cr spinel
    plagioclase + olivine + augite

Augite phenocrysts are rare and usually confined to rocks with abundant olivine and plagioclase. Oli-vine, spinel and calcic plagioclase are the first minerals to crystallize, followed by augite and then Fe–Ti oxides (Bender *et al.* 1978, Walker *et al.* 1979, Bryan 1983). The occurrence of olivine as the liquidus phase is consistent with models of MORB petrogenesis involving olivine fractionation from a more picritic primary magma (Section 5.11). Amphibole is exceedingly rare, being observed only in basalts with alkaline affinities and in cumulate gabbros. In the latter it is either a product of late-stage crystallization or hydrothermal altera-tion. Figure 5.29 shows a series of photomicro-graphs illustrating the petrographic variability of MORB. In some instances, the phenocryst miner-als appear highly embayed and are clearly out of equilibrium with the host magma. This lends

Figure 5.29 Photomicrographs illustrating the petrographic variability of MORB: (a) olivine phyric MORB from the FAMOUS area of the Mid-Atlantic Ridge (x40, crossed polars); (b) coarse-grained plagioclase phyric MORB from the FAMOUS area (x40, crossed polars); (c) olivine-clinopyroxene phyric MORB from Reykjanes Ridge (MAR) (x40, crossed polars); (d) fine-grained MORB from the Reykjanes Ridge, showing quench crystals of plagioclase (x100, crossed polars).

support to the importance of magma mixing in the evolution of MORB.

In general, basalts from normal and elevated (hot spot) ridge segments have different petrographic characteristics. In normal MORB, plagioclase is usually the dominant phenocryst phase, accompanied by olivine. Pyroxene is generally absent and bulk rock compositions may be modified by plagioclase and olivine accumulation (Section 5.10). In contrast, MORB from elevated ridge segments include both olivine and pyroxene phyric types. This may reflect both differing magma compositions and crystallization conditions in the two environments (Michael & Chase 1987).

The composition of olivine phenocrysts varies with the host magma composition, ranging from Fo_{73} in ferrobasalts to Fo_{91} in picrites (Fig. 5.30). It is generally euhedral in habit, becoming more anhedral in pyroxene-rich rocks. Moderate zoning of the olivine phenocrysts is common and in many cases the cores are too Mg-rich to be in equilibrium with the bulk rock, attesting to their derivation from a more mafic magma by magma mixing.

A spinel phase (Mg−chromite or Cr−spinel) is common in picritic and olivine-rich basalts, frequently occurring as tiny inclusions within olivine. This is rarely seen in plagioclase-rich basalts. Compositions of this spinel phase vary widely, with Al_2O_3 ranging from 12 to 30 wt.% and Cr_2O_3 from 25 to 45%. Extreme variations often occur within a single sample and are a consequence of the extreme sensitivity of the phase to fO_2 fluctuations (Fisk & Bence 1979).

Plagioclase compositions range from An_{88} to An_{40} and are uniformly orthoclase poor. As with the olivine phenocrysts they are freqently not in equilibrium with the bulk rock, again attesting to the importance of magma mixing processes. In general, there is a negative correlation between the An content of early formed plagioclase and the spreading rate. The fast spreading East Pacific Rise has plagioclase compositions in the range An_{56-88} whereas the slower-spreading Mid-Atlantic Ridge has more calcic plagioclase phenocrysts (An_{75-92}) (Hekinian 1982). This reflects the fact that EPR basalts are more evolved than MAR basalts.

MORB clinopyroxene phenocrysts are colourless to pale green (in thin section) diopsidic augites, of generally very restricted chemical composition, clustering around Wo_{35-40} En_{50} Fs_{10-15} in the pyroxene quadrilateral (Fig. 5.31). Subcalcic augite and Mg−pigeonite are rare.

Gabbroic rocks dredged from the ocean floor show considerable overlap in composition with mid-ocean ridge basalts. Mineralogically, they comprise plagioclase, olivine, clinopyroxene, orthopyroxene and accessory minerals such as sphene, hornblende, apatite and titanomagnetite. Orthopyroxene-bearing gabbros are not common and are usually highly altered. However, their existence is significant as orthopyroxene is not observed as a phenocryst phase in the erupted basalts (Section 5.11). Textures and modal proportions of the above minerals vary widely, as might be expected in rocks of an essentially accumulative origin.

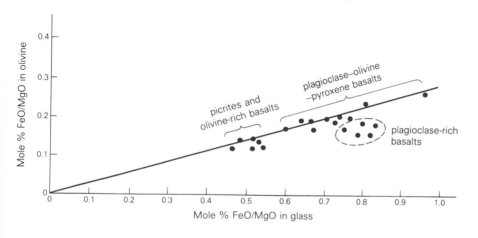

Figure 5.30 Correlation of molecular FeO/MgO ratio in olivine and coexisting basalt glass for MORB from the FAMOUS area of the Mid-Atlantic Ridge (after Hekinian 1982, Fig. 1.15, p. 42).

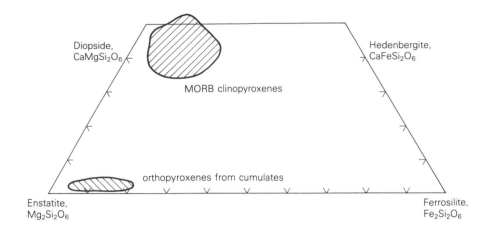

Figure 5.31 MORB pyroxene compositions projected into the pyroxene quadrilateral (after Basaltic Volcanism Study Project 1981, Fig. 1.2.5.5, p. 137).

5.10   Chemical composition of erupted magmas

5.10.1   Characteristic magma series

The majority of mid-ocean ridge basalts (MORB) are sub-alkaline and tholeiitic, according to the classification schemes given in Chapter 1 (see also Fig. 5.32). Alkali and transitional basalts occur only rarely, associated with seamounts, aseismic ridges and fracture zones. In terms of their major element chemistry, MORB are broadly similar to oceanic-island tholeiites (Ch. 9), island-arc tholeiites (Ch. 6) and continental flood tholeiites (Ch. 10) (Table 5.3). However, compared to such basalts they show characteristically low concentra-

tions of incompatible elements, including Ti and P and LIL elements (K, Rb, Ba) (Section 5.10.3). Low K_2O contents appear to be a particularly useful discriminant in distinguishing MORB from basalts erupted in other tectonic settings (Pearce 1976). Hawaiian (ocean-island) tholeiites typically have lower Al_2O_3 contents than MORB at comparable degrees of differentiation, suggesting a difference in the Al_2O_3 contents of the primary magmas. This could be related to differing source compositions, degrees of partial melting and residual source mineralogies or to the high-pressure fractionation of an Al-rich mineral in the evolution of oceanic-island tholeiites (Ch. 9). Available data on the geochemistry of MORB are strongly biased towards the Atlantic Ocean. Much of the discussion in the

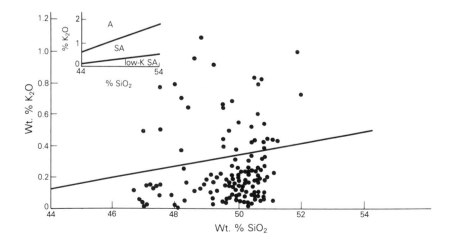

Figure 5.32 The variation of wt% K_2O versus wt% SiO_2 for basalts from 29°N to 73°N along the Mid-Atlantic Ridge. Most of the basalts fall in the low-K subalkalic field, although basalts from elevated ridge segments are richer in K_2O (data from Schilling *et al.* 1983). Boundaries between the alkalic (A), sub-alkalic (SA) and low-K subalkalic (low-K SA) fields from Middlemost (1975).

Table 5.3 Comparison of the major element goechemistry of MORB with that of a typical oceanic-island tholeiite, island-arc ᐟ tholeiite and continental flood tholeiite.

	MORB[a]			OIT[b]	IAT[c]	CFT[b]
	MAR	EPR	IOR			
SiO_2	50.68	50.19	50.93	50.51	51.90	50.01
TiO_2	1.49	1.77	1.19	2.63	0.80	1.00
Al_2O_3	15.60	14.86	15.15	13.45	16.00	17.08
FeO	9.85	11.33	10.32	9.59	9.56	10.01
Fe_2O_3	—	—	—	1.78	—	—
MnO	—	—	—	0.17	0.17	0.14
MgO	7.69	7.10	7.69	7.41	6.77	7.84
CaO	11.44	11.44	11.84	11.18	11.80	11.01
Na_2O	2.66	2.66	2.32	2.28	2.42	2.44
K_2O	0.17	0.16	0.14	0.49	0.44	0.27
P_2O_5	0.12	0.14	0.10	0.28	0.11	0.19

Data sources: [a] Melson *et al.* (1976); [b] Basaltic Volcanism Study Project (1981), Tables 1.2.6.2 & 1.2.3.2; [c] Jakes & White (1972).

MAR, Mid-Atlantic Ridge; EPR, East Pacific Rise; IOR Indian Ocean Ridge; OIT, oceanic-island tholeiite; IAT, island-arc tholeiite; CFT, continental flood tholeiite.

following sections thus relies heavily on Atlantic data sets, comparison with Pacific and Indian ocean MORB being made whenever possible.

5.10.2 Major elements

Early studies of MORB emphasized their compositional uniformity, but since the 1970s significant geochemical variations have been observed, suggesting that a variety of magmatic processes and a heterogeneous mantle source region are involved in their genesis (Le Roex *et al.* 1983, Schilling *et al.* 1983, Le Roex 1987).

SiO_2, TiO_2, Al_2O_3, Fe_2O_3, FeO, MnO, MgO, CaO, Na_2O, K_2O, P_2O_5 and H_2O can all be considered as major element oxides in the description of MORB geochemistry. For most mid-oceanic ridge segments, SiO_2 shows a remarkably narrow range of variation, from 47 to 51%, and therefore it cannot be used successfully as an index of differentiation. Only in propagating rift segments (e.g. Galapagos), transform fault zones and anomalous ridge segments (e.g. Iceland) does extensive fractionation produce more silicic differentiates.

Table 5.3 shows a representative range of MORB glass compositions from the Atlantic, Pacific and Indian oceans. When comparing analyses of MORB from different provinces it is important to use glass compositions whenever possible, as bulk rock analyses can differ from the original magmatic liquid compositions due to the accumulation of olivine and plagioclase. There appear to be no significant inter-ocean differences in major element chemistry, apart from a tendency for Pacific MORB to be slightly more Fe- and Ti-rich.

As SiO_2 is unsuitable as an index of differentiation, MgO content or M value ($100\ Mg/(Mg + Fe^{2+})$) is used instead to illustrate differentiation from primitive to more evolved compositions. Figure 5.33 shows the frequency of occurrence of M values for MORB glasses from the major ocean basins. Although there is a wide range of values there is a very obvious maximum in the data between 55 and 65. A value of 70 defines a basaltic magma in equilibrium with mantle olivine (Section 2.4), and thus the diagram reveals the rarity of primitive glass compositions amongst the spectrum of erupted MORB. Relatively fractionated magma compositions appear to be dominant, indicating that the primary MORB magmas must have

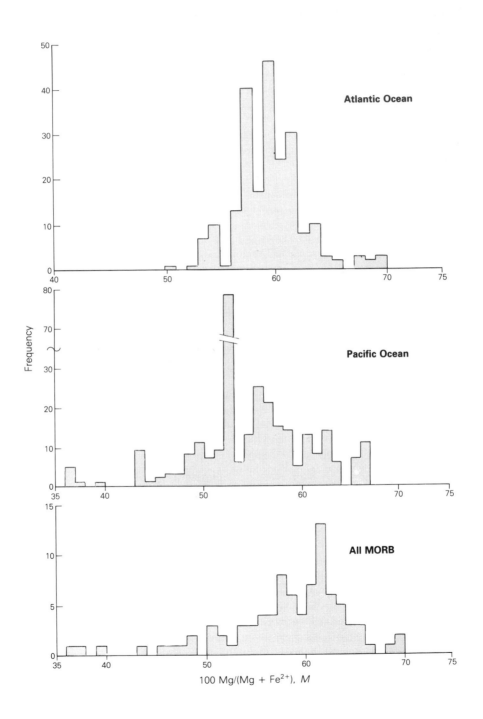

Figure 5.33 Frequency of occurrence of M values for MORB glasses from the major ocean basins: $M = 100Mg/(Mg + Fe^{2+})$ (after Wilkinson 1982, Figs 1–3).

undergone high-level fractionation after segregation from their mantle source.

Volcanics erupted along the strike of mid-ocean ridges can show important compositional changes which appear to correlate with topographic and structural features that characterize particular ridge segments (Hekinian 1982, Klein & Langmuir 1987). Topographic highs and volcanic platforms appear to be associated with 'hot spots' and have positive free-air gravity anomalies, high geothermal gradients and crustal thicknesses intermediate between oceanic and continental values, in addition to

erupting MORB of a rather distinctive trace element geochemistry (Section 5.10.3). On this basis MORB have been classified as normal (N-type, depleted), plume (P-type, enriched) and transitional (T-type) (Bryan *et al.* 1976, Sun *et al.* 1979, Schilling *et al.* 1983). However, despite significant variations in the trace element concentrations of basalts erupted along these different types of ridge segment, major element compositions remain remarkably uniform (Table 5.4). N-type basalts are recovered mostly from the Pacific and from the Atlantic south of 30°N, whereas P-type basalts

Table 5.4 Major (a) and trace element (b) geochemistry of average primitive (M= 60−70) normal, plume and transitional type MORB from the Mid-Atlantic Ridge (data from Schilling *et al.* 1983, Table 3).

	Normal MORB		Plume MORB		Transitional MORB	
	28−34°N	49−52°N	Azores	Iceland	34−38°N	61−63°N
(a)						
SiO_2	48.77	50.55	49.72	47.74	50.30	49.29
Al_2O_3	15.90	16.38	15.81	15.12	15.31	14.69
Fe_2O_3	1.33	1.27	1.66	2.31	1.69	1.84
FeO	8.62	7.76	7.62	9.74	8.23	9.11
MgO	9.67	7.80	7.90	8.99	7.79	9.09
CaO	11.16	11.62	11.84	11.61	12.12	12.17
Na_2O	2.43	2.79	2.35	2.04	2.24	1.93
K_2O	0.08	0.09	0.50	0.19	0.20	0.09
TiO_2	1.15	1.31	1.46	1.59	1.21	1.08
P_2O_5	0.09	0.13	0.22	0.18	0.14	0.12
MnO	0.17	0.16	0.16	0.20	0.17	0.19
H_2O	0.30	0.29	0.42	0.42	0.26	0.31
M value	66.5	64.1	64.9	62.2	62.8	63.9
(b)						
La	2.10	2.73	13.39	6.55	5.37	2.91
Sm	2.74	3.23	3.93	3.56	3.02	2.36
Eu	1.06	1.12	1.30	1.29	1.07	0.92
Yb	3.20	3.01	2.37	2.31	2.91	2.33
K	691	822	4443	1179	1559	572
Rb	0.56	0.96	9.57	2.35	3.50	1.02
Cs	0.007	0.012	0.123	0.025	0.042	0.013
Sr	88.7	106.4	243.6	152.5	95.9	86.0
Ba	4.2	10.7	149.6	36.0	39.8	14.3
Sc	40.02	36.47	36.15	39.49	42.59	41.04
V	262	257	250	320	281	309
Cr	528	278	318	330	383	374
Co	49.78	40.97	44.78	57.73	45.70	54.94
Ni	214	132	104	143	94	146
$(La/Sm)_N$	0.50	0.60	2.29	1.28	1.27	0.85
K/Rb	1547	869	475	498	465	560

come mostly from the Atlantic north of 30°N and from the Galapagos spreading centre.

The major element chemistry of suites of volcanic rocks is commonly used to model genetic relationships in terms of fractional crystallization processes. Several authors (Bryan *et al.* 1976, Bender *et al.* 1978, Rhodes & Dungan 1979) have pointed out the importance of cotectic crystallization of olivine and plagioclase at low pressures in controlling the bulk rock chemistry of MORB. This is clearly demonstrated in Figures 5.34 and 5.35, in which Atlantic MORB data exhibit a marked covariance between Al_2O_3 and MgO and TiO_2 and MgO, MgO being used as an index of differentiation. There is obviously a certain amount of scatter in the data attributable to phenocryst accumulation and magma mixing processes, but nevertheless the trends are clear. On the basis of such Harker diagrams the compositional variability of Atlantic MORB can be interpreted in terms of olivine plus plagioclase fractionation from primitive magmas with 10−11% MgO and 16% Al_2O_3.

Figure 5.35 shows the dominance of plagioclase over olivine in the fractionating assemblage, and both diagrams demonstrate the non-involvement of clinopyroxene as a major fractionating phase.

There is apparently a greater compositional variability of basalts erupted along the East Pacific Rise relative to those erupted along the Mid-Atlantic Ridge (Natland 1978), suggesting that Pacific MORB are in general more evolved. This has been attributed to the higher spreading rates along the EPR and the consequent existence of larger sub-axial magma reservoirs in which low-pressure fractional crystallization of magmas can occur.

One of the diagnostic characteristics of the tholeiitic magma series as defined in Chapter 1 is the marked trend of iron enrichment in the early stages of fractionation. Figure 5.36 shows the variation of FeO + Fe_2O_3 versus MgO for basalt data from 29 to 73°N along the MAR (Schilling *et al.* 1983). These data clearly show a pronounced trend of progressive iron enrichment with frac-

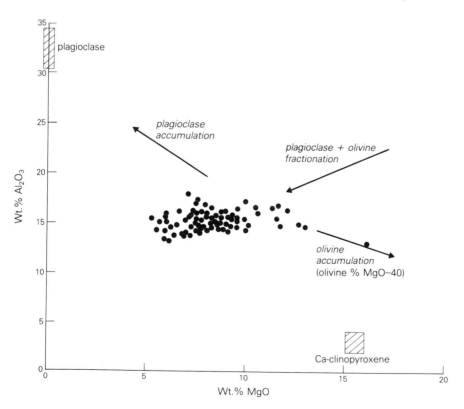

Figure 5.34 The variation of wt.% Al_2O_3 versus wt.% MgO for basalts from 29°N to 73°N along the Mid-Atlantic Ridge. The data can clearly be explained in terms of olivine + plagioclase fractionation. Clinopyroxene does not appear to be involved (data from Schilling *et al.* 1983).

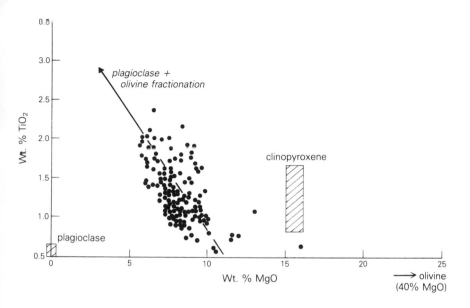

Figure 5.35 Variation of wt.% TiO_2 versus wt.% MgO for basalts from 29°N to 73°N along the Mid-Atlantic Ridge. The data array demonstrates the importance of plagioclase + olivine fractionation. Clinopyroxene is not a major fractionating phase (data from Schilling *et al.* 1983).

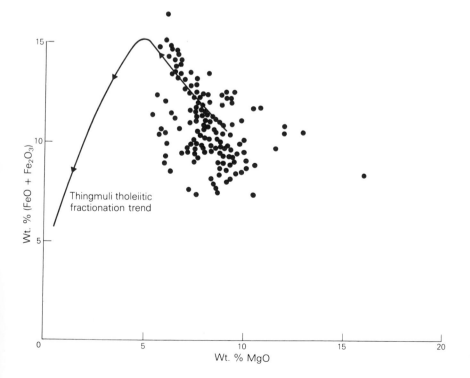

Figure 5.36 The variation of wt.% (FeO + Fe_2O_3) versus wt.% MgO for basalts from 29°N to 73°N along the Mid-Atlantic Ridge. Shown for comparison is the characteristic tholeiitic differentiation trend of increasing iron enrichment in the early stages of fractionation displayed by the volcanics of Thingmuli, Iceland (Carmichael 1964). Data from Schilling *et al.* (1983).

tionation, although there is considerable scatter. Shown for comparison is the differentiation trend for the Iceland tholeiitic volcano, Thingmuli (Carmichael 1964), which shows the complete evolutionary spectrum from basalt to rhyolite.

5.10.3 Trace elements

Large low-valency cations

Most MORB are depleted in large low-valency cations (Cs, Rb, K, Ba, Pb and Sr) relative to oceanic island and continental tholeiites (Basaltic Volcanism Study Project 1981). Additionally, for basalts erupted along topographically normal ridge segments, the larger ions are depleted to a greater extent than the smaller ions such that element ratios K/Rb, K/Ba and Sr/Rb are characteristically higher for MORB than for tholeiitic basalts generated in other tectonic environments (Table 5.5). Basalts erupted along elevated ridge sections adjacent to volcanic platforms associated with oceanic islands (e.g. Iceland, Azores and Galapagos) typically contain higher abundances of large cations, with the exception of Sr, and have much lower small/large cation ratios (Table 5.5). In this respect they have much closer affinities with oceanic island tholeiites.

With the exception of Sr, which is partitioned into plagioclase, most of the large low-valency cations are incompatible. Thus their abundance ratios should be essentially independent of the source mineralogy, the degree of partial melting and the extent of high-level fractional crystallization. The ratios of these elements in MORB should therefore reflect the ratios in their mantle source. The marked differences in K/Rb and K/Ba between normal and plume-type MORB shown in Table 5.5 must reflect significant differences in their source compositions, at least with respect to trace elements. However, it must be remembered that this group of elements is particularly susceptible to seawater alteration, and therefore only analyses of fresh glassy basalts should be used for comparative purposes.

Large high-valency cations

The group of large high-valency cations (Th, U, Zr,

Table 5.5 Typical large cation abundances (ppm) and ratios for oceanic tholeiites (Basaltic Volcanism Study Project 1981, Table 1.2.5.4,p. 144).

	MORB		Ocean-island tholeiite
	Normal	Plume	
K	1064	1854	1600−8300
Rb	1.0	4.5	5−12
Ba	12.2	55	70−200
Sr	127	105	150−400
K/Rb	1046	414	400
K/Ba	109	34	25−40
Sr/Rb	127	23	20−70

Hf, Nb and Ta) are called 'immobile elements', and these have been widely used in conjunction with other alteration-resistant elements (Ti, Y, P and Sr) to discriminate amongst basalts from different tectonic settings (Ch. 2). These elements tend to be depleted in N-type MORB relative to P-type and oceanic-island tholeiites (Basaltic Volcanism Study Project 1981). The Zr/Nb ratio serves as a particularly useful discriminant; N−type MORB have high ratios (>30) whereas P-type MORB have low ratios (∼10), similar to oceanic-island tholeiites. Zr/Nb ratios have been used to investigate regional heterogeneities in basalts erupted along the strike of the Mid-Atlantic Ridge (Wood et al. 1979).

Ferromagnesian elements (Cr, V, Sc, Ni and Co)

Crystal−liquid distribution coefficient data indicate that Ni and Co will partition into olivine during partial melting and fractional crystallization processes, while Sc, Cr and V will enter clinopyroxene. Thus the abundances of these elements should be useful indicators of petrogenetic processes. Despite controversy concerning the choice of appropriate olivine/basaltic melt partition coefficients, there is little doubt that Ni abundances in MORB are strongly controlled by olivine fractionation. Ni contents range from >300 ppm in primitive glassy basalts to 25 ppm in highly evolved basalts, and correlate well with MgO content (Fig. 5.37). Cr contents similarly show a marked reduction from 700 to 100 ppm with progressive fractionation. However, this is not considered to reflect significant clinopyroxene fractionation (which has already been negated on major element grounds) but the simultaneous crystallization of olivine- and Cr-rich spinel (Basaltic Volcanism Study Project 1981).

Rare earth elements

Figure 5.38 shows the range of chondrite-normalized REE patterns displayed by MORB.

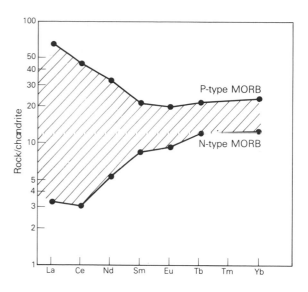

Figure 5.38 The range of chondrite-normalized REE patterns displayed by basalts from normal and plume ridge segments along the Mid-Atlantic Ridge. P-type MORB from 71° 33'N, $(La/Sm)_N = 3.04$; N-type MORB from 66° 51'N, $(La/Sm)_N = 0.4$ (data from Schilling *et al.* 1983).

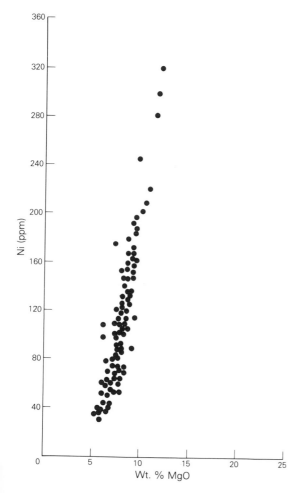

Figure 5.37 The variation of Ni content (ppm) versus wt.% MgO for basalts from 29°N to 73°N along the Mid-Atlantic Ridge. Ni content is clearly controlled by olivine fractionation (data from Schilling *et al.* 1983).

Typical N-type MORB have unfractionated heavy REE abundances and are strongly depleted in light REE. Primitive basalts have REE concentrations of 10× chondrite or less, whereas extremely differentiated basalts may contain up to 50× chondrite. Fractional crystallization involving olivine, plagioclase, clinopyroxene and spinel increases the total REE content of more evolved MORB, but does not produce any significant inter-element fractionations. Thus the characteristic shape of the primary basalt REE pattern will be maintained in the more evolved basalts. There is, however, a tendency for a negative Eu anomaly to develop as fractionation proceeds, because Eu is preferentially partitioned into plagioclase. In contrast, P-type MORB show relatively little tendency for light-REE depletion and in some instances are light-REE enriched. Generally, N-type MORB have $(La/Sm_N < 1$ whereas P-type have $(La/Sm)_N > 1$.

If partial melting is fairly extensive (>10%) the REE should not be fractionated from each other during partial melting and therefore ratios of REE (e.g. La/Sm, La/Yb and La/Ce) should reflect the ratios in the mantle source of the magmas. However, only the very light REE are truly incompatible and thus, of the above ratios, only La/Ce is likely to be diagnostic of source composition. Figure 5.39 shows the variation of $(La/Sm)_N$ with latitude along the Mid-Atlantic Ridge (Schilling *et al.* 1983). This shows that the division into N- and P-type MORB is clearly an artificial one. Lavas with the highest $(La/Sm)_N$, and hence the greatest degree of light-REE enrichment, are associated with the volcanic platform areas of the Azores, Iceland and Jan Mayen. Figure 5.40 shows that there is a good correlation between $(La/Sm)_N$ and the Zr/Nb ratio for Atlantic, Pacific and Indian ocean MORB, suggesting that binary mixing of end-member source components may be significant in determining their geochemical characteristics. Figure 5.41 shows the variation of La versus Ce for basalts sampled along a traverse along the Mid-Atlantic Ridge from 29 to 73°N. Also plotted for

comparison are the data of Humphris & Thompson (1983) for the Walvis Ridge. These data define a remarkably coherent trend which must reflect the La/Ce ratio of the Atlantic MORB source mantle.

Spiderdiagrams

Following Sun (1980), Figure 5.42 compares incompatible element abundances, normalized to primordial mantle values, for N- and P-type MORB and a typical oceanic-island tholeiite. The elements are ordered in a sequence of decreasing incompatibility, from the left to right, in a four-phase lherzolite undergoing partial fusion. Assuming that MORB are indeed produced by comparatively large degrees of partial melting, then their relative incompatible element abundances should be similar to those of their source. Partial melting of a chondritic mantle will produce magmas with spiderdiagram patterns variably enriched in the elements Rb to Nd, depending upon the degree of

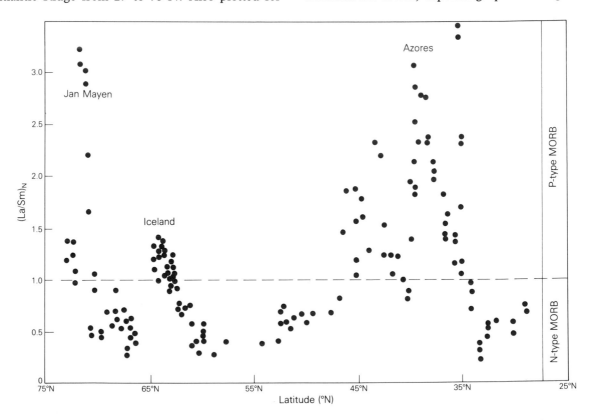

Figure 5.39 Variation in $(La/Sm)_N$ with latitude along the Mid-Atlantic Ridge (data from Schilling *et al.* 1983).

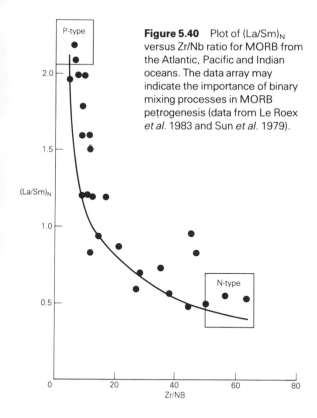

Figure 5.40 Plot of $(La/Sm)_N$ versus Zr/Nb ratio for MORB from the Atlantic, Pacific and Indian oceans. The data array may indicate the importance of binary mixing processes in MORB petrogenesis (data from Le Roex *et al.* 1983 and Sun *et al.* 1979).

melting. Thus the very obvious depletions in the most incompatible elements in N-type MORB must reflect similar depletions in their source mantle. This may be a long term phenomenon related to the continued extraction of continental crustal materials from the upper mantle throughout geological time. In contrast, MORB from plume ridge segments have a distinctive pattern enriched in the most incompatible elements, similar to oceanic-island tholeiites. This would seem to suggest that the sources of P-type MORB and oceanic-island tholeiites have similar geochemical characteristics. The implications of this for mantle dynamics will be considered further in Section 5.11.

5.10.4   Volatile contents

Estimates of the proportions of juvenile volatiles in any basalt are always subject to a certain degree of uncertainty due to the effects of degassing. Moore (1965) showed that vesicle formation and gas release in ocean-floor basalts is directly related to the depth of water in which extrusion takes place. Degassing appears to be inhibited by the hydro-static pressure of the water at depths greater than

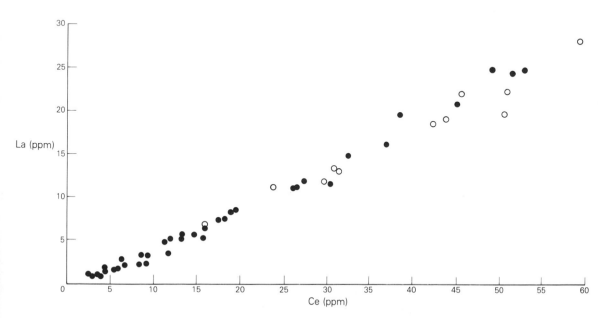

Figure 5.41   Graph of La (ppm) versus Ce (ppm) for basalts from 29°N to 73°N along the Mid-Atlantic Ridge (closed circles) and the Walvis Ridge (open circles) (data from Schilling *et al.* 1983 and Humphris & Thompson 1983).

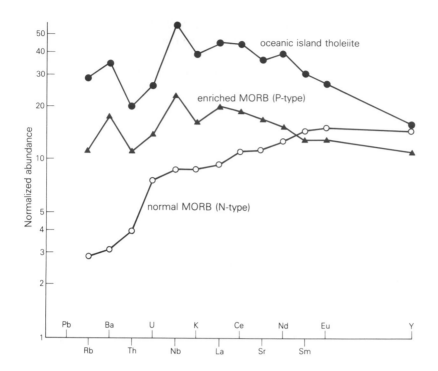

Figure 5.42 Spiderdiagrams showing the difference between normal and P-type MORB, and the similarity between P-type MORB and oceanic-island tholeiites (data from Sun *et al.* 1979).

200 m and therefore most MORB should retain near-primary volatile component characteristics. Concentrations of $0.2-1.0$ wt.% H_2O appear to be typical and correlate broadly with K_2O content. Byers *et al.*(1983), in a study of basalts and andesites from the Galapagos spreading centre, have shown that abundances of H_2O, Cl and F in rapidly quenched glasses increase progressively with fractionation (Table 5.6).

5.10.5   Radiogenic isotopes

Sr, Nd and Pb

Geochemical studies of MORB have resulted in their characterization as a remarkably coherent group of oceanic basalts, displaying only minor petrological and geochemical variations along and between different mid-oceanic ridge spreading centres when compared to the spectrum of oceanic-

Table 5.6   Volatile contents (wt.%) in the glassy rims of basalts and andesites from the Galapagos spreading centre (Byers *et al.* 1983).

M-value	Total volatiles	H_2O	CO_2	Cl	F
55	0.31	0.13	0.09	0.04	0.01
52	0.58	0.33	0.10	0.10	0.02
51	0.65	0.20	0.16	0.11	0.02
50	0.52	0.27	0.07	0.13	0.01
35	1.12	0.77	0.07	0.22	0.04
33	1.41	0.87	0.15	0.34	0.02
24	1.50	0.94	0.11	0.34	0.09
16	1.90	1.27	0.10	0.37	0.13

island basalts (Ch. 9). Nevertheless, MORB do show a significant range in $^{87}Sr/^{86}Sr$, $^{208}Pb/^{204}Pb$, $^{207}Pb/^{204}Pb$, $^{206}Pb/^{204}Pb$ and $^{143}Nd/^{144}Nd$ isotopic ratios, indicating that they are in fact derived from a heterogeneous mantle source. Variations in isotopic and trace element geochemistry have been variously attributed to the proximity of nearby hot spots, variations in magma-chamber dynamics or melt generation processes, fracture zone effects and large-scale mantle heterogeneities (Cohen & O'Nions 1982a, Zindler *et al.* 1982, Dupré & Allègre 1983, Allègre *et al.* 1984, Hamelin *et al.* 1984, Schilling 1985).

Some of the spread in the observed $^{87}Sr/^{86}Sr$ ratios may be attributable to seawater-alteration effects, but the Nd and Pb isotopic ratios are unlikely to be modified significantly. In general, N-type MORB have very restricted $^{87}Sr/^{86}Sr$ in the range 0.7024–0.7030, whereas P-type MORB are characterized by slightly more radiogenic compositions (0.7030–0.7035), overlapping with the oceanic-island basalt range (0.7030–0.7050).

Figures 5.43, 5.44 and 5.45 are plots of $^{143}Nd/^{144}Nd$ versus $^{87}Sr/^{86}Sr$, $^{207}Pb/^{204}Pb$ versus $^{206}Pb/^{204}Pb$ and $^{87}Sr/^{86}Sr$ versus $^{206}Pb/^{204}Pb$ for MORB

from the Atlantic, Pacific and Indian oceans compared to the spectrum of oceanic-island basalts (OIB). The Nd-Sr diagram (Fig. 5.43) clearly shows the restricted range in isotopic composition of MORB when compared to the entire oceanic basalt data array. Simplistic interpretations of the Nd–Sr 'mantle array' involve binary mixing between depleted and enriched regions of the Earth's mantle (Cohen & O'Nions 1982a), with MORB source mantle representing the depleted end-member. However, interpretation of the combined Pb–Sr–Nd data arrays requires more complex models (Dupré & Allègre 1980, Zindler *et al.* 1982). The MORB-source mantle has clearly been depleted in Nd with respect to Sm and Rb with respect to Sr over a large part of Earth history (>1Ga), and has conventionally been regarded as the geochemical complement of the incompatible enriched upper crust.

Hamelin *et al.* (1984), in a detailed study of the Pb–Sr isotopic variations in Atlantic and Pacific MORB, revealed good positive correlations between $^{207}Pb/^{204}Pb$ versus $^{206}Pb/^{204}Pb$ and $^{87}Sr/^{86}Sr$ versus $^{206}Pb/^{204}Pb$ (Figs 5.44 & 45). The trend on the Pb–Pb plot is subparallel to the oceanic basalt

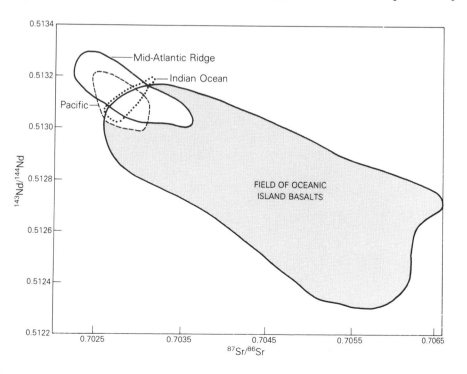

Figure 5.43 $^{143}Nd/^{144}Nd$ versus $^{87}Sr/^{86}Sr$ for MORB from the Atlantic, Pacific and Indian oceans, compared to the spectrum of oceanic-island basalts (after Staudigel *et al.* 1984, Fig. 4).

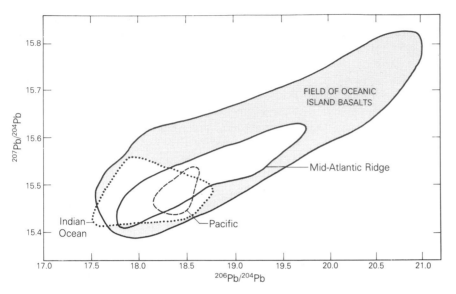

Figure 5.44 $^{207}Pb/^{204}Pb$ versus $^{206}Pb/^{204}Pb$ for MORB from the Atlantic, Pacific and Indian oceans compared to the spectrum of oceanic-island basalts (after Staudigel *et al.* 1984, Fig. 5).

data array, whereas the Sr–Pb MORB trend is mostly at variance with the oceanic-island basalt spectrum. The precise interpretation of these trends is still somewhat equivocal but they must represent the mixing of isotopically distinct components in the source region of MORB. This will be considered further in Ch. 9.

White & Schilling (1978) documented systematic variations in $^{87}Sr/^{86}Sr$ with latitude along the axis of

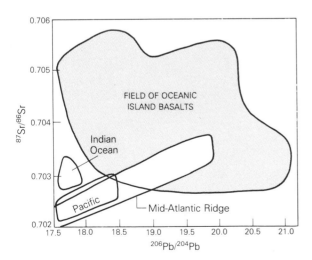

Figure 5.45 $^{87}Sr/^{86}Sr$ versus $^{206}Pb/^{204}Pb$ for MORB from the Atlantic, Pacific and Indian oceans compared to the spectrum of oceanic basalts (after Staudigel *et al.* 1984, Fig. 7; and Dupré & Allègre 1983).

the northern Mid-Atlantic Ridge which correlate well with topographic and other geochemical anomalies (Fig. 5.46). Such along-strike variations in isotopic characteristics can be explained in terms of mixing between an isotopically fairly homogeneous depleted upper mantle reservoir (the source of N-type MORB) and blobs of an isotopically heterogeneous more radiogenic mantle reservoir (the source of OIB and P-type MORB) injected at hotspot locations along the ridge axis. Detailed discussion of the geochemical characteristics of this latter reservoir will be deferred until Chapter 9. The isotopic heterogeneity of a particular ridge segment will depend upon internal blob heterogeneity, the proportions of blobs and depleted asthenosphere in the mixture and the efficiency of the mixing process. The 'blob' component appears to show distinct regional characteristics on a global scale, as evidenced by comparative studies of oceanic-island basalts which are inferred to sample blob material less diluted by MORB-source upper mantle. For example, OIB from the South Atlantic and Indian oceans are quite different isotopically from those from the North Atlantic and Eastern Pacific (Dupré & Allègre 1983, Hart 1984).

In general, Pacific MORB appear to show a much narrower range in Sr, Nd and Pb isotopic compositions than Atlantic MORB (White *et al.* 1987). This might be attributable to a greater

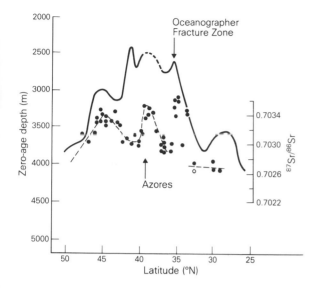

Figure 5.46 Superposition of the zero-age depth curve and variation in $^{87}Sr/^{86}Sr$ with latitude along the northern Mid-Atlantic Ridge (after Le Douaran & Francheteau 1981; $^{87}Sr/^{86}Sr$ data from White & Schilling 1978).

isotopic heterogeneity in the Atlantic MORB source compared to the Pacific. However, it seems more reasonable to postulate similar degrees of isotopic heterogeneity in all MORB source regions, with the greater degrees of melt production associated with the fast-spreading East Pacific Rise, effectively homogenizing the isotopic compositions.

5.11   Detailed petrogenetic model

The most primitive basalt compositions recorded amongst the MORB spectrum have 10% MgO, M values of 70, Ni contents of 300 ppm and highly magnesian olivine phenocrysts (Fo_{90-91}). Magmas with such chemical characteristics could be primary mantle partial melts (Bender *et al.* 1978) and thus there is no fundamental necessity to propose more picritic primary magma compositions. However, O'Hara (1968, 1973, 1982) has consistently argued for picritic primary magmas, which undergo extensive olivine fractionation en route to the surface, in his models of MORB petrogenesis. The high-MgO primary magma theory is supported by lherzolite melting experiments and by petrological studies of

ophiolite complexes. However, it is inconsistent with the observation that no glasses or aphyric rocks with MgO contents greater than 11% have ever been sampled from the ocean floor. Huppert & Sparks (1980) support a picritic primary magma hypothesis and explain the lack of high-MgO erupted basalts as a consequence of the fluid dynamics of such high-density liquids.

The vast majority of MORB samples are, however, more fractionated (evolved), attesting to the importance of olivine + plagioclase ± clinopyroxene fractionation in producing their observed geochemical characteristics. O'Hara (1977) has stressed the importance of magma mixing and fractional crystallization in high-level open-system magma chambers in the production of magma batches whose composition corresponds to some 'perched state'. His model explains the rarity of both primitive basalts and highly evolved differentiates in the MORB spectrum. Mixing and fractionation processes will obviously be enhanced in large sub-axial magma reservoirs beneath fast-spreading ridge segments, perhaps explaining why Pacific MORB appear more evolved than Atlantic MORB.

Two extreme types of basalt are erupted along mid-oceanic ridges:

(1) *Normal (N–type)*. LREE and incompatible element depleted. High K/Ba, K/Rb, Zr/Nb and low $^{87}Sr/^{86}Sr$.

(2) *Plume (P-type)*. Less depleted than N-type in LREE and incompatible elements with higher $^{87}Sr/^{86}Sr$ ratios. K/Ba, K/Rb, La/Ce and Zr/Nb ratios are lower than those of N-type MORB and comparable to those of oceanic-island tholeiites.

A continuous spectrum of intermediate types exists between these two end-members. N-type MORB appear to be derived from a depleted asthenospheric upper-mantle source, whereas P-type MORB are derived from a more enriched plume or hotspot component. Source heterogeneity is thus an important parameter in MORB petrogenesis combined with fractional crystallization, magma mixing, variable degrees of partial melting and variable residual source mineralogies.

Mg-rich MORB are multiply saturated with olivine + clinopyroxene + orthopyroxene at pressures in excess of 8−10 kbar (Bender *et al.* 1978, Green *et al.* 1979, Stolper 1980), corresponding to minimum depths of segregation from the mantle of 25−30 km. Thus MORB parent magmas should have last equilibrated within the spinel lherzolite stability field, which is consistent with their observed trace element geochemistry. The primary magmas clearly evolve by polybaric partial melting processes in ascending mantle diapirs, commencing at considerably greater depths, perhaps as much as 60 km. However, it is the last point of equilibration within the mantle before segregation which constrains the geochemical characteristics of these magmas.

Sr, Nd and Pb isotopic studies of MORB (Section 5.10.5) have revealed important source heterogeneities which can be explained in terms of present-day mixing processes beneath the ridge axis between depleted mantle material from the asthenosphere and blobs of hotspot or plume material coming from deeper levels (Allègre *et al.* 1984, Dupré & Allègre 1983). Such mixing processes can explain the observed negative correlation between Nd and Sr isotopic ratios, and the positive correlation between Sr and Pb isotopic ratios for Atlantic MORB. Figure 5.47 is a cartoon depicting such a mixing model. N-type MORB are derived by partial melting of the isotopically fairly homogeneous, well mixed, depleted upper-mantle

(a) N-type MORB

(b) P-type MORB

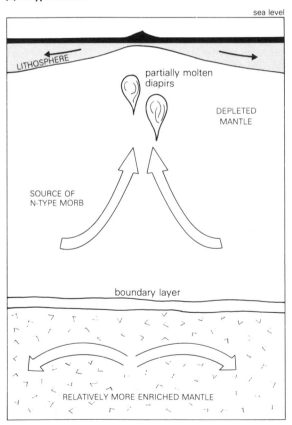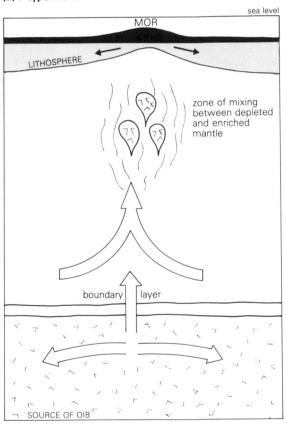

Figure 5.47 Models for the involvement of different source components in the origin of N- and P-type MORB. (a) N-type MORB are derived by partial melting of an isotopically fairly homogeneous, well mixed, depleted upper mantle reservoir. (b) P-type MORB are derived from sources containing variable amounts of a 'blob' component, derived from a lower isotopically heterogeneous reservoir mixed with the depleted N-type MORB source; this lower reservoir is also the source of OIB. (After Zindler *et al.* 1984, Fig, 12).

reservoir, whereas P-type MORB contain variable amounts of a blob component from the underlying isotopically heterogeneous reservoir, which is also the source of oceanic-island basalts (OIB). Parts of this lower reservoir may be almost primordial in composition, whereas other parts may represent recycled subducted slab components (Ch. 9).

The idea of hotspot injection of material beneath mid-ocean ridge axes originated in studies of the northern Mid-Atlantic Ridge by Schilling (1973). In the original model of J. T. Wilson (1973) hotspots were considered to be fixed, more or less continous, flows ascending from the lower mantle. However, the model shown in Figure 5.47 is more realistic, involving the ascent of discontinuous blobs from the lower-mantle source. On slow-

spreading ridges (e.g. the Mid-Atlantic Ridge) the hotspot signature is clearly visible in both bathymetry and geochemical characteristics, whereas on fast-spreading ridges (e.g. the East Pacific Rise) it may be diluted by the rapid supply of asthenospheric material. In the model, blobs reach high levels more or less efficiently mixed with asthenospheric material. Such blobs may be expected to be larger beneath slow-spreading ridges, and smaller and more numerous beneath fast-spreading ridges (Allègre et al. 1984; see also Figure 5.48).

Many studies have shown a correlation between mid-ocean ridge bathymetry and MORB isotopic and trace element geochemistry (Hart et al. 1973, White & Schilling 1978, Dupré & Allègre 1980, le Douaran & Francheteau 1981, Hamelin et al. 1984, Humphris et al. 1985, Hanan et al. 1986). Ridge-

(a) Slow-spreading regime

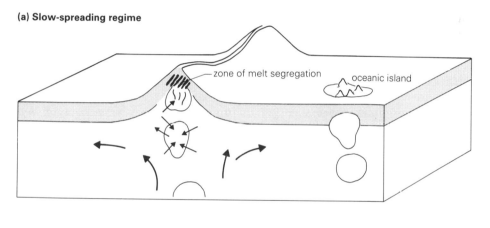

(b) Fast-spreading regime

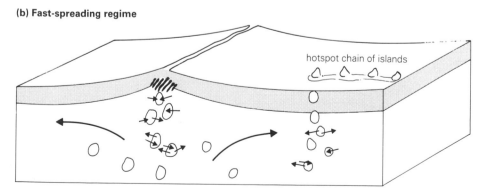

Figure 5.48 Injection of blobs of a lower more enriched mantle reservoir beneath the axis of (a) slow- and (b) fast-spreading mid-ocean ridges (after Allègre et al. 1984, Fig. 7).

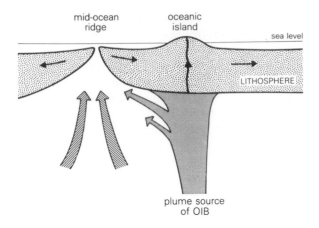

Figure 5.49 Migration of a mid-ocean ridge axis away from a hotspot, inducing a non-radial flow in the rising plume towards the ridge (after Schilling *et al.* 1985, Fig. 4).

centred hotspots appear to produce large-scale isotopic and trace elements gradients, unusually intense constructional volcanism and elevation anomalies (Schilling *et al.* 1985). If the ridge axis subsequently drifts away from the hot spot then the rising plume tends to develop a preferential non-radial flow towards the migrating ridge axis (Fig. 5.49). In such a case lava compositions are still gradational along the ridge axis, symmetrical about

the hotspot and the connecting channel, but the gradients are not as pronounced and the elevation anomaly is more subdued. Eventually, with continued migration of the ridge axis, the plume supply will be cut off and will cease to influence axial magmatic processes.

Figure 5.50 is a schematic illustration of the hypabyssal environment existing at the axis of a mid-ocean ridge spreading sufficently rapidly to maintain a dynamic sub-axial magma chamber of moderate dimensions. Here magmas may pond, fractionate, mix and precipitate mafic and ultramafic cumulates before erupting. The primary magma entering this plumbing system is either a picrite or a highly magnesian basalt. Initally, this may remain as a high-density layer at the base of the chamber underlying more fractionated (less dense) chamber magma until fractionation proceeds to the point of chamber rollover, and homogenization of the two layers occurs. If the chamber is horizontally stratified in this way then the roughness of the fault-block-controlled chamber roof can lead to the sampling of more primitive magmas at the centre of the rift and more differentiated magmas at the margins. This has been observed in the FAMOUS area of the Mid-Atlantic Ridge. Lavas emerging from such a periodically replenished, periodically

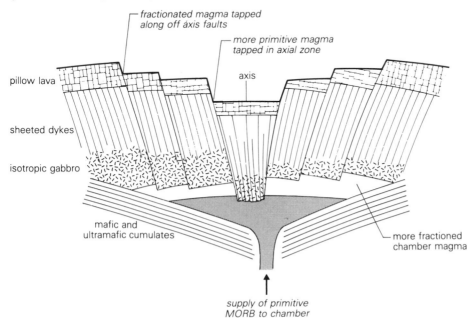

Figure 5.50 Hypabyssal environment at the axis of a mid-ocean ridge that is spreading sufficiently rapidly to maintain a dynamic sub-axial magma chamber.

tapped, continuously fractionating magma chamber may be far evolved from their parent magmas, displaying relatively constant major element compositions, but large variations in concentrations and concentration ratios of incompatible trace elements (O'Hara 1982). The chemical characteristics of such magmas obviously cannot be uniquely inverted to elucidate the nature of the parent magma and the geochemical characteristics of the source mantle. Thus, although MORB have conventionally been regarded as a 'window to the upper mantle', that window is by no means transparent.

Further reading

Basaltic Volcanism Study Project 1981. *Basaltic volcanism on the terrestrial planets*. New York: Pergamon Press.

Gass, I.G., S.J. Lippard & A.W. Shelton (eds) 1984. *Ophiolites and oceanic lithosphere*. Oxford: Blackwell Scientific.

Hekinian, R. 1982. *Petrology of the ocean floor*. Amsterdam: Elsevier.

Kennet, J.P. 1982. *Marine geology*. Englewood Cliffs, N.J: Prentice-Hall.

Le Roex, A.P. 1987. Source regions of mid-ocean ridge basalts: evidence for enrichment processes. In *Mantle metasomatism*, M.A. Menzies & C.J. Hawkesworth (eds). London: Academic Press.

Wilkinson, J.F.G. 1982. The genesis of mid-ocean ridge basalt. *Earth Sci. Rev.* 18, 1–57.

PART THREE

Magmatism at destructive plate margins

Destructive plate margins mark the sites of subduction of oceanic lithosphere into the Earth's mantle; one of the most significant phenomena in recent global tectonics. Most of the world's active volcanoes and earthquakes, including nearly all those with intermediate and deep foci, are associated with such descending lithospheric plates. To appreciate the scale on which subduction takes place it is only necessary to consider that both the Atlantic and Pacific oceans were created over the past 200 Ma by seafloor spreading (Ch. 5). As the Earth is not expanding, then an equivalent area of lithosphere must have been simultaneously subducted. At present-day subduction rates, an area equal to the entire surface of the Earth would be consumed in about 160 Ma (Toksoz 1975).

The overriding plate can be either of oceanic or continental lithosphere, resulting in different geometrical forms for the surface volcanism — oceanic island arcs and active continental margins respectively. Both have the following major characteristics (Thorpe 1982):

(a) Arcuate chains of islands or linear belts of volcanoes with a length of the order of hundreds to thousands of kilometres and a relatively narrow width (200–300 km).
(b) A deep oceanic trench (6000–11 000 m deep) on the oceanic side.
(c) Active volcanism in which there is an abrupt oceanward boundary to the volcanic zone, the volcanic front, usually parallel to and some 100–200 km from the oceanic trench.
(d) A dipping zone of seismicity, the Benioff zone, including shallow, intermediate and deep-focus earthquakes. This marks the plane

of descent of the oceanic lithosphere into the mantle.

(e)   A characteristic volcanic association which has been called the 'orogenic andesite' association (Gill 1981).

The chains of stratovolcanoes arranged in elongate zones above Benioff zones represent the most conspicuous volcanic features on Earth. Their eruptive products range in composition from basalt to rhyolite, with andesite being the most common magma type erupted sub-aerially. The range of igneous activity is extremely varied, making this one of the most complex magma generation environments which we will consider in this text.

The morphologies of the volcanoes with their flanking clastic debris reflect processes of rapid growth and rapid sub-aerial erosion. Volcanism is frequently highly explosive and caldera formation common, accompanied by the eruption of widespread pyroclastic flows and air-fall tuffs.

Evidence from the geological record suggests that island arcs and active continental margins have been important volcanic phenomena throughout much of geological time and have been the dominant agents for crustal growth.

CHAPTER SIX

Island arcs

6.1 Introduction

The oceanic island arcs represent the sites of subduction of one oceanic lithospheric plate beneath another (Fig. 6.1). Their characteristic features are linear or arcuate chains of islands forming the volcanic front, often flanked by marginal basins formed by seafloor spreading type processes behind the arc (Ch. 8). Sediments forming the upper layer of the oceanic crust are frequently scraped off the subducting plate as it descends, and form an accretionary wedge in the forearc region.

Figure 6.2 shows the distribution of the major oceanic island arc systems in the Pacific and Atlantic oceans and Indonesia. The western Pacific has major arcs extending from the New Zealand – Tonga system through New Britain – Papua-New Guinea and the Mariana−Izu Islands to the Japan−Kurile−Kamchatka system and the Aleutians. The Mariana−Izu and Japanese arcs are associated with marginal basins and with the exception of Japan all of these arcs appear to lack ancient continental basement. In contrast, the eastern Pacific has no island arcs but exhibits volcanism on an active continental margin extending from the western USA through Mexico and Central America to South America. This will be considered in further detail in Chapter 7. In the Atlantic, oceanic island-arc volcanism occurs in the Lesser Antilles and South Sandwich arcs. Subduction-related volcanism also occurs in the Mediterranean Aeolian and Aegean arcs and in an active continental collision zone extending from the Alps, through Turkey and Iran, to the Himalayas (this zone has been omitted from the general discussion here and in Chapter 7 because of its

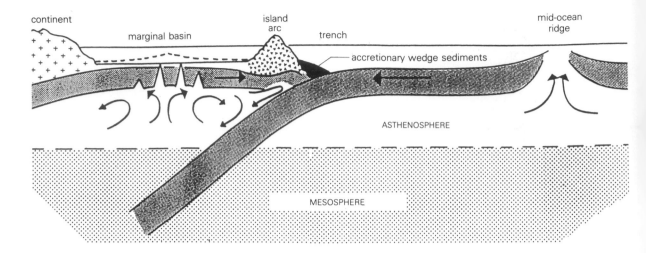

Figure 6.1   The formation and subduction of oceanic lithosphere. New oceanic lithosphere is created at the mid-oceanic ridge and a deep trench is formed where the lithospheric plate descends into the mantle. Secondary convection currents in the asthenosphere cause a small spreading centre, a marginal basin, to develop behind the arc.

tectonic complexities).

In this chapter emphasis will be placed upon the volcanic phenomena associated with island-arc magmatism. Intrusive igneous activity is obviously equally important in the evolution of the arc crust but, for the most part in young arcs, plutonic equivalents of the volcanic rocks are not exposed for direct study. Subduction-related plutonic associations will be considered further in Chapter 7.

6.2   Simplified petrogenetic model

The subduction-zone environment must undoubtedly be one of the most complex tectonic provinces on Earth, and many processes taking place there are as yet incompletely understood. Theoretically, oceanic island arcs should represent the products of the least complicated type of subduction-related magmatism, specifically one in which contamination of ascending magmas by continental crustal materials should be eliminated. It is generally agreed that the process of magma generation in this environment is a multistage multisource phenomenon (Hawkesworth & Powell 1980, Dupuy *et al.* 1982, Sekine & Wyllie 1982a,b,

Kay 1984, Wyllie 1984, Arculus & Powell 1986).

The process of subduction transports a cold oceanic lithospheric plate deep into the mantle (Figs 6.1 & 3). The plate is composed of the following components:

(a)   Variably depleted mantle lherzolite of the oceanic lithosphere (3 in Fig. 6.3).
(b)   Oceanic crust (2 in Fig. 6.3), comprising basalt and gabbro generated at a mid-ocean ridge which are hydrothermally metamorphosed to an uncertain extent and depth (Ch. 5).
(c)   Serpentinite bodies.
(d)   Oceanic sediments.

During subduction the cold crust is progressively heated by conduction of heat from the surrounding mantle and also possibly by frictional heating at the surface of the slab. With increasing pressure and temperature, prograde metamorphic reactions take place and the basaltic components of the oceanic crust are converted through greenschist and amphibolite facies mineralogies to eclogite (Fig. 6.3). The net effect of this metamorphism is to dehydrate an originally hydrous mineral assemblage, releasing H_2O as a separate fluid phase. The precise depth at which the various metamorphic

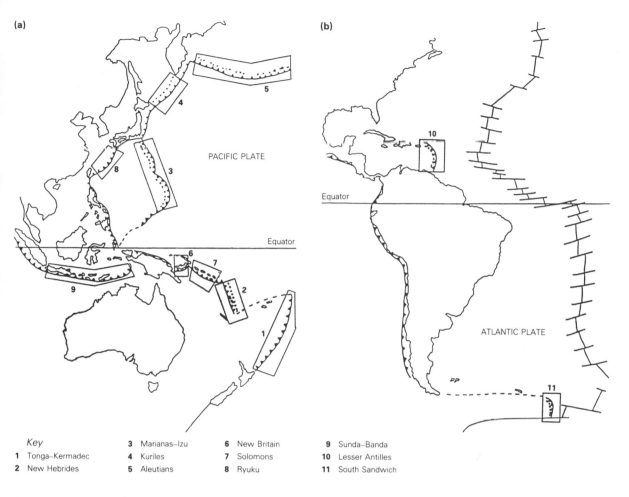

Key

1	Tonga–Kermadec	3	Marianas–Izu	6	New Britain	9	Sunda–Banda
2	New Hebrides	4	Kuriles	7	Solomons	10	Lesser Antilles
		5	Aleutians	8	Ryuku	11	South Sandwich

Figure 6.2 Distribution of the major currently active oceanic-island arc systems in the Pacific ocean and Indonesia (a) and the Atlantic ocean (b) (after Wilson & Davidson 1984).

transitions take place depends upon the thermal regime within the slab and thus will vary from arc to arc (Anderson *et al.* 1978,1980; Wyllie 1984). This will be considered further in Section 6.5.

Much controversy has centred upon the subsequent behaviour of this aqueous fluid, specifically whether it is retained as a grain boundary phase in the surface of the slab or released into the overlying mantle wedge as soon as it is formed. In Figure 6.3 the latter is assumed. The presence of such a fluid phase is of critical importance in models of island-arc magma genesis, as will be shown in Section 6.5.

Despite the very obvious correlation between the subduction of lithosphere and the generation of

magma in island arcs, the role of the subducted lithosphere is by no means a simple one. Early petrogenetic models favoured partial melting of the subducted oceanic crust in the generation of the voluminous andesitic magmas characteristic of this tectonic setting (Marsh & Carmichael 1974, Green & Ringwood 1968). However, more recent models have favoured multistage multisource phenomena, involving the mantle wedge to a much greater extent (Wilson & Davidson 1984, Wyllie 1984, Arculus & Powell 1986).

In general, magma generation in any environment will commence at sites where the temperature exceeds the solidus of the various rock types

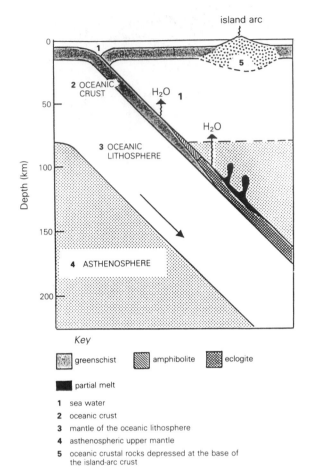

Figure 6.3 Potential source regions involved in island-arc magma genesis. Progressive metamorphism of the oceanic crust takes place upon subduction from greenschist through amphibole to eclogite facies. Dehydration reactions occur, releasing aqueous fluids into the mantle at shallow depths. At greater depths hydrous partial melting of eclogite produces H_2O-rich intermediate to acid partial melts which then rise into the mantle wedge (after Wyllie 1982).

present. The following must all be considered as potential sources for island-arc magmas:

(a)   The mantle wedge above the subducted slab. This consists of two components:

    (1)   A 40−70 km thick section of oceanic lithosphere which must be variably depleted due to the extraction of MORB at the ridge which generated it. It probably comprises lherzolites and harzburgites of a fairly refractory nature which are unlikely to melt readily, even in the presence of a fluid phase.

    (2)   A zone of asthenospheric upper mantle of varying thickness, depending upon the specific arc geometry. The lherzolites of this zone should be considerably more fertile than those of the overlying lithosphere (Ch. 3). It is a general observation (Gill 1981) that where the angle of subduction is so shallow that there is no wedge of asthenosphere above the slab then no surface volcanism occurs. This must have fundamental implications for models of island-arc magma genesis.

(b)   The oceanic crust. This also consists of two components:

    (1)   variably metamorphosed ocean-floor basalt, dolerite and gabbro;

    (2)   oceanic sediments ranging from pelagic clays and carbonate oozes to terrigenous clastic sediments.

The oceanic crust could become involved in the magma generation process in two distinct environments:

    (1)   the upper part of the subducted oceanic lithosphere;

    (2)   the base of the island-arc volcanic sequence, as all intra-oceanic arcs are built upon a foundation of foundered oceanic crust.

(c)   Sea water. This must ultimately provide the H_2O component which appears to be fundamental in island-arc magmatism. It is incorporated during hydrothermal alteration of the oceanic crustal layer during ocean-floor metamorphism (Ch. 5), and possibly also by direct circulation of sea water within the island-arc crust.

Models for magma generation in this tectonic setting have attempted to assess the relative importance of these different potential source regions in different island-arc systems, often with particular emphasis on the extent of involvement of sedimentary components (Kay 1980; Hole *et al.* 1984; Thirlwall & Graham 1984; Davidson 1985,

1906, Woodhead & Fraser 1985; Tera *et al.* 1986; White & Dupré 1986).

In general, we can consider that island-arc magmas could be generated by partial melting of any of the following sources:

(1) amphibolite, with or without aqueous fluid;
(2) eclogite, with or without aqueous fluid;
(3) lherzolite with aqueous fluid;
(4) lherzolite modified by reaction with hydrous siliceous magma derived by partial melting of the slab.

The role of aqueous slab-derived fluids and partial melts is a crucial one, which appears to distinguish magma generation in the subduction zone environment from all others. In Figure 6.3 partial melts of the subducted oceanic crust are depicted rising into the lherzolites of the mantle wedge, where they react and lose their chemical identity (Section 6.5, and see Sekine & Wyllie 1982a,b). Such slab-derived fluids (both aqueous fluids and partial melts) have the effect of lowering the mantle solidus, thereby promoting partial melting. They may thus be regarded as a catalyst for the bulk of island-arc magmatism rather than a prime source.

6.3   Structure of island arcs

Detailed knowledge of the physical structure of island arcs can provide important constraints for petrogenetic models (Gill 1981, Cross & Pilger 1982, Uyeda 1982, Jarrard 1986). Figure 6.4 shows the subdivision of a model island arc into trench, fore-arc, arc and back-arc regions, with their associated gravity and heat flow anomalies. The negative gravity anomaly near the trench is attributed to the presence of sediment wedges in the fore-arc and the positive anomaly to the cold dense subducted lithosphere beneath the arc. Heat flow is typically low in the fore-arc ($10-20°C$ km^{-1}) but rises abruptly at the volcanic front ($30-40°C$ km^{-1}) and remains high for distances of $200-600$ km behind the arc. This high heat flow can only be accounted for by the mass transfer of hot material (magma) to higher levels.

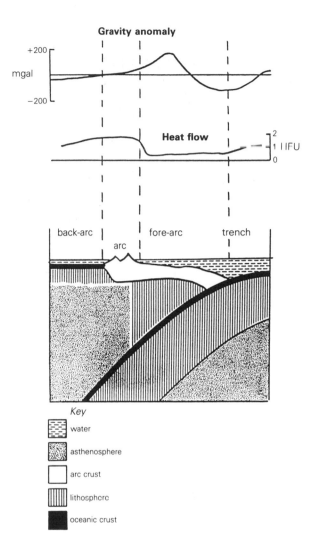

Figure 6.4   Schematic cross section through an island arc (after Gill 1981, Fig. 2.3, p. 25).

In Figure 6.4, lherzolites of the lithosphere and asthenosphere are differentiated by different ornaments, reflecting their differing seismic properties. Typical P-wave velocities of the lithosphere are $8.0-8.1$ km s^{-1}, whereas those of the asthenosphere are lower ($7.5-7.9$ km s^{-1}), this usually being attributed to the presence of a partial melt phase. The apparently vertical boundary between mantle with lithospheric characteristics and mantle with asthenospheric characteristics may thus reflect the rise of magma beneath the arc. Anomalously shallow asthenospheric mantle beneath the back-arc region may indicate the ascent of magmas formed by back-arc spreading processes (Ch. 8).

More detailed aspects of the crustal structure of island arcs, as revealed by seismic refraction studies, are shown in Figure 6.5. In general, arc crusts are less than 25 km thick, crustal thickness being proportional to the age of the subduction system and the rate of magma generation. Most arcs have 6−9 km of upper crust (V_p=5.0−5.7 km s^{-1}) underlain by 10−15 km of lower crust (V_p=6.5−7.0 km s^{-1}). These two layers are similar in velocity to layers 2 and 3 of the oceanic crust (Ch. 5). The Moho, or crust/mantle boundary, on these diagrams is depicted as a sharp line on the assumption that velocities in excess of 7.5 km s^{-1} represent mantle values. This is not necessarily correct, and in all probability the crust/mantle boundary in island arcs is actually a diffuse zone across which V_p steadily increases.

Crustal thickness plays an important role in constraining low-pressure fractionation of ascending magmas (Leeman 1983; see also Ch. 4). In general, in regions of thin crust mantle-derived magmas can ascend rapidly to the surface and can maintain near-primary characteristics. However, in more mature arcs with thickened crusts, the low-density crustal rocks act as a filter impeding the ascent of primary magmas and causing extensive low-pressure crystal fractionation in high-level magma chambers. This will be considered further in Section 6.6.

6.4 Earthquakes and magma genesis

The spatial association of volcanic activity and earthquakes in arc−trench systems is well documented, with volcanic arcs generally overlying earthquake foci 100−200 km deep (Gill 1981). However, despite this close correlation there is no consistent spatial relationship between large earthquakes and active volcanism, active volcanoes occurring in areas characterized by both their presence and absence.

Most of the world's large earthquakes (magnitude >7.0) occur along actively subducting plate margins, and the slip causing the earthquake has in general been presumed to reflect the relative motion of the two plates. The lack of a direct relationship between volcanic and seismic activity suggests that magma generation is not dependent upon the component of differential motion but is instead related to aseismic slip (Acharya 1981). The aseismic slip rate is defined as:

aseismic slip rate = rate of plate motion − seismic slip rate

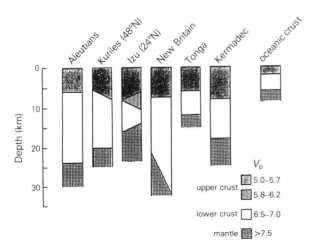

Figure 6.5 Crustal P-wave velocity structure in island arcs (after Gill 1981, Fig. 3.3, p. 48).

Locked segments of destructive plate boundaries have high seismic slip rates and generally low eruption rates. In contrast, if decoupling between the two plates is high, i.e. low seismic slip rate, a greater volume of magma appears to be generated (Fig. 6.6). Thus, to a first approximation, magma generation and earthquake activity appear to be antipathetic.

Several characteristics of the dipping seismic zone change beneath the volcanic front, implying that significant changes in the physical properties of the underthrust lithosphere occur here. Notably, a double seismic zone (Fig. 6.7), consisting of two 10−15 km thick bands, has been identified beneath several volcanic arcs and attributed to slab unbend-

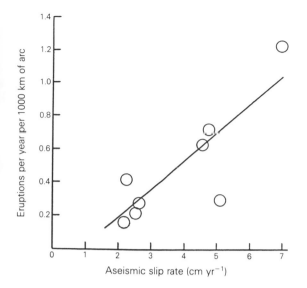

Figure 6.6 Relation between aseismic slip rate and number of eruptions per year per 1000 km of arc (after Acharya 1981, Fig. 2).

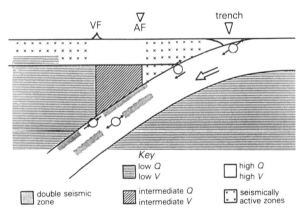

Figure 6.7 Schematic seismic characteristics of an island arc. AF is the *aseismic front*, arcwards of which no earthquakes occur in the asthenosphere part of the mantle wedge. VF is the *volcanic front*. At the AF intersection with the slab focal mechanisms change from interplate thrust to intra-plate compression down dip. It is suggested that the region of intermediate Q and seismic velocity (V) between AF and VF reflects water rising into the asthenospheric mantle wedge without causing partial melting. In this model the AF thus represents the onset of dehydration within the slab. Q is a dimensionless parameter which measures seismic wave attenuation. Low values of Q mean high seismic wave attenuation. Q decreases rapidly as temperature and degree of melting increase (after Anderson *et al.* 1980, Fig. 5).

ing (Gill 1981). Focal mechanisms for the upper band indicate downdip compression, whereas the lower band is characterized by downdip extension. Hasegawa *et al.* (1978) interpreted the uppermost plane to be the interface between the slab and the overlying mantle wedge, and the lower plane to be the base of the rheologically rigid upper portion of the lithosphere.

An interesting observation, valid for Japan at least (Anderson *et al.* 1980), is that there is an abrupt decrease in the frequency of large-magnitude earthquakes close to the volcanic front. This could be interpreted as delineating a region of lowered stresses due to high fluid pore pressures, and may mark the onset of dehydration within the slab.

6.5 Thermal structure and partial melting processes

Knowledge of the thermal structure of subduction zones is essential in understanding the complex magma generation processes responsible for island-arc volcanism and the distribution and characteristics of seismicity in and around the subducting slab. Innumerable thermal models have been presented in the literature (Anderson *et al.* 1978, 1980; Furlong *et al.* 1982) and yet none fully explains the range of physical characteristics observed. Any general model must include the effects of dehydration in the subducting oceanic crust, frictional heating along the upper surface of the slab and convection within the asthenospheric mantle wedge. Such a model is shown in Figure 6.8, which may be considered to represent broadly the correct configuration of isotherms. However, each subduction system must have its own unique thermal structure given the potential perturbations to the model caused by variations in the age of the subducted lithosphere, subduction rate and angle of subduction.

The temperature distribution in the slab and overlying mantle wedge is one of the decisive factors controlling the onset of partial melting, magma generation commencing at those sites where the temperature exceeds the solidus of the various

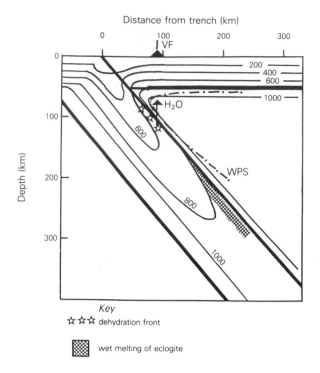

Key

☆☆☆ dehydration front

▨ wet melting of eclogite

Figure 6.8 Thermal model for the subduction of a 100 km thick oceanic lithospheric plate beneath another oceanic plate 50 km thick. Asthenospheric flow concentrates the isotherms in the corner of the wedge near the base of the arc lithosphere. The wet peridotite solidus (WPS) lies directly above the dehydration front in this model and thus partial melting of the asthenospheric part of the mantle wedge will occur. The shaded area represents the zone in which the temperature of the wet eclogite solidus is exceeded and thus partial melting of the subducted oceanic crust occurs. For this particular thermal model this obviously does not contribute to the arc volcanism (after Anderson *et al.* 1980, Fig. 2).

source materials present. Partial melting of any or all of the following materials could occur in the subduction-zone environment, given the right conditions.

6.5.1   Subducted oceanic crust

This may involve basic igneous rocks in the amphibolite or eclogite facies and metamorphosed subducted oceanic sediments, possibly in the presence of an aqueous fluid phase.

The mineralogy of the basic rocks (basalt, dolerite or gabbro) of the oceanic crust varies

during subduction as a function of pressure, temperature and vapour phase composition, changing from zeolite through blueschist or amphibolite facies to eclogite. This prograde metamorphism is accompanied by dehydration, which is thought to occur principally in the 80–125 km depth range. Some hydrous minerals such as biotite and a 14 Å chlorite may, however, persist to deeper levels before breaking down (Delaney & Helgeson 1978). Any sediments subducted will pass through the same sequence of metamorphic grades, their mineralogies reflecting their bulk chemical composition. H_2O released by the various dehydration reactions can either migrate upwards into the overlying mantle wedge as soon as it is formed or be transported to greater depths as an intergranular pore fluid. The behaviour of this fluid is of fundamental importance in models of island-arc magma genesis.

Figure 6.9 shows the minimum conditions (i.e. the solidus) necessary for partial melting of basic

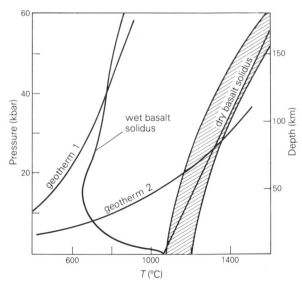

Figure 6.9 Solidi for basic rocks under anhydrous and water-saturated conditions. The shaded area represents the range of andesite liquidus temperatures. Geotherms 1 and 2 are two possible thermal gradients within the upper part of the subducted slab. Geotherm 1: a cold slab model in which the upper part of the slab is cooled by endothermic dehydration reactions. Geotherm 2: a warm slab model in which the upper surface of the slab is heated by frictional effects. (After Gill 1981, Fig. 8.1, p. 232).

igneous rocks in the subducted oceanic crust under both water-saturated and anhydrous conditions. The presence of water lower the solidus temperature by several hundred degrees. The precise depth at which the subducted oceanic crust will begin to partially melt can only be predicted if the following are known: (1) the geothermal gradient in the upper part of the slab, and (2) the amount of water in the upper part of the slab. For cold-slab models (geotherm 1) partial melting only begins at realistic depths for water-saturated conditions, whereas for warm-slab models (geotherm 2) it can begin at fairly shallow depths for a range of H_2O contents. Shown for comparison is the field of andesite liquidus temperatures (shaded). This suggests that if eruptive andesites are derived by direct partial melting of the subducted oceanic crust then this must occur under essentially anhydrous conditions (Gill 1981). For the thermal model shown in Figure 6.8 (similar to geotherm 1), wet partial melting of eclogite could only occur at depths greater than 150 km. However, the wet eclogite solidus lies virtually subparallel to the upper surface of the slab in this model, and thus whether or not the oceanic crust melts is highly sensitive to slight variations in the configuration of the isotherms.

To assess the partial melting behaviour of subducted sediments, solidi need to be determined for the range of compositions involved under water-saturated and anhydrous conditions. Unfortunately, few such data are available at present.

6.5.2   The mantle wedge

This involves partial melting of lherzolites modified by reaction with aqueous fluids, or hydrous siliceous partial melts derived from the subducted oceanic crust.

Initiation of partial melting in the mantle wedge is critically dependent upon the geothermal gradient and the amount of volatiles present. A wide body of data exists on the melting behaviour of lherzolite, much of which is contradictory (Mysen 1982; and see Ch. 3). Figure 6.10 shows that in the presence of H_2O and CO_2 solidus temperatures are lowered markedly, thus facilitating partial melting. Several different geothermal gradients are shown in

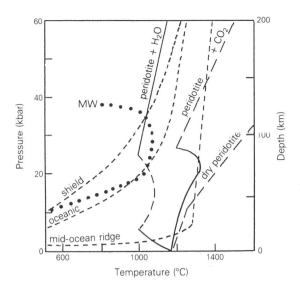

Figure 6.10 Peridotite (lherzolite) solidi in the presence of H_2O (0.4%) and CO_2 (5%) compared to the anhydrous (dry) solidus. Note that these volatile contents are not necessarily sufficient to generate a free vapour phase. The solid lines depict those portions of the solidi with a coexisting vapour phase. The short dashed lines are schematic shield, oceanic and mid-oceanic ridge geotherms. The dotted line MW represents the geothermal gradient in the mantle wedge beneath the arc in Figure 6.8 (after Wyllie 1981, Fig. 3).

this diagram for comparison. In the subduction-zone environment geothermal gradients in the mantle wedge have an unusual configuration (MW in Fig. 6.10), due to the cooling effects of the subducted slab (Wyllie 1981). For such a thermal model, mantle partial melting beneath the arc could only occur under water-saturated conditions.

It is generally accepted that partial melting of lherzolite under dry conditions produces liquids of basaltic or picritic composition, depending upon the degree of melting (Ch. 3). However, in the presence of H_2O and CO_2 the compositions of near-solidus liquids are changed (Fig. 6.11) (Mysen 1982, Wyllie 1982). Andesitic magmas could be generated directly by partial melting of hydrous lherzolite at depths less than 40 km. However, such melts would be water-saturated, containing some 15 wt.% H_2O at 40 km, and would therefore begin to crystallize and evolve vapour as soon as they began to rise. Clearly, such magmas could not

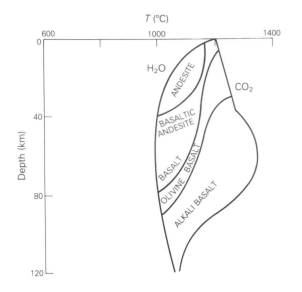

Figure 6.11 Compositions of near-solidus partial melts in the system lherzolite–H_2O–CO_2 (modified after Wyllie 1982, Fig. 6).

reach the surface without undergoing extensive crystal fractionation. This would also apply to more basic hydrous magmas generated at greater depths.

Despite the potentially continuous supply of volatiles from the subducted slab to the overlying mantle wedge, the total amount of H_2O in the wedge must still be low. Thus, although near-solidus partial melts would be water-saturated, all of the H_2O would dissolve in the melt within a few degrees of the solidus. With progressive fusion the partial melts would become more undersaturated with H_2O, and would change in composition towards basaltic or picritic liquids in equilibrium with volatile free peridotite. H_2O-undersaturated basalts of this type could be the parental magmas to the island-arc volcanic suites, with fractionation at higher levels yielding the more silica-rich members.

The bulk of the experimental data on the partial melting behaviour of mantle lherzolite has been obtained under anhydrous or H_2O–CO_2 vapour phase conditions. However, at the high pressures and temperatures of the subduction-zone environment, hydrothermal fluids rising from the subducted slab could contain considerable volumes of dissolved silicates. Reaction of the mantle wedge

with such fluids would lower its solidus temperature and, in addition, as shown by Sekine & Wyllie (1982a,b), the composition of near-solidus partial melts could change significantly from those shown in Figure 6.11. Specifically, there could be a marked expansion of the fields of basaltic andesite and andesite partial melts to greater depths.

Thus the possibility exists, according to experimental prediction, of generating the whole spectrum of magma compositions observed in island arcs by variable depth and degree of partial melting of a mantle source heterogeneously metasomatized by a silicate-rich fluid phase derived from the subducted slab. It is likely that there is a continuous spectrum of fluid compositions generated within the slab, ranging from aqueous silica-rich fluids to hydrous andesitic–dacitic partial melts. All of these would have a similar effect on the mantle wedge although, as shown in Section 6.11.2, they might imprint different trace element characteristics on subsequent partial melts.

Uncertainties attached to thermal models of subduction zones are so great at present that it is effectively impossible to make quantitative predictions as to which of the various source components will melt and at what depth. In principle, any one of the following possibilities may occur:

(1)   both subducted slab and mantle wedge melt;
(2)   the subducted slab dehydrates and the mantle wedge melts;
(3)   the slab alone melts.

Of these, (3) is the least likely model to account for the petrogenesis of the range of eruptive magma compositions in island arcs (Section 6.7).

6.6   Segregation, ascent and storage of magma

From the data presented so far it is apparent that the major site of magma generation in island arcs is in the asthenospheric part of the mantle wedge above the subducting slab. Partial melting may occur over a considerable depth range if lherzolite diapirs rise buoyantly towards the surface. At some unspecified depth, segregation of primary magmas will occur (Section 3.6) and these magmas will then rise towards the Earth's surface along a variety of ascent paths.

Several lines of evidence indicate the existence of high-level magma reservoir systems within the crust and upper mantle of island arcs in which the primary magmas may pond and fractionate. These include petrological evidence for low-pressure crystal fractionation, the occurrence of shallow volcanic tremor, ground-surface deformation and caldera formation, and the marked attenuation of S-waves. All the available evidence indicates that such magma chambers normally occur at depths less than 20−30 km and may extend to within a few hundred metres of the surface (Iyer 1984). Gill (1981) suggests that where shallow magma chambers exist (< 20 km deep) they usually underlie volcanoes with historic eruptions of andesite or dacite. In contrast, seismically identifiable magma reservoirs extending into the upper mantle normally underlie volcanoes whose most recent eruptions are of basalt or basaltic andesite.

Figure 6.12 is a schematic illustration of the magma storage system in a fairly mature island arc with a 30 km thick crust. Quite probably, there will be a series of interconnected reservoirs fed from the zone of magma generation, here indicated to be the asthenospheric mantle wedge. It is possible that such high-level storage systems are not present in very young island arcs and only become established once a certain crustal thickness has been reached. This may account for the predominance of basaltic magmas in immature arcs (Section 6.8).

Evidence for the importance of low-pressure fractional crystallization in the evolution of island-arc volcanic suites is most clearly found in Harker-type chemical variation diagrams, which frequently show linear patterns of compositional variation for sequences of volcanic rocks closely related in space and time (Fig. 6.13 & Section 6.11). Further evidence is provided by the frequent inclusion of cumulate plutonic xenoliths in island-arc volcanic rocks. These are coarse-grained fragments of layered igneous bodies, believed to represent the products of crystal accumulation on the floors of high-level magma chambers (Ch. 4). The xenoliths commonly

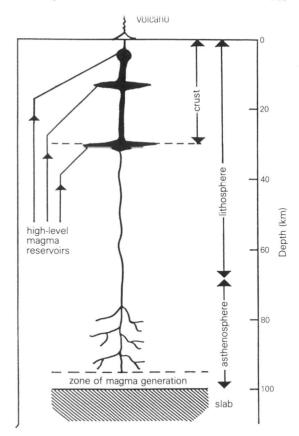

Figure 6.12 Magma reservoir systems beneath a mature island arc (after Gill 1981, Fig. 3.8, p. 58).

contain combinations of the following minerals (Fig. 6.14): olivine−clinopyroxene−orthopyroxene−plagioclase−amphibole−magnetite. The presence of plagioclase feldspar constrains their depth of crystallization to less than about 30 km, as plagioclase does not readily crystallize from basic melts at greater depths (Powell 1978). The frequent occurrence of amphibole is at first sight rather intriguing as it is uncommon as a phenocryst phase in the erupted magmas. It is clearly a stable crystallizing phase from volatile-rich basic magmas at depth, but becomes unstable and is subsequently resorbed as the magmas ascend towards the surface (Section 6.10). The rarity of primary magma compositions in island arcs (Section 6.11.3) also attests to the importance of crystal fractionation processes, but does not necessarily provide direct evidence for high-level magma reservoirs.

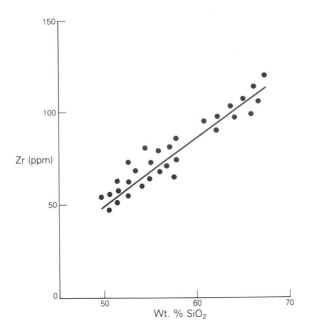

Figure 6.13 Variation of Zr versus SiO₂ for a hypothetical island-arc volcanic suite. Zr behaves incompatibly and increases in residual liquids with progressive crystal fractionation.

Figure 6.14 Anorthite—olivine cumulate plutonic xenolith from St Vincent, Lesser Antilles (x40, crossed polars).

6.7   Characteristic magma series

Attempts to classify the spectrum of magmas erupted in island arcs have frequently led to confusion because, unlike plants and animals,

magmas do not fall naturally into distinct species. Typically, the eruptive products have been subdivided into three major magma series, tholeiitic, calc-alkaline and alkaline (Ch. 1), each series spanning the compositional range from basalt to rhyolite. Much of the confusion over classification has arisen due to the use of different geochemical criteria by different authors. Following Gill (1981), two simple diagrams can be used to subdivide the range of magma compositions.

K_2O versus SiO_2 diagram

Using this Harker diagram (Fig. 6.15), island-arc volcanic suites can be subdivided into four distinct magma series:

(a)   low-K series;
(b)   calc-alkaline series;
(c)   high-K calc-alkaline series;
(d)   shoshonitic series.

Series (a) can be considered synonymous with the island-arc tholeiite series of Jakeš & Gill (1979), and (d) can also be referred to as an alkaline series. Basaltic members of the calc-alkaline series are sometimes referred to as high-alumina basalts. Within each of these series the relative proportion of basalt varies widely with respect to more evolved magma types (Baker 1982). Figure 6.16 shows the relative frequency of occurrence of basalts, andesites, dacites and rhyolites in the island-arc tholeiite and island-arc calc-alkaline series. The high-K calc-alkaline series is broadly similar to the calc-alkaline series in this respect, while the alkaline-shoshonitic series is dominantly basaltic but can be very variable.

$FeO*/MgO$ versus SiO_2 diagram

The low-K or tholeiitic magma series is characterized by marked Fe enrichment in the early stages of fractionation, in marked contrast to the calc-alkaline series in which total iron content decreases steadily with increasing SiO₂ (Miyashiro 1974). A variety of plots can be used to differentiate such trends, the simplest (Fig. 6.17) being FeO*/MgO

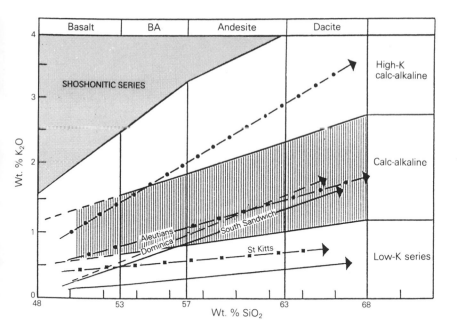

Figure 6.15   Plot of wt.% K_2O versus wt.% SiO_2 showing the major subdivisions of the island-arc volcanic rock suites. Individual trend lines are discussed in Section 6.11.1. Data sources: Aleutians, Marsh (1982); Dominica, Wills (1974); St Kitts, Smith *et al.* (1980); South Sandwich, Luff (1982). (After Basaltic Volcanism Study Project 1981, Fig. 1.2.7.1., p. 193).

versus SiO_2 ($FeO^\star$ = all Fe as FeO). A triangular diagram of $Na_2O + K_2O$ (A) − $FeO^\star$ (F) − MgO (M), commonly known as an AFM diagram, can also be used to show the divergent trends of the two magma series (Fig. 6.18). For basaltic rocks, a graph of alkali index versus Al_2O_3 differentiates tholeiitic from high-alumina (calc-alkaline) types (Fig. 6.19).

The differences between these four major magma series are reflected to a variable extent in their eruptive morphologies (Section 6.9). Island-arc tholeiite series volcanoes are characterized by eruptions of very fluid basalts and basaltic andesites, producing low-lying platforms around central vents. Pyroclastic rocks are uncommon and, in particular, pyroclastic flows are very rare. There is a far greater proportion of aphyric lavas than in the calc-alkaline series, and hydrous minerals such as amphibole and biotite are almost entirely absent, suggesting low volatile contents in the parent magmas. Typical examples of tholeiitic arcs are the South Sandwich Islands, Tonga, the Izu Islands and the northern Lesser Antilles.

In marked contrast, the principal rock type of the calc-alkaline series (including the high-K calc-alkaline series) is a two-pyroxene andesite with about 59% SiO_2. Eruptions tend to be more explosive than those of the island-arc tholeiite association, and pyroclastic fall and flow deposits are common. Andesitic magmas are much more viscous than basalt and produce steep-sided strato-volcanoes, often with volcanic domes or spines in the central crater area. The majority of lavas are highly porphyritic, with calcic plagioclase the most common phenocryst phase. Hydrous minerals such as amphibole and biotite are frequently present, reflecting (along with the explosive activity) the more volatile-rich nature of the magmas. Most of

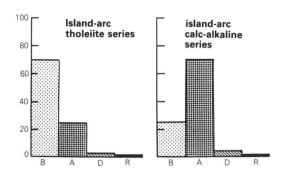

Figure 6.16   Relative volumes of basalt, andesite, dacite and rhyolite in the island-arc tholeiite compared to the island-arc calc-alkaline series (after Baker 1973).

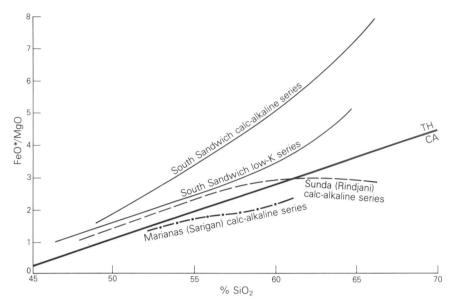

Figure 6.17   Plot of FeO*/ MgO versus SiO_2, used to differentiate tholeiitic (TH) from calc-alkaline (CA) suites; FeO* is all Fe as FeO (wt.%). The dividing line has the equation: FeO*/MgO = 0.1562 × SiO_2 − 6.685. Data sources: South Sandwich, Luff (1982); Marianas, Meijer & Reagan (1981); Sunda, Foden (1983).

the volcanic rocks of the circum-Pacific arcs, the Lesser Antilles and Indonesia fall into this category.

The shoshonitic series is much more variable but, on average, comprises 50% by volume basalt, 40% andesite and 10% dacite. Both sodic and potassic alkaline suites are known to erupt in island arcs, although sodic lavas are apparently restricted to specific tectonic settings such as:

(a) along or near the lateral edges of subduction zones where hinge faulting occurs, e.g. Grenada in the Lesser Antilles;

(b) where a fracture zone approximately perpendicular to the trench is being subducted.

Strictly speaking, the term 'shoshonitic' should only be applied to potassic alkaline suites. Typical examples of alkaline island-arc suites occur in Fiji, the Sunda arc of Indonesia, the Aeolian arc and Grenada in the Lesser Antilles.

In addition to the above-mentioned magma series, there also exists in island arcs an unusual group of high-MgO (>6 wt.% MgO) andesites called boninites. These are apparently restricted to fore-arc regions, which suggests that rather special conditions are required for their generation. The most extensive occurrences are in the present-day fore-arc of the Mariana−Izu system, outcropping

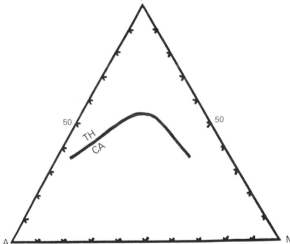

Figure 6.18   AFM diagram used to differentiate tholeiitic (TH) from calc-alkaline (CA) suites: A = Na_2O + K_2O; F = FeO + $0.9Fe_2O_3$; M = MgO. The solid line separates tholeiitic from calc-alkaline suites, using the criteria of Irvine & Baragar (1971).

in the Bonin Islands (Crawford *et al.* (1981). Here they occur stratigraphically above arc-related volcanics, and are associated with tholeiitic basalts produced during the development of a back-arc basin (Ch. 8). They appear to characterize the embryonic stages of magmatism associated with the

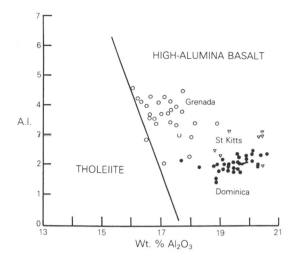

Figure 6.19 Alkali Index (A.I.) versus Al_2O_3, showing the subdivision between tholeiitic and high-alumina (calc-alkaline) basalts: A.I. $= [Na_2O + K_2O]/[(SiO_2 - 43) \times 0.17]$. The boundary line is from Middlemost (1975). Basaltic rocks from the Lesser Antilles island arc (St Kitts (low-K), Dominica (calc-alkaline) and Grenada (alkaline)) all display high-alumina characteristics (after Powell 1978).

Table 6.1 Classification of island arcs according to convergence rate, crustal thickness and characteristic magma series (after Gill 1981, Table 7.3, p. 220).

1	convergence rate >7 cm yr^{-1}, crust <20 km thick	
	A >40% of volcanoes calc-alkaline	Solomons, Aleutians
	B >80% of volcanoes tholeiitic	Tonga–Kermadec, Mariana–Izu, South Sandwich
2	convergence rate >7 cm yr^{-1}, crust 30–40 km thick	
	30–70% of volcanoes tholeiitic	New Britain, New Hebrides, Kuriles, Sunda (Java)
3	convergence rate <7 cm yr^{-1}, crust >30 km thick	
	<50% of volcanoes tholeiitic	Lesser Antilles, Ryuku

splitting of an island arc (Cameron *et al.* 1979; and see Ch. 8).

The classification schemes discussed in this section are essential for ease of communication and, using them, the oceanic-island arcs can be divided into distinct groups (Table 6.1). However, as far as petrogenetic models are concerned (Section 6.12), it should be remembered that they are essentially arbitrary subdivisions of a continuous spectrum of magma compositions.

6.8 Spatial and temporal variations in island-arc magmatism

Early models for island-arc magmatism, based on studies of the Japan arc, suggested that the erupted magmas should increase in alkalinity away from the trench (Fig. 6.20) (Kuno 1959, Dickinson & Hatherton 1967, Sugimura 1973). This led to the development of the so-called $K-h$ relationship, whereby the K_2O content (K) of the magmas at a fixed SiO_2 value was apparently correlated with the depth to the Benioff zone (h) (Dickinson 1975). However, it has become increasingly apparent that many arc systems do not follow this simple pattern. For example, in the Lesser Antilles the erupted magmas change in composition from tholeiitic to calc-alkaline to alkaline from north to south along the arc, at an approximately constant depth of 100 km to the Benioff zone. In the New Hebrides, Barsdell *et al.* (1982) have reported a reversed $K-h$ relationship with K_2O contents decreasing with increasing depth to the Benioff zone. As we shall see in Section 6.12, it is not surprising that there is no simple predictable pattern to island-arc magmatism. Source heterogeneity combined with variable depths and degrees of partial melting can result in the generation of a spectrum of primary magma compositions quite independently of the geometry of the subduction system. This, combined with a variety of crystal fractionation processes, can account for the complex array of eruptive magma compositions. However, tectonic setting should not be completely eliminated from the model as it can be demonstrated that alkaline island-arc magmas are frequently associated with anomalous environments such as hinge faults or fracture zones.

Figure 6.20 also shows a summary of apparent

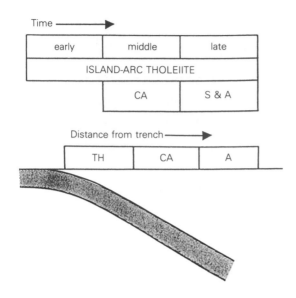

Figure 6.20 Spatial and temporal variations in the compositions of island-arc magmas.

temporal variations in the geochemistry of the erupted magmas. Very young arcs, e.g. South Sandwich, are typically characterized by eruptions of tholeiitic basalts, whereas Baker (1973) has suggested that as arcs mature volcanism progresses from an early tholeiitic phase to the eruption of dominantly calc-alkaline magmas. However, proof of this depends upon a detailed knowledge of the geology of island-arc basement complexes, which are mostly inaccessible below sea level. Precisely why there should be such a change in magma chemistry with time is not fully understood. It must be related to changing primary magma compositions, combined with varying conditions of low-pressure crystal fractionation. In the early stages of its development the arc builds up its submarine foundations upon a basement of oceanic crust which is at most 10 km thick (Ch. 5). Rising basaltic magmas will not be impeded in their ascent to the surface, and thus the resultant volcanism will be dominantly basaltic. As the arc develops, the thickening basalt pile will depress the oceanic crustal layer, which will also be accreting downwards by underplating of basic magma. Once the crust has thickened to some 20−25 km it may start to act as a density filter, arresting the ascent of

primary magmas which will then pond in high-level magma chambers. Subsequent crystal fractionation will result in the generation of lower-density andesitic derivatives, which can rise more easily towards the surface. This may explain the predominance of intermediate magma compositions in mature island arcs.

6.9 Surface volcanic features

The extremely varied styles of eruptive activity in island arcs cover almost the entire range of terrestrial volcanic phenomena. This diversity is accounted for by the wide range in magma chemistries, viscosities, yield strengths and volatile contents. In general, the eruptive style is governed by the SiO_2 and gas content of the magma. Coherent lava flows form by extrusion of gas-poor magma, whereas high volatile contents tend to produce highly explosive eruptions.

Basaltic magmas, having lower viscosities and lower yield strengths, form thin laterally extensive flows, whereas more SiO_2-rich magmas, with higher viscosities and yield strengths, form thick flows which do not travel far from the vent. In highly silicic magmas, the flows are so viscous that they pile up over the vent, forming domes or spines. The collapse of such gravitationally unstable features can lead to the formation of high-temperature pyroclastic flows, which are a characteristic feature of mature island arcs.

Island-arc volcanoes can be divided into two distinct types; (a) basalt − basaltic andesite volcanoes, and (b) andesite − dacite volcanoes. These have differing morphologies by virtue of the nature of their eruptive products. The basalt − basaltic andesite volcanoes commonly are of a low-angled shield type, whereas the andesite−dacite volcanoes are generally large conical composite edifices, the steep form of which is related to the slow extrusion of viscous andesitic lava and the formation of coarse debris flow deposits. Volcanoes of these two types can often be active contemporaneously less than tens of kilometres apart (Wills 1974). The origin of these different volcano types is as yet incompletely understood, but has been ascribed to differences in their high-level subvolcanic plumbing systems (Hawkesworth & Powell 1980).

A characteristic feature of island-arc magmas is their high volatile content (Gill 1981). This tends to promote highly explosive eruptions in which tephra is expelled long distances from the vent. Air-fall deposits are relatively well sorted and well bedded accumulations of ash and lapilli, deposited by gravitational fallout from the large eruption columns which accompany most explosive eruptions. In contrast, pyroclastic flows are generally poorly sorted and poorly stratified, being deposited from hot mixtures of gas and solid ejecta moving along the ground surface. Large-scale pyroclastic flows may be dispersed more than 100 km from the volcanic vent, and may be generated by the collapse of vertical eruption columns or gravitationally unstable summit domes or spines.

Eruptions of large volumes of magma from high-level magma chambers may be sufficient to cause collapse of the roof into the void, producing a caldera. These are a common feature of island-arc volcanoes and provide further evidence for the existence of high-level magma chambers.

6.10 Petrography of island-arc volcanic rocks

A characteristic feature of island-arc volcanic rocks is their highly porphyritic nature (Ewart 1982), magmas of the tholeiitic series being in general the least porphyritic (Figs 6.21–3). Figures 6.24–7 show, in graphical form, the important aspects of the phenocryst mineralogy of the four major magma series: tholeiitic, calc-alkaline, high-K calc-alkaline and shoshonitic. In these diagrams the solid bar indicates the dominant magma composition range for crystallization of a particular mineral, while the broken bar indicates composition fields in which that mineral crystallizes only sporadically. For example, in the low-K series olivine is a major phenocryst phase in magmas ranging in composition from basalt to andesite, but can continue crystallizing right into the dacite field.

In general, all the major ferromagnesian minerals (olivine, clinopyroxene, orthopyroxene and, to a

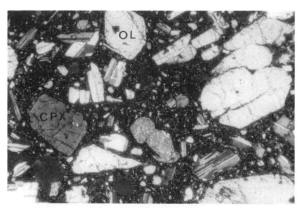

Figure 6.21 Highly porphyritic oceanite from the South Sandwich arc, with phenocrysts of olivine, clinopyroxene and plagioclase in a glassy groundmass (×40, crossed polars).

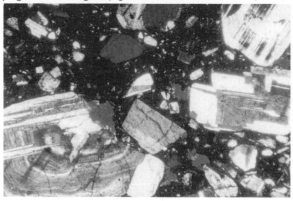

Figure 6.22 Dacite from St Lucia, showing strongly oscillatory zoned plagioclase feldspar phenocrysts (×40, crossed polars).

Figure 6.23 Porphyritic andesite from the South Sandwich arc with phenocrysts of plagioclase, clinopyroxene and orthopyroxene (×40, crossed polars).

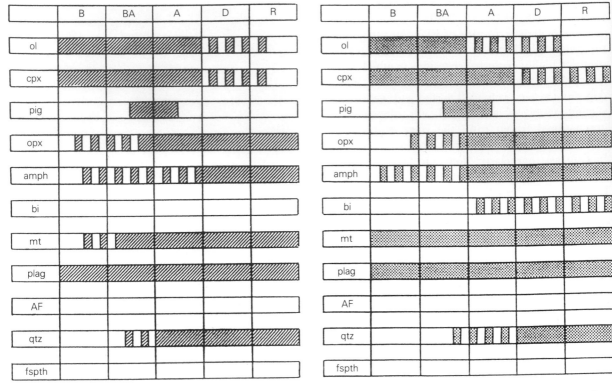

Figure 6.24 Major phenocryst mineralogy of the low-K or island-arc tholeiite series.

Figure 6.25 Major phenocryst mineralogy of the island-arc calc-alkaline series.

lesser extent, amphibole and biotite) tend to be relatively Mg-rich even in the dacites and rhyolites (Ewart 1982). This is shown for olivine and both Ca-rich and Ca-poor pyroxenes in Figure 6.28. Olivine compositions range from Fo_{70-90} in the basalt to basaltic andesite range to Fo_{50-90} in more andesitic compositions.

Augite is second only to plagioclase as a major phenocryst phase and is the most common groundmass pyroxene. Amongst the Ca-poor pyroxenes, orthopyroxene is a common phenocryst phase, occurring in rocks with widely varying SiO_2 contents. It is generally antipathetic in occurrence with olivine, amphibole and biotite. Clinoenstatite occurs rarely in the unusual group of high-Mg andesites called boninites (Ch. 8), while pigeonite occurs rarely as a phenocryst but commonly as a groundmass mineral in tholeiitic andesites.

Amphiboles in island-arc volcanic rocks are

strongly pleochroic hornblendes varying from green to brown in colour, and frequently showing intense opaque reaction rims due to low-pressure instability (Fig. 6.29). When reddish brown they are termed oxyhornblendes or basaltic hornblendes. Figure 6.30 shows their compositional range in terms of a plot of Ca + Na + K versus Si.

Biotite occurs frequently only in the more evolved members of the high-K calc-alkaline series.

Plagioclase is usually the most abundant phenocryst phase and generally displays complex oscillatory zoning. It is characteristically highly calcic (Fig. 6.31), but shows a wide range of compositions within each of the major magma series. The high Ca contents have been attributed to the high water contents of island-arc magmas. Quartz is the next most common felsic mineral, whereas alkali feldspar and feldspathoids are restricted to the high-K series.

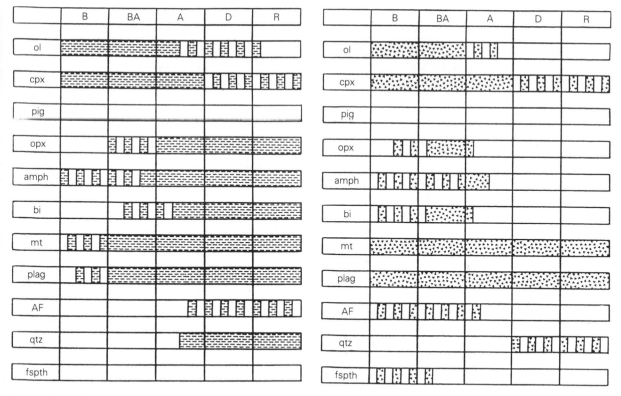

Figure 6.26 Major phenocryst mineralogy of the island-arc high-K calc-alkaline series.

Figure 6.27 Major phenocryst mineralogy of the island-arc shoshonitic series.

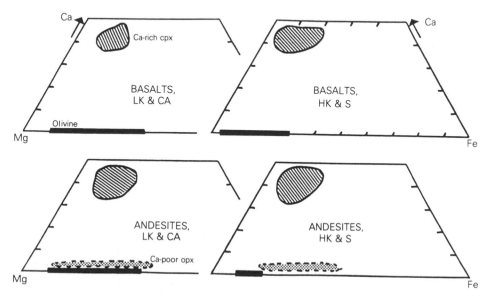

Figure 6.28 The pyroxene quadrilateral, showing the compositional ranges for olivine, Ca-rich clinopyroxene and Ca-poor orthopyroxene in basalts and andesites of the four major magma series (after Ewart 1982, Fig. 12, p. 64).

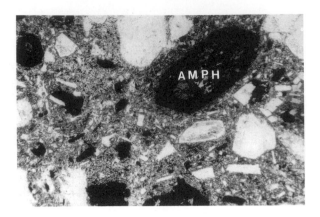

Figure 6.29 Hornblende andesite from the Lesser Antilles island arc, showing strongly resorbed amphibole phenocrysts (×40, crossed polars).

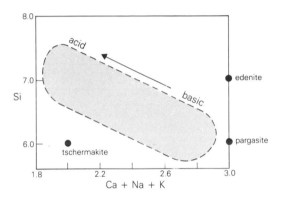

Figure 6.30 Compositional variation of amphiboles in island-arc volcanic rocks, plotted in terms of cations per formula unit (23 O) (after Ewart 1982, Fig. 15, p. 67).

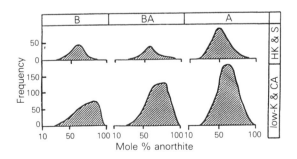

Figure 6.31 The variation in plagioclase composition between basalts (B), basaltic andesites (BA) and andesites (A) of the major magma series (after Ewart 1982, Fig. 9, p. 61).

Titanomagnetite is the normal Fe–Ti oxide phase occurring throughout the compositional range from basalt to rhyolite. Suppression of titanomagnetite crystallization during the early stages of evolution of the island-arc tholeiite series has been invoked to account for their characteristic trend of early iron enrichment. Coexisting ilmenite is generally absent, reflecting the low TiO_2 contents of island-arc magmas (Section 6.11).

Groundmasses can vary from glassy to microcrystalline, but generally contain the same minerals which also occur as phenocryst phases. Pigeonite is the characteristic groundmass pyroxene in tholeiitic rocks, in contrast to hypersthene in calc-alkaline rocks. However, these are often very difficult to identify using the optical microscope.

6.11 Chemical composition of erupted magmas

6.11.1 Major elements

SiO_2, TiO_2, Al_2O_3, Fe_2O_3, FeO, MnO, MgO, CaO, Na_2O, K_2O, P_2O_5 and H_2O can all be considered as major elements in the description of the geochemistry of island-arc magmas. In terms of these, the most obvious distinction between the major magma series is one of increasing total alkali content in the sequence tholeiitic–calc-alkaline – high-K calc-alkaline – shoshonitic, K_2O showing proportionately the greater increase. This has already been used as the basis for classification of island-arc volcanic suites (Section 6.7).

Figure 6.15, a plot of wt.% K_2O versus wt.% SiO_2, is one of a family of Harker diagrams (see Figure 6.32), which can be used to investigate the major element geochemical variations within suites of apparently co-genetic igneous rocks. K_2O generally behaves incompatibly within island-arc suites, and thus rocks which are genetically related should define linear trends with K_2O increasing progressively with increasing SiO_2. Similarly, in Figure 6.32, TiO_2, CaO and FeO + Fe_2O_3 show good negative correlations with silica for a suite of genetically related lavas from the Sunda arc of Indonesia, confirming the role of plagioclase and magnetite as major fractionating phases in the evolution of the magmas.

Shown in Figure 6.15 are an array of apparent evolutionary trends from different oceanic island-arc systems, the linearity of which strongly suggests the importance of low-pressure crystal fractionation in the evolution of the magmas. Of particular note are the divergent trends displayed by volcanics from the Aleutian and South Sandwich arcs. These can be most simply explained by variable conditions of low-pressure fractionation within individual arcs. In general, it is considered that

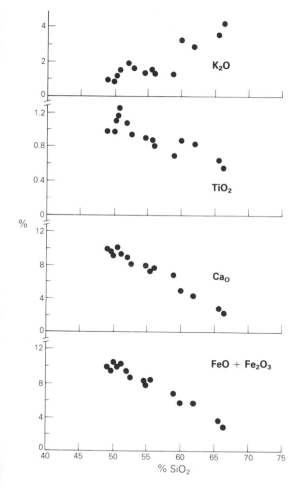

Figure 6.32 Harker diagrams showing the importance of low-pressure crystal fractionation in the evolution of island-arc volcanic suites. The data points shown are for the calc-alkaline lavas of the Rindjani volcano, Sunda arc (Foden 1983).

fractionation under reducing conditions suppresses the crystallization of magnetite, thus promoting iron enrichment in the early stages (tholeiitic trend). In contrast, under oxidizing conditions magnetite crystallizes from the outset, rapidly depleting residual liquids in iron (calc-alkaline trend; Osborn 1962, Miyashiro 1974). Taking the South Sandwich lavas as an example, we can see that in Figure 6.15 they define both a calc-alkaline and a low-K series. However, in terms of FeO^*/MgO versus SiO_2 (Fig. 6.17) both series display tholeiitic characteristics. This highlights one of the major sources of confusion in attempting to classify magmas as tholeiitic or calc-alkaline: magmas with higher K_2O contents than the low-K series can still fractionate along Fe-enrichment trends. Such trends are probably more closely related to differing aO_2/aH_2O conditions during high-level fractionation than to any fundamental differences in the chemistry of the parent magmas. An important point which should not be overlooked is that linear trends on variation diagrams such as Figures 6.15 & 6.32 are frequently composites of data from several distinct volcanic centres, sometimes spanning a range of ages. In such cases linearity does not necessarily imply a direct genetic relationship, and Sr isotopic analysis of the samples can often reveal significant isotopic heterogeneity (Section 6.11.5).

If crystal fractionation processes are indeed fundamental to the generation of the spectrum of erupted magma compositions in island arcs, then studies of basaltic magmas which might be parental to the suites become of paramount importance. This will be considered further in Section 6.11.3. Table 6.2 shows some typical major element analyses of island-arc basalts compared to those from mid-ocean ridge and intra-oceanic plate settings. These data reveal a remarkable similarity in terms of most major elements between all oceanic basalts. Only TiO_2 appears to be distinctive, being characteristically low in island-arc basalts. Tables 6.3−5 compare the major element chemical variations within the tholeiitic, calc-alkaline and high-K calc-alkaline magma series respectively.

6.11.2   Trace elements

Major element studies of island-arc volcanic suites

Table 6.2 A comparison of the major element composition of basaltic rocks from island arcs, mid-oceanic ridges and oceanic islands.

	Average Atlantic MORB[a]	Island-arc tholeiitic basalt[b]	Island-arc calc-alkaline basalt[b]	Island-arc high-K calc-alkaline basalt[b]	Hawaiian alkali basalt[c]	Hawaiian tholeiite basalt[d]
SiO_2	50.67	49.20	49.40	51.00	44.50	49.20
TiO_2	1.28	0.52	0.70	0.93	2.15	2.57
Al_2O_3	15.45	15.30	13.29	13.6	14.01	12.77
FeO*	9.67	9.00	10.15	8.11	12.51	11.40
MnO	(0.15)	0.18	0.20	0.14	0.19	0.17
MgO	9.05	10.1	10.44	12.50	10.12	10.00
CaO	11.72	13.00	12.22	7.92	10.63	10.75
Na_2O	2.51	1.51	2.16	2.67	2.47	2.12
K_2O	0.15	0.17	1.06	2.37	0.53	0.51
P_2O_5	0.20	0.06	0.20	0.59	0.42	0.25

() = estimated.
Data sources: [a] Best (1982); [b] Perfit *et al.* (1980); [c] Clague & Frey (1982); [d] Hughes (1982).

Table 6.3 Major and trace element characteristics of the island-arc tholeiite series lavas from the South Sandwich arc (Luff 1982).

	Basalts		Basaltic andesite	Andesite	Dacite
%					
SiO_2	45.80	50.49	57.10	62.39	64.81
TiO_2	0.17	0.70	0.92	1.00	0.91
Al_2O_3	20.05	19.44	16.15	14.12	13.92
Fe_2O_3	8.53	9.83	10.15	9.76	9.06
MnO	0.16	0.17	0.18	0.18	0.18
MgO	10.49	4.26	3.38	2.13	1.62
CaO	13.44	11.66	8.47	6.16	5.42
Na_2O	0.98	2.53	3.56	4.06	4.38
K_2O	0.04	0.14	0.32	0.44	0.49
P_2O_5	0.10	0.11	0.15	0.17	0.17
ppm					
Rb	1	2	8	9	9
Sr	101	121	116	110	107
Ba	7	36	66	103	115
Cr	177	55	51	37	30
Ni	27	5	2	0	0
Y	3	14	27	35	38
Zr	10	21	51	71	83

frequently reveal the importance of crystal fractionation of parental basalt magmas in the generation of the more evolved magma types (Meijer & Reagan 1981, Foden 1983). In this section we will focus our attention on the trace element characteristics of island-arc basalts in an attempt to characterize more fully the nature of the partial melting processes involved in their genesis. Table 2.1 summarizes those features of the trace element geochemistry of the basalts which have petrogenetic significance.

Trace element abundances in oceanic island-arc basalts (Table 6.6), are conventionally compared with those of N-type MORB (Ch. 5), as this is a relatively well understood magma type also derived from the oceanic upper mantle. Typically, the arc basalts are characterized by selective enrichment of incompatible elements of low ionic potential (Sr, K, Rb, Ba ± Th) and low abundances of elements of high ionic potential (Ta, Nb, Ce, P, Zr, Hf, Sm, Ti, Y, Yb, Sc and Cr) relative to N-type MORB (Basaltic Volcanism Study Project 1981, Pearce 1982). The low abundances of most incompatible elements are highly significant in terms of the source of island-arc basalts, as MORB itself is thought to be derived from a chemically depleted source (Ch. 5).

Island-arc basalts in general have low Ni contents, which suggests that they are not primary magma (Table 2.1) and have undergone olivine fractionation en route to the surface. This would tend to increase the incompatible element concentrations from those of the primary magmas, even further enhancing the observation that island-arc basalts are highly depleted in certain incompatible elements.

The low ionic potential elements are those most readily mobilized by a fluid phase, and their enrichment in island-arc basalts has been attributed to metasomatism of their mantle source region by hydrous fluids derived from the subducted oceanic crust. The low abundances of the high ionic potential elements have been variously attributed to (Pearce 1982):

(1)   high degrees of partial melting of the mantle source;

(2)   stability of minor residual phases (e.g. rutile, zircon and sphene) in the mantle source, which preferentially concentrate a range of trace elements;

(3)   remelting of an already depleted mantle source.

Following Sun (1980), Figures 6.33 & 6.34 compare incompatible element abundances, normalized to primordial mantle values, for basaltic magmas erupted in both arc and non-arc settings. The elements are ordered in sequence of decreasing incompatibility from left to right in a four-phase lherzolite undergoing partial fusion (Ch. 2). Compared to the range of basaltic magmas generated at mid-ocean ridges and oceanic islands, island-arc basalts have a highly distinctive 'spiked' trace element pattern or spiderdiagram (Ch. 2), regardless of whether they belong to the tholeiitic, calc-alkaline or shoshonitic magma series. Marked spikes occur at Sr, K, Ba and to a lesser extent U.

Table 6.4   Major and trace element characteristics of the island-arc calc-alkaline series lavas from the South Sandwich arc (Luff 1982).

	Basalt	Andesite	Dacite
%			
SiO_2	49.24	60.23	66.29
TiO_2	0.55	1.12	0.67
Al_2O_3	20.16	14.13	13.46
Fe_2O_3	9.39	11.71	8.71
MnO	0.17	0.22	0.17
MgO	4.80	2.13	1.00
CaO	11.61	6.12	4.18
Na_2O	2.71	3.78	4.26
K_2O	0.27	1.18	1.66
P_2O_5	0.31	0.26	0.23
ppm			
Rb	4	29	39
Sr	198	137	122
Ba	58	243	332
Cr	54	28	26
Ni	11	3	2
Y	14	39	50
Zr	23	126	178

Table 6.5 Major and trace element characteristics of the island-arc high-K calc-alkaline series lavas from the Sunda arc (Rindjani volcano; Foden 1983).

	Basalts		Basaltic andesite	Andesite	Dacite
%					
SiO_2	48.32	50.20	55.49	61.69	66.22
TiO_2	0.69	1.13	0.91	0.86	0.56
Al_2O_3	10.53	18.11	18.45	16.85	16.86
Fe_2O_3	1.53	1.62	1.39	0.96	0.53
FeO	7.81	8.28	7.07	4.88	2.69
MnO	0.17	0.18	0.16	0.17	0.10
MgO	14.02	5.63	3.10	1.90	1.14
CaO	14.38	9.71	7.47	4.42	2.52
Na_2O	1.50	3.67	4.09	4.82	4.96
K_2O	0.90	1.21	1.60	3.00	4.26
P_2O_5	0.15	0.25	0.28	0.47	0.15
ppm					
Rb	21	20	35	69	113
Sr	452	452	433	405	216
Ba	—	—	—	—	—
Cr	510	219	27	27	12
Ni	151	71	2	1	2
Y	13	20	30	45	40
Zr	53	82	134	244	310
Nb	2	4	2	—	10

Table 6.6 Trace element concentration in island-arc basalts, compared to other oceanic basalt magma types.

	N-type[a] MORB	E-type[a] MORB	Within plate[b] tholeiitic	Back-arc[c] tholeiitic	Island-arc[a] tholeiitic	Island-arc[a] calc-alkaline	Within plate[a] alkalic
Rb	1.0	3.9	7.5	6	4.6	14	22
Ba	12	68	100	77	110	300	380
K	1060	1920	4 151	3569	3240	8640	9 600
Nb	3.1	8.1	13	8	0.7	1.4	53
La	3.0	6.3	(9)	7.83	1.3	10	35
Ce	9.0	15.0	31.3	19.0	3.7	23	72
Sr	124	180	290	212	200	550	800
Nd	7.7	9.0	(19)	13.1	3.4	13	35
Zr	85	75	149	130	22	40	220
Sm	2.8	2.5	5.35	3.94	1.2	2.9	13
Ti	9300	8060	13 369	8753	3000	4650	20 000
Y	29	22	26	30	12	15	30
Th	0.20	0.55	—	—	0.25	1.1	3.4
U	0.10	0.18	—	—	0.10	0.36	1.1

() = estimated value.
Data sources: [a] Sun (1980); [b] Pearce (1982); [c] Hawkesworth *et al.* (1977).

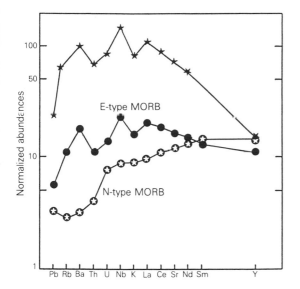

Figure 6.33 Incompatible element abundances, normalized to primordial mantle values, in mid-ocean ridge tholeiite (MORB) and oceanic-island alkali basalts (after Sun 1980).

There is a pronounced trough at Nb which is of particular interest in considering the petrogenesis of island-arc alkali basalts as alkali basalts from non-arc settings characteristically have very high Nb contents.

The significance of the relative magnitudes of the

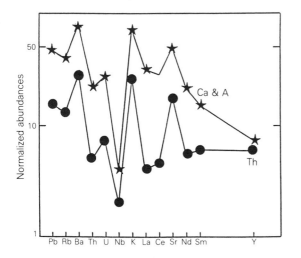

Figure 6.34 Incompatible element abundances, normalized to primordial mantle values, in island-arc basalts (after Sun 1980).

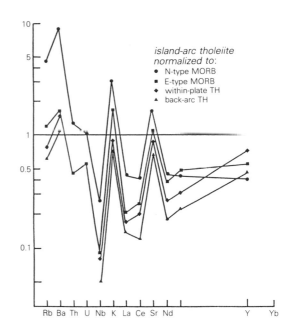

Figure 6.35 Incompatible element abundances in an island-arc tholeiitic basalt, normalized to a range of tholeiitic basalts from other tectonic settings (data from Table 6.7).

various spikes has been the subject of considerable debate (Dupuy *et al.* 1982, Hole *et al.* 1984, Thompson *et al.* 1984, Arculus & Powell 1986). However, it is important to realize the rather arbitrary nature of Sun's normalization factors (Sun 1980). Figure 6.35 shows the typical island-arc tholeiite from Figure 6.34, normalized to a range of tholeiitic basalts generated in non-arc environments (Table 6.6). The spiked trace element pattern still remains as a distinctive feature. However, it is apparent that when compared to tholeiitic basalt types other than MORB, island-arc tholeiites are actually depleted in the whole range of trace elements.

Rare earth element patterns (Fig. 6.36), for island-arc basalts show a wide spectrum from light-REE depleted to flat to strongly light-REE enriched. These broadly correspond to the major subdivisions between the different magma series based on K_2O content. For example, island-arc tholeiitic basalts typically have light-REE depleted patterns, whereas the calc-alkaline basalts are light-REE enriched. In general, the relative magni-

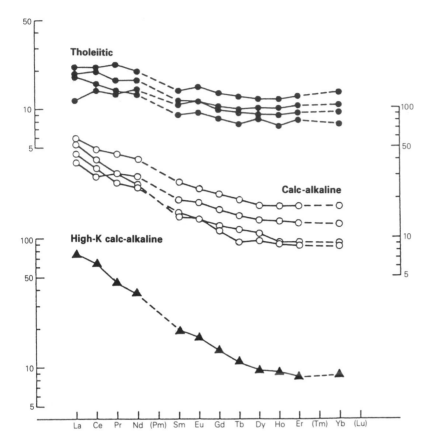

Figure 6.36 Chondrite-normalized rare earth element abundances in island-arc basalts.

tude of the Sr, K and Ba spikes appears to correlate with the degree of light-REE enrichment (Fig. 6.34).

The normalized incompatible element abundance patterns for island-arc basalts indicate that they cannot have been derived from MORB source or ocean-island tholeiite source mantle through the involvement of normal mantle phases, i.e. olivine, orthopyroxene, clinopyroxene, garnet, spinel and plagioclase (Ch. 3). Their characteristic spiked trace element signature appears to require the addition of a component rich in Sr, Ba, K, Pb and light REE, presumably derived from the subducted lithospheric slab, to the lherzolites of the mantle wedge. This component could either be a hydrous fluid or a water-saturated silicate melt. Variations in parental basalt composition from tholeiitic to calc-alkaline to alkaline must then reflect the relative proportions of this LIL (Large-Ion Lithophile) enriched component and normal mantle lherzolite

phases entering subsequent partial melts.

Figure 6.37 is a plot of Ba/La versus La/Sm (Arculus & Powell 1986), which clearly reveals the characteristic high Ba/La ratios of island-arc basalts compared to MORB and intra-plate basalts. The high ratio appears to be the consequence of the enrichment of the mantle wedge by Ba-rich subduction-zone fluids, with much of the Ba being derived from subducted oceanic sediments (Hole *et al.* 1984).

Figure 6.38 is a particularly useful trace element variation diagram devised by Pearce (1983) to highlight the nature of the subduction-zone component. The elements plotted all behave incompatibly ($D \ll 1$) during most partial melting and fractional crystallization events, with the exception of Sr which may be concentrated in plagioclase, Y and Yb in garnet and Ti in magnetite. Element abundances are normalized to those in a typical mid-ocean ridge basalt (MORB), and the elements are ordered

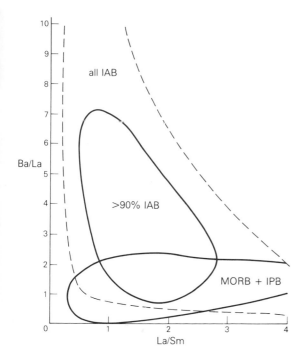

Figure 6.37 Plot of Ba/La versus La/Sm for island-arc basalts (IAB), compared to MORB and intra-plate basalts (IPB) (after Arculus & Powell 1986).

placed from the MORB line at 1.0 simply as a consequence of a different degree of partial melting, or subsequent crystal fractionation, than the MORB sample chosen for the normalization factor. The shaded area above the line gives an indication of the contribution to the magma due to the subduction-zone component added to the mantle wedge.

Comparing the patterns for typical island-arc calc-alkaline and tholeiitic basalts shows that Sr, K, Rb, Ba and Th show an even greater degree of enrichment in the former. In addition, Ce, P and Sm are also enriched. However, the immobile elements Ta, Nb, Zr, Hf, Ti, Y and Yb still define a relatively flat trend parallel to the MORB pattern,

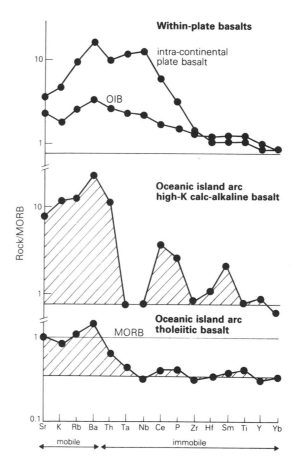

Figure 6.38 MORB-normalized trace element patterns for tholeiitic and high-K calc-alkaline island-arc basalts, compared to within-plate basalts (data from Pearce 1983).

in the diagram in terms of their mobility in an aqueous fluid phase and their relative incompatibility. Sr, K, Rb and Ba are classed as mobile and plot at the left of the pattern, whereas elements Th to Yb are generally immobile. The elements are arranged so that the incompatibilities of both mobile and immobile elements increase from the outside to the centre of the pattern.

For oceanic-island arc tholeiitic basalts, the part of the pattern from Ta to Yb (i.e. the immobile elements) lies parallel to but at a lower level than MORB, which plots as a horizontal line at 1.0 as the data are MORB-normalized. In contrast, Sr, K, Rb and Ba (and to a lesser extent Th) are enriched above this level. A line drawn through the Ta–Yb part of the pattern and extrapolated towards Sr should represent what the magma composition would have been without an element input from the subduction zone, assuming that it was derived by partial melting of MORB source mantle (depleted asthenosphere). This horizontal line may be dis-

presumably reflecting the pre-subduction characteristics of the mantle wedge. Ce, P and Sm are much more likely to be transported in a partial melt than an aqueous fluid, and this may reflect a fundamental difference in the petrogenesis of magmas of the tholeiitic and calc-alkaline series (Hawkesworth & Powell 1980).

Shown for comparison in Figure 6.38 are patterns for a typical oceanic-island basalt (Ch. 9) and a within-continental-plate basalt (Ch. 10). In general, all basalts erupted in within-plate settings (both oceanic and continental) are enriched in most incompatible elements compared to MORB and show 'humped' MORB-normalized trace element patterns. This suggests similarities between the sources of continental intra-plate basalts (?the subcontinental lithosphere) and oceanic-island basalts (OIB). Indeed, McKenzie & O'Nions (1983) have suggested that recycled continental lithosphere provides a component for some OIB magmatism.

It is appropriate at this stage to reconsider the significance of the spiked spiderdiagram pattern in Figure 6.34. From the above discussion it is clear that the marked trough at Nb may not actually reflect a real depletion in Nb as its concentration is close to that in MORB. The apparent sharpness of the trough is in fact a consequence of the marked enrichment of the adjacent elements U and K in the spiderdiagram. Nevertheless, in diagrams such as Figure 6.38, the elements Ta to Yb frequently plot well below the MORB line and it is possible therefore that stabilization of minor residual phases in the mantle wedge could be responsible, although such phases would have to retain all these elements equally (Saunders *et al.* 1980, Briqueu *et al.* 1984, Thompson *et al.* 1984, Arculus & Powell 1986).

Comparison of the spiderdiagram patterns for island-arc basalts with those of MORB and oceanic-island alkali basalts (Figs 6.33 & 4), reveals that both groups show the same range in overall slope of the REE pattern (La, Ce, Sm) from light-REE depleted to light-REE enriched. Thompson *et al.* (1984) consider that this is because the convecting mantle wedges above subducted slabs contain variable proportions of MORB-source and OIB-source components, to which are added fluids

derived from the subducted slab. Thus they regard low-K oceanic island-arc tholeiites as the hydrous subduction-related equivalents of MORB, whereas calc-alkali basalts and shoshonites are generated from subduction-modified OIB source components. This seems an attractive proposition meriting further study, although we must exercise some caution as it is easy to get involved in rather circular arguments. If correct, this would require that, for the generation of calc-alkaline and shoshonitic magmas, the subduction-zone fluid has a strong negative Nb anomaly which, when superimposed upon the Nb-enriched OIB source component, would produce arc basalts with a negative Nb anomaly. Saunders *et al.* (1980) consider that subduction-zone fluids could have such a signature if Ta and Nb were retained in a titaniferous mineral phase such as ilmenite or sphene in the subducted oceanic crust.

The metasomatism of the mantle wedge by slab-derived fluids could fundamentally alter both its mineralogy and melting relationships. Sekine & Wyllie (1982b) have attempted to model these effects by experimentally investigating the system granite−peridotite−H$_2$O. They have successfully shown that, for example, olivine can be eliminated as a residual mantle phase during partial melting, facilitating the generation of basaltic andesite and andesite as primary partial melts. On the basis of these experiments alone, the presence of residual phases in the metasomatized mantle wedge, which could preferentially concentrate a range of trace elements, must be considered a distinct possibility.

6.11.3   Identification of primary magmas

In those arcs in which basalt forms a significant percentage of the eruptive products, candidates for primary magmas are sought amongst those basalts with high MgO contents (>6% MgO), 250−300 ppm Ni and 500−600 ppm Cr (Perfit *et al.* 1980). Generally, such compositions are rare in island-arc volcanic suites. Nevertheless, petrological and geochemical evidence frequently suggests the development of the basalt−andesite−dacite−rhyolite spectrum by low-pressure fractionation of such primary magmas.

In some arcs, however, individual volcanic centres are composed dominantly of andesitic—dacitic material. A major problem in these cases is whether the primary magma is also andesitic or basaltic andesitic in composition, implying a different source region or magma generation process. Hawkesworth & Powell (1980), in a detailed isotopic study of contrasting basic and andesitic centres from Dominica, Lesser Antilles, concluded that both types probably had a similar basaltic parent. However, further studies are required before such conclusions can be considered generally applicable. If correct, this emphasizes the importance of detailed geochemical studies of the most primitive basaltic compositions in island-arc volcanic suites as petrogenetic indicators.

6.11.4   Volatile contents

Juvenile volcanic gas is that originally contained in a magma, of which most is lost during the eruption and crystallization of the resultant lava. H_2O appears to be the predominant volatile species with lesser amounts of CO, CO_2, H_2S, SO_2, HCl and H_2 in order of decreasing abundance. H_2O/Cl ratios are lower in island-arc magmas than in those from other tectonic settings, and the Cl enrichment could reflect the involvement of a sea water component derived from the subducted oceanic crust in the magma generation process (Perfit *et al.* (1980).

Tholeiitic-arc magmas are generally considered to have low volatile contents. However, for calc-alkaline magmas high but unspecified water contents have been assumed, based upon the frequency of explosive eruptions, the presence of hydrous minerals as phenocrysts and the high An content of the plagioclase phenocrysts.

The amount of water that can dissolve in a magma increases greatly with pressure, as shown in Figure 6.39 for a typical basalt and an andesite respectively. Ascending magmas are unlikely to be water-saturated at depth but they will eventually rise to a level at which saturation and hence vesiculation occur. The relative ease with which the entrained gas can escape from the magma will control the explosivity of subsequent volcanic eruptions.

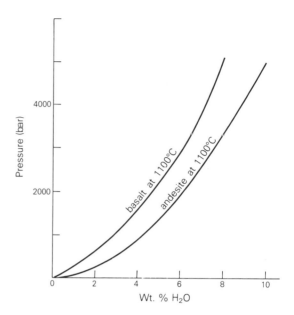

Figure 6.39   Maximum solubility of water in basaltic and andesitic magmas at 1100°C, as a function of pressure (after Hughes 1982, Fig. 6.1, p. 150).

Island-arc lavas have relatively high Fe_2O_3/FeO ratios, suggesting that the magmas are more hydrous and more oxidized than magmas from other tectonic settings. The oxidation state during low-pressure fractionation is believed to play a fundamental role in constraining whether magmas fractionate along tholeiitic trends of early iron enrichment (reducing conditions) or calc-alkaline trends of no iron enrichment (oxidizing conditions). Powell (1978) calculated the aO_2 during the crystallization of cumulate plutonic xenoliths (Section 6.6) from the Lesser Antilles island arc as being above that defined by the synthetic NNO oxygen buffer, i.e. significantly oxidizing conditions.

6.11.5   Radiogenic isotopes

Sr—Nd

Recent Nd and Sr isotope studies have significantly increased our understanding of the origins of island-arc magmas. In particular, they have demonstrated the involvement of both the subducted oceanic crust (basalt + sediments) and the overlying mantle wedge (DePaolo & Johnson 1979, Hawkesworth & Powell 1980, McCulloch & Perfit 1981, White & Patchett 1984, Davidson 1986, White & Dupre 1986). Unfortunately, these data cannot reveal the nature of the involvement of the subducted oceanic crust, i.e. whether the crustal component is derived from melting of the slab or merely from fluids released during dehydration.

$^{143}Nd/^{144}Nd$ and $^{87}Sr/^{86}Sr$ data are plotted for a range of volcanic rocks from oceanic island arcs in Figure 6.40, for comparison with the fields for mid-ocean ridge basalts (MORB; see Ch. 5), and ocean-island basalts (OIB; see Ch. 9), which together define a broad negative trend, the mantle array (Ch. 3). Island-arc volcanic rocks show considerable overlap with the OIB field and could therefore be derived from partial melts of enriched mantle

sources unmodified by subduction-zone components. When the mantle array was first defined in the 1970s, it was considered that most island-arc volcanic rocks showed small but variable displacements from it in the direction of increasing $^{87}Sr/^{86}Sr$. This was generally attributed to the involvement of a component with a high $^{87}Sr/^{86}Sr$ ratio, derived from the subducted oceanic crust, in the magma generation process (DePaolo & Johnson 1979, Hawkesworth & Powell 1980, Perfit et al. 1980). However, as more high-quality isotopic data have become available, the OIB field has expanded considerably (Ch. 9) and such simplistic interpretations are no longer tenable.

Two-arc systems, the Sunda−Banda arc of Indonesia and the Lesser Antilles arc in the Atlantic ocean, show considerable deviations from the MORB−OIB array in Figure 6.40. These data may be most easily accounted for by involving terrigenous sedimentary components with high $^{87}Sr/^{86}Sr$ and low $^{143}Nd/^{144}Nd$ ratios in the petrogenesis of the magmas. Samples with the lowest $^{87}Sr/^{86}Sr$ ratios in the Sunda−Banda arc plot within the OIB field (Whitford & Jezek 1982), which could imply that the mantle wedge beneath this arc had an enriched composition relative to MORB source mantle prior to subduction. In contrast, low $^{87}Sr/^{86}Sr$ samples from the Lesser Antilles overlap with the MORB field, which suggests the existence of depleted lherzolite (MORB source mantle) above the subducted slab. However, caution must be exercised in interpreting these data in this way, as location of samples within the mantle array does not necessarily indicate that they were derived from partial melts of unmodified mantle lherzolite. Mixing of a mantle component lying within the MORB field with an appropriate slab-derived component could very easily produce displacements of the bulk magma source along the mantle array (Fig. 6.41). In general, Nd and Sr isotopic data alone are insufficient to identify pre-subduction mantle compositions and must be combined with other isotope systems, e.g. Pb.

The Marianas, New Britain, Aleutian and South Sandwich island arcs are, isotopically at least, some of the simplest examples of oceanic island-arc magmatism. Their Sr−Nd isotopic data overlap the

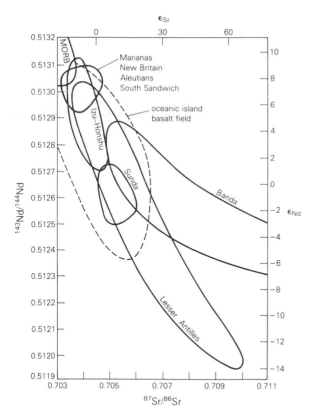

Figure 6.40 Variation of $^{143}Nd/^{144}Nd$ versus $^{87}Sr/^{86}Sr$ for oceanic island-arc volcanic rocks compared to the spectrum of oceanic basalts (MORB + OIB) (after Arculus & Powell 1986, Fig. 4, p. 5916).

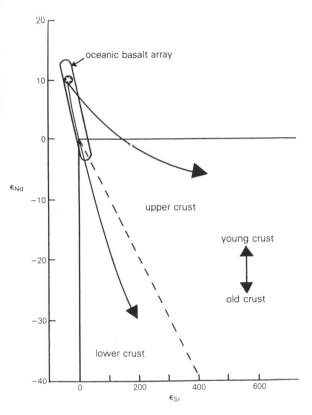

Figure 6.41 Plot of ε_{Nd} versus ε_{Sr} to show how contamination with materials having a continental crustal signature can produce variable displacements from the oceanic basalt array. For example, if subducted sediments derived from an old lower crustal terrane imprinted their isotopic characteristics upon fluids emanating from the slab, then the metasomatized mantle wedge could have a range of ε_{Nd}–ε_{Sr} values lying within the oceanic basalt array, depending upon the volume of metasomatizing fluid present. Equally, such a trend could be produced by contamination of ascending magmas by terrigenous sediments in the base of the island-arc crust. In contrast, involvement of a sedimentary component derived from a young upper crustal terrane can produce marked deviations from the array in the direction of increasing ε_{Sr} (modified from DePaolo & Wasserburg 1979).

MORB field, indicating that the arc basalts are generated from depleted asthenospheric mantle sources similar to those which melt to produce mid-ocean ridge basalts. However, the influx of fluids from the subducted oceanic crust is still required in their petrogenesis to account for their distinctive trace element geochemistries (Section 6.11.2).

A fundamental problem in the interpretation of the petrogenesis of island-arc volcanic suites which appear to carry an isotopic fingerprint of crustal contamination (e.g. the Lesser Antilles and the Banda arc) is the site at which such contamination occurs. It is now generally accepted that continental crustal material is recycled into the mantle at subduction zones via hydrothermally altered subducted oceanic crust and its veneer of oceanic sediments. The extent to which sediment subduction occurs is still a matter for speculation (Scholl *et al.* 1980), although most workers agree that some portion of the sedimentary layer must resist scraping off at the trench and must therefore be carried to greater depths. Several authors (Cohen & O'Nions 1982b, Hofmann & White 1982, White & Hofmann 1982) have considered sediment subduction as a mechanism for the localized enrichment of the mantle in incompatible elements, ultimately providing source components for OIB magmatism (Ch. 9).

If sediments are subducted they are most likely to partially melt at depth and thus carry their distinctive isotopic signature into the overlying mantle wedge in a melt phase. Such a process may be described as *source contamination*. Ascending mantle-derived magmas may additionally assimilate sedimentary material intercalated within the arc crust, thereby acquiring a signature of continental crustal contamination. This may be described as *high-level* or *crustal contamination*. Unfortunately, it is not possible to resolve the site at which contamination occurs (source or crustal or both) using radiogenic isotopes alone and stable isotope systems, particularly oxygen, must be considered. In Figure 6.42 the array of Lesser Antilles Sr–Nd isotopic data can be seen to intersect the MORB field at one end and the field of Atlantic sediments at the other. This data array could therefore be interpreted in terms of either source contamination by partial melts of subducted terrigenous sediments or high-level crustal contamination of the arc magmas.

Pb

Studies of the Pb isotopic characteristics of island-arc magmas have proved particularly useful in identifying the involvement of sedimentary compo-

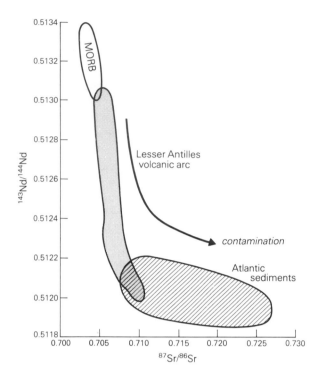

Figure 6.42 $^{143}Nd/^{144}Nd$ versus $^{87}Sr/^{86}Sr$ for volcanic rocks from the Lesser Antilles island arc (Davidson 1986), relative to the field of Atlantic sediments (White *et al.* 1985) and MORB.

nents in their petrogenesis. This is because the concentration of Pb in oceanic sediments is high, and its isotopic composition very distinctive compared to mantle lherzolite and its partial melts, making them highly sensitive to relatively small degrees of contamination (<10%). Figure 6.43 is a plot of $^{207}Pb/^{204}Pb$ versus $^{206}Pb/^{204}Pb$, which shows that, in general, island-arc magmas are characterized by higher $^{207}Pb/^{204}Pb$ ratios than MORB (White 1985), showing significant overlap with the fields for oceanic sediments. Arcs which plot within the MORB−OIB field in Figure 6.40 (Marianas, Aleutians, South Sandwich and Sunda) show variable displacements from the MORB field in Figure 6.43, indicating that even these have a sedimentary component involved in their petrogenesis. Data for the Lesser Antilles island arc extend to both higher and lower $^{207}Pb/^{204}Pb$ ratios than sediments sampled from the adjacent Atlantic plate (White *et al.* 1985), which has led Davidson

(1986) and Thirlwall & Graham (1984) to postulate that the magmas have become contaminated at crustal levels by a sedimentary component intercalated within the crust of the Caribbean plate, which may have a different provenance and thus different isotopic characteristics from the Atlantic sediments. White & Dupré (1986) suggest that subducted oceanic sediment contributes to the magma source of most, if not all, arcs. It is more obvious, however, in the Lesser Antilles because the sediments being subducted have much more radiogenic Pb isotopic compositions than most oceanic sediments, because they contain a terrigenous component derived from erosion of an Archaean shield area.

Figure 6.44 is a plot of $^{87}Sr/^{86}Sr$ versus $^{206}Pb/^{204}Pb$ for volcanic rocks from the Lesser Antilles, which clearly reveals the involvement of two distinct source components in the petrogenesis of the magmas, MORB source mantle and a terrigenous sedimentary component with $^{206}Pb/^{204}Pb$ ~ 20.0 and $^{87}Sr/^{86}Sr$ >0.710. This sedimentary component projects to the high-$^{206}Pb/^{204}Pb$ side of the Atlantic sediment field and may thus support Davidson's (1986) model for contamination by terrigenous sediments intercalated in the arc crust. However, it must be stressed that source versus crustal contamination models cannot be differentiated using radiogenic isotopes alone.

Be isotope data

The isotope ^{10}Be is produced by cosmic ray induced spallation reactions on oxygen and nitrogen in the upper atmosphere. It is transported via rain and snow to the Earth's surface, where it becomes preferentially incorporated into pelagic sediments which may subsequently be subducted and become incorporated into the source of island-arc magmas. Thus ^{10}Be may be a potentially useful tracer for the involvement of such sediments in the petrogenesis of subduction-related magmas. It has a half-life of 1.5×10^6 years, which is ideal for this purpose as it is long enough to function as a tracer for subduction but short enough to disappear from recycling on a longer timescale.

Tera *et al.* (1986) have measured the concentration of ^{10}Be in basalts and andesites from several

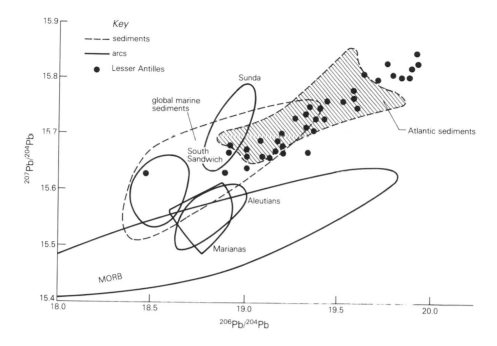

Figure 6.43 Variation of ^{207}Pb/^{204}Pb versus ^{206}Pb/^{204}Pb for oceanic island-arc volcanic rocks compared to MORB and oceanic sediments. Fields for global marine sediments, MORB, Atlantic sediments, South Sandwich and Aleutian arcs from White & Dupré (1986); Marianas data from Woodhead & Fraser (1985); Sunda arc data from Whitford & Jezek (1982); data for Lesser Antilles volcanic rocks from Davidson (1986) and White & Dupré (1986).

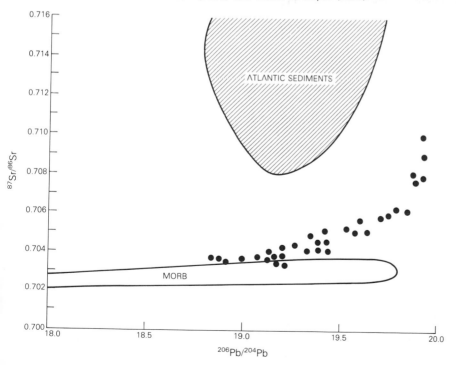

Figure 6.44 Variation of ^{87}Sr/^{86}Sr versus ^{206}Pb/^{204}Pb for volcanic rocks from the Lesser Antilles. These data reveal the apparent contamination of the magmas with a sedimentary component with ^{206}Pb/^{204}Pb ~20, which is slightly more radiogenic than analysed Atlantic sediments. Fields of Atlantic sediments and MORB from White & Dupré (1986); Lesser Antilles data from Davidson (1986) and White & Dupré (1986). (After White & Dupré 1986, Fig. 4.)

island-arc volcanoes and compared them with basaltic rocks from other environments. Figure 6.45 shows that volcanic rocks from non-arc settings typically have very low ^{10}Be concentrations, whereas arc-related magmas show a much wider range. Provided that it can be shown that the ^{10}Be has not been introduced by surface contamination, carried into magma chambers by ground water or generated *in situ* by radioactive decay of ^7Li, then its presence in the arc rocks seems to provide strong evidence for the involvement of subducted oceanic sediments. However, incorporation of ^{10}Be into the magma source region requires that several conditions be met (Tera *et al.* 1986). First, an adequate amount of ^{10}Be must be supplied to the trench, and the uppermost ^{10}Be-rich sediments must be subducted rather than accreted. Secondly, such sediments must be subducted and recycled into the mantle on a timescale of less than about 10 Ma. Failure of any of these conditions may prevent ^{10}Be becoming incorporated into the arc magma sources. Thus, while the presence of ^{10}Be in arc lavas provides unequivocal evidence for the role of sediment subduction, its absence does not prove the converse. Considerably more data need to be obtained on recent volcanic rocks before this method may be used to provide a reliable petrogenetic constraint on sediment subduction.

U–Th isotope data

Uranium has three naturally occurring isotopes, ^{238}U, ^{235}U and ^{234}U. Thorium exists primarily as one isotope, ^{232}Th, which is itself radioactive, although five other isotopes occur in nature as short-lived intermediate daughter products of ^{238}U, ^{235}U and ^{232}Th. The daughters of ^{238}U, which include ^{230}Th, are not in radioactive equilibrium with their parent in young volcanic rocks, the disequilibrium probably resulting from chemical fractionation during magma generation in the upper mantle. This is shown in Figure 6.46, a plot of activity ratios of $(^{230}Th/^{232}Th)$ versus $(^{238}U/^{232}Th)$. The equiline on this diagram defines the isotope ratios for equilibrium conditions.

All recent oceanic lavas plot to the left of the equiline, which has fundamental implications for partial melting processes in the upper mantle (Allègre & Condomines 1982). Partial melting fractionates the Th/U ratio of the system, whereas crystal fractionation does not, and in general produces a magma with a Th/U ratio greater than its mantle source, i.e. the melt is preferentially enriched in Th relative to U. In contrast, magmas generated in subduction-zone environments fall to

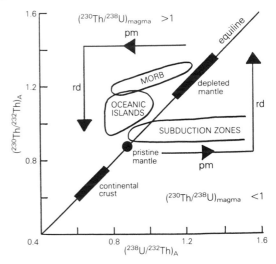

Figure 6.46 Plot of activity ratios of $(^{230}Th/^{232}Th)$ versus $(^{238}U/^{232}Th)$ for young volcanic rocks. The equiline defines the equilibrium ratios. Vectors show the evolution of the isotopic ratios during partial melting (pm) and subsequent radioactive decay (rd) (after Allègre & Condomines 1982).

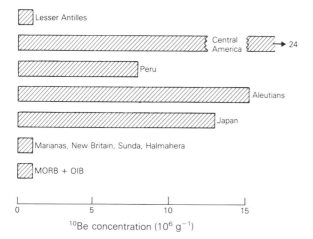

Figure 6.45 ^{10}Be concentrations in subduction-related magmas compared to MORB and OIB. (Lesser Antilles data from White & Dupré 1986; remaining data from Tera *et al.* 1986.)

the right of the equiline and therefore must be preferentially enriched in U relative to Th. This could suggest the importance of vapour phase transport in the generation of island-arc magmas, with U possibly being derived from subducted oceanic sediments.

Application of U–Th disequilibrium data to island-arc petrogenesis is still at the development stage, but clearly has significant potential.

6.11.6   Stable isotopes

Oxygen

The analysis of $^{18}O/^{16}O$ ratios in island-arc volcanic rocks is a powerful tool for tracing the recycling of materials ultimately derived from the continental crust in subduction-zone magmatism (James 1981; Davidson 1985, 1986) because of the large differences in oxygen isotopic composition between crustal rocks and rocks derived from the mantle (Fig. 6.47). Rocks which have reacted with the atmosphere or hydrosphere at low temperatures are typically richer in $\delta^{18}O$ than those from the mantle, but nevertheless significantly poorer than those sedimentary rocks derived by erosion of the continental crust.

By using a combination of $^{18}O/^{16}O$ and $^{87}Sr/^{86}Sr$ data it is possible in some instances to assess which

of the following processes have been responsible for the observed isotopic characteristics of island-arc magmas:

(a) *Crustal contamination:* Contamination of mantle-derived magmas through assimilation of, or isotopic equilibration with, materials having a continental crustal signature in the base of the arc crust.

(b) *Source contamination:* Reflux of subducted continentally derived sedimentary material into the mantle source of the arc magmas.

These two contamination processes are fundamentally different. Crustal contamination involves the physical and chemical reaction of the magma with the crustal rocks it intrudes and, being an endothermic process, it causes the magma to fractionate. Source contamination, on the other hand, involves closed system partial melting of a mixture of crustal and mantle materials within the mantle.

In terms of Figure 6.48, the effects of these two processes will be quite different. For source contamination the trace element concentrations in the unmodified mantle are so low that the trace element concentrations and Sr and Nd isotopic ratios of the modified mantle will be dominated by the highly enriched slab-derived fluid or partial melt, even if the proportion of the latter involved in small. The oxygen isotopic composition will approximate a simple linear function of the bulk proportion of slab-derived fluid (melt) to the total mantle material with which it equilibrates and of the $\delta^{18}O$ of the two end-members. Figure 6.48 shows that a suite of magmas generated by source contamination processes should lie on a straight or concave downwards mixing curve. In contrast, for the case of crustal contamination, the Sr contents of the magma and the assimilated crust should be approximately the same or greater in the magma. Therefore the mixing curves shown in Figure 6.48 should be convex. Thus, theoretically, a combined Sr–O isotopic study of a suite of genetically related lavas should reveal which, if any, of these two contamination processes has been involved in their genesis. Obviously, the existence of contamination by continental crustal components must have been estab-

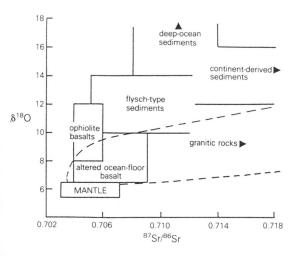

Figure 6.47 $\delta^{18}O$ versus $^{87}Sr/^{86}Sr$ for common igneous and sedimentary rocks (after James 1981, Fig. 2).

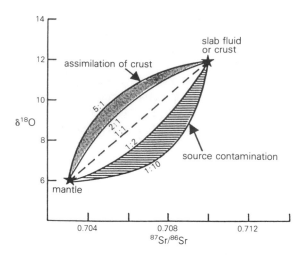

Figure 6.48 Theoretical two-component mixing curves for $\delta^{18}O$ versus $^{87}Sr/^{86}Sr$. Ratios shown on each curve denote the proportion of Sr in the mantle or mantle-derived end-member to the proportion of Sr in the crustal contaminant or slab-derived fluid (after James 1981, Fig. 6).

lished previously by combined Sr−Nd−Pb isotopic studies.

Unfortunately, there is very little oxygen isotope data available for island-arc volcanic suites. Margaritz *et al.* (1978) showed a good hyperbolic correlation between $^{87}Sr/^{86}Sr$ and $\delta^{18}O$ in the lavas of the Banda arc, Indonesia, which were shown in Section 6.11.5 to be significantly contaminated by continental crustal materials. This correlation is concave and thus consistent with a model of source contamination. In contrast, Davidson (1985, 1986) has demonstrated a convex relationship between $\delta^{18}O$ and $^{87}Sr/^{86}Sr$ for volcanic rocks from Martinique, Lesser Antilles, which he interprets in terms of intracrustal contamination of the magmas.

6.12   Detailed petrogenetic model

It is now generally accepted that the basaltic magmas erupted in island arcs have to be generated by partial melting of the mantle wedge above the subducted lithospheric slab (Green 1982; Mysen 1982; Wyllie 1982, 1984; Arculus & Powell 1986).

In most circumstances it is the asthenospheric part of this mantle wedge which is most likely to melt, the overlying lithospheric mantle already having been rendered significantly refractory by previous partial melting events associated with the generation of mid-ocean ridge basalt. This asthenospheric mantle component could show substantial pre-subduction heterogeneity, being a mixture of MORB-source and OIB-source mantle components. In addition, it is a source which can be potentially replenished by convective overturn in the mantle wedge related to instabilities generated by the subduction of cold oceanic lithosphere.

If no additional components were involved in the magma generation process, then the major and trace element characteristics of the resultant partial melts should be broadly similar to the range of mid-ocean ridge and ocean-island basalts. However, the distinctive trace element characteristics of arc basalts require distinctive source characteristics unique to the subduction-zone environment. These characteristics can most readily be achieved by metasomatism of the lherzolite of the mantle wedge by fluids ascending from the subducted oceanic lithosphere.

Satisfactory interpretation of the trace element and Sr, Nd and Pb isotope geochemistry of island-arc lavas requires the involvement of a component derived from the subducted oceanic crust. This could be either a hydrous fluid or a partial melt, and could have variable Sr, Nd and Pb isotopic ratios depending upon the extent of ocean-floor metamorphism of the subducted crust and the extent of involvement of subducted sedimentary components. Wyllie (1982) stressed that aqueous fluids derived from dehydration reactions within the subducted oceanic crust must contain considerable proportions of dissolved silicates at the pressures involved. It is therefore likely that there is a continuous spectrum of fluid compositions generated within the slab, ranging from aqueous SiO_2-rich fluids to hydrous andesitic partial melts.

Basaltic island-arc magmas ranging in chemistry from low- to high-K and showing variable displacements from the MORB−OIB arrays in Sr, Nd and Pb isotope systems, all show the same characteristic spiked trace element pattern (Fig. 6.34). This suggests that the trace element and isotope systems can become variably decoupled during the magma generation process. It is suggested here that all island-arc magmas are ultimately related to a parental partial melt of mantle lherzolite, metasomatized to varying degrees by an isotopically heterogeneous slab-derived fluid.

In most of the arc systems studied so far, evidence for the involvement of sedimentary components suggests that, at most, it is limited to a few per cent. This may provide indirect evidence for the non-subductability of oceanic sediments. However, Sr, Nd and Pb isotopic data for the Sunda−Banda and Lesser Antilles arcs indicate substantial involvement of terrigenous sediments derived from an old continental crustal terrane.

Such sediments may represent clastic wedges developed at passive continental margins during the early stages of ocean opening. A major and as yet largely unresolved problem is the site at which such sediments become involved in the magma generation process. Thus far, it has been assumed that it is the slab-derived fluids which carry the continental crustal signature. However, many authors have refuted the physical possibility of significant sediment subduction and thus, indirectly, the involvement of partial melts of subducted sediments to account for the isotopic data. Alternatively, mantle-derived magmas could become contaminated by terrigenous sediments *in situ* in the base of the island-arc crust. The results of such contamination would be isotopically indistinguishable from the

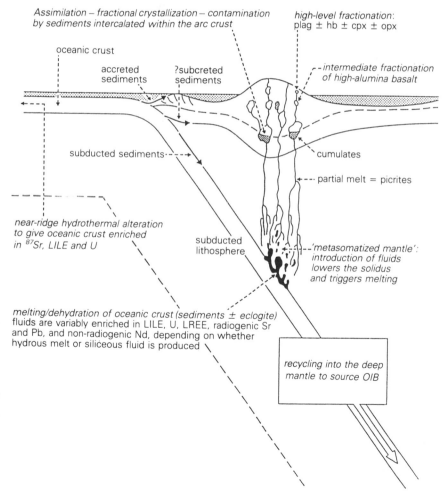

Assimilation − fractional crystallization − contamination by sediments intercalated within the arc crust

high-level fractionation: plag ± hb ± cpx ± opx

oceanic crust

accreted sediments

?subcreted sediments

intermediate fractionation of high-alumina basalt

subducted sediments

cumulates

partial melt = picrites

near-ridge hydrothermal alteration to give oceanic crust enriched in ^{87}Sr, LILE and U

subducted lithosphere

'metasomatized mantle': introduction of fluids lowers the solidus and triggers melting

melting/dehydration of oceanic crust (sediments ± eclogite) fluids are variably enriched in LILE, U, LREE, radiogenic Sr and Pb, and non-radiogenic Nd, depending on whether hydrous melt or siliceous fluid is produced

recycling into the deep mantle to source OIB

Figure 6.49 Summary of the magma generation processes in an ocean−ocean collision zone (after Davidson 1984).

source contamination model. Combined Sr−O isotopic studies do, however, provide a means of distinguishing between the two.

There must be substantial circulation of sea water in the crust during the early submarine stages of arc development. This provides a potential reservoir of hydrothermally altered volcanic rocks enriched in $^{87}Sr/^{86}Sr$, which could contaminate magmas stored in high-level magma chambers during the later stages of arc development. Isotopically, the effects of such contamination might be difficult to distinguish from those of aqueous fluids ascending from the subducted oceanic crust. Oxygen isotope studies may be of crucial importance in resolving the effects of such high-level contamination.

The compositions of evolved magmas, such as andesite, erupted in island arcs are sufficiently diverse that no single magma-generation process is capable of explaining the range of chemical characteristics. Crystal fractionation of plagioclase−olivine−clinopyroxene−orthopyroxene−magnetite ± amphibole from basaltic parent magmas is probably the major process (Powell 1978). However, from the experiments of Sekine & Wyllie (1982a,b) generation of primary magmas more SiO_2-rich than basalt from a highly metasomatized mantle source remains a possibility. In addition, high-level contamination of basaltic magmas by terrigenous sediments in the base of the arc crust could generate more acidic magmas.

The chemistry of the unusual group of high-MgO andesites called boninites indicates that they are derived from an ultra-depleted harzburgite source. However, even these lavas display the characteristic island-arc enrichment in K, Ba, Rb and Sr, which suggests that an incompatible element enriched fluid was added to the harzburgite prior to the melting event. This metasomatized harzburgite could represent the lithospheric part of the mantle wedge, which does not normally become hot enough to melt.

Figure 6.49 presents a summary of the various processes which may operate during magma generation in an ocean−ocean collision zone. This provides a basis for consideration in the next chapter of the more complicated type of subduction-related magmatism in which the overriding plate is a continental one.

Further reading

Best, M.G. 1982. *Igneous and metamorphic petrology*. New York: W.H. Freeman; see Chapter 3.

Gill, J.B. 1981. *Orogenic andesites and plate tectonics*. Berlin: Springer-Verlag.

Hughes, C.J. 1982. *Igneous petrology*. Amsterdam: Elsevier; see Chapter 11.

Thorpe, R.S. (ed.) 1982. *Andesites*. Chichester: Wiley.

CHAPTER SEVEN

Active continental margins

7.1 Introduction

In Chapter 6 we considered the simplest type of subduction-related magmatism produced as a consequence of the subduction of one oceanic plate beneath another. Now we shall focus our attention on the more complex case in which the overriding plate is a continental one. Magmas generated in this tectonic environment occur along the west coast of the Americas, Japan, Sumatra, Alaska, New Zealand and the Aegean (Fig. 7.1).

Since the early days of plate tectonics, the South American Andes have been cited as the type example of an ocean−continent collision zone, or active continental margin (Mitchell & Reading 1969), and much of the discussion in the following sections will be based on Andean data. A volcanic arc developed upon an uplifted surface of Precam-

brian and Palaeozoic rocks along much of the Pacific margin of the Americas by the late Triassic or early Jurassic (Dalziel 1986) and volcanic activity has been essentially continuous to the present day along different segments of the plate margin. However, compared to North America, the South American continental margin has been a comparatively simple active margin since Triassic times and consequently it may be utilized to develop general models for a variety of geological processes which may then be applied to more complex tectonic situations.

The orogenic andesite association characteristic of island arcs (Ch. 6) also typifies the volcanism of active continental margins and, in many respects, is broadly similar, although the passage of magmas through thick continental crust produces added complexities. Although it was once considered that

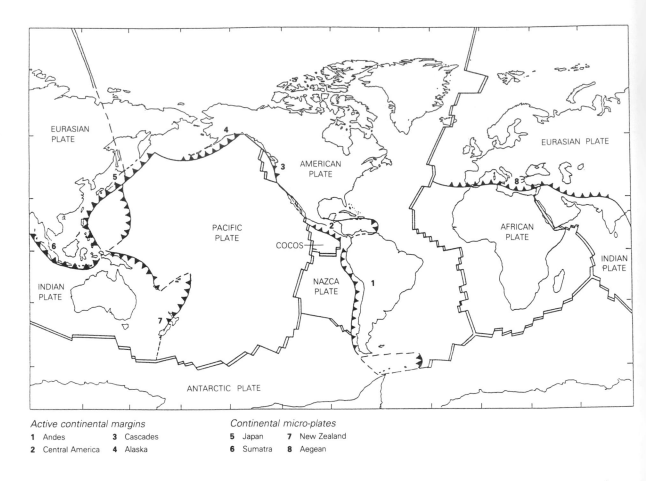

Active continental margins
1 Andes   3 Cascades
2 Central America 4 Alaska

Continental micro-plates
5 Japan   7 New Zealand
6 Sumatra   8 Aegean

Figure 7.1 Location of the major active continental margins and subduction systems involving continental micro-plates.

such margins were dominated exclusively by calc-alkaline rocks (Baker 1982), it is clear that the four main magma series recognized in island arcs (low-K, calc-alkaline, high-K and shoshonitic) are all represented. Additionally, alkaline lavas are often closely associated with the calc-alkaline volcanics, but generally form a separate zone of activity to the landward side of the volcanic belt (Thorpe *et al.* 1982). These may not necessarily be directly related to the subduction system being formed in an extensional regime similar to that of back-arc basins (Ch. 8).

One of the most conspicuous differences between the island-arc and continental-margin calc-alkaline series is the greater abundance of more silica-rich magmas (dacites and rhyolites) in the latter. Much

of this additional volume of acid rock occurs as pyroclastic flow material (ignimbrite) and appears to have a particular association with zones of thickened continental crust. It is therefore a distinct possibility that these acid magmas are derived, at least in part, by partial melting of the continental crust.

Chemically, the most distinctive features of the continental-margin volcanic suites compared to those erupted in oceanic island arcs are the higher concentrations of K, Sr, Rb, Ba, Zr, Th and U, higher K/Rb and Fe/Mg ratios and a much wider range of $^{87}Sr/^{86}Sr$, $^{143}Nd/^{144}Nd$ and Pb isotopic compositions. These characteristics must be largely explained in terms of crustal involvement in the petrogenesis of the magmas, although the distinctive geochemical characteristics of the subcontinental mantle wedge may also be important (Section 7.7).

In any destructive plate margin environment (oceanic or continental) the nature and distribution of magmatic activity in the overriding plate is directly linked to the geometry of the subducted slab (Pilger 1984). This, in turn, is a function of the convergence rate of the lithospheric plates, the age of the subducted lithosphere and the presence of features such as aseismic ridges, oceanic island and seamount chains, oceanic plateaus and microcontinents in the underthrust plate. The latter, by virtue of their increased crustal thickness, are more buoyant and tend to resist subduction, frequently becoming accreted on to the plate margin when they collide with the landward plate. Numerous such accreted terrains have now been recognized on the North American continental margin (Uyeda 1982, Nur & Ben Avraham 1983), making it a more complex example of an active continental margin than the South American Andes.

The most complex case of subduction-related magmatism occurs where two continental plates approach and collide by subduction of the intervening ocean, e.g. the Alpine–Himalayan system. The consequent suture zone becomes an area of thickened crust characterized by complex tectonic and magmatic activity and uplift. After collision, calc-alkaline andesites and dacites may be erupted, followed by alkaline volcanism as extensional tectonic regimes develop as a consequence of the rapid uplift (Harris et al. 1986). Houseman et al. (1981) have suggested that, during a collision orogeny, the thickened subcontinental mantle root may become detached and sink, to be replaced by hot asthenospheric mantle which then partially melts as it rises to produce the post-orogenic magmas. However, the transition from subduction related to intra-plate characteristics may not become apparent immediately due to the interaction of the rising magmas with the hot, thickened continental crust.

The Andean Cordillera of South America extends for 10 000 km along the western margin of the continent, from the Caribbean Sea to the Scotia Sea, making it the longest sub-aerial mountain chain on Earth. A significant feature of the present-day subduction system is its segmentation into shallow dipping ($<10°$) and more steeply dipping ($\sim30°$) zones, with active volcanism occurring only in association with the steeply dipping segments (Figs. 7.2 & 7.3). This seems at first surprising, as the rate of convergence of the Nazca plate and the South American plate is practically uniform (~10 cm yr^{-1}) along the whole convergence zone (Wortel 1984). The cause of the anomalously shallow dipping segments has been attributed to the subduction of buoyant aseismic ridges, the Nazca Ridge and the Juan Fernandez Ridge, within the Nazca plate. Barazangi & Isacks (1979) have attributed the absence of active volcanism in north and central Peru and central Chile (Fig. 7.2), where the Nazca plate is subducting at shallow angles, to the displacement of the asthenospheric mantle wedge and the direct superposition of the two lithospheric plates (Fig. 7.3). The moderate angle of subduction ($\sim30°$) characteristic of most of the plate boundary has been attributed to the combined effects of rapid plate convergence, overriding of the trench by the South American plate and the relative youth of the subducting Nazca plate (Cross & Pilger 1982).

As shown in Figure 7.2, active volcanism within the Andes is divided into three zones (Thorpe et al. 1982), a northern volcanic zone (NVZ) extending from 5°N to 2°S in Colombia and Ecuador, a central volcanic zone (CVZ) extending from 16°S to 28°S in southern Peru, northern Chile, Bolivia and Argentina, and a southern volcanic zone (SVZ) in southern Chile and Argentina. In each of these zones volcanism has occurred episodically since the Mesozoic. Table 7.1 summarizes the physical and geochemical characteristics of each of these volcanic zones. The lavas of the NVZ are dominantly basaltic andesites and andesites, which have mineralogical and major element characteristics similar to island-arc volcanic suites. In general, the lavas of the SVZ are similar but slightly more basic, with high-alumina basalt and basaltic andesite being the most common rock types. The lavas of the CVZ are characteristically intermediate to acid in composition and show a marked increase in K_2O content (at constant $SiO_2\%$) with increasing depth to the Benioff zone, calc-alkaline volcanics grading east-

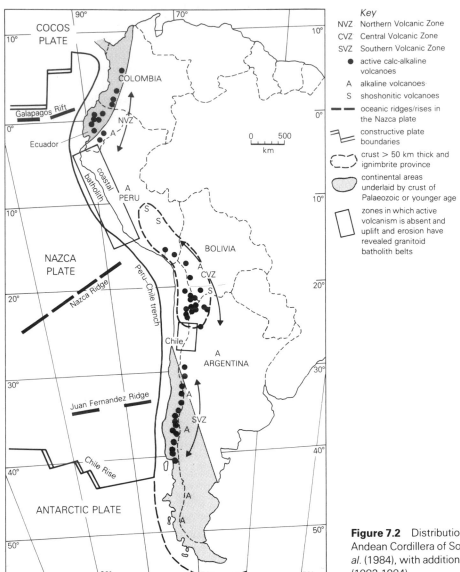

Figure 7.2 Distribution of active volcanoes along the Andean Cordillera of South America (after Harmon *et al.* (1984), with additional data from Thorpe *et al.* (1982, 1984).

wards into shoshonites.

Andean magmas result from a complex interplay of partial melting and fractional crystallization processes within the mantle, and contamination and fractional crystallization processes within the crust. Significantly, one of the most obvious differences between the northern, central and southern volcanic zones is the occurrence of Precambrian basement beneath the CVZ, but only much younger Mesozoic—Cenozoic crust beneath the NVZ and SVZ. In terms of magma—crust interaction models (Leeman 1983), it would be expected that the volumetric proportions of erupted rock types and their geochemical characteristics should be strongly correlated with the thickness and chemical characteristics of the crust through which the rising magmas have passed. This is clearly true for the Andes (Section 7.7), supporting

(a)

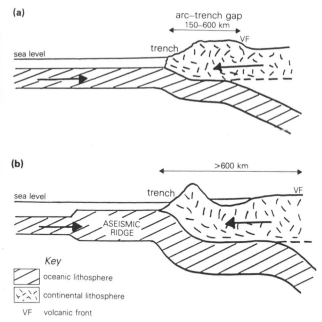

arc–trench gap
150–600 km

trench

VF

sea level

(b)

>600 km

trench

VF

sea level

ASEISMIC
RIDGE

Key

oceanic lithosphere

continental lithosphere

VF   volcanic front

Figure 7.3 Schematic illustration of the effect of subduction of an aseismic ridge on the angle of subduction. (a) A fast convergence rate and the subduction of young oceanic lithosphere results in a fairly shallow angle of subduction of ~30°, with the active volcanic front occurring at distances of 150–600 km from the trench. (b) Subduction of the thickened lithosphere of an aseismic ridge results in the direct superposition of the continental and oceanic lithospheric plates over a much greater distance, increasing the distance between the volcanic front and the trench to >600 km, and in some instances completely eliminating active volcanism. (After Cross & Pilger 1982, Fig. 1, p. 547.)

petrogenetic models involving interaction of mantle-derived magmas and their crustal wall rocks to the extent that some of the more acid magmas may actually be crustal remelts.

Many Andean volcanic rocks carry a trace element and isotopic signature of the continental crust through which they have passed (Section 7.7). However, as discussed in Chapter 6, destructive plate margin magmas can also inherit such a signature from the subduction of continentally derived terrigenous sediments. Unfortunately, in most cases it is impossible to separate these two effects. Opinions are conflicting as to the amounts of continentally derived detritus currently being

subducted beneath the Andes. Shepherd & Moberley (1981) consider that the lack of a substantial accretionary wedge in the Peru–Chile trench suggests that all material derived from the continent has either been subducted or removed laterally, whereas Uyeda (1982) attributes this to a low rate of sediment supply to the trench.

One of the characteristic features of the Andean tectonic setting is the close spatial association of calc-alkaline volcanic and plutonic rocks, the latter now generally accepted as the root zones of former active volcanoes. The intrusive rocks range in composition from gabbro, though diorite, tonalite and granodiorite to granite, and show similar compositional ranges to the volcanic rocks (Section 7.7), strengthening the precept of a genetic relationship. Collectively, the intermediate to acid intrusives are known as granitoids.

Studies of subduction-related plutonic rocks have tended to focus on the western Americas, but crystalline plutonic rocks of calc-alkaline affinity also outcrop in many of the island arcs of the western Pacific, Indonesia, the Aleutians and the Caribbean. Island arcs do not develop on ancient continental crust but on a foundation of oceanic crust, and evolve by thickening of the volcanic pile by the combined effects of volcanism and plutonism. Eventually, more mature arcs develop continental crustal-like profiles (Ch. 6) and, during the Phanerozoic, New Zealand, Japan and Central America have evolved to this intermediate stage between immature arcs with thin crust and active continental margins with Precambrian basements.

The cessation of active volcanism in those segments of the Andes where the angle of subduction has decreased to less than 10° is related to periods of extensive uplift and erosion, revealing enormous linear batholith belts paralleling the continental margin. The largest of these is the Coastal Batholith of Peru, which is over 1600 km long and 60 km wide and comprises more than 1000 plutons emplaced over a 60 Ma timespan from 100 to 37 Ma. Most of these plutons were emplaced by permissive cauldon subsidence and stoping within 3–4 km of the surface (Pitcher & Cobbing 1985). Tonalites and granodiorites (granitoids) predominate, associated with swarms of basaltic andesite

Table 7.1 Tectonic and geological characteristics of the late Cenozoic volcanic zones of the Andes (after Harmon *et al.* 1984, Table 1, p.805).

	SVZ (45−33°S)	CVZ (26−18°S)	NVZ (2°S−5°N)
dip of seismic zone	<25°	*c.* 25−30°	*c.* 20−30°
depth to seismic zone	*c.* 90 km	*c.* 140 km	*c.* 140 km
maximum crustal elevation	2000−4000 m	5000−7000 m	4000−6000 m
crustal thickness	30−35 km	50−70 km	40 km
crustal age	Mesozoic−Cenozoic	Precambrian−Palaeozoic	Cretaceous−Cenozoic
composition of volcanics	basalt with minor andesite and dacite	andesite−dacite with dacite−rhyolite ignimbrites	basaltic andesite to andesite
SiO_2 (wt.%)	50−69	56−66	53−63
K_2O (wt.%)	0.4−2.8	1.4−5.4	1.4−2.2
	medium-K series	high-K series	medium-K series
$\delta^{18}O$	5.2−6.8	6.8−14.0	6.3−7.7
$^{87}Sr/^{86}Sr$	0.7037−0.7044	0.7054−0.7149	0.7036−0.7046
$^{206}Pb/^{204}Pb$	18.48−18.59	17.38−19.01	18.72−18.99
$^{207}Pb/^{204}Pb$	15.58−15.62	15.53−15.68	15.59−15.68
$^{208}Pb/^{204}Pb$	38.32−38.51	38.47−39.14	38.46−38.91

dykes. The batholith follows a remarkably linear course over most of its length, suggesting a fundamental deep control to its emplacement, postulated to be a deep structural lineament channelling the magmas upwards. Similar granite batholith belts are exposed along the whole of the western continental margin of the Americas from Alaska to the Antarctic, and their emplacement history spans most of Mesozoic and Tertiary time. All are intruded within and flanked by subduction-related volcanic rocks, and many are within regions of thickened Precambrian crust.

Subduction-related volcanism is one of the main mechanisms for the growth of the continental crust, either by lateral accretion of island arcs or by the vertical addition of intermediate composition intrusives and extrusives in active continental margins. Thorpe *et al.* (1981) have estimated that, for the Andes, plutonic rocks volumetrically exceed volcanic rocks by a factor of at least 10, attesting to their fundamental role in crustal growth. Genetically, the volcanic and plutonic suites must be related, although many authors have found it difficult to reconcile the dominantly andesitic composition of the lavas with the more silicic granitoids of the batholith.

7.2   Simplified petrogenetic model

In Section 6.2 it was considered that magma generation in the subduction-zone environment is a multistage multisource phenomemon. Much of the preceding discussion about petrogenetic processes in island arcs is equally relevant to the active continental margin environment and therefore will not be reiterated here. However, in active continental margins, passage of the magmas through thick sections of continental crust produces added complexities.

Figure 7.4 shows a schematic cross section of an active continental margin which might be appropriate for the CVZ of the Andes. Processes in the subducted oceanic lithosphere should be exactly the same as those described in Section 6.2 for the case of ocean−ocean collision. Upon subduction, the cold oceanic lithosphere is heated by the combined effects of friction and conduction and, as a consequence, the oceanic crustal layer undergoes a series of metamorphic transformations from greenschist through amphibolite to eclogite facies. Prograde metamorphic reactions generally involve dehydration, and the resulting hydrous fluids are released into the mantle wedge where they lower the solidus

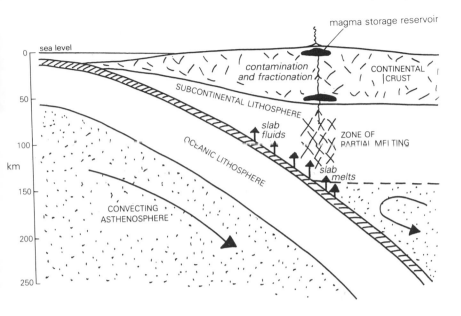

Figure 7.4 Schematic cross section of an active continental margin.

and promote partial melting. If the solidus temperature of the subducted crust is exceeded a hydrous intermediate-acid partial melt may be generated which similarly metasomatizes the mantle wedge and causes partial melting. A significant difference between the model shown in Figure 7.4 and that shown in Figure 6.3 lies in the thickness of the continental lithosphere. This is of the order of 140 km beneath the central Andes (Barazangi & Isacks 1979) compared to 70–80 km for typical oceanic lithosphere (Ch. 5). Additionally, the continental crust in this profile is 50 km thick, compared to 10 km for average oceanic crust.

In the island-arc tectonic setting (Ch. 6) it is generally agreed that for volcanism to occur there must be a wedge of more fertile asthenospheric mantle above the slab (Gill 1981). In Figure 7.4 a similar wedge has been shown for the active continental margin case, although its role must be considered more equivocal. This is because, unlike the oceanic lithosphere, which is variably depleted due to magma generation events at the mid-ocean ridge, the continental lithosphere may be considerably metasomatized and enriched, particularly if it has formed part of a stable continental root for a considerable period of time. Thus slab-derived fluids could initiate partial melting in the subcontinental lithosphere, adding further complexity to

the isotope and trace element geochemistry of the magmas. Indeed, Pearce (1983) considers that enriched subcontinental mantle (lithosphere) plays a dominant role in the petrogenesis of all basalts generated in active continental margin tectonic settings, rather than the convecting asthenosphere.

Any mantle-derived magma passing through a 50 km thick section of continental crust must inevitably interact with it, and therefore assimilation and fractional crystallization (AFC) processes (Ch. 4) should be important in the petrogenesis of these magmas. Of fundamental importance in this respect is the age and geochemical characteristics of the crust. Ancient Precambrian basement gneisses have distinctive isotope geochemical signatures which should readily fingerprint contaminated magmas (Section 7.7.5), whereas younger greywacke type sediments may differ little isotopically from the mantle-derived magmas and therefore if these are the crustal contaminant the process is much harder to identify.

In general, it seems reasonable to assume, as in Chapter 6, that the primary mantle-derived magmas are basaltic in composition, although generation of more siliceous magmas from the metasomatized mantle wedge remains a distinct possibility. Low-pressure crystal fractionation of such magmas, combined with crustal contamination, can then account for the spectrum of more evolved rock types observed at the surface.

7.3.   The structure of active continental margins

Four main interdependent variables control the geometry of subduction zones (Cross & Pilger 1982, Jarrard 1986):

(1)   the relative plate convergence rate;
(2)   the direction and rate of absolute upper plate motion;
(3)   the age of the subducting plate;
(4)   the subduction of aseismic ridges, oceanic plateau or intra-plate island/seamount chains.

Their effects may be additive, or one variable may cancel the effect of another.

In general, low-angle subduction results from combinations of rapid absolute upper plate motion towards the trench, relatively rapid plate convergence, subduction of low-density oceanic lithosphere and the subduction of young oceanic lithosphere. The consequences are either a landward displacement of the magmatic arc or a cessation of subduction-related magmatism, and the development of a compressional tectonic regime within and behind the arc. In contrast, steeper subduction results from combinations of slow or retrograde absolute upper plate motion, slow relative rates of plate convergence and subduction of old dense oceanic lithosphere. This induces the development of a magmatic arc closer to the trench and extensional tectonics within and behind the arc.

Seismology has thus far been the only geophysical method available for examining the shape, continuity and physical characteristics of subduction zones. In Figure 7.5 profiles of the top of the Benioff zone for three oceanic island arcs are compared (Marianas, Lesser Antilles and New Hebrides) with those of three segments of the Andes in Peru, central Chile and northern Chile. All of the oceanic examples show considerably steeper profiles, which may be explained in terms of the greater age of the subducting plate and the

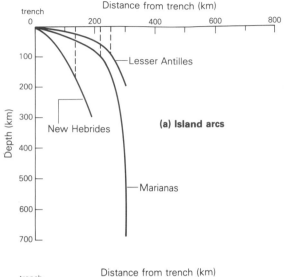

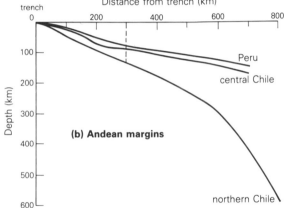

Figure 7.5   Profiles of the top of the Benioff zone for three intra-oceanic island arcs (a) compared with three segments of the Andes (b). The vertical dashed line indicates the location of the volcanic front in each case (after Jarrard 1986, Fig. 2, p. 223).

slow rate of plate convergence. Comparing the three Andean profiles, those from Peru and central Chile are characterized by a very shallow dip (<10°) and lack active volcanism. These are among the most shallow dipping of all modern subduction zones (Jarrard 1986) in which the slab appears to flatten and travel horizontally along the lower surface of the South American lithosphere. The shallow angle of subduction has been attributed to the subduction of buoyant aseismic ridges in the underthrust Nazca plate (Cross & Pilger 1982). In contrast, the moderate angle of subduction (~30°) characteristic of much of the plate margin may be attributed to the combined effects of rapid convergence, overriding of the trench by the South American plate and the relative youth of the subducting Nazca plate.

During the evolution of the Andean subduction system over the past 250 Ma there have doubtless been numerous changes in the angle of subduction and the consequent location of the volcanic front. For example, between 50 and 25 Ma the age of the subducting oceanic plate progressively decreased (Pilger 1981), resulting in a gradual decrease in the subduction angle and consequent eastward shift in the loci of magmatic activity. Additionally, a major reorganization of plate motions at 25 Ma resulted in an increased convergence rate normal to the Chilean Andes, with a consequent shallowing of the subduction zone.

As suggested in the introduction, Andean magmatism results from a complex interplay of crystal/melt equilibria in the mantle, and contamination and fractional crystallization processes in the crust. To understand the nature of the crustal contamination process it is important to have as much detailed information as possible about the crustal structure. Seismic refraction surveys provide information about the thickness and velocity structure and enable identification of upper and lower crustal layers (Fig. 7.6). In this diagram crustal profiles for Andean segments in Colombia and northern Chile are compared with profiles for the Cascades, Alaska and New Zealand and for Tonga, an immature island arc. It is important to realize that the seismic data can only give information about the broad lithology of the crustal profile, not the age. Thus lower crustal rocks with a V_p of 6.5–7.0 beneath Tonga are undoubtedly much younger than those with a similar range of V_p beneath northern Chile. Upper crustal rocks with V_p 5.0–6.2 comprise sedimentary rocks, volcanics and young intermediate-acid plutonic rocks. In contrast, the lower crust with V_p of 6.5–7.4 may be considered to be made up of high-grade crystalline metamorphic rocks in the granulite and amphibolite facies.

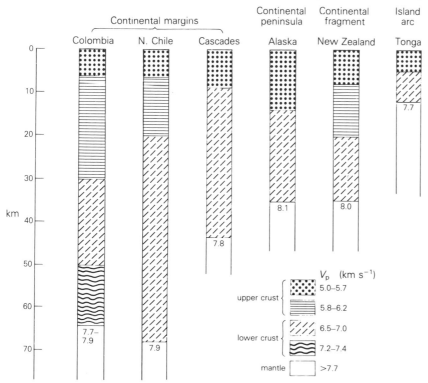

Figure 7.6 P-wave velocity (V_p) crustal profiles for a variety of active continental margin tectonic settings, compared with that for an immature intra-oceanic island arc (after Gill 1981, Fig. 3.3, p. 48).

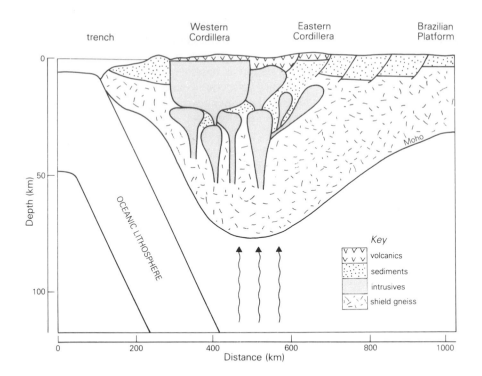

Figure 7.7 Schematic cross section through the central Andean active continental margin to show the crustal structure (×5 vertical exaggeration). Arrows indicate the direction of magma and volatile streaming from the downgoing plate (after Brown & Mussett 1981, Fig. 9.6, p. 168).

Figure 7.7 depicts the crustal structure of the central Andean plate margin. This shows that the bulk of the upper crust in the active volcanic belt comprises a young granitoid batholith overlain by intermediate composition volcanic rocks. The high-grade ancient metamorphic rocks of the Brazilian shield pass beneath the batholith and outcrop on the Pacific coast. Clearly, in this region of thickened continental crust, magmas must pass through in excess of 50 km of high-grade Precambrian gneiss before reaching the high-level magma storage reservoirs. Therefore, it is not in the least surprising that the CVZ magmas carry a strong imprint of continental crustal contamination in comparison to magmas generated in the NVZ and SVZ, where ancient gneissic basement complexes are absent.

Figure 7.8 is a schematic cross section of the Peru–Chile subduction zone, showing the distribution of zones of moderate and high seismicity and the variation of the seismic attenuation factor, Q. The upper zone of the subducted oceanic lithosphere is characterized by high seismicity and high Q, while the overlying mantle wedge is

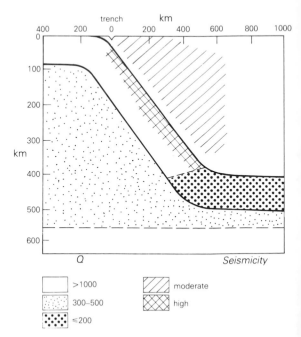

Figure 7.8 Schematic cross section through the Peru–Chile subduction zone, showing the distribution of zones of moderate and high seismicity and the variation of the seismic attenuation factor Q (after Condie 1982, Fig. 6.12, p. 112).

moderately seismically active and also has high Q (1000–3000). Sacks (1983) suggests that this indicates the existence of a thickened zone of subcontinental lithosphere (~350 km) above the slab. Low values of Q (<200) indicate high seismic attenuation and the probable presence of a partial melt phase. Below a depth of about 400 km the subducting slab appears to flatten and to develop low Q characteristics suggestive of partial melting.

7.4 Thermal structure and partial melting processes

The thermal structures of subduction zones are not well defined (Section 6.5) and a variety of models have been proposed (Anderson *et al.* 1978, 1980; Furlong *et al.* 1982). The subducted oceanic crust may be relatively warm as a consequence of frictional heating, or it may be significantly cooled by endothermic dehydration reactions. Similarly, the mantle wedge may be relatively cool, chilled by the subducting slab, or it may be heated by induced convection (Toksöz & Hsui 1978, Wyllie 1984).

Figure 7.9 shows the schematic thermal structure of an active continental margin with a 50 km thick crust for two extreme thermal models which may be described as cool (A) and warm (B) respectively. The consequence of induced convection in the mantle wedge (model B) is to raise the 750°C isotherm into the base of the continental crust, thereby increasing the likelihood of crustal melting. Additionally, the 1000° and 1250°C isotherms are raised higher in the mantle and thus this latter thermal regime will be more conducive to extensive mantle partial melting. Figure 7.10 shows the same thermal models with the approximate location of the solidi in the presence of H_2O for the major sources which may contribute to partial melts in the subduction-zone environment; the subducted oceanic crust, the continental crust and the mantle wedge. The line D–D' marks the onset of significant dehydration within the slab, broadly coincident with the greenschist/amphibolite facies boundary. Hydrous fluids are shown streaming into the mantle wedge and the base of the continental crust where, under subsolidus conditions, they may promote the growth of extensive metasomatic amphibole (hatched regions). The dotted line in each diagram shows the maximum depth of stability of amphibole. Partial melting occurs where any of these three major sources are raised above their

(a) Cool mantle wedge, chilled by the subducting slab

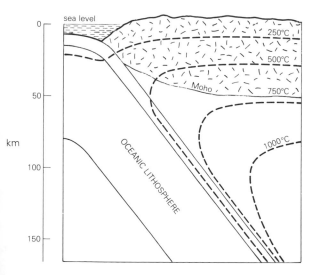

(b) Warm mantle wedge, heated by induced connection

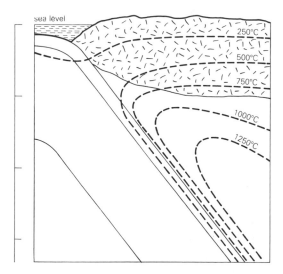

Figure 7.9 Schematic thermal structure of an active continental margin (after Wyllie 1984, Fig. 8, p.449).

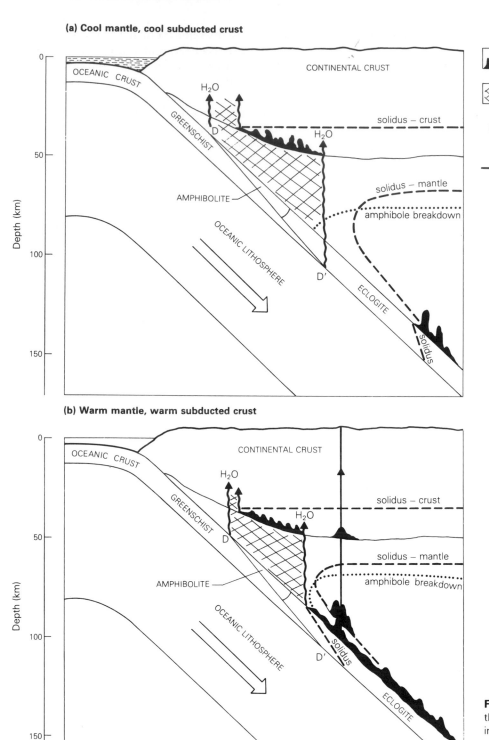

Figure 7.10   Location of the sites of partial melting in active continental margins for two different thermal models (after Wyllie 1984, Fig. 9, p.450).

respective solidus temperature, particularly where H_2O streams into regions to the high-temperature side of the solidus.

For both thermal models, aqueous fluids penetrating the base of the continental crust may promote partial melting. However, only in thermal model B does extensive mantle partial melting occur, triggered by partial melts ascending from the subducted oceanic crust. These mantle partial melts then rise into the base of the continental crust, where mixing with anatectic crustal melts may occur. Model B may be considered to be generally applicable to the Andean tectonic setting, and thus we can see that the erupted volcanics may contain a contribution from each of the three potential magma sources. If the subducted oceanic crust contains a significant proportion of continentally derived terrigenous sediment, it may be difficult to resolve geochemically components with a continental crustal signature inherited from the subduction zone from those introduced by high-level crustal contamination. Oxygen isotope studies may be particularly useful in this respect (Section 7.7.6).

Many workers have suggested the importance of crustal melting in the generation of the vast sheets of ignimbrite characteristic of the CVZ of the Andes (Gill 1981). Figure 7.11 shows the range of compositions of partial melts which might be derived from continental gneisses in the presence of H_2O (Wyllie 1984). There is clearly a very narrow temperature interval for the existence of H_2O-saturated rhyolitic liquids close to the solidus, except for pressures less than 2 kbar, and at all crustal depths very high temperatures are required to derive andesitic liquid compositions. Increasing pressure produces liquids with lower SiO_2 contents and at the base of a 40−50 km thick crust the near-solidus partial melt may be syenitic (Huang & Wyllie 1981). Thus it is possible that a range of the more acidic Andean magma compositions could be generated by direct partial melting of the continental crust.

7.5 Magma storage in the crust

Evidence for the existence of shallow magma reservoirs in the crust beneath active volcanoes is provided by the following:

(1) geophysical data;
(2) petrological evidence for the role of low-pressure crystal fractionation in the geochemical evolution of the magmas;
(3) the existence of plutons underlying eroded volcanic complexes.

Geophysical techniques for detecting magma bodies are based on the dramatic decrease in density and seismic velocity, and increase in seismic attenuation and electrical conductivity, which occur at the onset of partial melting in rocks. Seismicity beneath volcanoes is caused by magma-induced tectonic stresses, and if extensive zones of partial melt are present (i.e. magma chambers) earthquake stresses cannot accumulate. Thus zones of seismic quiescence may indicate the locations of crustal magma reservoirs. Iyer (1984) has reviewed the geophysical evidence for the locations, shapes, sizes and internal structure of magma bodies beneath selected regions of Quaternary volcanism including Alaska and Kamchatka (continental

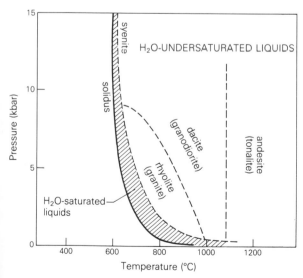

Figure 7.11 Compositions of liquids generated by partial melting of continental gneisses in the presence of H_2O (2%). The shaded field depicts the conditions under which H_2O-saturated liquids can occur (after Wyllie 1984, Fig. 6, p.445).

peninsulas) and New Zealand (continental fragment), all of which may broadly be considered as examples of active continental margins. Unfortunately, these data are limited and similar high-quality seismic data for the Andean margin are lacking altogether. For Kamchatka there is evidence for magma storage bodies in the depth range 30–90 km, with dimensions between 8 and 40 km across and up to 30 km thick. However, in some instances much shallower reservoirs occur, within 10 km of the surface, fed by conduits extending into deep-seated mantle magma reservoirs. Marked low-velocity zones have been recorded at depths of 10 and 35 km in the crust of the central Andes, and have been interpreted by Ocala & Myer (1972) as potential zones of magma storage.

In active continental margins, volcanic and plutonic rocks, ranging in composition from basalt (gabbro) to rhyolite (granite), frequently display good linear correlations on Harker variation diagrams (Section 7.7), suggestive of the derivation of the more acid magmas by fractional crystallization of olivine, plagioclase, pyroxene, magnetite and amphibole mineral assemblages from basaltic parent magmas. In suites of volcanic rocks for which Sr–Nd–Pb isotopic data suggest little crustal contamination, these data may be interpreted as reflecting liquid lines of descent. However, most Andean magmas, particularly those erupted in the CVZ, have geochemical characteristics reflecting the combined processes of assimilation and fractional crystallization (Ch. 4) which, in general, will tend to blur coherent linear trends on Harker diagrams. Nevertheless, there still appears to be abundant geochemical evidence for low-pressure crystal fractionation trends, providing strong supporting evidence for the existence of high-level crustal magma reservoirs.

One of the most useful lines of evidence in elucidating the structure of high-level magma chambers beneath active volcanoes is to examine the plutonic root zones of deeply dissected volcanic belts, the granitoid batholiths. Previous to the 1970s, the plutonic and volcanic phases of Andean magmatism tended to be treated as separate unconnected phenomena, based on the incorrect assumption that the batholiths were dominantly granitic as opposed to the intermediate composition of the volcanic belt. However, it is now well established that the study of individual plutons comprising the batholith can provide invaluable information about high-level (<10 km) magma storage reservoirs.

Mesozoic and Cenozoic batholiths are exposed in the mobile belts of the western Americas, attesting to the continuity of subduction-related magmatism along the whole of the continental margin from the late Triassic. The Coastal Batholith of Peru is some 1600 km long by 60 km wide and up to 15 km thick, elongated parallel to the present trench. It is composed of over 1000 plutons intruded over a 70 Ma period from 100 to 30 Ma. The plutonic rocks are spatially coincident with two groups of volcanic rocks, the 100 Ma Casma group and the early Tertiary Calipuy group which overlies an erosion surface cut through the batholith. The batholith is divided into five segments (Fig. 7.12), exhibiting recognizably distinct groups of plutonic rocks which may be related to discontinuities in the underlying subduction system at their time of formation, similar to the segmentation of the presently active volcanic zone.

The plutonic rocks of the batholith comprise 16% by volume gabbro and diorite, 58% tonalite and granodiorite, 25.5% adamellite and 0.5% granite (Hughes 1982). This clearly indicates that the term 'granite batholith' is a misnomer as intermediate composition rocks predominate. For much of its length the coastal batholith occupies the axis of an early Cretaceous marginal basin (Pitcher et al. 1985), although to the south it penetrates old crystalline basement. The batholithic magmas appear to have been channelled along the same deep-seated suture along which the marginal basin opened. Magmatism within the batholith was distinctly episodic (Fig. 7.13), with quiescent periods often longer than 15 Ma between intrusive phases (Beckinsale et al. 1985).

At the present rather shallow (<5 km) level of erosion the batholith comprises arrays of intersecting plutons, forming complexes with a surprisingly regular spacing of 120 km. Pitcher et al. (1985) relate this 120 km spacing to the location of separate melt cells at depth. At such high crustal levels magmas are hydrostatically emplaced by a com-

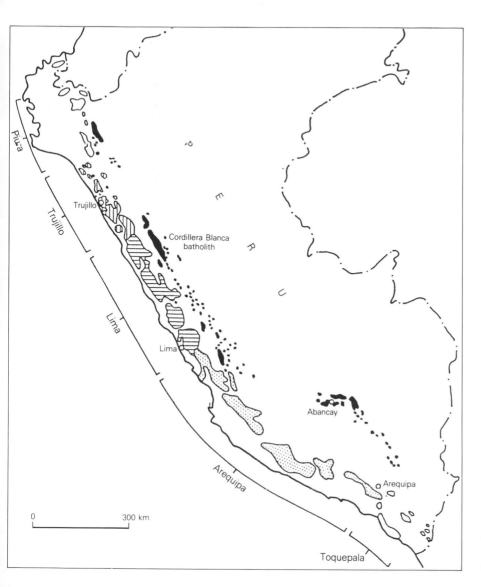

Figure 7.12 The segmentation of the Cretaceous Coastal Batholith of Peru (shown by various ornaments or blank); Also shown in black is a belt of Cenozoic stocks and batholiths paralleling, but to the landward side of, the Coastal Batholith (after Pitcher & Cobbing 1985, Fig. 3.2, p. 22).

bination of roof lifting and cauldron subsidence (Pitcher 1979, Pitcher *et al*. 1985) and the shapes of plutons are controlled by magma-induced fracture patterns (Bussell 1976). In three dimensions, the roofs of the plutons are flat with rapid turn-downs into steep sides, forming a box-like shape. Such plutons may be simplistically represented as the magmatic filling of a cavity above a down-dropped block of pre-existing country rock. The majority of plutons have a circular outcrop pattern, with some degree of elongation along the structural grain of the country rocks.

Figure 7.14 shows a schematic cross section of the batholith showing the nested belljar like plutons intruded into a basement of pre-Cretaceous rocks overlain by volcanics of the 100 Ma Casma group, which may represent the earliest phase of volcanic activity associated with the oldest plutons of the batholith. The cross section shows the youngest phases of plutonism venting to the surface to produce the Tertiary volcanic cover of the Calipuy group (Cobbing *et al*. 1981).

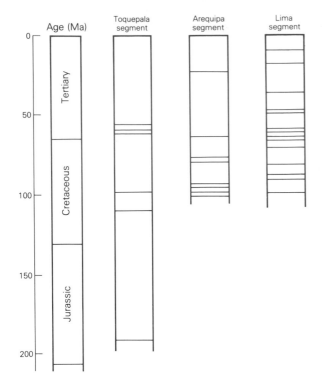

Figure 7.13 Major intrusive phases of the Toquepala, Arequipa and Lima segments of the Coastal Batholith of Peru (after Beckinsale *et al.* 1985, Fig. 16.10, p. 198).

7.6 Petrographic characteristics of the volcanic and plutonic rocks

In Section 6.10 the petrographic characteristics of the four major magma series erupted in oceanic island arcs (tholeiitic, calc-alkaline, high-K calc-alkaline and shoshonitic) were described. Chemically similar magmas erupted in active continental margin tectonic settings are virtually identical and thus the reader is referred to Section 6.10 for the relevant information, and to Ewart (1982) for a more detailed synthesis. In this section emphasis is placed on the petrography of plutonic rocks of the calc-alkaline series, as these form the bulk of the exposed granitoid batholith belts. Much of this information is also relevant to the study of subduction-related plutonic rock associations exposed in the more mature oceanic island-arc systems.

Figure 7.15 shows the distribution of the major rock-forming minerals in rocks, ranging in composition from gabbro to granite from the Coastal Batholith of Peru (Mason 1985). The mineralogy and textures of these rocks reflects a history of magmatic crystallization in high-level subvolcanic magma chambers. However, as with all slowly cooled plutonic rocks there is abundant evidence

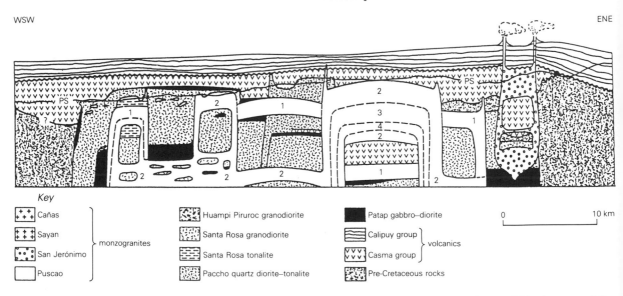

Key

+++ Cañas		
+++ Sayan	monzogranites	
+++ San Jerónimo		
Puscao		

Huampi Piruroc granodiorite

Santa Rosa granodiorite

Santa Rosa tonalite

Paccho quartz diorite–tonalite

Patap gabbro–diorite

Calipuy group ⎫
               ⎬ volcanics
Casma group ⎭

Pre-Cretaceous rocks

0   10 km

Figure 7.14 Cross section of the Coastal Batholith of Peru, showing the nested belljar-shaped plutons. PS is the trace of the present topography (after Bussell & Pitcher 1985, Fig. 15.4, p. 169).

	Gabbro	Diorite	Tonalite	Grano-diorite	Granite
olivine	▓				
cpx	▓	▓			
pigeonite	▓				
opx	▓		▓		
amphibole	▓	▓	▓	▓	
biotite		▓	▓	▓	▓
magnetite	▓	▓	▓	▓	
plagioclase	▓	▓	▓	▓	▓
alkali feldspar			▓	▓	▓
quartz			▓	▓	▓

Figure 7.15 Distribution of the major rock-forming minerals in calc-alkaline plutonic rock suites (after Mason 1985).

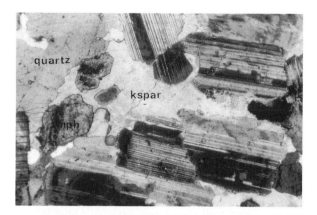

Figure 7.16 Characteristic textural features of a calc-alkaline granodiorite from Chile. (a) Multiply twinned plagioclase with interstitial quartz and k-feldspar (×40, crossed polars). (b) Intergowth of amphibole, biotite and magnetite (×40, ordinary light).

for the growth of subsolidus minerals such as biotite, amphibole and chlorite due to the interaction of the solid rocks with high-temperature hydrothermal fluids. Figure 7.16 shows some of the characteristic textural features of a calc-alkaline granodioritic plutonic rock.

The major rock-forming minerals are plagioclase, alkali feldspar, quartz, pyroxene, amphibole, biotite and magnetite. Sphene and apatite are common accessory minerals, even in the more basic rocks, while allanite occurs quite frequently in the highly differentiated granites.

Pyroxene. The dominant pyroxene phase is an augite or calcic augite, joined by hypersthene in the intermediate composition range. Inverted pigeonite occurs in some of the gabbros, and Mason (1985) suggest that it may be a high-pressure phenocryst phase in the more basic magmas. The occurrence of calcic augite and hypersthene is considered to reflect relatively high water fugacities during crystallization.

Amphibole. Hornblende is one of the major mafic minerals crystallizing from magmas ranging from basic to acid in composition. This is in marked contrast to its occurrence in calc-alkaline volcanic suites, in which it occurs infrequently and often in a highly resorbed state. The abundance of hornblende in the plutonic rocks reflects the increased stability of amphibole at depth in the crust. Crystals are generally euhedral or subhedral, indicating early crystallization, and change in colour from brown through green–brown to green with increasing differentiation of the magma. The colour changes appear to correlate with progressively decreasing TiO_2 contents. In some rocks an original green–brown hornblende may be patchily replaced by green hornblende and associated sphene. This is most probably a solid-state reaction product in the presence of a hydrothermal fluid phase. Early formed amphiboles in the basic rocks are tschermakitic hornblendes, whereas in the acid rocks later formed amphiboles tend to be actinolitic hornblendes.

Biotite. Biotite is a common mafic mineral in many granitoid rock types, appearing late in the crystallization sequence of the more basic rocks but early in the more acid intrusives, where it may form well developed crystals. $Mg/(Mg + Fe^{2+})$ ratios vary from 0.38 to 0.61 proportional to those in the host rock. Biotite may be quite commonly altered to chlorite as a consequence of interaction with late-stage hydrothermal fluids.

Plagioclase. Plagioclase is the major rock-forming mineral in nearly all the plutonic rocks, ranging in composition from An_{93} to An_{10}. The crystals often show complex oscillatory zoning similar to that observed in plagioclase phenocrysts in andesitic lavas. This is a characteristic feature of the intermediate to acid rocks. Fine-scale myrmekite (plagioclase–quartz intergrowth) is common in all rock compositions, but particularly so in the more basic rocks.

Alkali feldspar. The amount of alkali feldspar present in the plutonic rocks varies in a regular manner with the bulk rock composition. In more basic rocks it tends to occur interstitially, whereas in the more acid rocks it forms larger 'pools'. Some of the granitoids contain K-feldspar megacrysts which are generally considered to have been produced by late-stage K-rich metasomatism (subsolidus). Orthoclase is by far the most common type of K-feldspar in the granitoids, while microcline occurs only in some of the most differentiated rocks. The degree of ordering in the K-feldspar seems to be mainly controlled by the concentration of volatile components in the melt, with microcline crystallization being favoured by the most volatile-rich conditions. Exsolution textures are ubiquitous, although the alkali feldspar observed in basic rocks normally lacks exsolution lamellae and is probably a cryptoperthite. Vein perthites are dominant in the intermediate and acid rocks, while patch perthites are most common in the most evolved rocks. Parsons (1978) has suggested that magmatic water might be the prime catalyst in causing perthite coarsening. Granophyric intergrowths are characteristic of the most highly differentiated rocks which formed from the most volatile-rich magmas. These are considered to have formed from the rapid crystallization of quartz and alkali feldspar as a consequence of a sudden reduction in vapour pressure due to loss of volatiles from the system (Mason 1985).

Magnetite. Magnetite is the major opaque oxide phase throughout the spectrum of basic to acid magmas, with ilmenite occurring only rarely. Both phases tend to exhibit high degrees of subsolidus re-equilibration.

7.7 Chemical composition of the magmas

7.7.1 Charactistic magma series

The four major magma series recognized in oceanic island arcs (low-K, calc-alkaline, high-K calc-alkaline and shoshonitic; see Section 6.7) also occur in active continental margin tectonic settings. Their classification is based upon the same K_2O versus SiO_2 diagram, and the reader is referred to Section 6.7 for further details. However, in comparison with island-arc volcanic suites (Fig. 7.17), low-K series magmas are poorly represented, while high-K and shoshonitic magmas are more common, particularly at the acid end of the spectrum. These high-K characteristics may reflect increasing degrees of crustal contamination in the active margin magmas. Additionally, suites of alkaline volcanic rocks may occur to the landward side of the volcanic front, ranging from mildly alkaline basalts to leucite basanites and their derivatives. These magmas are not necessarily subduction-related, and may be generated as a consequence of extensional tectonics in a back-arc region.

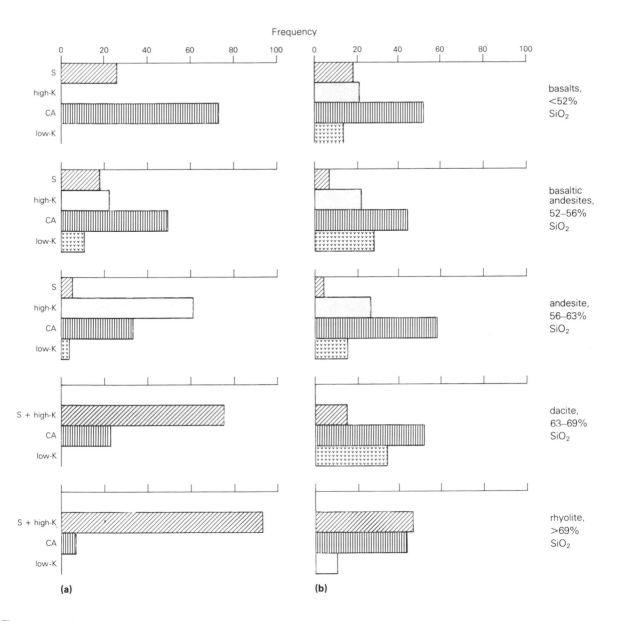

Figure 7.17 Comparison of the relative frequency of occurrence of rocks of the low-K, calc-alkaline (CA), high-K calc-alkaline and shoshonitic (S) series in (a) the Andes and (b) the oceanic island arcs of the south-west Pacific (data from Ewart 1982.)

Figure 7.18 compares the frequency distribution of basalts, basaltic andesites, andesites, dacites and rhyolites, irrespective of magma series, in the Andes with that in the island arcs of the south-west Pacific (Ewart 1982). This clearly reveals the greater abundance of intermediate and acid magmas erupted in the active continental margin tectonic setting which, as stated previously, may be a consequence of crustal contamination.

7.7.2 Major elements

SiO_2, TiO_2, Al_2O_3, Fe_2O_3, FeO, MnO, MgO, CaO, Na_2O, K_2O, P_2O_5 and H_2O can all be

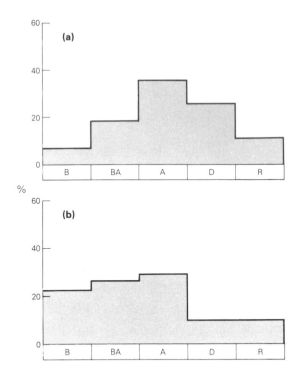

Figure 7.18 Frequency distribution of basalts (B), basaltic andesites (BA), andesites (A), dacites (D) and rhyolites (R) in the Andes (a) compared with that in the island arcs of the south-west Pacific (b). (Data from Ewart 1982.)

considered as major elements in the description of the geochemistry of active continental margin magmas. In terms of these, the most obvious distinction between the major magma series is one of increasing total alkali content in the sequence tholeiitic − calc-alkaline − high-K calc-alkaline − shoshonitic, K_2O showing proportionately the greater increase. This has already been used in Section 6.7 as the basis for the classification of island-arc volcanic suites.

Figure 7.19 is a plot of wt.% K_2O versus wt.% SiO_2 for recent volcanic rocks from the northern (NVZ), central (CVZ) and southern (SVZ) zones of the Andes. Volcanics from the NVZ and SVZ have medium-K or calc-alkaline characteristics and are restricted to SiO_2 contents <63% (i.e. dacites and rhyolites are lacking). In contrast, volcanics from the CVZ have generally high-K characteristics, spanning the complete compositional range from

basalt to rhyolite. Figure 7.20 is a comparable plot for plutonic rocks from the Arequipa and Lima segments of the Coastal Batholith of Peru, showing that there is total overlap between the compositions of Andean volcanic and plutonic rocks. In island-arc volcanic suites K_2O behaves essentially incompatibly, and thus genetically related suites of rocks define positive linear trends in plots of K_2O versus SiO_2 (Section 6.7). While this is also broadly true for the active continental margin volcanic and plutonic suites, there is a considerably greater degree of scatter which may be attributable to the effects of crustal contamination.

Suites of rocks related by fractional crystallization processes and unmodified by extensive crustal contamination should also define coherent linear trends on all types of Harker diagram. For example, Figure 7.21 shows the variation of wt.% MgO, CaO and Al_2O_3 versus % SiO_2 for plutonic rocks from the Lima and Arequipa segments of the Coastal Batholith of Peru. The data define remarkably good linear trends, consistent with the fractionation of ferromagnesian minerals and plagioclase from parental basalts, bearing in mind the difficulty of obtaining true liquid compositions by analysis of plutonic rocks because of the effects of crystal accumulation. However, caution must be exercised in interpreting such trends as true liquid lines of descent until the isotopic homogeneity of all members of the plutonic suites is verified (Section 7.7.5). Shown for comparison in Figure 7.22 is a plot of wt.% MgO and wt.% K_2O versus wt.% SiO_2 for volcanic rocks from the Tertiary Calipuy group of Peru, which overlies an erosion surface cut through the Coastal Batholith. Coherent trends are still visible, although they are rather more noisy than the plutonic data. Both volcanic and plutonic suites show typical calc-alkaline differentiation trends with total iron content decreasing progressively as the SiO_2 content increases due to the early crystallization of magnetite.

Table 7.2 shows average major element analyses of Andean volcanic rocks compared to those from the island arcs of the south-west Pacific. These data clearly show that magmas erupted in the Andean region are enriched in K_2O, Na_2O, TiO_2 and P_2O_5 and depleted in CaO, compared to their island-arc

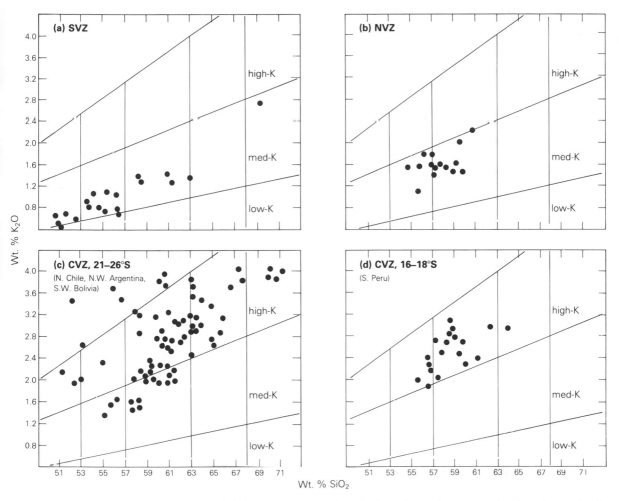

Figure 7.19 Plots of wt.% K_2O versus wt.% SiO_2 for young volcanic rocks from the northern, central and southern volcanic zones of the Andes. The boundaries between the low-, medium- and high-K fields are those of Peccerillo & Taylor (1976) (after Harmon *et al.* 1984, Fig. 2, p. 810).

counterparts (Ewart 1982). In Table 7.3 average compositions of basaltic andesites (52–56% SiO_2) from the NVZ, CVZ and SVZ of the Andes are compared. The CVZ basaltic andesites appear slightly richer in TiO_2 and K_2O than those from the NVZ and SVZ, but otherwise the analyses are broadly similar. In Table 7.4 a typical basalt from the SVZ is compared with an alkali basalt erupted in an extensional tectonic setting to the east of the volcanic front in the CVZ. The alkali basalt is much poorer in SiO_2 and therefore the two analyses are not directly comparable. However, it is evident that

the alkali basalt has much higher concentrations of TiO_2 and P_2O_5 and the whole range of incompatible trace elements and lower Al_2O_3. Table 7.5 shows whole-rock analyses of plutonic rocks from the Lima segment of the Coastal Batholith of Peru, for comparison with the volcanic data.

7.7.3 Trace elements

In Section 6.7 it was demonstrated that island-arc basalts are characterized by selective enrichment of elements of low ionic potential (Sr, K, Rb, Ba ±

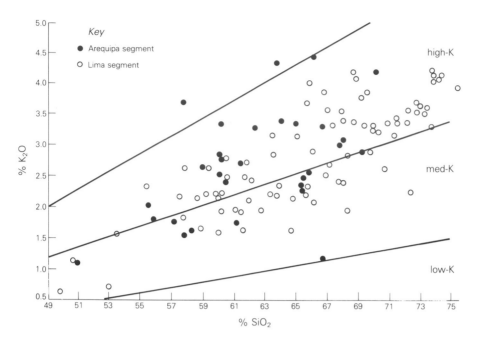

Figure 7.20 Plot of % K_2O versus % SiO_2 for plutonic rocks from the Arequipa and Lima segments of the Coastal Batholith of Peru (data from Pitcher *et al.* 1985).

Th) and low abundances of elements of high ionic potential (Ta, Nb, Ce, P, Zr, Hf, Sm, Ti, Y, Yb, Sc and Cr) compared to N-type MORB. The enrichment in low ionic potential elements has been attributed to metasomatism of the mantle source of arc basalts by fluids released from the subducted slab. In contrast, the relative depletion in high ionic potential elements has been variably attributed to higher degrees of partial melting and to the stability of residual mantle phases (Pearce 1982).

Figure 7.23 shows chondrite-normalized trace element abundance patterns (spiderdiagrams) for basaltic andesites from the northern, central and southern volcanic zones of the Andes. More primitive basaltic compositions would normally be used for such a diagram, but unfortunately data are unavailable. Compared to the equivalent diagram for oceanic island-arc basalts (Fig. 6.34), they clearly show the same distinctive spiked pattern with peaks at K, Sr and Th and a marked trough at Nb. It appears that such patterns must be a characteristic of all subduction-related magmas, attesting to the involvement of subduction-zone fluids enriched in Sr, K, Rb, Ba and Th in their petrogenesis.

Figure 7.24 shows a MORB-normalized trace element variation diagram (Pearce 1983) for the least enriched of the two CVZ basaltic andesites shown in Figure 7.23. Comparing this with the patterns for intra-plate and island-arc basalts in Figure 6.37, we can see that the immobile elements Ta, Nb, Zr, Hf, Ti, Y and Yb define a pattern (dashed line) more akin to that of intra-plate basalts than to MORB. Following Pearce (1983) it is suggested therefore that the mantle source of this magma was enriched subcontinental lithosphere (as opposed to depleted asthenosphere in the case of island-arc basalts) to which mobile elements (Sr, K, Rb, Ba, and to a lesser extent Ce and Sm) had been added by a subduction-zone fluid. Shown for comparison in Figure 7.24 is the trace element pattern for an alkali basalt erupted to the east of the CVZ in an extensional tectonic regime. This shows a typical intra-plate signature (Fig. 6.37) and may be derived by partial melting of the subcontinental lithosphere, possibly with some contamination by the continental crust. Thus Figure 7.24 clearly attests to the involvement of subcontinental lithosphere as a major source component in the petrogenesis of Andean volcanic rocks.

A difficult question to resolve is how patterns such as those in Figure 7.24 reflect involvement of

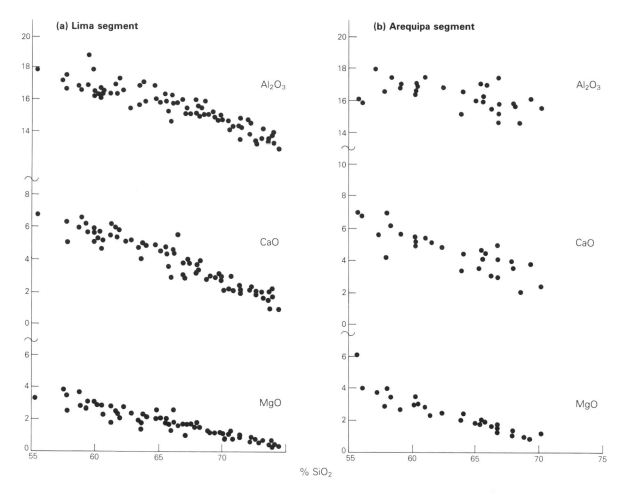

Figure 7.21 Variation of Al$_2$O$_3$, CaO and MgO versus SiO$_2$ for plutonic rocks from the Lima and Arequipa segments of the Coastal Batholith of Peru (data from Pitcher *et al.* 1985).

continental crustal materials. This is particularly important in the petrogenesis of the CVZ Andean magmas, as we shall see in Section 7.7.5. The trace element signature of crustal contamination is particularly difficult to predict, given the great range of crustal rocks which could be involved and the likelihood that the contaminant will be a partial melt of one of these rocks rather than the bulk rock itself. Figure 7.25 illustrates the types of trace element patterns that might result from selective contamination of a basalt (with MORB-normalized abundances of 0.5) with 50% partial melts of diorite and greywacke crustal rocks respectively, in the proportion 4 : 1, basalt : contaminant (Pearce

1983). This is obviously an extreme case, as the addition of such a large volume of acidic partial melt would change the composition of the basalt to that of a basaltic andesite or andesite. Of particular significance is the fact that crustal contamination by these components does not appreciably add elements of the group Ta to Yb, or indeed Sr. Ba and Th are the most enriched elements in both cases. Contamination effects will obviously be easier to detect in basalts with originally flat MORB-normalized trace element patterns. For intra-plate basalts (Fig. 6.37) with originally 'humped' shaped patterns, crustal contamination effects would be very much more difficult to discern. Thus for

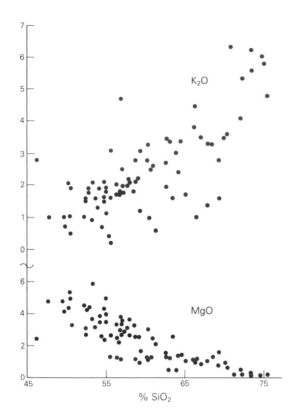

Figure 7.22 Variation of % K_2O and % MgO versus SiO_2 for volcanic rocks form the Calipuy group of Peru (data from Pitcher *et al.* 1985).

Andean magmas, generated from enriched sub-continental lithosphere sources, quantification of the role of high-level crustal contamination using trace element geochemistry alone may prove to be a near-impossible task.

In studying suites of subduction-related basalts, bivariate diagrams based on trace element ratios have been found to be useful in separating subduction-related from mantle components in the petrogenesis of the magmas (Pearce 1982). Figure 7.26 is such a diagram showing the variation of Th/Yb versus Ta/Yb (Pearce 1982, 1983). Yb is the denominator in both of these ratios, and this has the effect of largely eliminating variations due to partial melting and fractional crystallization processes, allowing attention to be focused on source composition as a major petrogenetic variable. Mid-ocean

ridge basalts (MORB) and uncontaminated intra-plate basalts plot within a well defined band with a slope of unity, as mantle enrichment events appear to concentrate Ta and Th equally. In contrast, island-arc and active continental margin basalts are displaced to higher Th/Yb ratios, presumably reflecting the influence of subduction-zone fluids enriched in Th in their petrogenesis. The fact that active continental margin basalts plot above the enriched end of the oceanic mantle array in Figure 7.26 would seem to provide strong support for the involvement of subcontinental lithosphere in their petrogenesis. Contaminated continental intra-plate basalts also plot in a similar position, however, and thus once more we are faced with the problem of distinguishing between the effects of subduction-zone fluids and those of near-surface crustal contamination in producing the observed trace element characteristics of the magmas.

Tables 7.2–5 include trace element data for volcanic and plutonic rocks from the Andes. In general, the active continental margin magmas appear to show greater degrees of enrichment of a whole range of incompatible trace elements compared to oceanic island-arc basalts, which may reflect the combined effects of derivation from an enriched mantle source and crustal contamination.

7.7.5 Radiogenic isotopes

Nd–Sr isotopes

Isotopic compositions of Sr, Nd and Pb provide some of the most useful information for elucidating magmatic processes at convergent plate boundaries, because the various source components involved have such contrasting isotopic signatures. Figure 7.27 shows the variation of $^{143}Nd/^{144}Nd$ versus $^{87}Sr/^{86}Sr$ for volcanic rocks from the northern, central and southern volcanic zones of the Andes (Hawkesworth *et al.* 1982, James 1982, Thorpe *et al.* 1984, Hickey *et al.* 1986) compared with fields for MORB, OIB and oceanic island arcs. Data for the NVZ in Ecuador and Colombia and for the SVZ are displaced to the low $^{143}Nd/^{144}Nd$ side of the MORB field, falling within the field of oceanic island basalts (OIB). Clearly, these data cannot be

Table 7.2 Average major and trace element compositions of Andean volcanic rocks, compared with those from the island arcs of the south-west Pacific (SWP) (data from Ewart 1982).

	Basalt		Basaltic andesite		Andesite	
	Andes	SWP	Andes	SWP	Andes	SWP
%						
SiO_2	51.05	50.07	53.90	54.19	59.89	59.09
TiO_2	1.14	0.85	1.27	0.83	0.95	0.73
Al_2O_3	18.57	16.23	17.50	17.07	17.07	16.83
Fe_2O_3	3.42	3.23	3.13	3.25	3.31	2.82
FeO	5.48	6.75	5.39	5.68	3.00	4.16
MnO	0.16	0.18	0.15	0.16	0.12	0.13
MgO	5.54	7.84	5.35	5.24	3.25	3.83
CaO	8.87	10.82	7.68	9.08	5.67	7.05
Na_2O	3.98	2.51	3.67	2.92	3.95	3.41
K_2O	1.42	1.24	1.62	1.30	2.47	1.70
P_2O_5	0.38	0.28	0.35	0.26	0.31	0.23
ppm						
Rb	49.9	29.1	45.4	30.3	75.4	41.2
Ba	345	364	676	402	886	479
Sr	608	628	644	561	648	516
Zr	162	69.7	179	105	195	138
La	16.3	11.6	24.6	20.2	38.0	25.4
Ce	41.6	25.9	51.3	36.4	66.8	44.0
Y	31.0	19.7	25.4	23.3	12.2	24.7
Yb	2.29	1.54	2.32	1.57	1.94	1.94
Cu	30.0	121	49.6	105	40.0	51.8
Ni	57.9	104	67.4	44.9	38.6	34.4
Co	29.6	43.0	30.5	29.7	18.6	21.3
Cr	67.9	273	202	110	48.4	87.4
V	187	300	220	235	125	154
Nb	—	5.3	12.5	6.5	—	6.3
Pb	—	7.2	—	8.0	—	9.9
Hf	2.9	1.3	3.67	1.75	5.46	2.7

accounted for simply by partial melting of a depleted asthenospheric mantle wedge (MORB source mantle) enriched in radiogenic Sr by slab-derived fluids, as is the case for many intra-oceanic island arcs (Hawkesworth & Powell 1980, Wilson & Davidson 1984). Instead, petrogenetic models could involve partial melting of a subduction-modified enriched mantle source (subcontinental lithosphere; Pearce 1983) or contamination of primary magmas derived from subduction-modified MORB source mantle with a continental crustal component. However, extensive crustal contamination of the NVZ and SVZ lavas appears to be ruled out by combined Sr−O isotopic studies (Section 7.7.6; see also James 1982, Harmon *et al.* 1984) and thus their isotopic compositions may give a good indication of the isotopic characteristics of the subduction-modified mantle wedge. In contrast, the CVZ lavas are characterized by much more varied isotopic compositions with higher $^{87}Sr/^{86}Sr$ and lower $^{143}Nd/^{144}Nd$. These data unequivocally require contamination of mantle-derived magmas by the continental crust (Hawkesworth *et al.* 1982, James 1982, Harmon *et al.* 1984, Thorpe *et al.* 1984). This is consistent with the observation made in Section 7.2 that the CVZ is

Table 7.3 Major and trace element analyses of basaltic andesites from the northern (NVZ) central (CVZ) and southern (SVZ) active volcanic zones of the Andes.

	NVZ		CVZ		SVZ	
	a	b	West [a]	East [a]	c	a
%						
SiO_2	55.72	55.50	54.22	52.41	54.35	54.88
TiO_2	0.89	0.81	0.95	2.02	0.93	1.33
Al_2O_3	16.89	15.20	16.02	16.25	18.16	16.50
Fe_2O_3	8.72	0.90	8.46	9.27	8.50	3.76
FeO	—	6.00	—	—	—	6.56
MnO	0.10	0.13	0.13	0.14	0.14	0.19
MgO	5.12	8.32	7.66	6.03	5.60	3.28
CaO	7.51	7.56	7.88	6.93	8.46	7.25
Na_2O	3.86	3.35	3.14	3.95	3.35	4.44
K_2O	1.14	1.16	1.19	2.50	0.72	0.84
P_2O_5	0.23	0.17	0.20	0.49	0.17	0.21
H_2O	—	—	0.10	—	—	0.28
CO_2	—	—	—	—	—	—
ppm						
Cr	—	515	120	144	96	12
Ni	—	166	81	82	47	40
Rb	18	23	32	63	18.2	17
Sr	640	495	501	633	557	485
Y	13	17	21	25	15	—
Zr	110	79	115	238	80	—
Nb	6	5	11	34	1.9	—
Ba	—	729	367	509	224	265
La	13.4	11.6	15.7	39.3	9.8	10.0
Ce	27.0	23.5	35.0	84.2	24.1	25.2
Nd	16.7	15.27	18.7	40.4	14.5	14.6
Sm	3.9	3.64	4.0	7.7	3.01	2.9
Tb	0.4	0.52	0.6	0.9	0.47	0.7
Yb	1.0	1.37	1.7	2.1	1.59	2.4
Hf	2.7	2.03	3.2	5.8	1.7	1.9
Ta	0.5	0.25	0.5	2.9	—	—
Th	2.6	2.63	2.5	6.8	2.0	2.0

Data sources: [a] Thorpe *et al.* (1984); [b] Marriner & Milward (1984); [c] Hickey *et al.* (1986).

characterized by much thicker crust with a substantial Precambrian basement.

Also shown for comparison in Figure 7.27 are Nd-Sr isotopic data for Cenozoic plateau basalts from Patagonia (Hawkesworth *et al.* 1979). These have been erupted in an extensional tectonic regime similar to that of a marginal basin (Ch. 8) to the east of the Andean Cordillera. It is possible that these magmas, which range in composition from tholeiites through alkali basalts to leucite basanites, have largely escaped contamination by the continental crust as they show primitive characteristics with high MgO contents (6–11%). If this is correct then their isotopic characteristics may reflect those of the subcontinental lithospheric mantle, thus indirectly supporting the largely uncontaminated nature of the NVZ and SVZ magmas.

Combined Nd–Sr data are not available for the

Table 7.4 Comparison of the geochemical characteristics of an alkali basalt erupted to the east of the volcanic front in the CVZ (Thorpe *et al.* 1984) and a calc-alkaline basalt erupted in the SVZ (Hickey *et al.* 1986).

	SVZ calc-alkaline basalt	CVZ alkali basalt
%		
SiO_2	50.30	43.49
TiO_2	0.85	2.34
Al_2O_3	18.88	13.43
Fe_2O_3	9.56	13.19
MnO	0.15	0.18
MgO	5.91	9.95
CaO	10.59	12.30
Na_2O	2.95	3.12
K_2O	0.44	1.42
P_2O_5	0.14	0.74
ppm		
Sc	32	—
V	219	—
Cr	112	—
Co	34	—
Ni	50	127
Zn	80	—
Ga	17	—
Y	16	27
Zr	59	190
Hf	1.4	4.5
Ta	—	3.5
Nb	2.0	47
Th	0.9	5.4
Rb	7.7	24
Cs	0.58	—
Ba	146	—
Sr	437	871
La	6.09	47.5
Ce	15.3	96.1
Nd	9.3	49.0
Sm	2.36	8.8
Eu	0.92	—
Tb	0.42	1.1
Yb	1.60	2.0
Lu	0.26	—

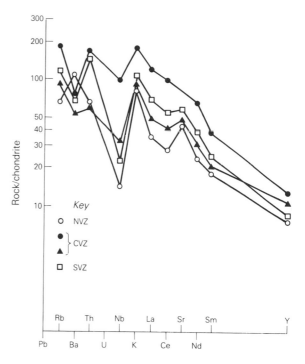

Figure 7.23 Spiderdiagrams for basaltic andesites from the northern (NVZ), central (CVZ) and southern (SVZ) active volcanic zones of the Andes. Data from Thorpe *et al.* (1984) and Hickey *et al.* (1986). Normalization factors from Sun (1980).

plutonic rocks of the Coastal Batholith of Peru and therefore direct comparison of volcanic and plutonic suites is not possible. However, Figure 7.28 shows the available Sr isotopic data (Beckinsale *et al.* 1985) for three segments of the batholith, Arequipa, Lima and Toquepala. The Lima seg-

ment is emplaced along the axis of a marginal basin (Section 7.5) into country rocks consisting of 'new' crust; lavas, dykes, sills and basic plutons. It is probable that all of the magmas which built up this segment were derived by crystal fractionation of mantle-derived parental basalts with comparatively little crustal involvement (Atherton & Sanderson 1985), as evidenced by their low $^{87}Sr/^{86}Sr$ ratios. However, it is important to realize that the young crust into which this segment of the batholith is emplaced will be isotopically similar to the primary magmas and thus crustal contamination may in fact be extensive but not detectable isotopically. In contrast, the Arequipa and Toquepala segments are emplaced partly into a craton, the Arequipa massif, composed of Precambrian gneisses, Upper Palaeozoic and Mesozoic sediments and 400–440 Ma intrusive igneous rocks. This crustal assemblage provides an array of possible sources of

Table 7.5 Analyses of plutonic rocks from the Lima segment of the Coastal Batholith of Peru (data from Pitcher *et al.* 1985).

	Gabbro	Diorite	Granodiorite	Granite
%				
SiO_2	49.84	58.65	69.04	75.58
TiO_2	0.94	0.81	0.42	0.22
Al_2O_3	24.92	16.84	15.03	13.35
Fe_2O_3	1.27	2.76	1.37	0.90
FeO	4.03	4.63	1.77	0.43
MnO	0.13	0.15	0.07	0.05
MgO	2.65	3.66	1.21	0.69
CaO	10.58	6.01	2.85	1.41
Na_2O	2.73	2.85	3.49	3.96
K_2O	0.64	2.16	4.07	3.90
P_2O_5	0.12	0.17	0.10	0.03
ppm				
Ba	259	564	741	595
Ce	28	37	38	34
Co	15	20	8	3
Cr	11	16	6	3
Hf	—	—	—	—
La	7	14	18	14
Nd	17	19	19	15
Ni	7	10	7	14
Pb	9	15	14	12
Rb	19	70	159	144
Sc	26	25	10	5
Sr	431	352	237	104
Th	1	8	21	16
V	163	196	65	18
Y	15	25	22	21
Zn	48	83	28	24
Zr	27	120	191	85

radiogenic Sr and it seems probable that crustal contamination of mantle-derived magmas can explain the observed variations in $^{87}Sr/^{86}Sr$ initial.

Pb isotopes

Figure 7.29 shows the variation of $^{207}Pb/^{204}Pb$ versus $^{206}Pb/^{204}Pb$ for volcanic rocks from the northern, central and southern volcanic zones of the Andes (a) and for plutonic rocks from the Coastal Batholith of Peru (b). These data define broadly linear trends which are quite different from the trend defined by oceanic basalts (MORB + OIB), the Northern Hemisphere Reference Line (NHRL) (Hart 1984, see also Ch. 9). Data for both volcanic and plutonic rocks plot to the high-

$^{207}Pb/^{204}Pb$ side of the NHRL, similar to the Dupal group of oceanic islands and to oceanic island-arc volcanics. Pb isotopic data for the volcanic and plutonic rocks show extensive overlap, supporting the contention that the plutonic rocks do indeed represent the eroded root zones of former active volcanoes. Additionally, these data do not appear to support extensive involvement of depleted asthenospheric mantle, similar to the source of Nazca plate MORB, in the petrogenesis of the magmas. Instead, the data arrays appear to define a mixing line between an enriched mantle component (?the subcontinenetal lithosphere) and Precambrian gneissic crustal rocks. Precambrian basement gneiss with very low $^{206}Pb/^{204}Pb$ clearly appears to have been involved in the petrogenesis of the

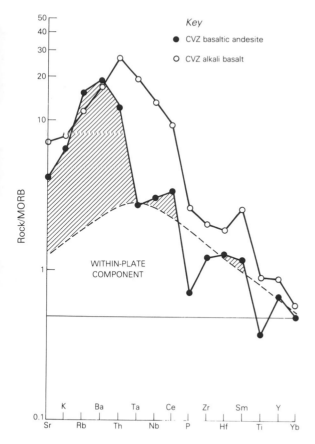

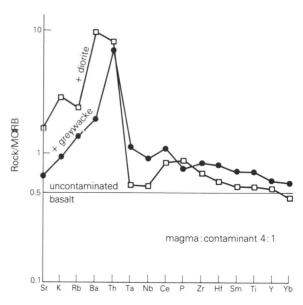

Figure 7.25 MORB-normalized trace element diagram to show the effects of crustal contamination by mixing a basalt magma (with MORB-normalized concentrations of 0.5) with 50% partial melts of greywacke and diorite crustal rocks respectively in the proportions 4:1 magma:contaminant (after Pearce 1983, Fig. 7).

Figure 7.24 MORB-normalized trace element diagram (after Pearce 1983), showing a typical CVZ basaltic andesite and an alkali basalt erupted in an intra-plate setting to the east of the active volcanic zone. For the subduction-related basaltic andesite the dashed line indicates the within-plate component (subcontinental lithosphere), while the shaded area indicates those elements enriched in the sources by subduction-zone fluids (data from Thorpe *et al.* 1984).

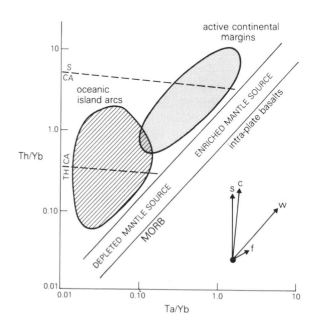

Figure 7.26 Th/Yb versus Ta/Yb plot to show the difference between subduction-related basalts and oceanic basalts derived from depleted sources (MORB) and enriched sources (OIB). Uncontaminated intracontinental plate basalts should plot in the enriched mantle source region. Vectors shown indicate the influence of subduction components (S), within-plate enrichment (W), crustal contamination (C) and fractional crystallization (F). Dashed lines separate the boundaries of the tholeiitic (TH), calc-alkaline (CA) and shoshonitic (S) fields (after Pearce 1983, Fig. 9).

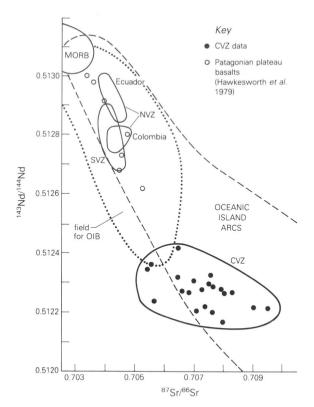

Figure 7.27 Plot of ^{143}Nd/^{144}Nd versus ^{87}Sr/^{86}Sr for volcanic rocks from the northern (NVZ), central (CVZ) and southern (SVZ) active volcanic zones of the Andes. Data from Hawkesworth *et al.* (1982), James (1982), Thorpe *et al.* (1984) and Hickey et al. (1986). Field of oceanic island-arc volcanic rocks from Figure 6.46 and field of oceanic-island basalts (OIB) from Figure 9.23.

magmas forming the Arequipa and Toquepala segments of the Peruvian coastal batholith. This is in good agreement with the crustal models of Couch *et al.* (1981) and Jones (1981), which show a thick Precambrian crustal layer in southern Peru beneath the Arequipa and Toquepala segments, and an extremely thin one beneath the Lima segment.

It is probable that some of the scatter in the volcanic and plutonic ^{207}Pb/^{204}Pb−^{206}Pb/^{204}Pb data arrays is a consequence of multicomponent mixing involving a depleted MORB source mantle component and Pb derived from subducted oceanic sediments, in addition to the subcontinental lithosphere component. For example, the Pb isotopic composition of the NVZ and SVZ lavas could be

modelled in terms of the introduction of Pb derived from subducted continentally derived sediments (via subduction-zone fluids) into a MORB-source mantle wedge.

Segments of the CVZ between 16−18°S and 21−26°S have very distinctive Pb isotopic compositions which may be related to different crustal contaminants. This is more clearly revealed in Figure 7.30, a plot of ^{87}Sr/^{86}Sr versus ^{206}Pb/^{204}Pb. Data from the two segments define remarkably good linear trends pointing in the direction of different crustal contaminants. The 16−18°S data project towards the isotopic composition of 2000 Ma Precambrian basement gneisses (Charcani gneiss), whereas the 21−26°S data can be explained in terms of contamination of mantle-derived magmas by late Precambrian − Palaeozoic metamorphic and granitoid intrusive rocks. Both trends project back to the fields of NVZ and SVZ magmas, the isotopic characteristics of which may therefore indicate those of the subduction-modified mantle wedge. Again, these data appear to confirm the involvement of an enriched mantle source, the subcontinental lithosphere, rather than MORB-

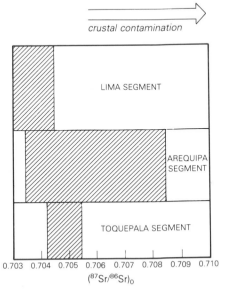

Figure 7.28 Variation of ^{87}Sr/^{86}Sr initial ratio for plutonic rocks from the Lima, Arequipa and Toquepala segments of the Coastal Batholith of Peru (data from Beckinsale *et al.* 1985).

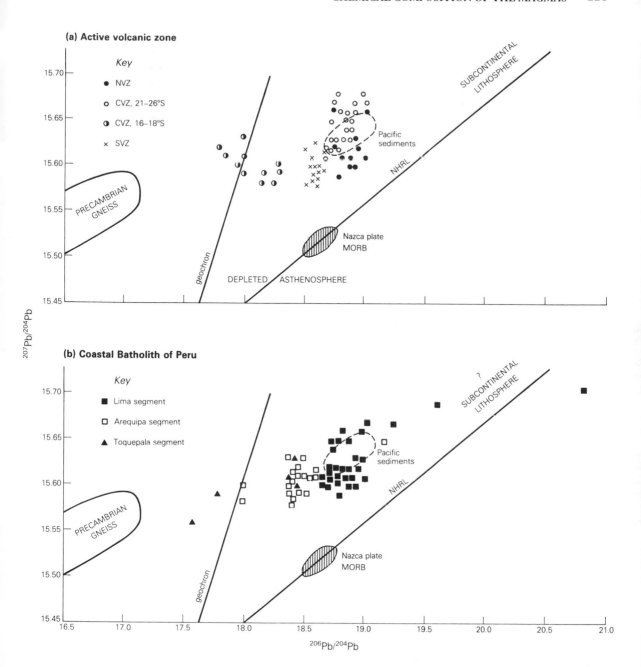

Figure 7.29 ^{207}Pb/^{204}Pb versus ^{206}Pb/^{204}Pb for (a) volcanic and (b) plutonic rocks from the Andes. (a) Data from James (1982), Harmon *et al.* (1984) and Hickey *et al.* (1986); (b) data from Mukasa (1986). Fields for southern Peru Precambrian gneiss, Nazca plate MORB and Pacific sediments from Harmon *et al.* (1984); Northern Hemisphere Reference Line (NHRL) from Hart (1984).

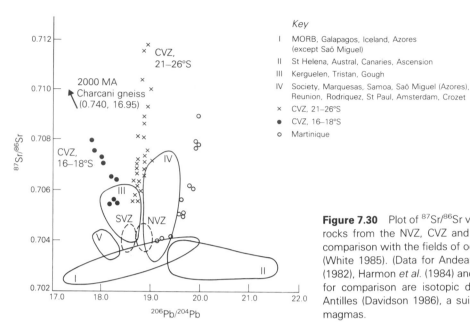

Figure 7.30 Plot of $^{87}Sr/^{86}Sr$ versus $^{206}Pb/^{204}Pb$ for volcanic rocks from the NVZ, CVZ and SVZ of the Andes, to show comparison with the fields of oceanic basalts (MORB + OIB) (White 1985). (Data for Andean volcanic rocks from James (1982), Harmon *et al.* (1984) and Hickey *et al.* (1986).) Shown for comparison are isotopic data from Martinique, Lesser Antilles (Davidson 1986), a suite of contaminated island-arc magmas.

source mantle, in the petrogenesis of Andean magmas.

Also shown in Figure 7.30 are Sr–Pb isotopic data for volcanic rocks from the Lesser Antilles island arc (Davidson 1986). These data also define a steep trend, but in this case point to a contaminant with much higher $^{206}Pb/^{204}Pb$ than the Andean crust. Davidson has explained this apparent continental crustal contamination trend in an oceanic island-arc tectonic setting in terms of contamination of mantle-derived magmas by terrigenous sediments intercalated in the arc crust. However, White & Dupré (1986) favour a source contamination model to explain these data (Ch. 6).

7.7.7 Stable isotopes

Oxygen

As considered in Section 6.11.6, the analysis of oxygen isotopes is a powerful tool for tracing the involvement of continental crustal materials in magma genesis because of the large differences in $\delta^{18}O$ between crustal rocks and rocks derived from the mantle (James 1981). Figure 7.31 shows the variation of $\delta^{18}O$ with $^{206}Pb/^{204}Pb$ for volcanic rocks from the northern, central and southern

volcanic zones of the Andes (James 1982, Harmon *et al.* 1984). $\delta^{18}O$ is lowest in the rocks of the SVZ, ranging from 5.2 to 6.8‰, indistinguishable from the oxygen isotopic composition of fresh MORB and OIB (Kyser *et al.* 1982). It thus seems reasonable to assume that the SVZ lavas represent essentially uncontaminated magma compositions, the isotopic characteristics (Sr, Nd, Pb, O) of which reflect those of the subduction-modified mantle wedge. The NVZ lavas are relatively homogeneous in terms of $\delta^{18}O$ and overlap with the low-$\delta^{18}O$ end of the field for CVZ rocks from 21 to 26°S, which show evidence of contamination by a high-$\delta^{18}O$ crustal component. CVZ rocks from 16 to 18°S define a completely different trend, projecting towards a Precambrian gneissic component with moderate $\delta^{18}O$. The different trends of the 16–18°S and 21–26°S segments of the CVZ mirror those in Figure 7.30, the Sr–Pb isotope diagram, clearly supporting the involvement of different crustal contaminants, old crust with high Rb/Sr, low U/Pb and moderate $\delta^{18}O$, and young crust with high Rb/Sr, high U/Pb and high $\delta^{18}O$ respectively. The NVZ lavas appear to have undergone slight contamination by a young crustal component broadly similar to that involved in the petrogenesis of the CVZ lavas from 21 to 26°S.

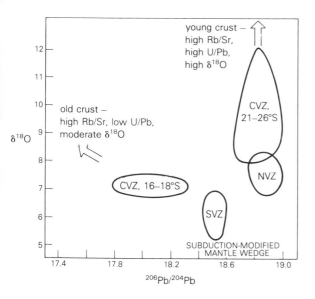

Figure 7.31 Plot of $\delta^{18}O$ versus $^{206}Pb/^{204}Pb$ for volcanic rocks from the northern (NVZ), central (CVZ) and southern (SVZ) volcanic zones of the Andes (after Harmon *et al.* 1984, Fig. 9, p. 818).

7.8   Detailed petrogenetic model

Most recent studies (Hawkesworth & Powell 1980, Perfit *et al.* 1980, Arculus & Johnson 1981, Kay 1984, Wilson & Davidson 1984, Arculus & Powell 1986) have attributed the main features of island-arc basalt geochemistry to variable contributions from two main source components; the asthenospheric mantle wedge overlying the subducting slab of oceanic lithosphere and a metasomatic component, either a hydrous fluid or a partial melt, derived from the subducted oceanic crust. In the active continental margin tectonic setting two additional components are involved, the crust and mantle portions of the continental lithosphere, making this one of the most complex magma generation environments on Earth.

It is generally accepted that the continental lithosphere is thicker than the oceanic lithosphere, and in Andean-type margins it is probable that much if not all of the mantle wedge overlying the subducted slab has lithospheric characteristics. This subcontinental mantle wedge may have very

different trace element and isotope geochemical characteristics from the underlying depleted asthenosphere, particularly if it has formed part of a stable continental keel for several billion years. Specifically, it may be heterogeneously trace element enriched due to the migration of partial melts generated during previous intra-plate magmatic events (Ch. 3). Addition of slab-derived fluids to such enriched mantle will induce partial melting if solidus temperatures are exceeded, and the resultant magmas should have distinctive trace element geochemistries (Section 7.7.3).

Individual subduction systems differ in significant ways, and therefore it is unrealistic to expect any simple general model to explain all the characteristics of all arcs, both oceanic and continental. For example, there may be significant pre-subduction heterogeneity in the mantle wedge and the geochemistry of slab-derived fluids may vary as a consequence of variable degrees of submarine alteration of the oceanic crustal layer, and variations in the proportions and geochemical characteristics of any sediments that may be subducted. Nevertheless, there is one characteristic feature which appears to be common to all instances of subduction-related magmatism − the transfer of Sr, K, Rb, Ba, Th ± Ce, P and Sm to the mantle wedge by partial melt or fluid-transfer processes associated with the dehydration of the subducted slab (Anderson *et al.* 1980, Hawkesworth & Powell 1980, Wilson & Davidson 1984). This provides the critical link between the physical process of subduction and arc magmatism.

Once primary magmas have been generated by partial melting of the subduction-modified mantle wedge, they must subsequently rise through a thick section of continental crustal rocks, up to 70 km in the case of the CVZ of the Andes. Crustal contamination seems inevitable and the subsequent geochemical evolution of the magmas must be dominated by assimilation − fractional crystallisation processes (AFC) (DePaolo 1981). Thus active continental margin magmas should in general have distinctive Sr, Nd, Pb and O isotopic signatures, reflecting the nature of the specific crustal component with which they have interacted. This may be upper or lower crust, young crust or ancient

Precambrian crust, each of which will have different isotopic characteristics. Where magmas rise through young crust Sr, Nd and Pb isotopic data may give the misleading impression that the magmas are uncontaminated. This is because young crustal rocks can have isotopic characteristics quite close to those of the mantle-derived magmas, particularly so if they represent island-arc sequences newly accreted to the continental margin. In such a situation, while AFC processes may have operated, the isotopic composition of the magmas is not modified significantly.

In addition to the effects of high-level crustal contamination, subduction-related magmas may also inherit an isotopic signature from the continental crust via the subduction of terrigenous sediments. This has been clearly demonstrated for some intra-oceanic island-arc magmas (Section 6.11), the isotopic compositions of which could not have been modified by direct interaction with the continental crust. In an Andean tectonic setting it is probable that contamination in both of these environments contributes towards the overall continental crustal fingerprint, but isotopic and trace element data do not allow us to separate these effects.

Primitive basaltic magmas generated in the mantle wedge rise, because they are less dense, to depths at which there is a zero density contrast between the magma and the wall rock. In oceanic island arcs this may be only a few kilometres from the surface, whereas in a continental margin environment it is most likely to be in the deep crust close to the Moho (crust/mantle boundary). The continental crust, by virtue of its lower density than the oceanic crust, thus acts as a filter causing the subduction-zone magmas to stagnate, become contaminated and fractionate at much deeper levels. The comparative rarity of basaltic lavas in continental margin arcs may thus reflect their inability to rise through the continental crust, rather than a lack of basaltic primary magmas.

Young immature intra-oceanic island arcs are characterized by relatively high proportions of tholeiitic mafic volcanic rocks, the trace element and isotopic compositions of which reflect derivation from depleted asthenospheric mantle with minor additions of slab-derived material (Perfit et al. 1980, Arculus & Johnson 1981, Gill 1981). In contrast, more mature island arcs and continental margin arcs underlain by thicker crust erupt greater proportions of more silicic volcanic rocks. Additionally, in these arcs, although tholeiitic, calc-alkaline and shoshonitic series volcanics are all represented, calc-alkaline and shoshonitic types predominate. This may reflect the combined effects of more enriched mantle sources and crustal contamination in the petrogenesis of the magmas.

As an oceanic island arc evolves with time, repeated influx of magma causes the crust to thicken and thus the depth of stagnation of primitive basaltic magmas to increase (Leeman 1983). Thus, in some instances, the changeover from dominantly tholeiitic to calc-alkaline arc magmatism may not necessarily reflect any fundamental differences in the primary magma chemistry, but simply differences in fractionation conditions. For example, the evolution of basaltic magmas fractionating at shallow depths outside the stability field of amphibole will be dominated by anhydrous assemblages involving plagioclase, olivine, orthopyroxene, clinopyroxene and magnetite, and the magmas may consequently evolve along a tholeiitic liquid line of descent. However, at greater depths crystal fractionation of hydrous basic magmas will be dominated by amphibole, which has been postulated by many authors to be fundamental in producing calc-alkaline magma chemistries (Eggler & Burnham 1973, Cawthorn & O'Hara 1976, Allen & Boettcher 1978). However, Hawkesworth & Powell (1980) suggested, for the Lesser Antilles island arc, that tholeiitic and calc-alkaline magmatism was triggered by the release of hydrous fluids and partial melts respectively from the subducted slab, and thus that the parental magma compositions of the two series do differ. This remains a matter for further detailed study.

In a region of particularly long-lived arc magmatism the thermal effects of basaltic magma influx into the base of the crust become important (Patchett 1980) and may eventually cause crustal anatexis (partial melting). As considered in Section 7.4, partial melting of lower crustal gneisses could produce silicic magmas, and many authors have attributed the ignimbrite eruptions of the central Andes to such a mechanism. Additionally, mantle-derived magmas may mix with such crustal melts while simultaneously undergoing crystal fractionation (DePaolo 1981).

The Andean active continental margin has provided a particularly useful natural laboratory in which to study the interaction between subduction-related magmas and the continental crust, as it shows marked variations from north to south in the subduction-zone geometry, the volume and provenance of subducted sediments and in the thickness, age and composition of the overriding continental crust. The uniformity of the volcanic front relative to the position of the Peru–Chile trench along the whole length of the Andean Cordillera implies an intimate association between volcanism and subduction, which is most easily attributed to the role of slab-derived fluids. Chemical and isotopic data for the most primitive basalts erupted in all three active volcanic zones (NVZ, CVZ and SVZ) suggest that magma genesis is initiated as slab-derived fluids, enriched to variable extents in incompatible elements of low ionic potential, radiogenic Sr derived from sea water and radiogenic Pb derived from subducted sediments, invade the mantle wedge. Partial melting of this enriched peridotite source region to different degrees produces the primitive mafic magmas which are parental to the range of rock types observed in all these volcanic zones. The Sr, Nd, Pb and O isotopic characteristics of the NVZ and SVZ lavas have been interpreted in terms of derivation from a subduction-modified enriched mantle source with very little crustal contamination (Thorpe et $al.$ 1981, 1984; Harmon et $al.$ 1984; Déruelle et $al.$ 1983). In contrast, the CVZ lavas are more evolved, with higher $^{87}Sr/^{86}Sr$ and $\delta^{18}O$ and lower $^{143}Nd/^{144}Nd$ ratios, indicating contamination by the continental crust. There appears to be a particularly good correlation between the chemistry of the Andean volcanic rocks and the crustal thickness and age (Table 7.1). Thus CVZ lavas have the most obvious continental crustal fingerprint as these have risen through the greatest thickness of Precambrian basement gneisses.

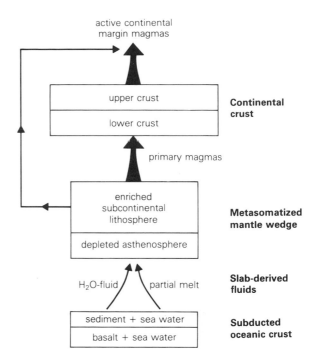

Figure 7.32 Flow diagram to summarize the source components involved in the petrogenesis of active continental margin magmas.

Figure 7.32 summarizes the processes and source components involved in the petrogenesis of active continental margin magmas.

Further reading

Gill, J.B. 1981. $Orogenic$ $andesites$ and $plate$ $tectonics$. Berlin-Heidelberg; Springer-Verlag, 390 pp.

Moorbath, S. & R.N Thompson (eds) 1984. The relative contributions of mantle, oceanic crust and continental crust to magma genesis. $Phil$ $Trans$ $R.$ $Soc.$ $Lond.$ **A310**, 437–780.

Pitcher, W.S., M.P. Atherton, E.J. Cobbing & R.D. Beckinsale (eds) 1985. $Magmatism$ at a $plate$ $edge$. Glasgow: Blackie 328 pp.

Thorpe, R.S. (ed.) 1982. $Andesites:$ $orogenic$ $andesites$ and $related$ $rocks$. Chichester: Wiley 724 pp.

CHAPTER EIGHT

Back-arc basins

8.1 Introduction

Back-arc or marginal basins are semi-isolated basins or series of basins lying behind the volcanic chains of island-arc systems (Karig 1971). It is generally accepted that these are extensional features produced by seafloor spreading type processes broadly similar to those occurring at mid-oceanic ridges (Ch. 5; see also Saunders & Tarney 1979, 1984; Crawford *et al.* 1981; Taylor & Karner 1983; Jarrard 1986). An extensional origin is supported by the high heat flow characteristic of such basins (Sclater *et al.* 1972, Hawkins 1974) and by the occurrence of sets of magnetic lineations similar to those observed in normal oceanic crust. These were first described from the East Scotia Sea, an active back-arc basin behind the South Sandwich island arc in the South Atlantic (Barker 1972),

and have subsequently been documented from the Lau, Mariana, south Fiji and west Philippine basins. Taylor & Karner (1983) have listed all known Neogene back-arc basins and compared their characteristics to those of normal oceanic spreading centres.

Figure 8.1 shows the main currently active back-arc basins of the Pacific and Atlantic oceans. These are associated with the Tonga, Kermadec, Mariana, New Hebrides, New Britain and South Sandwich island arcs. Additionally, the Ryukyu (Japan), Izu−Bonin and north Sulawesi subduction zones exhibit incipient or very slow back-arc spreading (Jarrard 1986). Table 8.1 lists the duration of spreading and the total opening rate for these different back-arc spreading systems. In each of these back-arc basins, except the Andaman which may have originated by more complex

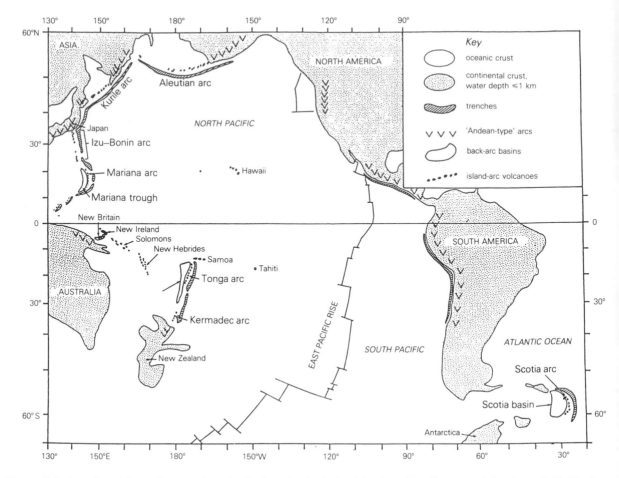

Figure 8.1   Location of the main currently active back-arc basins in the Atlantic and Pacific oceans (after Stern 1982, Fig. 1, p. 478).

processes, the spreading direction is approximately perpendicular to the associated trench, suggesting an intimate relationship with the dynamics of the subduction system.

Back-arc basins are essentially an oceanic phenomenon, although extensional tectonic regimes to the landward side of the volcanic front in active continental margins (Ch. 7) may be considered similar. However, in such cases no new oceanic crust is generated. For example, active thinning of the continental crust is occurring in a diffuse zone behind the Hellenic arc in the Aegean Sea (Jarrard 1986). However, this extension is not simple back-arc spreading, as the back-arc region is complicated by the collision of a Turkish plate with the Eurasian plate (McKenzie 1978a). Alkaline

Table 8.1   Duration of spreading and total opening rate for currently active back-arc basins (data from Jarrard 1986).

Arc	Associated back-arc basin	Age of inception of spreading (Ma)	Total opening rate (cm yr^{-1})
New Britain	Bismarck Sea	3.5	13.2
New Hebrides	Fiji Plateau	10	7.0
Tonga	Lau Basin	6	7.6
Marianas	Mariana Trough	6–7	4.3
Andaman	Andaman Sea	13	3.7
South Sandwich	East Scotia Sea	8	5–7

volcanic rocks erupted to the east of the Andean Cordillera (Ch. 7) are generated in a similar tectonic regime. In this chapter attention is focused on the oceanic examples, as subduction-related intracontinental plate extension can be more effectively considered under the headings of Chapters 10 & 11.

All areas of active oceanic back-arc extension overlie steeply dipping subduction zones (Cross & Pilger 1982) and, in general, it appears that extensive back-arc spreading only occurs where the subducting lithosphere is old (> 80 Ma) and consequently cold and dense (Furlong et al. 1982). Additionally, there appears to be some correlation with arcs in which the relative vector of migration of the overriding plate is away from the trench. Molnar & Atwater (1978) noted that subduction systems in the western Pacific are characterized by steep-angled subduction of old dense lithosphere and back-arc extension, whereas those in the eastern Pacific are characterized by the subduction of young buoyant lithosphere at relatively shallow angles (~30°) and back-arc compressional tectonics.

Seafloor spreading in back-arc basins clearly differs from that at normal mid-oceanic ridges in that spreading is always closely associated with subduction. Petrogenetic models for the formation of back-arc basin magmas must therefore consider whether there are any fundamental differences in magma generation processes between the two environments in terms of source compositions, depth and degree of partial melting and the role of volatiles. The tectonic setting of back-arc spreading is obviously one in which there is potential for the involvement of fluids from the subducting slab, depending upon the specific arc geometry. This may affect the magma generation process and produce basalts with geochemical characteristics transitional to those arc basalts. Nevertheless, many studies have suggested that back-arc basin basalts form by partial fusion of mantle sources analogous to those involved in the generation of normal or slightly enriched MORB (Pineau et al. 1976, Hawkesworth et al. 1977, Saunders & Tarney 1979) and that, despite their association with island arcs, fluids derived from the subducted slab are not necessarily involved in their petrogenesis.

Despite the considerable body of data on mid-ocean ridge basalts (MORB) (Ch. 5), there are unfortunately few sets of geochemical data available for back-arc basin basalts for comparative purposes. Basaltic magmas erupted in back-arc basin tectonic settings vary from low-K tholeiites, the major element chemistry of which is essentially identical to MORB, to subalkaline basalts with slightly higher alkali contents. However, in terms of their trace element geochemistry back-arc basin basalts may differ from MORB as a consequence of the involvement of subduction-zone fluids in their petrogenesis. In addition to basalts, an unusual group of high-MgO andesites, called boninites, are in some instances associated with the fore-arc regions of island arcs with a history of back-arc spreading (Cameron et al. 1979). These have been given the status of a separate island-arc magma series, the boninite series, by Meijer (1980).

Many ophiolite complexes (Ch. 5) are now regarded as the obducted floors of back-arc basins, rather than as tectonically emplaced slivers of true oceanic crust (Dewey 1976, Saunders et al. 1979). Crawford et al. (1981) consider that the Bay of Islands complex in Newfoundland (Suen et al. 1979) and the Tortuga−Sarmiento complex of southern Chile (Stern 1980) are examples of magmatism in an extensional back-arc basin associated with a subduction system.

8.2 Simplified petrogenic model

Models for the evolution of island arc − back-arc basin systems are essentially based on those of Karig (1971) for the Tonga−Lau and west Philippine − Mariana regions. Figure 8.2 shows a series of schematic cross sections depicting the development of a back-arc basin, based on Karig's ideas and those of Crawford et al. (1981). Section (a) shows a typical oceanic island-arc tectonic setting which might exist prior to the inception of a back-arc basin. Fluids or partial melts ascending from the subducted oceanic lithosphere metasomatize the asthenospheric mantle wedge, which then partially melts to produce the arc basalts. Section (b) depicts the onset of rifting and the development

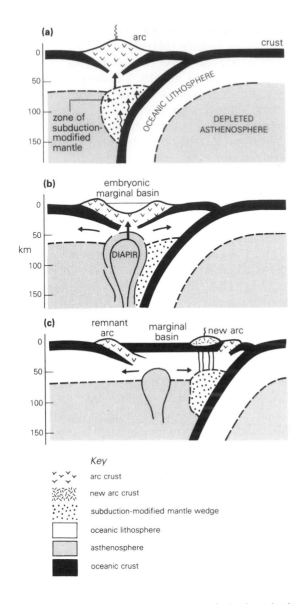

Key

y v y v   arc crust

new arc crust

subduction-modified mantle wedge

oceanic lithosphere

asthenosphere

oceanic crust

Figure 8.2 Models for the development of a back-arc basin. (a) Normal island-arc magmatism: subduction-zone fluids metasomatize the asthenospheric mantle wedge, which then partially melts to produce the arc magmas. (b) Ascending diapirs of asthenospheric (MORB-source) mantle rise beneath the arc and interfere with arc magma generation processes. Arc magmatism ceases and partial melts of the diapir feed an embryonic marginal basin. (c) The marginal basin develops, rupturing the old arc as it spreads. Eventually, subduction-related magmatism is re-established upon a foundation of older arc rocks. (after Crawford *et al.* 1981, Fig. 3, p. 351).

of an embryonic back-arc basin, generated as a consequence of the diapiric upwelling of deep asthenospheric mantle beneath the arc axis, which partially melts as a consequence of adiabatic decompression to produce MORB-like back-arc basin basalts. Arc volcanism appears to cease around the time that the back-arc opening commences (Crawford *et al.* 1981), implying that the rising mantle diapirs interfere with the process of arc magmatism, effectively isolating the arc volcanoes and their plumbing systems from the arc magma sources in the mantle. Eventually the arc crust ruptures (section (c)) into two blocks and a true back-arc basin develops. The block furthest from the trench sinks and becomes a remnant arc, while that adjacent to the trench migrates oceanwards as the basin extends. Taylor & Karner (1983) suggest that the initial rifting may either split the arc as depicted in Figure 8.2 or it may occur to either side of the arc.

From the early stages of basin opening, arc-related volcanism ceases on both the subsiding remnant block and on the migrating fore-arc block (Crawford *et al.* 1981). However, after a period of extension arc magmatism may recommence (section (c)) and a new magmatic arc develops partly on the rifted-off fore-arc block of the old arc. The sequence of events shown in Figure 8.2 may occur several times along an active destructive plate boundary, producing a complex sequence of magmatic events, particularly in the fore-arc region. While the back-arc spreading centre remains close to the trench, the possibility exists that slab-derived fluids may influence the chemistry of the back-arc basalts. However, as spreading proceeds the influence of such fluids progressively diminishes and the back-arc basalts approach the composition of light-REE depleted N-type MORB (Tarney *et al.* 1977).

In terms of Figure 8.2, partial melting and extensional tectonics are related to the adiabatic decompression of ascending mantle lherzolite beneath the arc axis. However, there is no general consensus as to why such upwelling should occur. Toksöz & Bird (1977) suggested that the descending slab exerts a viscous drag on the asthenosphere, causing complementary convective circulation in the mantle wedge behind the arc. Aternatively, Oxburgh & Parmentier (1977) have proposed that the subducted oceanic crust and mantle components of the lithosphere may segregate at depth into dense eclogite and buoyant harzburgite, the latter rising behind the arc to induce back-arc spreading.

Regardless of the details of the model, it is clear that there are potentially more variable source components available for the production of back-arc basalts than beneath a mid-oceanic ridge. These include a variety of depleted and more fertile peridotites from the oceanic lithosphere and underlying asthenosphere, which may be secondarily enriched by subduction-zone fluids during the early stages of basin opening. In terms of the model shown in Figure 8.2, the generation of boninite-series magmas appears to occur at the point when arc volcanism ceases and back-arc spreading is initiated (Crawford et al. 1981). They are thus considered to characterize the embryonic stages of the magmatism, which results in the splitting of an island arc and the generation of a back-arc basin. Boninite magmas have geochemical characteristics (Section 8.4) suggestive of derivation by the partial melting of highly depleted mantle sources (harzburgites) and Crawford et al. (1981) postulate that they originate from the envelope of the rising MORB-source mantle diapirs. This envelope may comprize variably depleted peridotites and harzburgites of the oceanic lithosphere, some of which may have been previously metasomatized by subduction-zone fluids.

8.3   Petrography of the volcanic rocks

Few petrographic data are available for back-arc basin basalts. Saunders & Tarney (1979) present brief descriptions of East Scotia Sea basalts and, in the absence of other data, we have to regard these as representative. The basalts are sparsely porphyritic with large, commonly resorbed, phenocrysts of olivine and plagioclase and microphenocrysts of plagioclase, olivine, Ca-rich clinopyroxene, chromite and titanomagnetite set in a fine-grained, sometimes glassy groundmass. Mineralogically, they appear similar to mid-ocean ridge basalts (Ch. 5) and their textures are typical of basalts quenched in a submarine environment. Plagioclase compositions vary from An_{67} to An_{90}, with the groundmass phases being more sodic than the associated phenocrysts. Olivine compositions are remarkably homogeneous in the range Fo_{86-88} close to the range of mantle olivines (Fo_{98-92}).

Boninites are glassy olivine, orthopyroxene and clinopyroxene phyric lavas, characterized by an absence of feldspar in all but most evolved varieties. Olivine and polysynthetically twinned clinoenstatite, often showing a reaction relation, are common in the more primitive end-members, while strongly differentiated samples lack olivine but may contain hornblende and calcic plagioclase phenocrysts (Crawford et al. 1981).

8.4   Chemical composition of the erupted magmas

8.4.1   Major elements

SiO_2, TiO_2, Al_2O_3, Fe_2O_3, FeO, MnO, MgO, CaO, Na_2O, K_2O and P_2O_5 can all be considered major elements in the description of the geochemistry of back-arc basin basalts and boninite-series magmas.

Few detailed studies have been made of suites of back-arc basin volcanic rocks and the discussion in this section is based upon the data of Saunders & Tarney (1979) for the East Scotia Sea, the back-arc basin associated with the young tholeiitic South Sandwich island arc in the South Atlantic (Fig. 8.1). Dredge samples from the East Scotia Sea are all basaltic, ranging from 49 to 54% SiO_2, and include fairly primitive basalts with 7–8% MgO which are sometimes associated with more fractionated basalts (4–5% MgO) in a single dredge site. Table 8.2 presents major and trace element geochemical data for the more primitive basalts and for a tholeiitic basalt from the South Sandwich arc (Luff 1982), and enriched and depleted MORB compositions from the South Atlantic (Humphris et al. 1985) for comparative purposes. The Scotia Sea basalts are similar to both MORB and island-arc

Table 8.2 Major and trace element analyses of back-arc basin basalts from the East Scotia Sea (Saunders & Tarney 1979), a typical tholeiitic basalt from the associated South Sandwich island arc, and depleted and enriched South Atlantic MORB compositions (Humphris *et al.* 1985). South Sandwich island arc data from Luff (1982).

	East Scotia Sea basalts				South Sandwich arc tholeiite	South Atlantic MORB	
	1	2	3	4		Enriched	Depleted
%							
SiO_2	50.36	50.70	51.00	53.84	51.03	51.22	50.40
TiO_2	1.46	1.29	1.02	0.61	0.82	1.62	1.26
Al_2O_3	16.36	16.58	17.90	14.51	15.76	15.87	17.20
Fe_2O_3	9.07	8.43	7.54	9.24	11.76	10.66	10.24
MnO	0.16	0.16	0.14	0.17	0.21	0.17	0.17
MgO	7.36	7.67	7.37	7.71	6.30	7.51	7.72
CaO	10.84	11.12	10.70	10.80	10.72	11.04	11.98
Na_2O	3.39	3.15	2.72	1.79	2.40	2.69	2.51
K_2O	0.43	0.34	0.57	0.24	0.18	0.38	0.08
P_2O_5	0.20	0.19	0.15	0.08	0.13	0.18	0.10
ppm							
Cr	270	270	196	295	100	229	267
Ni	64	63	66	42	14	127	127
Rb	6	5	8	4	3	8	2
Ba	77	49	83	55	57	74	20
Sr	212	193	195	123	133	151	117
Zr	130	107	84	40	29	117	91
Hf	2.9	2.6	2.0	0.98	—	—	—
Nb	8	3	3	1	1	7.2	2
Ta	1.1	0.78	0.82	0.36	—	—	—
Zn	68	68	69	71	79	74	67
Ga	15	17	13	12	—	—	—
La	7.83	—	5.71	—	1.59	6.87	2.63
Ce	19.0	16.1	13.3	6.45	4.99	21.4	9.1
Nd	13.1	11.1	8.98	4.55	4.39	13.5	7.7
Sm	3.94	3.29	2.76	1.46	1.62	4.69	2.77
Eu	1.44	1.25	1.00	0.56	0.69	1.59	1.24
Gd	4.87	—	3.49	1.99	2.36	6.8	4.5
Dy	5.24	4.69	3.87	3.38	2.88	6.4	—
Er	3.20	3.03	2.48	1.58	1.91	—	—
Yb	3.02	2.87	2.45	1.59	1.84	3.60	3.19
Y	30	29	24	14	18	37	34

tholeiites in terms of their major element chemistry, although in general they tend to have slightly higher alkali contents.

Figure 8.3 is a plot of wt.% K_2O versus wt.% SiO_2 for the East Scotia Sea basalts, which shows that although some have low-K characteristics similar to MORB tholeiitites (Ch. 5) others have higher K_2O contents, falling within the sub-alkalic field. Figure 8.4 is a plot of Alkali Index (A.I.) versus wt.% Al_2O_3 for the more primitive basalts

with MgO contents greater than 7%, a figure which was used in Chapter 6 (Fig. 6.19) to distinguish tholeiitic from high-alumina arc basalts. This clearly shows that the more K_2O-rich primitive basalts also have high alumina characteristics, perhaps suggesting the involvement of subduction-modified mantle sources in their petrogenesis.

As the range of silica contents of the basaltic rocks is so restricted, SiO_2 is not useful as an index of differentiation in variation diagrams and so the

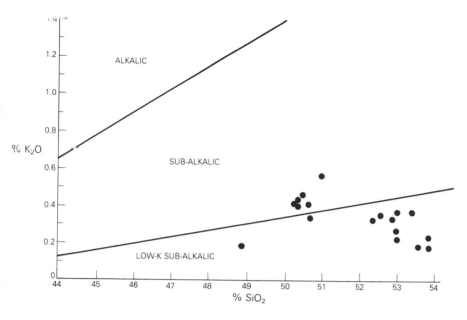

Figure 8.3 Plot of wt.% K_2O versus wt.% SiO_2 for back-arc basin basalts from the East Scotia Sea. Data from Saunders & Tarney (1979). Field boundaries from Middlemost (1975).

incompatible trace element Zr is used instead. Figure 8.5 is a plot of wt.% MgO versus Zr (ppm) which reveals a spectrum of primitive basalt types (1–4 in Table 8.2) with MgO contents greater than 7%. These could be generated by progressively greater degrees of partial melting of a homogeneous source, as indicated by their decreasing total REE contents from basalt 1 to basalt 4 (Section 8.4.2). Also shown in Figure 8.4 is a coherent fractionation trend for basalts from a single dredge site (open symbols) involving parental basalts slightly more MgO-rich than basalt 4. Figure 8.6 is a plot of wt.% TiO_2 versus Zr, which shows a remarkably coherent linear correlation for all the Scotia Sea basalts, both primitive and fractionated. This suggests that TiO_2 behaves incompatibly during both partial melting and fractional crystallization processes.

Boninites are high-SiO_2 (>55%) high-MgO (>9%) lavas characterized by high compatible trace element contents (Ni = 70–450 ppm, Cr = 200–1800 ppm) and very low TiO_2 contents (<0.3%) (Crawford *et al.* 1981, Hickey & Frey 1982). Meijer (1980) has suggested that they should be classified as a separate boninitic magma series because associated rocks range widely in MgO content (4–25%) as a consequence of low-pressure fractionation of orthypyroxene (Jenner 1981). The

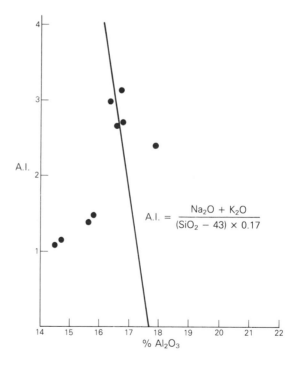

$$A.I. = \frac{Na_2O + K_2O}{(SiO_2 - 43) \times 0.17}$$

Figure 8.4 Plot of Alkali Index (A.I.) versus wt.% Al_2O_3 for primitive East Scotia Sea back-arc basin basalts with >7% MgO. Data from Saunders & Tarney (1979). Field boundary from Middlemost (1975). Numbered samples are those from Table 8.2.

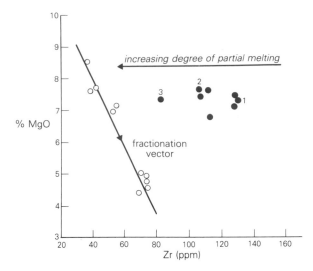

Figure 8.5 Plot of wt.% MgO versus Zr (ppm) for back-arc basin basalts from the East Scotia Sea. A suite of basalts from a single dredge haul, which appear to represent a fractionation sequence, are plotted as open circles. Data from Saunders & Tarney (1979). Numbered samples are those from Table 8.2.

geochemical characteristics of boninites indicate derivation from strongly depleted mantle sources under hydrous conditions (Hickey & Frey 1982). Table 8.3 presents major and trace element data for a suite of boninitic rocks from the Bonin Islands, Japan, (Hickey & Frey 1982) and an analysis of an island-arc andesite from the South Sandwich arc (Luff 1982) for comparative purposes. Despite their high SiO_2 contents, it is obvious that the boninite magmas with their high MgO and low Al_2O_3, TiO_2 and alkali contents are not fractionated magmas like the island arc andesite, but near-primary partial melts generated under unusual conditions.

8.4.2 Trace elements

Trace element data for East Scotia Sea basalts, a tholeiitic basalt from the associated South Sandwich arc and depleted and enriched Atlantic MORB compositions are presented in Table 8.2. Compared to N-type MORB, the back-arc basalts show relative enrichment in the large low-valency cations K, Rb, Ba and Sr, which have previously been considered (Ch. 6) to be those mobile elements

transported into the source of island-arc basalts by subduction-zone fluids. However, compared to enriched MORB compositions only K and Sr appear to be significantly enriched. Clearly, some caution must be exercised when comparing absolute trace element abundances in this way, as some of the apparent trace element enrichment of basalts 1, 2 and 3 may be a consequence of their generation by smaller degrees of partial melting than basalt 4 of an essentially homogeneous mantle source. Trace element ratios are perhaps more reliable petrogenetic indicators as far as mantle source compositions are concerned, and Table 8.4 compares K/Rb, K/Ba, Rb/Sr and Zr/Nb ratios for back-arc basalts with those of MORB and a typical oceanic island tholeiite. N-type MORB have high K/Rb, K/Ba and Zr/Nb and low Rb/Sr ratios compared to the back-arc basalts, which seem to have more affinity with enriched MORB and oceanic island tholeiites generated from less depleted mantle sources.

Chondrite-normalized REE patterns are shown in Figure 8.7 for comparison with those of tholeiitic basalts from the adjacent South Sandwich island arc. The back-arc basin basalt patterns are essentially flat at 6–30 times chondrite, with a slight tendency towards light-REE enrichment. They lie within the same range as the tholeiitic arc basalts and also within the spectrum of MORB REE

Figure 8.6 Plot of wt.% TiO_2 versus Zr (ppm) for back-arc basin basalts from the East Scotia Sea. Data from Saunders & Tarney (1979). Numbered samples are those from Table 8.2.

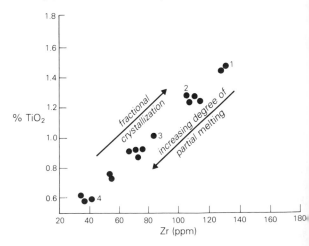

Table 8.3 Major and trace element abundances in boninite lavas from the Bonin Islands, Japan (data from Hickey & Frey 1982): listed for comparison is an analysis of a South Sandwich island-arc andesite (Luff 1982).

	Boninites					Island-arc andesite
	1	2	3	4	5	
%						
SiO_2	57.23	58.43	58.46	59.69	69.56	50.50
TiO_2	0.12	0.15	0.10	0.29	0.33	0.72
Al_2O_3	10.61	11.35	13.37	14.44	13.26	17.52
Fe_2O_3	—	—	—	1.67	0.87	7.10
FeO	8.80	8.57	8.27	6.73	5.16	—
MnO	—	0.12	—	0.23	0.12	0.14
MgO	12.27	11.40	9.39	5.71	1.65	3.43
CaO	9.69	7.76	8.11	8.38	4.80	7.55
Na_2O	0.87	1.74	1.59	2.28	3.27	3.11
K_2O	0.33	0.51	0.70	0.51	0.95	0.92
P_2O_5	—	—	—	0.07	0.04	0.19
ppm						
Sc	45.1	37.4	36.2	35.9	22.0	—
V	145	164	174	—	—	—
Co	46.1	41.7	37.3	31.6	12.4	25
Cr	888	832	538	208	<5	40
Ni	111	205	140	—	—	7
Rb	7.5	10.5	12.2	11.1	20.0	21
Sr	58.7	68.3	97.2	85.7	113.5	204
Ba	20.2	28.2	30.0	27.9	55.8	200
Y	2	5	4.9	8	7	24
Zr	11	19	25.4	30	44	91
Hf	0.31	0.54	0.69	0.88	1.34	—
La	0.71	0.95	1.27	1.13	1.82	6.24
Ce	1.62	2.16	2.57	2.69	3.96	16.76
Nd	0.97	1.47	1.65	1.95	2.69	10.37
Sm	0.266	0.429	0.426	0.623	0.769	3.00
Eu	0.107	0.150	0.146	0.231	0.268	0.77
Tb	0.078	0.109	0.099	0.160	0.188	—
Yb	0.480	0.663	0.591	0.894	1.08	2.99
Lu	0.084	0.115	0.103	0.149	0.186	—

patterns (Ch. 5, Fig. 5.38) but, unlike the arc basalts, do not display Eu anomalies. From a consideration of the REE patterns it is possible that basalts 1–4 in Table 8.2 could be generated by progressively increasing degrees of partial melting of a homogenous source with a flat REE pattern (1–2 times chondrite).

Figure 8.8 is a MORB-normalized trace element variation diagram (Pearce 1983) similar to that used in Chapter 6 (Fig. 6.38) to characterize subduction-zone fluid components added to the mantle source of island-arc basalts. Plotted are the most light-REE enriched (1) and the most light-REE depleted (4) back-arc basin samples from Table 8.2, which may represent the extremes of partial melting of a homogeneous source, and a typical tholeiitic basalt from the South Sandwich arc. The patterns reveal certain similarities between the geochemistry of back-arc and arc basalts, particularly for basalt 4. However, basalt 1 shows more of an affinity with intra-plate basalts (Fig. 6.38) having a characteristic 'humped' pattern. Basalt 4 appears to have

Table 8.4 Trace element characteristics of East Scotia Sea back-arc basin basalts, compared to MORB and oceanic-island tholeiites. Data from Table 6.5, Saunders & Tarney (1979) and Basaltic Volcanism Study Project (1981).

	East Scotia Sea back-arc basin basalts	MORB		Oceanic-island tholeiite
		N-type depleted	P-type enriched	
K/Rb	400–800	1046	414	400
K/Ba	40–60	109	34	25–40
Rb/Sr	0.025–0.04	0.0079	0.043	0.01–0.05
Zr/Rb	16–54	37	6	6–15

Nb–Yb characteristics in common with the arc tholeiites, but an unusually high Ta content more akin to that of intra-plate basalts. The trace element geochemistry of these back-arc basalts is clearly complex and their petrogenesis may involve depleted MORB source mantle (Ch. 5), more enriched OIB source mantle (Ch. 9) and subduction-zone components.

Table 8.3 represents trace element data for a suite of boninitic lavas from the Bonin Islands, Japan, and for comparison an andesite from the South Sandwich island arc. The high Ni and Cr contents of the boninites combined with their high MgO contents suggest that these are near primary partial melts despite their high SiO_2 contents. Boninites display unusual broad 'V'-shaped chondrite-normalized REE patterns (Fig. 8.9), which are rarely observed in other types of volcanic rock. However, such patterns are found in metasomatized harzburgite xenoliths in alkali basalts (Frey 1982), which may therefore be comparable to the boninite source. The unusual shape of the REE patterns may be a consequence of the metasomatism of a strongly light-REE depleted source by a light-REE enriched fluid or partial melt.

Figure 8.10 shows MORB-normalized trace element variation diagrams (Pearce 1983) for a boninite from the Bonin islands (4 in Table 8.3) with 5.71% MgO and a more MgO-rich sample (12.6% MgO) from Cape Vogel, Papua New Guinea. Compared to MORB the boninites have higher K,

Rb and Ba contents but significantly lower heavy REE and high field strength element concentrations. Island-arc basalts show broadly similar patterns (Fig. 6.38), suggesting that subduction-zone fluids also play a fundamental role in boninite petrogenesis. To account for the very low abundances of Ti, Y and Yb, the source peridotite must either have been depleted by a previous partial melting event or these elements must be retained in the source in residual mineral phases such as clinopyroxene, amphibole or garnet during partial melting. However, experimental studies (Green 1973, 1976) which suggest that boninites are formed by high degrees of partial melting (>30%), leaving a refractory residue of olivine and orthopyroxene, tend to argue against the role of residual phases.

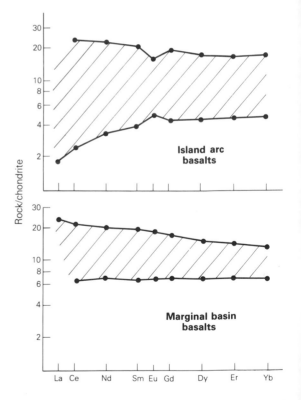

Figure 8.7 Chondrite-normalized REE patterns for basalts from the South Sandwich island arc, compared with those from the associated marginal basin, the East Scotia Sea (after Hawkesworth *et al.* 1977, Fig. 2, p. 257).

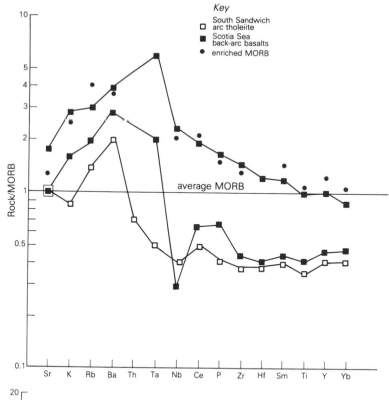

Figure 8.8 MORB-normalized trace element variation diagram for back-arc basalts from the East Scotia Sea (Saunders & Tarney 1979) and for a typical tholeiitic basalt from the associated South Sandwich island arc (Pearce 1983). Also shown are data for an enriched MORB composition from the South Atlantic (Humphris et al. 1985). Numbered samples are those from Table 8.2.

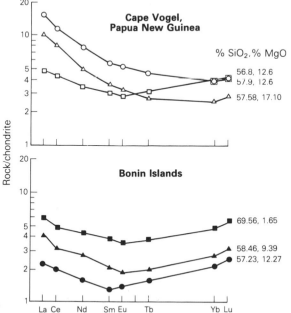

Figure 8.9 Chondrite-normalized REE patterns for boninite series magmas from Cape Vogel, Papua New Guinea, and the Bonin Islands, Japan (data from Hickey & Frey 1982).

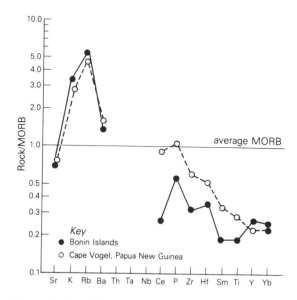

Figure 8.10 MORB-normalized trace element variation diagrams for boninites from Cape Vogel, Papua New Guinea, and the Bonin Islands, Japan (analysis 4 in Table 8.3) (data from Hickey & Frey 1982).

8.4.3   Radiogenic isotopes

Sr–Nd

Back-arc basin basalts are normally characterized by lower $^{87}Sr/^{86}Sr$ ratios than volcanics from the associated island arc (Stern 1982) and this has been considered to reflect the role of $^{87}Sr/^{86}Sr$-enriched fluids from the subducted slab in the petrogenesis of the arc magmas (Hawkesworth *et al.* 1977). Figure 8.11 shows a compilation of Sr isotopic data for the Mariana arc and its associated back-arc basin, the Mariana Trough (Woodhead & Fraser 1985), which clearly shows the lower $^{87}Sr/^{86}Sr$ of the back-arc basin basalts. The South Sandwich arc (0.70376–0.70423) and the East Scotia Sea back-arc basin (0.70281–0.70336) (Luff 1982) show a similar pattern. Stern (1982) has calculated that circum-Pacific intra-oceanic island-arc lavas have a mean $^{87}Sr/^{86}Sr$ of 0.70335, whereas associated back-arc basins have a mean of 0.70311. In the North Pacific the mean $^{87}Sr/^{86}Sr$ for island arcs (0.70335) and the back-arc basins (0.70287) is less radiogenic than for counterparts in the South Pacific (0.70374, 0.70336). Stern (op. cit.) considers that the western Pacific island arcs show the same regional isotopic variations as observed for the volcanic rocks of the intra-plate oceanic islands (Ch. 9). Thus both arcs and oceanic islands show enrichments in radiogenic Sr near the Dupal anomaly (Section 9.7.4) associated with Samoa, and become progressively depleted away from it. This seems reasonable, as it would be expected that regional isotopic variations in the mantle should be sampled equally by intra-plate and destructive plate margin volcanism.

Unfortunately, there are very few combined Nd–Sr isotopic analyses of back-arc basin basalts. Figure 8.12 shows the available data for the East Scotia Sea – South Sandwich arc system (Hawkesworth *et al.* 1977). Both arc and back-arc basin

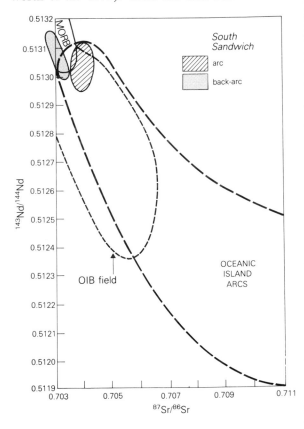

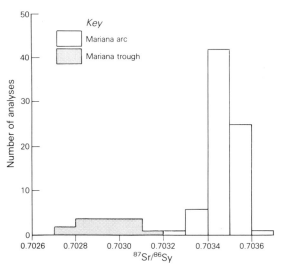

Figure 8.11 $^{87}Sr/^{86}Sr$ isotopic data for basalts from the Mariana island arc and its associated back-arc basin, the Mariana Trough (after Woodhead & Fraser 1985, Fig. 2, p. 1927).

Figure 8.12 Plot of $^{143}Nd/^{144}Nd$ versus $^{87}Sr/^{86}Sr$ to compare the isotopic composition of basalts from the South Sandwich island arc with those from the associated marginal basin, the East Scotia Sea (data from Hawkesworth *et al.* 1977, and Luff 1982). Fields for MORB, OIB and oceanic island arcs are from Figure 6.40.

basalts have similar $^{143}Nd/^{144}Nd$ ratios overlapping with the lower end of the MORB field. However, the arc basalts have slightly higher $^{87}Sr/^{86}Sr$ ratios and plot as a distinct group.

Hickey & Frey (1982) present Nd isotopic data for boninites from the Mariana fore-arc and the Bonin islands further to the north, but unfortunately no Sr isotopic data. The Mariana fore-arc boninites have $^{143}Nd/^{144}Nd$ ratios of 0.51295–0.51296, close to those of the Mariana arc lavas (0.51295–0.51305). However, boninites from the Bonin Islands show a much wider range of isotopic compositions (0.51262–0.51293; Fig. 8.13). In terms of this diagram, these clearly span a considerable range of the oceanic island basalt (OIB) field

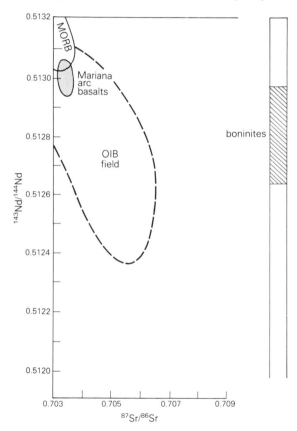

Figure 8.13 Plot of $^{143}Nd/^{144}Nd$ versus $^{87}Sr/^{86}Sr$ to show the Nd isotopic composition of boninites from the Mariana fore-arc and Bonin Islands in comparison to the Mariana arc basalts. Boninite data from Hickey & Frey (1982) and Mariana arc data from Hawkesworth (1982). MORB and OIB fields as in Figure 8.12.

and may be compared with the Nd isotopic compositions of lherzolite samples from the subcontinental lithosphere (Fig. 9.24). From the available trace element geochemical data (Section 8.4.2) it was concluded that boninites are derived from depleted mantle sources, possibly refractory harzburgites within the oceanic lithosphere, which seems at variance with the isotopic data. To account for the range of Nd isotopic compositions observed in boninites, such harzburgite sources would have to have been metasomatically re-enriched during their residence time in the oceanic lithosphere, possibly independently of the recent subduction-related metasomatic event which enriched the boninite source in Sr, Ba and alkalis.

Pb isotopes

As considered in Chapter 6, Pb isotopic studies of suites of volcanic rocks are particularly useful in resolving the relative roles of different source components in the petrogenesis of the magmas. Figure 8.14 is a plot of $^{207}Pb/^{204}Pb$ versus $^{206}Pb/^{204}Pb$ showing fields for South Sandwich and Mariana arc basalts compared to MORB. Only two Pb isotopic analyses are available from the Mariana Trough (Meijer 1976) and these reveal the distinction between the arc and back-arc magma sources, already demonstrated in terms of Sr isotopic compositions in Fig. 8.11. The Mariana trough basalts have identical Pb isotopic compositions to Pacific MORB, whereas the Mariana arc samples (Woodhead & Fraser 1985) have incorporated radiogenic Pb from subducted oceanic sediments (Ch. 6).

8.5 Detailed petrogenetic model

Several factors may contribute to the geochemical variability of back-arc spreading centre basalts, including variable degrees of partial melting, varying P_{H_2O} and P_{O_2} conditions during partial melting, heterogeneity of the mantle source along the basin axis and high-level crystal fractionation in magma chambers beneath the axis of the spreading centre. Additionally, the influence of fluids derived from the subducting plate may be significant. It is

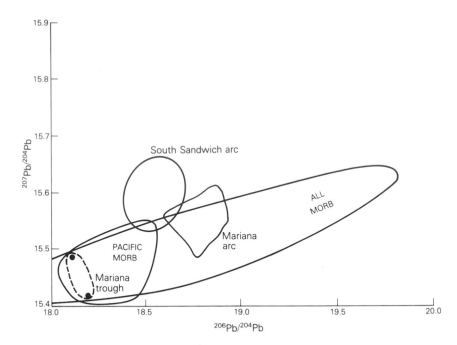

Figure 8.14   Plot of $^{207}Pb/^{204}Pb$ versus $^{206}Pb/^{204}Pb$ to show the different isotopic compositions of Mariana arc and Mariana Trough basalts. Mariana arc data from Woodhead & Fraser (1985) and Meijer (1976). Mariana Trough data from Meijer (1976). Fields for MORB and South Sandwich island arc basalts are from Figure 6.43.

important to assess these effects before comparisons are made between the geochemistry of back-arc basalts and those erupted at normal mid-ocean ridge spreading centres.

On the basis of their major element geochemistry, back-arc basin basalts fall within the MORB spectrum, although they generally appear to have greater affinity with enriched MORB compositions rather than normal or N-type MORB (Ch. 5). However, major element chemistry is not a particularly useful discriminant as all oceanic tholeiitic basalts (MORB, ocean-island tholeiites and island-arc tholeiites) tend to be very similar. The trace element geochemistry of the back-arc basalts may be complex, showing both MORB-like and arc-like characteristics. This might be anticipated, as the back-arc basin environment is clearly one in which subduction-modified mantle components may become involved in the magma generation process, producing spreading centre basalts with geochemical characteristics transitional to those of arc basalts. In general, it seems likely that the influence of subduction-zone fluids should be greatest during the early stages of basin opening and then diminish as the basin widens (Tarney *et al.* 1977), although this will depend upon the geometry of the particu-

lar subduction system. Saunders & Tarney (1979) have suggested that the Scotia Sea spreading centre basalts are more volatile-rich than normal MORB, with the possibility that dehydration of the subducted slab may have contributed to higher P_{H_2O} conditions during partial melting. This is consistent with their trace element geochemistry, which indicates the involvement of subduction-modified mantle sources in their petrogenesis.

Potential source components in the back-arc region may include both depleted and more fertile peridotites from the oceanic lithosphere, in addition to the relatively fertile lherzolite of the ascending asthenospheric mantle diapirs. The unusual geochemical characteristics of the boninite series of magmas suggests their derivation from strongly depleted harzburgite sources, possibly from the oceanic lithosphere, which had experienced metasomatism by an incompatible element enriched fluid prior to the magma generation event. Their enrichment in K, Ba and Rb appears to suggest that this is a subduction-related metasomatism. However, Nd isotopic studies of boninites (Hickey & Frey 1982) reveal a wide range in $^{143}Nd/^{144}Nd$ ratios, which may reflect enrichment events over a much longer period of time (perhaps

100 Ma), while the harzburgites, generated as the residuum from mid-ocean ridge partial melting processes, resided within the oceanic lithosphere. Normally, such refractory harzburgites would not be expected to participate in subsequent magma generation events because of their high solidus temperatures. Metasomatism by hydrous subduction zone fluids will clearly lower their solidus but, nevertheless, temperatures at shallow depths within the oceanic lithosphere are probably still too low for partial melting to occur. The boninites of the Mariana fore-arc occur stratigraphically above arc-related volcanics, and are associated with tholeiitic basalts generated during the early stages of back-arc extension. Crawford *et al.* (1981) have suggested that partial melting of refractory metasomatized harzburgite may be a consequence of the thermal pulse associated with the early stages of back-arc extension, which would raise temperatures in the oceanic lithosphere significantly. Experimental studies (Green 1973, Tatsumi 1981) have confirmed that relatively high degrees (>30%) of hydrous partial melting of peridotite at 30−60 km depth can produce liquids

with the high-SiO_2 and high-MgO characteristics of boninites, leaving a refractory residue of olivine and orthopyroxene.

Further reading

Crawford, A J., L. Beccaluva & G. Serri 1981. Tectono-magmatic evolution of the west Philippine − Mariana region and the origin of boninites. Earth Planet Sci. Lett.54, 346-56.

Hickey, R. L. & F. A. Frey 1982. Geochemical characteristics of boninite series volcanics: implications for their source. *Geochim. Cosmochim. Acta* 46, 2099−115.

Karig, D. E. 1971. Origin and development of marginal basins in the western Pacific. *J. Geophys. Res.* 76, 2542−61.

Saunders, A. D. & J. Tarney 1979. The geochemistry of basalts from a back-arc spreading centre in the East Scotia Sea. *Geochim. Cosmochim. Acta* 43, 555−72.

Taylor, B & G. D. Karner 1983. On the evolution of marginal basins. *Rev. Geophys.* 21, 1727−41.

PART FOUR

Magmatism within plates

As we have seen in Parts 2 and 3, present-day igneous activity is largely confined to plate boundary tectonic settings. Nevertheless, significant volcanism also occurs within plate interiors, both continental and oceanic, which is difficult to relate directly to the processes of plate tectonics. The Hawaiian islands provide a spectacular example of oceanic intra-plate magmatic activity (Ch. 9), lying at the end of a segmented chain of extinct volcanic islands and seamounts, which J. T. Wilson (1963) explained in terms of the motion of the Pacific plate over a stationary thermal anomaly or hot spot within the upper mantle. Morgan (1971) considered that such hot spots mark the locus of ascent of deep-mantle plumes, which impinge randomly on the base of both oceanic and continental plates. Partial melting in the rising plume occurs as a consequence of adiabatic decompression, and may generate magmas ranging from tholeiite to alkali basalt and nephelinite in composition, depending upon the depth and degree of partial melting and the composition and mineralogy of the mantle source. It has been suggested that major periods of kimberlite magmatic activity may be associated with hot spot activity within continental plates (Ch. 12). The present-day global distribution of mantle hot spots, which may be regarded as a secondary mode of mantle convection, is non-uniform, with particularly large concentrations within the African plate.

However, while a significant proportion of intra-plate volcanic activity may be associated with hot spots it is difficult to explain all occurrences in this way. Some intracontinental plate volcanic provinces are clearly associated with extensional tectonics and rifting, e.g. the East African

Rift, the Rhine Graben, and the Basin and Range province (Ch. 11), and in these the magmatism may be a passive response to lithospheric thinning, caused by differential stresses within the plate. In some oceanic environments volcanic activity may also be induced by intra-plate stress. There appears to be a complete spectrum between passively induced continental rift zones and those in which hot spot activity appears to be responsible for asthenospheric upwelling which stretches the lithosphere. As the rate of crustal extension in such intra-plate rifts increases, eventually culminating in continental fragmentation and the generation of a new ocean basin, the geochemical characteristics of the magmatic activity change from alkalic to sub-alkalic with the production of large volumes of transitional to tholeiitic flood basalts. Such flood basalt provinces appear to be the precursors to the formation of new ocean basins (Ch. 10).

Highly potassic magmatism occurs rarely within continental plates in a variety of tectonic settings, including zones of extension postdating active phases of subduction or continental collision and hot spot activity (Ch. 12). The geochemical characteristics of such magmas suggest that they are derived from metasomatically enriched mantle sources which probably reside within the sub-continental lithosphere.

CHAPTER NINE

Oceanic islands

9.1 Introduction

In Parts 2 and 3 it was established that some 90% of present-day volcanic activity is concentrated within or adjacent to zones of plate divergence or convergence. Nevertheless, within the ocean basins numbers of seamounts and volcanic islands occur at locations far from adjacent plate boundaries, and these are examples of intra(within)-plate volcanism.

Seamounts are small submarine volcanic structures, morphologically similar to sub-aerial shield volcanoes, which either never grow above sea level or, if they do, are subsequently eroded and subside. Those that emerge close to sea level in the Tropics frequently become capped with coral reef deposits and then subside to form structures called guyots. Although widespread throughout all the major ocean basins, they are most abundant in the Pacific, particularly the northeastern sector. Batiza (1982) has suggested that many of these structures are preferentially located along fracture zones, which would provide conduits for the upward passage of magma. In the Pacific it is estimated that there are between 22 000 and 55 000 (Batiza op. cit.), of which only about 2000 are presently volcanically active. Locally, seamount lavas may account for some 5–25% of the oceanic crust.

In contrast, oceanic-island volcanoes are immense structures rising up to 10 000 m above the base level of the adjacent ocean floor, with dimensions greater than those of the largest mountains on the continents. Little is known about the structure and composition of their submarine foundations as direct sampling has, thus far, been restricted to their emergent volcanic cones. Most consist of several overlapping volcanic centres, indicating migration of the focus of activity with time. In slow-spreading ocean basins such as the Atlantic, volcanic islands tend to occur singly or in small groups, sometimes associated with submarine aseismic ridges (Ch 5; and see Fig. 9.1), whereas in fast-spreading oceans such as the Pacific they commonly occur in linear chains (Fig. 9.2).

In addition to seamounts and volcanic islands there are other topographic highs within the ocean basins, known as oceanic plateaux, some of which may also have an intra-plate volcanic origin. These areas are several hundred square kilometres in dimension, and are elevated more than 1000 m above the surrounding ocean floor. Most are aseismic, with thick sedimentary covers, and overall crustal thicknesses greater than that of the adjacent crust. In general, they exhibit weak or no magnetic lineations, suggesting that they are not formed as typical ocean crust (Nur & Ben-Avraham 1982). Some are clearly foundered continental fragments, but the majority are regions of thickened crust formed in a variety of tectonic settings, including triple junctions, fracture zones, hot spots and subduction zones. Such plateaux cover about 10% of the present-day ocean floor, with particular concentrations in the western Pacific and Indian oceans.

A striking feature of the Pacific ocean basin are the linear chains of volcanic islands and seamounts, which are much younger than the oceanic crust upon which they sit (Fig. 9.2). These chains tend to be subparallel and almost perpendicular to the magnetic anomaly patterns and several, possibly all, are marked by a progressive increase in the age of the volcanoes away from the East Pacific Rise. In 1963, J. T. Wilson proposed a model for the origin of such linear volcanic features, which is still in current use as a general explanation for the occurence of intra-plate volcanism. The model involves a fixed magma source in the mantle, a hot spot or mantle plume, over which the oceanic plate moves (Fig. 9.3). Magmas rising from this plume feed surface volcanism, and a submarine volcano develops which may eventually become an emergent cone. As the volcano is carried progressively away from the centre of the hot spot, by the motion of the lithospheric plate, the magma supply is cut

off and volcanism ceases. Ultimately, this creates a chain of extinct volcanoes moving away from the hot spot in the direction of seafloor spreading. Such an age progression is clearly documented for the Hawaiian chain (Fig. 9.2). Several of the Pacific island chains (Hawaiian−Emperor, Gambier−Tuamotu, Austral−Cook and Gilbert−Marshall) change direction from SE−NW to slightly west of north (Fig. 9.2), which Morgan (1972a,b) explained in terms of a dramatic change in the direction of spreading of the Pacific plate 40−50 Ma ago.

While the hotspot model for oceanic-island volcanism seems just as attractive now as it did in the 1960s, it is unlikely to be the explanation for all occurrences of oceanic intra-plate volcanism, and other models must be considered. For example, Turcotte & Oxburgh (1978) related the origin of linear island chains to the development of propagating fractures caused by intra-plate stresses as the moving lithospheric plates adapt to latitudinal variations in the shape of the Earth. Such a model cannot account for the geometry of all the subparallel island chains in the Pacific Ocean basin, but may be a viable explanation for certain localized occurrences of intra-plate volcanism. In other instances, thermal stresses, induced by the contraction of the cooling oceanic lithosphere, may lead to the development of fractures which tap magmas from deeper levels. However, as cooling is most rapid in newly formed oceanic lithosphere, this might only be expected to be significant in the younger parts of oceanic plates.

Throughout this chapter we shall adopt a Wilson-type hot spot model to explain the global occurrence of oceanic intra-plate volcanism, bearing in mind that there are no convincing explanations for the anomalous features of hot spots in terms of simple plate-tectonic theory. Following Wilson, they are envisaged as rising plumes of deep-mantle material, the dimensions, temperature, rates of ascent and chemical characteristics of which are impossible to deduce by direct geological and geophysical observation. They may be regarded as a secondary mode of mantle convection, the depth of origin of which is a matter for considerable speculation. The overall circulation

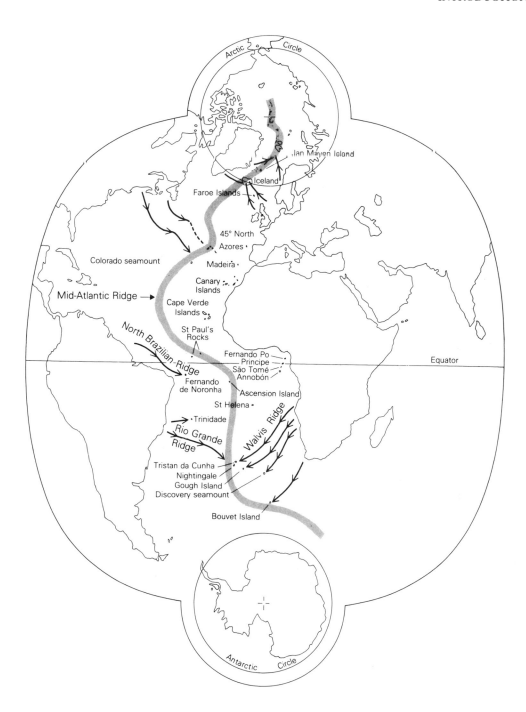

Figure 9.1 Distribution of volcanic islands within the Atlantic Ocean basin. The shaded band denotes the position of the Mid-Atlantic Ridge. Lines with arrows denote the positions of lateral aseismic ridges (after Burke & Wilson 1976, p. 35).

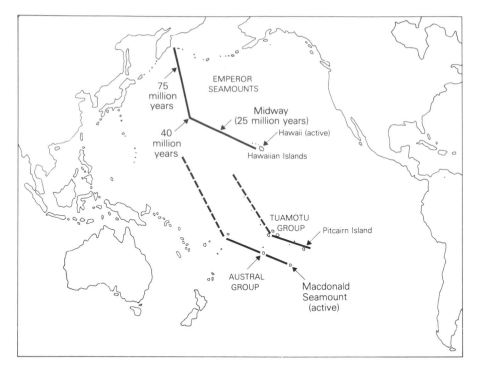

Figure 9.2 Distribution of linear volcanic island chains in the Pacific Ocean. Dashed lines indicate linear seamount chains extending to the north from the Austral and Tuamotu groups, subparallel to the Emperor seamount chain, the age progression of which is not adequately documented (after Burke & Wilson 1976, p. 34).

pattern of the mantle is poorly understood and debate continues as to whether convection is mantle-wide or occurs in two layers (Ch. 3). Assuming a simple two-layer convecting mantle, the most likely place for plumes to originate would be the boundary layer between the upper and lower convecting zones, which may correspond to the 670 km seismic discontinuity. However, Moberly & Campbell (1984), in a palaeomagnetic study of the Hawaiian−Emperor chain, concluded that hotspot volcanism is more prevalent during normal-polarity intervals of the Earth's magnetic field, and thus they linked plume dynamics to processes operating at far deeper levels − in the Earth's core.

Burke & Wilson (1976) identified a global pattern of 122 hot spots, active during the past 10 Ma, within both the oceanic and continental plates (Fig. 9.4). They defined 53 oceanic hot spots which have a tendency to be located close to mid-oceanic ridges. When a hot spot is located on an actively spreading ridge, chains of extinct volcanic islands and seamounts are formed on both sides of the ridge, extending away from the hot spot, in some instances forming lateral aseismic ridges (Ch. 5). Within the Atlantic ocean (Fig. 9.1), paired aseis-

mic ridges or hot spot tracks have a 'V'-shaped pattern which can be related to a northerly vector of motion during the opening of the ocean. In contrast, when a hot spot occurs within the interior of an oceanic plate, a single linear chain of islands and seamounts results, the dimensions of which reflect the velocity of plate motion over the hot spot.

If a continental plate comes to rest over a hot spot the upwelling may eventually rupture the continental lithosphere and initiate the formation of a new ocean basin (Ch. 11). Such a situation may have happened 120 Ma ago when the remains of the Gondwanaland supercontinent fractured along the line of the present Mid-Atlantic Ridge (Ch. 10). When the Atlantic ocean was born, hot spots such as Tristan da Cuhna may already have been active volcanoes lying along the continental rift that ruptured to form the new ocean basin (Morgan 1983). Tristan da Cuhna is no longer a ridge-centred hot spot due to the westward migration of the MAR some 30 Ma ago, although it continues to influence the chemistry of magmas erupted along the axis of the adjacent ridge segment (Section 9.7).

The Hawaiian island − Emperor seamount

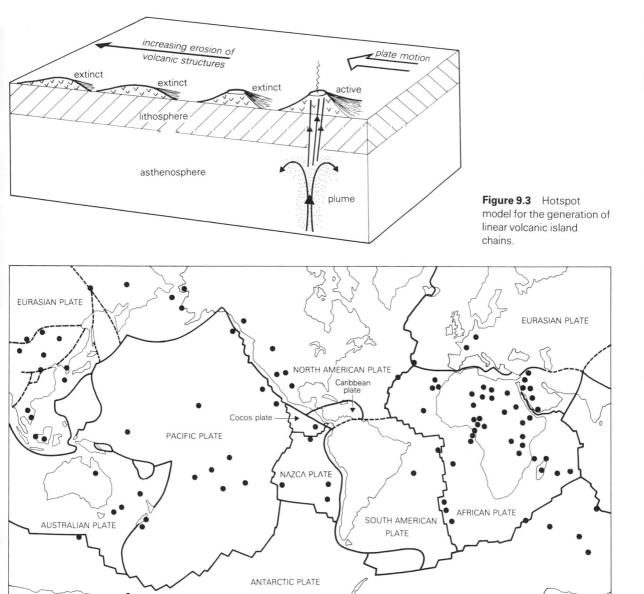

Figure 9.3 Hotspot model for the generation of linear volcanic island chains.

Figure 9.4 Distribution of hot spots within the oceanic and continental plates (after Burke & Wilson 1976, p. 37).

chain (Fig. 9.2) is perhaps the best studied example of a truly intra-oceanic plate hot spot located well away from the constructive plate boundary (East Pacific Rise). The Hawaiian archipelago is a dominantly submarine mountain chain, over 2000 km in length, elongated WNW−ESE approximate-ly parallel to the present spreading direction of the Pacific plate. Hawaii itself is the only currently active volcanic island, built of five coalescing shield volcanoes within a timespan of only 1 Ma (Mac-donald 1968). Of these, only two volcanoes are actually active, Kilauea and Mauna Loa. The individual shield volcanoes rise from a submarine pedestal of volcanic rocks 5 km thick lying on downwarped older oceanic crust. Mauna Loa, rising to 4170 m above sea level, and the slightly higher extinct volcano Mauna Kea, have the highest constructional reliefs above base level of any other volcano or mountain on Earth, attesting to the enormous volumes of magma produced at the hot spot. Shaw *et al.* (1980), using K−Ar geochronological data on basalts sampled from the Hawaiian chain, have calculated its propagation rate relative to the focus of the hot spot as approximately 6 cm yr^{-1} over the peroid 74−1.4 Ma BP. This is close to the average spreading rate of the East Pacific Rise (6−7 cm yr^{-1}), and suggests that the hot spot is effectively fixed in the mantle with respect to the overriding plate. When compared to Hawaii, most other oceanic islands show considerably longer periods of volcanic activity at a single site (>10 Ma versus <1 Ma at Hawaii), which may be related to their location in ocean basins that are spreading less rapidly than the Pacific.

In the ocean basins, volcanic rocks belonging to both alkalic and sub-alkalic magma series are recognized. Tholeiitic basalts (MORB) are the dominant rocks of the ocean floors and these have already been discussed in Chapter 5. Both tholeiitic and alkalic basalts (and their derivatives) have been documented within the eruptive sequences of oceanic-island volcanoes. These have distinctive trace element and isotopic geochemistries when compared to MORB, indicating derivation from different mantle source regions (Section 9.7). Some oceanic intra-plate volcanoes, e.g. Hawaii (Pacific) and Reunion (Indian Ocean), are dominantly composed of tholeiitic basalts but the vast majority are alkalic lavas and pyroclastics. However, it is impossible to assess the relative volumes of tholeiitic and alkalic magmas erupted in oceanic islands due to the restriction of sampling to the emergent volcanic cones. Volcanic rocks have not yet been sampled from the submarine pedestals of any oceanic island, because of the technical difficulties of drilling so close inshore.

In general almost all oceanic islands situated on mature oceanic crust are characterized by an evolutionary sequence from an early less alkalic voluminous shield-building stage to later more alkalic phases, which often postdate prolonged periods of dormancy. In the case of Hawaii, the shield-building lavas are dominantly tholeiitic, becoming transitional to alkalic with time, followed by late-stage post-erosional eruptions of highly alkalic lavas such as basanites, nephelinites and melilitites (Clague 1987). Traditionally, this temporal evolution of chemistry has been explained by the early eruptives being the products of large degrees of mantle partial melting at shallow depths, and the late-stage alkali basalts resulting from smaller degrees of melting at greater depths. If the Hawaiian eruptive pattern is generally applicable to all oceanic-island volcanoes, then much of their submerged portions could be composed of tholeiitic basalts, though this remains to be tested by direct drilling.

Two distinct differentiation trends are observed amongst the alkalic oceanic-island volcanic suites. By far the most common is an undersaturated evolutionary sequence from parental alkali basalts to nepheline-bearing phonolitic residua. More rarely, the parental alkali basalts evolve along an oversaturated trend to produce quartz-bearing alkali rhyolites (pantellerites and comendites). Whether the parental basalts fractionate along under- or oversaturated trends is likely to be a complex function of primary magma composition and high-level fractionation conditions (pressure, temperature, P_{O_2}, P_{H_2O} etc.).

9.2 Simplified petrogenetic model

The origin of volcanic oceanic islands and seamounts has been a controversial subject throughout the development of plate tectonic theory. The hotspot or plume model, first proposed by J.T. Wilson (1963) and developed by Morgan (1971, 1972a,b), still seems the most convincing explanation, although it is undoubtedly far more complex than these early workers envisaged.

Figure 9.5 presents a summary of the processes responsible for the generation of magma at both mid-plate and ridge-centred hot spots. It depicts a two-layer mantle with an upper convecting de-

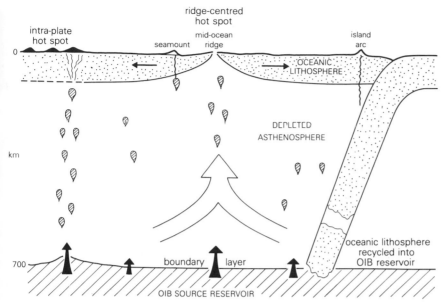

ridge-centred
hot spot

intra-plate
hot spot

mid-ocean
seamount   ridge

island
arc

0

OCEANIC
LITHOSPHERE

km

DEPLETED
ASTHENOSPHERE

700

boundary   layer

oceanic lithosphere
recycled into
OIB reservoir

OIB SOURCE RESERVOIR

Figure 9.5 Simplified model of oceanic intra-plate magmatism. The model depicts a two-layer mantle in which the lower OIB source reservoir is composed of a mixture of near-primordial mantle and recycled subducted oceanic crust. Diapirs upwell from the interface between the OIB reservoir and the overlying depleted asthenosphere (MORB-source reservoir) and partially melt as they rise. Depleted asthenosphere caught up in the flow will also start to partially melt, and these melts will mix with partial melts of the OIB-source mantle. At the ridge-centred hot spot the OIB-source component is significantly more diluted by the depleted asthenosphere component than at the intra-plate hot spot.

pleted layer (the source of MORB) and a lower, independently convecting, oceanic-island basalt (OIB) source layer. Plumes originate at the boundary between these two mantle layers and are expressed as a series of upwelling diapiric pods of the OIB source reservoir. Adiabatic decompression induces partial melting in the diapirs and also in the adjacent MORB source mantle caught up in the plume flow. Mixing of these different melt fractions will be inevitable, and the magma which ultimately segregates to feed the oceanic-island volcanism will have a geochemical signature reflecting source component mixing. Ridge-centred hot spots will, in general, show a more dilute plume component signature due to the greater extent of partial melting in the upwelling depleted asthenosphere beneath the ridge.

Morgan's (1972a,b) original model was based on the idea that the plume component (lower mantle layer) is relatively primordial mantle material which has not been involved in the magma generation events associated with the formation of the continental crust. The distinctive trace element and isotope geochemistry of MORB lends considerable support to such a 'chemical plume' concept, and isotope geochemical studies confirm that OIB must be derived from source components that have been isolated from the MORB source reservoir for

several Ga. However, as we shall see in Section 9.7.4, detailed Sr, Nd and Pb isotopic studies reveal that the concept of primordial mantle plumes as a source of OIB is a gross oversimplification. Many of their geochemical characteristics appear to require a source component derived from ancient recycled oceanic crust (Hofmann & White 1982). The OIB source reservoir in Figure 9.5 must therefore be envisaged as a mixture between relatively primordial mantle and other components, including those derived from subducted lithospheric slabs (Zindler & Hart 1986).

One of the major problems in understanding intra-plate volcanism is caused by the intangible nature of the proposed mantle plumes. We have no direct geophysical or geological methods to prove that they actually exist, and thus we have no information about their size, shape or velocity of ascent. Morgan (1972b) estimated that the plume beneath Hawaii must be of the order of 150 km in diameter, with an ascent velocity of several metres per year, based upon the pattern of volcanism within the Hawaiian chain, but this clearly cannot be generalized to all oceanic-island volcanoes. Originally, the system of mantle hot spots was thought to define a fixed reference frame, but Olson (1987) and Molnar & Stock (1987) have demonstrated that they must move relative to each other at rates of $10-20$ mm yr^{-1}.

Plumes must be considered as a secondary mode of mantle convection, with the motion of the lithosphere and underlying asthenosphere constituting the primary mode. It is difficult to deduce their dynamics without a knowledge of the depth in the mantle from which they arise. Parmentier *et al.* (1975) conducted numerical experiments on hypothetical mantle plume flows and concluded that they must be driven by base heating. This would be consistent with the two-layer convection model shown in Figure 9.5, in which the plumes originate at the boundary layer between the two convecting regimes. Many workers believe that this boundary layer corresponds to the depth of the 670 km seismic discontinuity (Houseman 1983a,b; Knittle *et al.*1986), which may mark a change in the average chemical composition of the mantle. However, Moberly & Campbell (1984) suggest that a major component of the Hawaiian plume flow originates near the core/mantle boundary, based upon a correlation between periods of intense volcanism and normal polarity intervals of the Earth's magnetic field. Thus the depths from which plumes originate remains a matter for further speculation and are probably variable on a global scale.

Hofmann & White (1982) proposed a model for OIB magmatism involving the recycling of subducted oceanic crust into the OIB source reservoir. Such recycling is necessary to explain many of the isotope geochemical characteristics of OIB (Section 9.7.4). They suggest that the oceanic crust becomes detached from the remainder of the subducted lithosphere and accumulates at some depth in the lower mantle, where it may remain stored for several Ga. Eventually, it becomes unstable as a consequence of internal heating and rises diapirically, providing a source component for OIB magmatism. These authors originally proposed storage at the core/mantle boundary, but Sekine *et al.* (1986) postulate shallower depths, possibly close to the 670 km discontinuity.

The magmas erupted in oceanic islands and seamounts belong to both the alkalic and sub-alkalic (tholeiitic) magma series (Section 9.7.1) and have trace element and isotope geochemical characteristics which clearly distinguish them from MORB. Basaltic magmas are dominant in general, and the origins of the more evolved magmas can be simply explained in most instances in terms of low-pressure fractional crystallization of parental basalts. In the model depicted in Figure 9.5, mantle-derived basaltic magmas (both tholeiitic and alkalic) continually rise through the lithosphere and are stored in shallow magma reservoirs, at $2-30$ km depth, where fractional crystallization and magma mixing occurs.

The chemical composition of primary basaltic magmas generated within the rising mantle plume will depend upon a variety of factors including:

(a) the composition and mineralogy of the source mantle;
(b the degree of partial melting of the source and the mechanism of partial melting (see Ch. 3);
(c) the depth of segregation of the magma.

Detailed discussion of these various parameters is, of necessity, deferred until the appropriate sections, but the general conclusions to be drawn are as follows:

Source. The geochemical characteristics of OIB indicate that they are partial melts of multi-component sources involving near primordial mantle, ancient recycled oceanic crust (basalt + sediments), depleted asthenosphere (MORB source mantle), depleted oceanic lithosphere and recycled subcontinental lithosphere.

Degree of partial melting. There is considerable controversy concerning the degrees of partial melting required to generate the spectrum of OIB chemistries. Calculations based upon trace element data generally predict unrealistically small degrees of melting ($2-10\%$) as a consequence of the assumptions involved (O'Hara 1985). On the basis of experimental melting studies (Ch. 3), it would seem reasonable that tholeiitic OIB should be generated by comparable degrees of partial melting to MORB ($20-30\%$), whereas alkalic OIB may represent smaller degrees of melting (5−15%), possibly at greater depths. However, variations in source composition and residual mineralogy may be one of the overriding factors in controlling alkalic versus tholeiitic chemistry.

Depth of segregation of magmas. The geochemical characteristics of all primary magmas are largely derived from their last point of equilibration with their mantle source − at the depth of segregation. Figure 9.6 shows a hypothetical thermal model of a mantle plume which predicts the locus of the major zone of partial melting close to the base of the lithosphere. This implies depths of magma segregation of 100 km or less. An upper limit may be provided by the data of Eissler & Kanamori (1986), who consider that intense earthquake swarms at depths of 50−60 km beneath the Hawaiian volcano Kilauea reflect the movement of magma upwards into the high-level reservoir system. If magmas do indeed segregate at depths of 100 km, it seems unlikely that they can traverse the whole of the lithosphere without further re-equilibrating with it, thus adding further complexites to their geochemical signatures.

9.3   Crustal structure of oceanic islands

Seismic studies can provide important constraints on the velocity and density structure of the crust and upper mantle beneath an active volcano, and thus are of great value in developing detailed petrogenetic models. Unfortunately, the deep structure of oceanic islands and seamounts is, in general, poorly understood. Much of the available data comes from the Hawaiian islands and these will be used as a model in this section.

It has been known since the earliest gravity measurements that the Hawaiian Islands are associated with large-amplitude free-air gravity anomalies. These are explained, in terms of flexure models of isostasy, as a consequence of the huge volcanic load which downwarps the oceanic lithosphere. Figure 9.7 shows a vertical cross section of the Hawaiian ridge in the vicinity of Oahu, deduced from combined gravity and seismic studies (Watts *et al.* 1985). This represents the results of the first detailed seismic experiment to be carried out over a mid-plate seamount chain. The section shows the flexing of the oceanic crustal layer and the marked increase in the thickness of the crust beneath the Hawaiian chain (15−20 km), in contrast to the

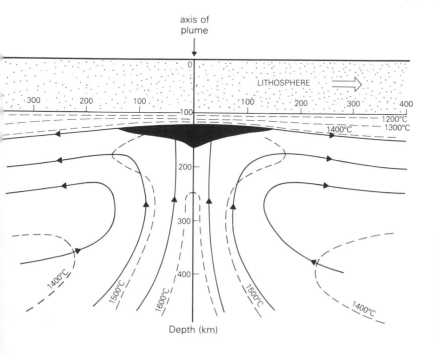

Figure 9.6   Structure of a thermal plume impinging on the base of a rigid 100 km thick oceanic lithospheric plate. The zone of partial melting is shaded black. Dashed lines are isotherms, and solid lines with arrows are mantle flow lines (after Parmentier *et al.* 1975, Fig. 4).

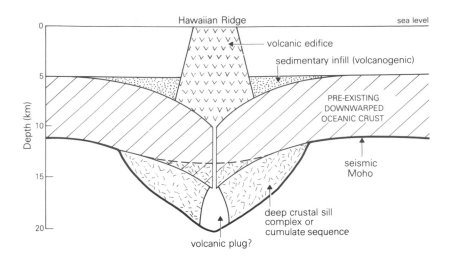

Figure 9.7 Schematic model of the crustal structure of the Hawaiian Ridge near the island of Oahu (after Watts *et al.* 1985, Fig. 6).

adjacent oceanic crust (5−6 km). If, as shown, the oceanic crustal layer maintains its normal thickness beneath the ridge then a high-density body is required beneath it to compensate for the low density of the extrusive volcanic complex. This may be interpreted as either a deep-crustal sill complex or a body of mafic and ultramafic cumulates.

9.3.1   Crustal magma reservoirs

It is possible to map out the zones within the structure of an active volcano which transmit and store magma on the basis of the spatial distribution of magma-related earthquake hypocentres (Koyanagi *et al.* 1976), assuming that such events are induced by the hydraulic opening of fractures via magmatic pressure build-up. Using these data, models can be developed to describe the magmatic plumbing system beneath a large volcanic structure. Unfortunately, detailed earthquake data are only available for the Hawaiian volcanoes Kilauea and Mauna Loa and models for high-level magma storage based on them may not be generally applicable, particularly so in the case of alkalic oceanic-island volcanoes which may have much deeper reservoir systems.

Figure 9.8 shows a three-dimensional model of the Kilauea magma plumbing system based on the distribution of earthquake hypocentres (Ryan *et al.* 1981). There is an aseismic zone approximately 3

km in diameter, extending from 3 to 6 km depth, which is interpreted as the zone of primary magma storage. This high-level reservoir system is fed by a primary conduit, extending to at least 14 km depth, which is probably a region of intense interconnected fractures along which magma percolates. Tholeiitic magma rising through the primary conduit into the high-level reservoirs is capable of deforming the summit region, as indicated by changes in surface elevation and tilt and the occurrence of shallow earthquakes above the reser-

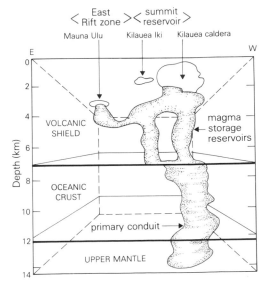

Figure 9.8 Subvolcanic magma plumbing system beneath the Hawaiian volcano Kilauea (after Ryan *et al.* 1981, Fig. 11).

voir (Ryan *et al.*1981). The centre of summit inflation migrates with time (Swanson *et al.* 1976) suggesting that the near-surface reservoir is not symmetrical and is likely to be an interconnected network of dykes, sills and small irregularly shaped magma bodies. A major eruptive sequence empties the near-surface reservoirs and is followed by a short period of repose (~260 days) during which the system refills. Such rapid refilling tends to suggest that magma withdrawal from the summit effectively unloads a plumbing system in hydraulic connection with the source region, triggering the ascent of new magma batches from depth (Dzurisin *et al.* 1984).

9.4 Partial melting processes

Partial melting in the mantle beneath oceanic islands occurs in response to adiabatic decompression of plume and asthenospheric components in the ascending flow. In order to understand such melting processes it is necessary to establish:

(1) the compositions (major and trace element, Sr, Nd, Pb isotopes) of primary oceanic-island basalts, both tholeiitic and alkalic;
(2) the mineralogy and chemical composition of the mantle source(s);
(3) the degree of partial melting;
(4) the mechanism of partial melting (batch, fractional etc.);
(5) the depth of beginning of melting and, more importantly, the depth of segregation of the magmas;
(6) the importance of mixing of different mantle sources, e.g. plume source, depleted asthenosphere and oceanic lithosphere.

The remoteness of most oceanic islands from continental plates and the relative thickness of the oceanic crustal layer upon which they are constructed should reduce the likelihood of near-surface crustal contamination of the magmas. Thus the geochemical characteristics of the most primitive basalts should reflect, albeit indirectly, the chemistry and mineralogy of the source, the

temperature, pressure and oxygen fugacity during partial melting, the degree of partial melting and the degree of subsequent fractional crystallization during storage in high-level magma reservoirs. Magmas with near-primary characteristics (high MgO, high Ni, high Cr etc.) are erupted, somewhat infrequently, in most oceanic islands, and these compositions can be used as an indication of the primary magma compositions unmodified by low-pressure fractional crystallization processes. Alternatively, primary magma compositions can be calculated employing fractionation-correction procedures which involve adding mineral phases, presumed to have fractionated during ascent, back into the lava composition, e.g. Feigenson *et al.* (1983), Wright *et al.* (1975) and Wright & Doherty (1970).

An important problem in the study of oceanic volcanoes which erupt both tholeiitic and alkalic basalts is whether the two primary magma types are derived by differing degrees of partial melting of a homogeneous mantle source or of different sources. Similar incompatible element ratios (K/Ba, K/Rb and Zr/Nb) and similar Sr, Nd and Pb isotopic ratios in alkali and tholeiitic basalts from some oceanic islands are consistent with a petrogenetic model involving a single relatively homogeneous source (Feigenson *et al.* 1983). However, for many of the Hawaiian volcanoes, late-stage and post-erosional alkalic basalts have distinctly different trace element and isotopic characteristics from the underlying shield tholeiites, indicating the involvement of different mantle sources (Section 9.8).

On the basis of the simplified petrogenic model proposed for oceanic-island volcanism (Section 9.2), it seems inevitable that most OIB are generated as mixtures of melt fractions derived from different source components, for example a plume component and a depleted asthenosphere component from the envelope of the plume. This automatically invalidates trace element partial melting calculations to deduce the degree of melting and source mineralogy, as these inherently assume a single stage melting event. Such calculations (e.g. Budahn & Schmitt 1985) tend to predict unrealistically small degrees of partial melting (2–10%) of a near-chondritic source to produce Hawaiian tholeiitic basalts. O'Hara (1985) has argued most convincingly that, even in the simplest case of partial melting of a homogeneous source, mixing of melts formed by variable degrees of partial melting in an ascending mantle plume generates trace element enriched magmas which could only be generated by ultra-low degrees of melting in a single-stage batch melting event.

Delineation of the depths at which partial melting occurs beneath oceanic islands is theoretically possible, using seismic methods to define zones of marked attenuation of P and S waves. Unfortunately, few such studies have been attempted. Ellsworth & Koyanagi (1977) modelled the 3-dimensional seismic structure beneath Hawaii and concluded that large volume concentrations of magma are absent from the lower crust and upper mantle to depths of at least 40 km. However, studies of the deep seismicity beneath Hawaii suggest that upper mantle rocks fracture in response to magmatic pressures to depths of at least 60 km (Koyanagi & Endo 1971), possibly corresponding to the minimum depth of magma segregation.

Experimental melting studies probably still give one of the best indications of the degrees of partial melting necessary to generate the range of primary magma compositions observed. However, as discussed in Chapter 3, these also have their limitations. Takahashi & Kushiro (1983) investigated the anhydrous partial melting behaviour of a spinel lherzolite xenolith from a Hawaiian alkali basalt and reached similar conclusions to Jaques & Green (1980) (Ch. 3), that near-solidus partial melts at pressures below 15 kbar are tholeiitic, becoming picritic as the degree of melting increases. At pressures between 15 and 25 kbar, near-solidus partial melts are alkali olivine basalts, becoming tholeiitic as the degree of melting increases, and at pressures greater than 25 kbar they are alkali picrites, changing to tholeiitic picrite as the degree of melting increases. These experimental studies are important because they indicate that both alkalic and tholeiitic primary magmas can be generated from the same mantle source, simply by varying the depth and degree of melting. However, Takahashi & Kushiro (op. cit) confirmed that

highly silica deficient magmas such as basanite and nephelinite, which characterize the post-erosional phase of Hawaiian volcanism, could not be generated from a spinel lherzolite source under anhydrous conditions. Wendlandt & Mysen (1980) consider that such magmas can only be generated in the presence of a CO_2-rich fluid phase.

9.5   High-level magma storage

9.5.1   Low-pressure fractional crystallization processes

The nature of crustal magma reservoirs beneath active oceanic-island volcanoes has already been considered in Section 9.3. For the Hawaiian volcano Kilauea these have been shown to consist of an interconnected complex of dykes, sills and small magma chambers at depths less than 7 km. Clearly, the Hawaiian data cannot be generalized to all oceanic-island volcanoes, particularly those erupting alkalic magmas. Nevertheless, it seems inevitable that all of the larger volcanic edifices, both tholeiitic and alkalic, should have high-level magma chamber systems in which fractional crystallization, crustal contamination and magma mixing occurs.

Direct evidence for the importance of low-pressure fractional crystallization in the geochemical evolution of oceanic-island volcanic suites is provided by the frequent occurrence of good linear correlations on major and trace element variation diagrams (Section 9.7). Additionally, alkalic basalts quite commonly bring to the surface a variety of low-pressure cumulate xenoliths which appear to be fragments of layered plutonic rocks from the walls and floors of the deeper-level magma chambers. These are generally basic (gabbro) to ultrabasic (dunite, pyroxenite and wehrlite) in composition, although both syenite and granite xenoliths have been recorded. In general, they vary between 2 and 25 cm in diameter and show marked variations in the modal proportions of the constituent minerals and in grain size. Layering may be present in some specimens, with feldspathic layers alternating with more melanocratic ones. The gabbroic cumulates generally consist of varying proportions of olivine, plagioclase feldspar, Ca-rich clinopyroxene, opaque oxides and amphibole, an identical mineral assemblage to that which usually crystallizes from the erupted basalts (Section 9.6). Only the occurrence of amphibole is in some instances anomalous, being an important constituent of the cumulate xenoliths but rare in the extrusive volcanic rocks. However, this may be easily explained in terms of the changing stability of amphibole in the near-surface environment (Section 9.6).

Occasionally, gabbroic xenoliths entrained in oceanic-island alkali basalts contain orthopyroxene as a cumulus phase, which is inconsistent with an origin by crystal accumulation from a fractionating alkali basalt (e.g. Harris 1983; Ascension). These may be cumulates from tholeiitic basalts erupted during early submarine volcanic phases or, alternatively, they may be fragments of the depressed oceanic crustal layer upon which the volcanic island is built.

The post-erosional phases of Hawaiian alkalic volcanism bring a variety of ultramafic xenoliths to the surface. Some are cumulates clearly related to the main tholeiitic shield-building stage, whereas others are tectonites which may not be (Leeman et al. 1980). The xenoliths include spinel lherzolites, rare garnet lherzolites and garnet pyroxenites which appear to have equilibrated at depths of 60–80 km, i.e. within the zone of magma segregation as deduced from seismic studies (Section 9.3). Some of these lherzolite xenoliths may be close to the actual source mantle composition for Hawaiian volcanism, but most are variably depleted residua. The garnet pyroxenites may represent high-pressure cumulates or pods of alkali basaltic magma which crystallized at high pressure. It seems inevitable that oceanic-island alkali basalts should transport fragments from the vicinity of their mantle source to the surface. However, in general, the xenolith suites have not been extensively studied and thus the mantle samples may have easily been overlooked.

9.5.2 Crustal contamination

As indicated in Section 9.3, the high-level subvolcanic magma chamber systems of mature oceanic-island volcanoes are likely to be located within sequences of previously erupted volcanic rocks, progressing upwards as the volcanic edifice grows. Early submarine phases of the volcanism may have undergone significant alteration by interaction with sea water, and thus chamber magmas assimilating their roof rocks could be digesting seawater-altered basalt, which might potentially alter their $^{87}Sr/^{86}Sr$ and $^{18}O/^{16}O$ ratios. The extent of such near-surface contamination is difficult to assess, given the marked isotopic variability of OIB (Section 9.7. 4.). Oxygen and hydrogen isotopic studies should provide the most reliable criteria, although few such studies have been attempted (Sheppard & Harris 1985). Additionally, it is possible for magmas to become contaminated by oceanic sediments associated with the depressed oceanic crustal layer beneath the island. Such contamination may be potentially identified by combined Sr−O isotopic studies (James 1981).

9.5.3 Magma mixing

In any long-lived magmatic plumbing system, two or more unlike magmas may mix together forming a hybrid daughter product. Such processes must be fundamental in the petrogenesis of all oceanic-island volcanoes with well developed magma storage reservoirs. In rapidly refluxed chamber systems, mixing causes the erupted magma compositions to correspond to some average 'perched state'. This may explain why Hawaiian tholeiitic basalts display such remarkable compositional uniformity, at least with respect to major elements.

If the two parental magmas could be identified and analysed then any blended daughter product should have an intermediate composition, falling on a straight line between the parents on a Harker variation diagram (Fig. 9.9). However, the example depicted is a particularly extreme case and, in general, the effects of magma mixing will merely mimic the effects of fractional crystallization. In some instances, lavas carrying corroded xenocrysts

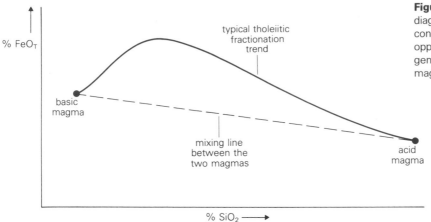

Figure 9.9 Schematic Harker variation diagram to show the geochemical consequences of magma mixing, as opposed to crystal fractionation, in the generation of intermediate-composition magmas.

may be considered to have originated by mixing of two porphyritic magmas, but this is not conclusive as the xenocrysts could also reflect assimilative disaggregation.

9.6 Petrography of oceanic-island volcanic rocks

It is clearly beyond the scope of this work to describe the petrographic variability of the entire range of volcanic rocks formed in the oceanic-island environment. Instead, emphasis will be placed upon the petrographic differences between tholeiitic and alkalic basalts, followed by a brief comparison of alkalic volcanic suites from the Atlantic oceanic islands of Gough, Tristan da Cunha, St Helena and the Azores.

Table 9.1 presents a summary of the major petrographic differences between alkalic and tholeiitic basalts. In general, oceanic-island tholeiites (OIT) are similar to MORB (Ch. 5) but may contain orthopyroxene in addition to olivine, spinel, Ca-rich clinopyroxene and Fe–Ti oxides (Basaltic Volcanism Study Project 1981). In some instances, there is evidence of a reaction relationship between olivine and orthopyroxene.

Spinel is common in both tholeiitic and alkalic basalts, frequently occurring as a co-liquidus phase with olivine. It is highly variable in composition, correlating with its relative time of crystallization. Early-formed spinels have the highest MgO, Cr_2O_3

and Al_2O_3 contents, whereas later-formed spinels have lower Cr_2O_3 and Al_2O_3 and higher Fe contents. The Cr_2O_3 content of spinel is sensitive to fO_2 (Hill & Roeder 1974) and high Cr_2O_3 contents are consistent with reducing conditions during crystallization. In general, there are systematic compositional differences between the spinels in tholeiitic and alkalic basalts, with the former having higher Cr_2O_3 contents.

Olivine in tholeiitic basalts occurs only as a phenocryst phase and, as such, has a narrow compositional range (Fo_{90-70}). In contrast, in alkali basalts it occurs as both phenocrysts and in the groundmass, and consequently shows a considerably wider composition range (Fo_{90-35}), often within a single sample.

The pyroxene mineralogy of tholeiitic basalts can be complex, involving both Ca-rich clinopyroxene (augite) and orthopyroxene (hypersthene) as phenocrysts, sometimes accompanied by low Ca-clinopyroxene (pigeonite) reaction rims on olivine phenocrysts. All three pyroxene phases may coexist in the groundmass. In contrast, alkali basalts contain only a single pyroxene phase, a brown titaniferous augite. In some highly phyric basalts clinopyroxene megacrysts occur which have high $Mg/(Mg + Fe^{2+})$ ratios (0.85–0.90). These may be high-pressure phenocrysts from more picritic primary magmas or possibly accidental xenocrysts. Figure 9.10 and Table 9.2 show the compositional ranges of the various pyroxene minerals in Hawaiian tholeiitic and alkalic basalts. Ca-rich

Table 9.1   Summary of the petrographic differences between tholeiitic and alkali basalts (after Hughes 1982, Table 9.5, p.297).

Tholeiitic basalts	Alkali basalts
(a) Phenocrysts	
infrequent large *olivine* phenocrysts, commonly unzoned — may show reaction rims of orthopyroxene	medium-sized *olivine* phenocrysts common — often strongly zoned with more iron-rich rims
Orthopyroxene phenocrysts may occur	*Orthopyroxene* absent
Plagioclase phenocrysts often appear early in the crystallization sequence:	*plagioclase* phenocrysts less common, appear later in the crystallization sequence:
olivine <plagioclase <augite	olivine <augite <plagioclase
phenocrysts of pale brown *augite*	*titaniferous augite* phenocrysts, strongly zoned with purplish brown rims
(b) Groundmass	
groundmass usually relatively fine-grained, with an intergranular texture	groundmass relatively coarse, with textures ranging from intergranular to ophitic
no groundmass *olivine*	groundmass *olivine*
groundmass pyroxene is variable; subcalcic augite or augite ± pigeonite	only one species of Ca-rich clinopyroxene in the groundmass (titansalite)
no alkali feldspar or analcite in the groundmass	interstitial alkali felsdspar and analcite may occur in the groundmass
interstitial glass relatively common	interstitial glass rare or absent
(c) Associated rocks	
ultramafic xenoliths very rare	ultramafic xenoliths fairly common, dunite and wehrlite predominating
associated accumulative rocks are picrites (oceanites), rich in olivine phenocrysts	associated accumulative rocks are ankaramites, rich in olivine and augite phenocrysts

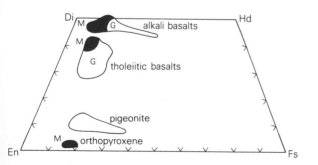

Figure 9.10 Pyroxene compositions in Hawaiian tholeiitic and alkalic basalts (after Basaltic Volcanism Study Project 1981, Fig. 1.2.6.19, p. 184). M = megacryst, G = ground-mass.

clinopyroxenes (both phenocrysts and groundmass) are significantly subcalcic in the tholeiitic basalts compared to those in alkali basalts.

Plagioclase is, in general, a more common phenocryst phase in tholeiitic than in alkalic basalts, occurring slightly earlier in the crystallization sequence (Table 9.1). Compositions vary widely (An $_{85-50}$) depending on the order of appearance and, for a given anorthite content, plagioclases in alkali basalts have higher K_2O contents.

Hydrous phases such as amphibole and biotite are conspicuously absent from tholeiitic volcanic suites, perhaps attesting to lower concentrations of volatiles in the magmas. In contrast, kaersutitic

Table 9.2 Pyroxene compositions in Hawaiian tholeiitic and alkalic basalts (data from Basaltic Volcanism Study Project 1981)

	CPX megacryst in Kilauea tholeiite	CPX megacryst in Kohala alkali basalt	OPX megacryst in Mauna Loa tholeiite	Pigeonite rim to olivine
%				
SiO_2	51.5	50.2	55.0	54.6
Al_2O_3	3.57	3.56	1.67	2.36
TiO_2	0.79	1.25	0.35	0.37
FeO_T	6.21	7.90	12.7	10.8
MgO	16.7	15.0	28.4	26.6
CaO	19.7	20.6	2.30	5.22
Na_2O	0.12	0.33	0.0	0.0
Cr_2O_3	0.91	0.29	0.22	0.35
MnO	0.10	0.11	0.21	0.17
Cations per six oxygens				
Si	1.896	1.870	1.945	1.942
Al^{iv}	0.104	0.130	0.055	0.058
Al^{vi}	0.051	0.027	0.015	0.004
Ti	0.022	0.035	0.009	0.010
Cr	0.026	0.009	0.006	0.010
Fe^{3+}	0.000	0.048	0.015	0.000
Fe^{2+}	0.191	0.198	0.362	0.321
Mn	0.003	0.003	0.006	0.005
Mg	0.916	0.835	1.500	1.410
Ca	0.777	0.822	0.087	0.199
Na	0.009	0.024	0.000	0.000

amphibole is a relatively common minor phase in many alkalic basalts. It is interesting to note that low-pressure cumulate xenoliths entrained within such basalts frequently contain abundant cumulus amphibole (Section 9.5.1). This apparent discrepancy can be explained in terms of the instability of amphibole at very shallow depths (<1–2km). At subvolcanic magma chamber levels (10–20 km), it is stable and is an early-crystallizing phenocryst in water-rich basaltic magmas. As the magmas rise towards the surface, amphibole passes out of its stability field and is rapidly resorbed back into the magma, thus occurring only infrequently as ragged phenocrysts in the erupted basalts.

The phenocryst and groundmass mineralogy of four alkalic volcanic suites from Atlantic oceanic islands is compared in Table 9.3. Gough, Tristan da Cunha and St Helena all follow undersaturated evolutionary trends, whereas the Azores alkali basalts evolve to quartz-saturated comenditic residua. Olivine, Ca-rich clinopyroxene, plagioclase, amphi-

bole, magnetite and apatite are common phenocryst minerals in magmas ranging from alkali basalt to phonolite and comendite. Alkali feldspar only becomes an important phenocryst phase in the more siliceous differentiates, where it is sometimes joined by biotite. The ferromagnesian minerals become more iron-rich as the SiO_2 content of the magma increases and, in addition, the clinopyroxene becomes an increasingly sodic, green pleochroic, aegirine–augite. Nepheline occurs rarely as a phenocryst phase in the more evolved members (phonolites and trachytes) of the alkalic-undersaturated series.

In contrast to the relatively simple phenocryst mineralogies, groundmass mineralogies in alkalic volcanic suites can be highly complex. In general, the phenocryst minerals continue to crystallize in the groundmass, where they are joined by alkali feldspar and feldspathoids (sodalite, leucite and nepheline) in the undersaturated suites. Note that both alkali feldspar and feldspathoids are conspi-

Table 9.3 Mineralogy of oceanic-island alkalic magma series.

		(a) phenocrysts									
		ol	cpx	plag	AF	mt	ap	bi	amph	ne	sph
Gough	basalt	x	x	x							
	trachyandosite	x	x	x		x	x				
	trachyte	x	x	x	x	x	x				
Tristan da Cunha	ankaramite	x	x	x		x					
	basalt	x	x								
	trachybasalt	x	x	x		x					
	leucite–trachybasalt	x	x						x		
	trachyandesite	x	x	x		x		x	x		
	trachyte		x	x		x	x		x	x	x
St Helena	ankaramite	x	x	x		x					
	basalt	x	x	x		x					
	trachybasalt	x		x		x	x				
	trachyandesite	x		x		x	x		x		
	trachyte	x				x	x		x		
	phonolite	x	x		x				x		
Azores	basalt	x	x	x							
	hawaiite	x	x	x		x			x		
	mugearite	x	x	x		x	x		x		
	trachyte	x	x	x		x	x	x	x		
	comendite	x	x	x		x	x	x	x		

(Continued on p. 262)

cuously absent from the groundmass of the over-saturated Azores magmas. An additional groundmass mineral in the more siliceous differentiates from St Helena is aenigmatite ($Na_2Fe_5TiSi_6O_{20}$), characterized by blood red to black pleochroism.

Figure 9.11 shows photomicrographs comparing the petrographic characteristics of oceanic-island alkalic and tholeiitic basaltic volcanic rocks.

9.7 Chemical composition of erupted magmas

9.7.1 Characteristic magma series

Amongst the volcanic rocks of the ocean basins, representatives of both tholeiitic and alkalic magma series are recognized. The tholeiitic basalts erupted at the system of mid-oceanic ridges (MORB) have already been discussed in Chapter 5. Oceanic-island tholeiitic basalts differ from MORB in terms of their trace element and isotope geochemistry (Sections 9.7.3 & 9.7.4) and must clearly be derived from different mantle sources, albeit probably by similar degrees of partial melting as they have similar major element chemistries. Alkali basalts and more evolved alkaline magmas dominate the upper flanks and crests of most oceanic islands and seamounts. However, by comparison with the eruptive sequence of many Hawaiian volcanoes (Section 9.2), the submarine portion of these volcanic edifices may be tholeiitic, although there is no direct evidence to substantiate this. On the Indian Ocean island of Reunion, alkali basalts are indeed underlain by tholeiitic basalts, whereas in the Galapagos (Pacific) tholeiitic and alkalic basalts are erupted contemporaneously. The shield volcanoes of Hawaii and Iceland are dominated by tholeiitic basalts but, in both, late-stage volcanic activity is invariably alkalic. Iceland has already been considered in Chapter 5 as an example of a

Table 9.3   Mineralogy of oceanic-island alkalic magma series (continued).

(b) Groundmass

		ol	cpx	plag	AF	mt	ap	sod	amph	leuc	bi	ne	aenig	qtz
Gough	basalt	x	x	x	x	x	x							
	trachyandesite			x	x	x								
	trachyte	x			x	x								
Tristan da Cunha	ankaramite		x	x		x	x							
	basalt		x	x		x		x	x	x				
	trachybasalt		x	x		x	x		x					
	leucite–trachybasalt	x	x	x		x		x	x	x	x			
	trachyandesite		x		x	x								
	trachyte		x		x	x		x			x			
St Helena	ankaramite	x	x	x	x	x								
	basalt	x	x	x	x	x		x			x			
	trachybasalt	x	x	x	x	x	x			x				
	trachyandesite		x	x	x	x	x							
	trachyte		x	x	x	x	x							
	phonolite		x		x	x						x	x	
Azores	basalt	x	x	x		x								
	hawaiite		x	x		x								
	mugearite		x	x		x								
	trachyte		x	x		x	x		x					
	comendite		x	x		x	x		x					x

ol, olivine; cpx, clinopyroxene (Ca-rich); plag, plagioclase; AF, alkali feldspar; mt, magnetite; ap, apatite; bi, biotite; amph, amphibole; ne. nepheline; sph, sphene; sod, sodalite; leuc, leucite; aenig, aenigmatite; qtz, quartz.
Data sources: Gough, Le Roex (1985); Tristan da Cunha, Baker *et al.* (1964); St Helena, Baker (1969); Azores, White *et al.* (1979).

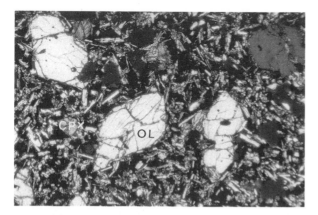

Figure 9.11   Photomicrographs comparing the petrographic characteristics of oceanic-island tholeiitic and alkali basalts. (a) Tholeiitic basalt from Hawaii with phenocrysts of olivine, set in a fine-grained matrix of plagioclase, clinopyroxene and glass (isotropic) (×40, crossed polars). (b) Alkali basalt from the Azores containg a glomeroporphyritic aggregate of Ti–augite, plagioclase and magnetite, set in a fine-grained matrix (×40, ordinary light).

ridge-centred hot spot (anomalous ridge segment) and thus will not be considered further in this chapter.

Within oceanic-island volcanic suites belonging to the alkalic magma series two distinct differentiation trends are recognized:

(a) *Undersaturated:* with the ultimate differentiation products being nepheline-bearing phonolites.

(b) *Oversaturated:* with the ultimate differentiates being quartz-bearing alkali rhyolites (comendite or pantellerite).

The undersaturated evolutionary trend is by far the most common and occurs in the following oceanic-island suites:

Atlantic Tristan da Cunha, Gough, Canary Islands, St Helena, Trinidade, Fernando de Noronha

Pacific Tahiti

Indian Kerguelen

The Atlantic oceanic islands of Ascension and the Azores are well documented examples of alkalic magmas evolving to quartz-saturated residua. In some of the larger oceanic islands such as Hawaii, highly alkaline basaltic magmas, including nephelinites, melilite nephelinites and basanites, succeed the eruption of tholeiitic and alkalic basalts, often following a significant period of dormancy (Section 9.2) (Clague 1987).

Figure 9.12 is a plot of $Na_2O + K_2O$ versus SiO_2, which has been used in Chapter 1 to differentiate between members of the alkalic and sub-alkalic (tholeiitic) magma series. The solid line separating the two series is schematic, and follows the boundary of Macdonald & Katsura (1964) at the basaltic end of the spectrum. Shown for comparison are data for three oceanic-island volcanic suites; Iceland (tholeiitic), Ascension (mildly alkalic —

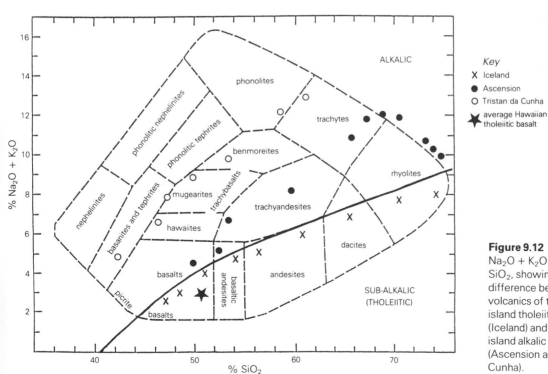

Figure 9.12 Weight. % $Na_2O + K_2O$ versus wt.% SiO_2, showing the difference between volcanics of the oceanic island tholeiitic series (Iceland) and the oceanic island alkalic series (Ascension and Tristan da Cunha).

Table 9.4   Representative analyses of volcanic rocks from Ascension Island, Atlantic (data from Harris 1983).

	Basalt	Hawaiite	Trachyandesite	Trachyte	Comendite
%					
SiO_2	48.24	51.42	59.42	66.95	74.05
TiO_2	3.15	2.61	1.34	0.38	0.13
Al_2O_3	16.33	15.66	17.04	15.40	12.44
FeO	11.70	11.04	6.79	4.21	2.53
MnO	0.19	0.21	0.27	0.15	0.06
MgO	5.10	5.30	2.22	0.33	0.04
CaO	8.37	8.60	4.38	0.82	0.22
Na_2O	4.01	3.67	5.38	6.71	5.53
K_2O	1.86	1.36	2.45	4.87	4.60
P_2O_5	1.02	0.43	0.66	0.12	0.02
	99.97	100.30	99.95	99.90	99.70
ppm					
Nb	—	42	95	—	205
Zr	—	219	488	—	871
Y	—	36	55	—	113
Rb	—	31	54	—	147
Sr	—	388	413	—	1.34

Table 9.5   Representative analyses of volcanic rocks from Tristan da Cunha (data from Baker *et al.* 1964).

	Basalt	Trachybasalts			Trachyte
%					
SiO_2	42.43	46.36	48.54	53.90	58.00
TiO_2	4.11	3.54	2.98	1.77	1.20
Al_2O_3	14.15	16.19	18.00	19.00	19.50
Fe_2O_3	5.84	3.66	3.78	3.37	1.70
FeO	8.48	6.94	5.18	3.05	2.20
MnO	0.17	0.18	0.18	0.18	0.10
MgO	6.71	4.57	3.32	1.68	1.00
CaO	11.91	9.45	8.49	6.25	3.30
Na_2O	2.77	3.97	4.74	5.04	6.50
K_2O	2.04	3.15	3.38	4.53	5.30
P_2O_5	0.58	1.42	1.18	0.74	—
ppm					
Nb	35	110	160	160	130
Zr	200	300	400	350	350
Y	15	40	50	45	35
Rb	110	160	220	200	350
Sr	700	1400	1100	1200	650
Ba	700	1000	950	1100	1000

óversaturated) and Tristan da Cunha (strongly alkalic (potassic) − undersaturated).

9.7.2   Major elements

SiO_2, TiO_2, Al_2O_3, Fe_2O_3, FeO, MnO, MgO, CaO, Na_2O, K_2O, P_2O_5 and H_2O can all be considered major element oxides in the description of the geochemistry of oceanic-island volcanic suites. It is clearly beyond the scope of this work to discuss the geochemical characteristics of the entire spectrum of oceanic-island volcanic rocks, and thus a small number of the better documented islands have been selected for comparative purposes.

Major and trace element data are presented in Tables 9.4 and 9.5 for volcanic rocks from Ascension and Tristan da Cunha, spanning the complete SiO_2 range from basalt to comendite (alkali rhyolite) and basalt to trachyte respectively. These

may be compared with data for typical Hawaiian tholeiitic and alkalic basalts and average N-type MORB given in Table 9.6. Oceanic-island basalts (both alkalic and tholeiitic) are, in general, variably enriched in TiO_2 and K_2O relative to MORB and have lower Al_2O_3 contents. This most probably reflects differing source compositions and residual mineralogies during partial melting.

Figure 9.13 shows the variation of wt.% K_2O versus SiO_2 for volcanic suites from the Atlantic oceanic islands of Ascension, St Helena, Tristan da Cunha, Gough, Azores and Iceland. Apart from the ridge-centred hotspot lavas of Iceland, all of these suites are alkalic, ranging from mildly alkalic on Ascension to highly alkalic (potassic) on Tristan da Cunha and Gough. K_2O appears to behave incompatibly and the data arrays define good linear trends which may be interpreted as liquid lines of descent, produced by low-pressure fractional crystallization

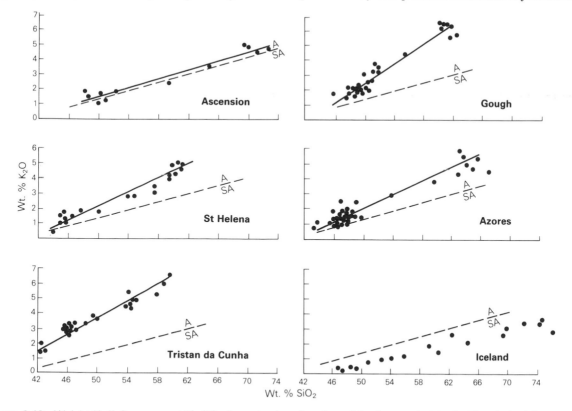

Figure 9.13 Weight.% K_2O versus wt.% SiO_2 for volcanic suites from Atlantic oceanic islands. The dashed line divides tholeiitic (sub-alkalic) and alkalic suites (after Middlemost 1975). Data sources: Ascension, Harris (1983); St Helena, Baker (1969); Tristan da Cunha, Baker et al. (1964); Iceland, Carmichael (1964); Gough, Le Roex (1985); Azores, White et al. (1979).

Table 9.6   Major and trace element geochemistry of typical tholeiitic and alkalic basalts from Hawaii, compared to N-type MORB (data from Basaltic Volcanism Study Project 1981). MORB data from Schilling *et al.* (1983).

	Tholeiites		Alkali basalts		MORB
	Kilauea	Mauna Loa	Hualalai	Kohala	
%					
SiO_2	50.51	51.63	46.37	47.52	48.77
Al_2O_3	13.45	13.12	14.18	15.95	15.90
Fe_2O_3	1.78	2.58	4.09	7.16	1.33
FeO	9.59	8.48	8.91	5.30	8.62
MgO	7.41	8.53	9.47	5.18	9.67
CaO	11.18	9.97	10.33	8.96	11.16
Na_2O	2.28	2.21	2.85	3.56	2.43
K_2O	0.49	0.33	0.93	1.29	0.08
MnO	0.17	0.17	0.19	0.19	0.17
TiO_2	2.63	1.94	2.40	3.29	1.15
P_2O_5	0.28	0.22	0.28	0.64	0.09
H_2O	—	—	—	1.16	0.30
ppm					
La	13.4	7.58	18.8	38.0	2.10
Ce	35.5	21.0	43.0	85	—
Sm	6.14	4.40	5.35	11.8	2.74
Eu	1.88	1.60	1.76	3.5	1.06
Yb	1.98	1.98	1.88	3.08	3.20
Rb	9.2	4.9	22	26	0.56
Sr	371	273	500	650	88.7
Ba	150	75	300	340	4.2
Hf	4.39	3.34	3.00	8.5	—
Zr	115	119	166	351	—
Nb	17	8	16	36	—
Y	25	23	21	39	—
Th	1.27	0.50	1.20	2.9	—
Pb	5	6	1	5	—

of a range of parental alkali basalt magmas. Many of these volcanic suites have comparatively limited ranges of Sr, Nd and Pb isotopic compositions (Section 9.7.4), which would tend to support this. However, island groups such as the Azores show a wide dispersion in isotopic data, implying the existence of a variety of parental basalts with differing isotopic but very similar major and trace element geochemistries, each of which may fractionate to produce more SiO_2-rich magmas. Thus the K_2O-SiO_2 trend shown in Figure 9.13 for the Azores cannot actually represent a true liquid line of descent and simply shows an average fractionation trend.

In many of these oceanic-island volcanic suites there is an apparent bimodality to the data, with a scarcity of intermediate composition magmas, the so-called 'Daly Gap'. This is most likely to be a function of incomplete sampling but may also reflect density filtration of magmas as they rise through the oceanic-island crust. All of the alkalic suites, apart from Ascension and the Azores, shown in Figure 9.13 follow undersaturated evolutionary trends, with phonolites as the ultimate differentiation products. In contrast, the more mildly alkalic basalts of Ascension and the Azores fractionate towards quartz-bearing residua (alkali rhyolite, comendite). Such an oversaturated fractionation trend is comparatively rare in the oceanic environment and its origins are incompletely understood.

Figure 9.14 is a plot of wt.% $Na_2O + K_2O$ versus SiO_2 for basalts from the five major volcanic centres on the island of Hawaii; Kohala, Hualalai, Mauna Kea, Mauna Loa and Kilauea. Basalts from Kohala (Feigenson *et al.* 1983) show a continuous transition from tholeiitic to alkalic types upwards in the stratigraphic sequence. This could reflect decreasing degrees of partial melting of a relatively homogeneous source, although this would have to be substantiated by a constancy of Sr, Nd and Pb isotopic compositions in all basalts (Hofmann *et al.* 1987, Lanphere & Frey 1987).

The importance of low-pressure fractional crystallization in the production of the observed linear K_2O-SiO_2 trends in Figure 9.13 may be further substantiated by a consideration of Figures 9.15 and 9.16. These show a series of Harker diagrams for volcanic rocks from the Atlantic oceanic islands of Ascension (oversaturated) and Gough (undersaturated). Segmented linear correlations indicate the importance of different fractionating mineral assemblages in the evolution of the two suites, with points of marked inflection marking the appearance of a new major crystallizing phase.

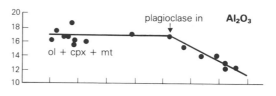
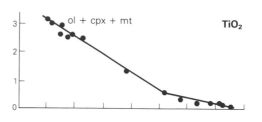
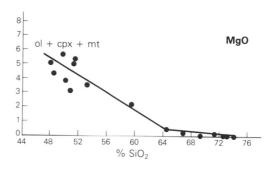

Figure 9.15 Variation of wt.% Al_2O_3, TiO_2 and MgO versus % SiO_2 for volcanic rocks from Ascension (data from Harris 1983).

9.7.3 Trace elements

Large low-valency cations

The group of large low-valency cations (Cs, Rb, K, Ba, Pb and Sr) are, in general, enriched in oceanic-island basalts relative to MORB, with alkali basalts showing the greatest levels of enrichment (Table 9.6). The abundances of these elements are controlled by the source composition and residual mineralogy, the degree of partial melting and the extent of subsequent fractional crystallization. Only Sr and Ba are preferentially incorporated into early-crystallizing minerals (calcic plagiocase) and thus as a group they behave incompatibly. Their

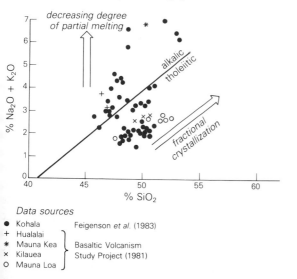

Data sources
● Kohala Feigenson *et al.* (1983)
+ Hualalai
✱ Mauna Kea ⎫ Basaltic Volcanism
× Kilauea ⎬ Study Project (1981)
○ Mauna Loa ⎭

Figure 9.14 Weight.% $Na_2O + K_2O$ versus wt.% SiO_2 for alkalic and tholeiitic basalts from the island of Hawaii. The dividing line between alkalic and tholeiitic basalt fields is from Macdonald & Katsura (1964).

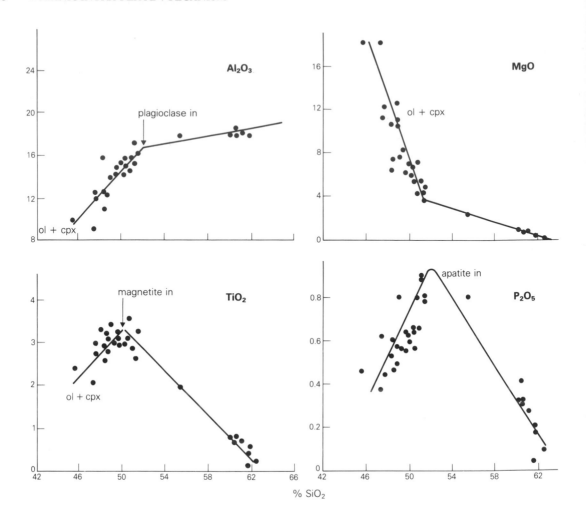

Figure 9.16 Weight.% Al_2O_3, TiO_2, MgO and P_2O_5 versus %SiO_2 for volcanic rocks from Gough (data from Le Roex 1985)

variation within suites of oceanic-island volcanic rocks is mostly a function of low-pressure fractional crystallization.

Figure 9.17 shows the variation of Ba and Sr concentrations (in ppm) with SiO_2 for the alkalic volcanic suite of Gough island. Both elements behave incompatibly until the onset of significant plagioclase fractionation from magmas with about 54% SiO_2, whereupon their abundances decrease dramatically. The concentrations of Ba and Sr in oceanic-island tholeiites are comparable with those of plume-type MORB (Ch. 5)

Variations in K/Ba ratio are sensitive indicators of source heterogeneity and the low ratios in

oceanic-island basalts relative to MORB (Table 9.7), clearly reflects their derivation by partial melting of different mantle sources.

Large high-valency cations

The group of large high-valency cations (Th, U, Ce, Zr, Hf, Nb, Ta and Ti) behave incompatibly and are preferentially concentrated in OIB relative to MORB (Tables 9.6 & 7). The Zr/Nb ratio is characteristically low in oceanic-island basalts (<10) compared to N-type MORB ($\geqslant$30) and can be used to demonstrate mixing of mantle sources. Figure 9.18 shows the variation of Y/Nb versus Zr/Nb for MORB suites erupted in the vicinity of

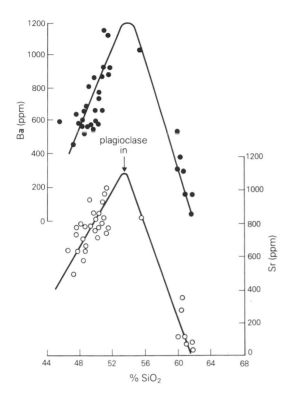

Figure 9.17 Variation of Ba and Sr (ppm) versus wt.% SiO_2 for the Gough Island volcanic suite (data from Le Roex 1985).

Figure 9.19 shows the variation of Nb/Zr (reciprocal of Zr/Nb) in MORB erupted along the axis of the MAR from 0 to 48°S. Hot spots associated with the oceanic islands of Ascension, St Helena, Tristan da Cunha and Gough produce positive Nb/Zr spikes, clearly attesting to the importance of mixing between MORB and OIB source components (Humphris *et al.* 1985).

Transition metals (Cr, Ni)

Nickel is a sensitive indicator of olivine fractionation from basaltic magmas because of its large mineral/melt partition coefficient. Oceanic-island volcanic suites frequently display good correlations between Ni and MgO (Fig. 9.20), indicating the importance of olivine fractionation/accumulation. Cr also tends to correlate with MgO, possibly due to the concurrent crystallization of olivine and a Cr-rich spinel phase. In general, oceanic-island alkali basalts are depleted in Ni and Cr relative to oceanic-island tholeiites and MORB, perhaps attesting to significant high-pressure fractional crystallization en route to the surface.

Rare earth elements

REE patterns of oceanic-island basalts are characterized by varying degrees of light-REE enrichment relative to the heavy REE. Figure 9.21 shows a range of chondrite-normalized REE patterns for tholeiitic and alkalic basalts from Hawaii and alkalic basalts from the Azores, compared to N- and P-type MORB. MORB have unfractionated heavy-REE abundances compared to OIB. The high light-REE abundances of oceanic tholeiites relative

the Tristan da Cunha (Humphris *et al.* 1985) and Bouvet (Le Roex *et al.* 1983, 1985) mantle hotspots. These clearly influence the geochemistry of MORB erupted along the adjacent segments of the southern Mid-Atlantic Ridge and the data arrays appear to be rather simply explained in terms of two-component mixing, between a slightly heterogeneous depleted asthenosphere and a low Zr/Nb plume component.

Table 9.7 Comparison of selected trace element abundances in MORB, oceanic-island tholeiites and oceanic-island alkali basalts (data from Basaltic Volcanism Study Project 1981).

	MORB	OIT	OIAB
% K_2O	< 0.1−0.3	0.2−1.0	1−7
Ba (ppm)	5−50	70−200	200−1400
Sr (ppm)	90−200	150−400	400−4000
Rb (ppm)	< 5	5−12	15−400
Zr (ppm)	15−150	100−300	200−1000
Nb (ppm)	1−15	5−25	20−160
K/Ba	20−160	25−40	~28

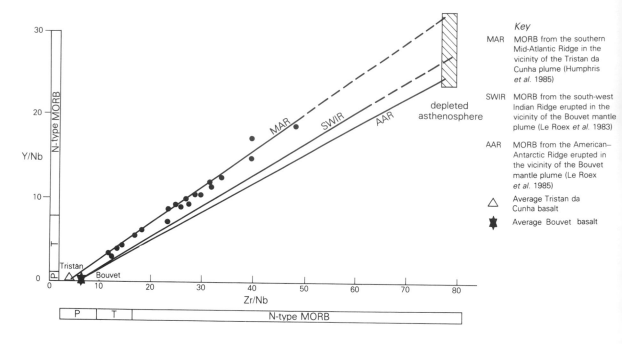

Figure 9.18 Y/Nb ratio versus Zr/Nb ratio for MORB erupted in the vicinity of mantle hotspots (individual data points for the SWIR and AAR have been omitted for clarity). Bars parallel to the axes of the diagram indicate the normal compositional ranges for normal (N), transitional (T) and plume (P) type MORB.

to N-type MORB are a consequence of their derivation from relatively undepleted mantle sources. The even greater enrichment of light REE in the alkali basalts, together with their relative depletion in heavy REE, is consistent with their derivation by relatively small degrees of partial melting of a source in which garnet remains as a residual phase.

Crystal fractionation involving olivine, plagioclase, clinopyroxene and magnetite increases the total REE content of more evolved OIB but does not produce any significant inter-element fractionations. Thus the characteristic shape of the primary basalt REE pattern will be maintained, while absolute abundances increase. Substantial plagioclase fractionation should, however, lead to the development of negative Eu anomalies.

If partial melting is fairly extensive (>10%) the REE should not be fractionated from each other during melting, and therefore ratios of REE (La/Sm, La/Yb, Ce/Yb and La/Ce) should reflect ratios in the mantle source. However, in reality only the

very light REE are truly incompatible and thus only La/Ce is likely to be diagnostic of source composition.

Whereas low-pressure fractional crystallization is unable to modify Ce/Yb ratios significantly, high-pressure eclogite (garnet + clinopyroxene) frac-

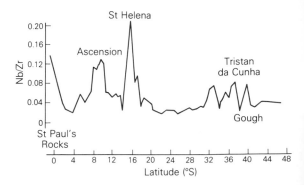

Figure 9.19 Variation of Nb/Zr ratio with latitude in MORB erupted along the southern Mid-Atlantic Ridge from 0 to 48°S. Note the spike highs corresponding to the location of off-axis mantle plumes (oceanic islands) – Tristan da Cunha, Gough, Ascension and St Helena (after Humphris *et al.* 1985, Fig. 2).

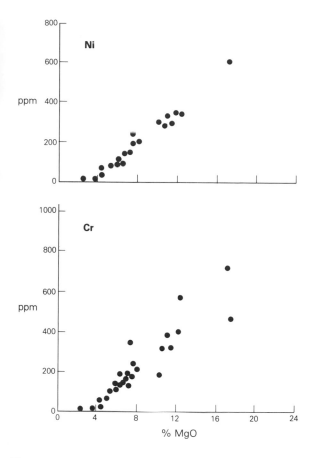

Figure 9.20 Variation of Ni and Cr content (ppm) versus wt.% MgO for basalts from Gough Island (data from Le Roex 1985).

tionation may do because garnet has high partition coefficients for the heavy REE. By evoking eclogite fractionation it is theoretically possible to account for most of the variations in Ce/Yb ratios in MORB and OIB. Nevertheless, the observed differences in Sr, Nd and Pb isotopic ratios (Section 9.7.4) cannot be explained in terms of derivation of all oceanic basalts from a homogeneous mantle source by crystal-liquid fractionation processes, and it is now generally accepted that the sub-oceanic mantle is chemically and isotopically heterogeneous. The source mantle for OIB and P-type MORB appears to be light-REE enriched, whereas the N-type MORB source is light-REE depleted.

Spiderdiagrams

Following Sun (1980), incompatible element abundances, normalized to primordial mantle values, for a typical tholeiitic basalt, alkali basalt and melilite nephelinite from Hawaii are compared in Figure 9.22. Shown for comparison are typical patterns for N- and P-type MORB (Ch. 5). The elements are ordered in a sequence of decreasing incompatibility from left to right in a four-phase lherzolite undergoing partial fusion. The spider-diagram patterns for all three basalt types have closely similar configurations, with significant troughs at K and Th and peaks at Ba and Nb, comparable to P-type MORB. This suggests that components in the sources of P-type MORB and OIB have similar geochemical characteristics.

9.7.4   Radiogenic isotopes; Sr, Nd, Pb and He

In Section 9.2 a model for oceanic-island magmatism was proposed, involving plumes of deep-mantle material rising up through the depleted asthenosphere. Mixing between these two components seems inevitable and, as discussed in Chapter 5, this is one way of explaining the isotopic and trace element heterogeneity of MORB (Schilling 1975, Sun *et al.* 1979, Hamelin *et al.* 1984). The influence of the plume source of OIB on the geochemistry of MORB erupted at adjacent mid-oceanic ridges is now well established, and is particularly well demonstrated by correlations between isotopic compositions of Sr, Nd and Pb and the bathymetry of the ridges (Allègre *et al.* 1984, Hamelin *et al.* 1984, Humphris *et al.* 1985, Schilling 1985, Schilling *et al.* 1985, Klein & Langmuir 1987).

Studies of the Sr–Nd–Pb isotope geochemistry of MORB erupted along the Mid-Atlantic Ridge and the East Pacific Rise have revealed good linear correlations in plots of $^{207}Pb/^{204}Pb$ versus $^{206}Pb/^{204}Pb$, $^{87}Sr/^{86}Sr$ versus $^{206}Pb/^{204}Pb$ and $^{87}Sr/^{86}Sr$ versus $^{143}Nd/^{144}Nd$ (Hamelin *et al.* 1984; and see Ch. 5) These have been interpreted in terms of mixing of two different mantle components beneath the ridges. One component is the convecting depleted asthenosphere, with less radiogenic Pb

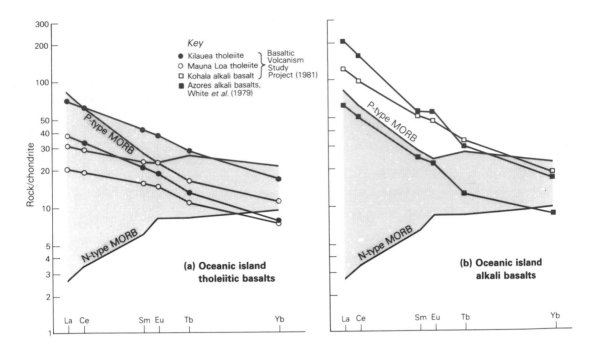

Figure 9.21 Chondrite-normalized REE abundances. Shown for comparison are typical N- and P-type MORB REE patterns.

and Sr and more radiogenic Nd isotopes, hereafter termed depleted mantle (DM). The other component comprises blobs of the source material of OIB, characterized by more radiogenic Pb and Sr and less radiogenic Nd. Such a model is supported by the observation that isotopic data for oceanic islands from the North Atlantic and East Pacific extend the MORB trends in all three diagrams, (Figs 5.43–5). In this section an attempt is made to investigate the isotopic characteristics of the OIB source component in more detail, and to assess its relative homogeneity or heterogeneity compared to the depleted asthenosphere on a global scale.

To a first approximation, the differentiation of the Earth's mantle may be described in terms of unidirectional transport of material from some portion of an initially homogeneous (?primordial) mantle into the lithosphere. This implies that the isotopic heterogeneities in Sr, Nd and Pb in young uncontaminated oceanic basalts are the result of mixing between regions of the mantle that have undergone differing degrees of depletion over the course of geological time (Cohen & O'Nions

1982a). Such a simplistic model originally formed the basis for interpretation of the broad correlation of $^{143}Nd/^{144}Nd$ versus $^{87}Sr/^{86}Sr$ for oceanic basalts (Ch. 2; and see O'Nions et al. 1977, DePaolo 1979), in which one end-member was the depleted asthenospheric source of N-type MORB and the other mantle material close to primordial composition (bulk Earth), which had never experienced any major magma-extraction events. However, as more Nd and Sr data were obtained for oceanic basalts, it became apparent that such a model was too simplistic.

Nd–Sr isotopic data for basalts from the Azores (Atlantic), Society Islands, Samoa and Marquesas (Pacific) and Kerguelen (Indian Ocean) (Fig. 9.23), deviate from the originally defined trend of the mantle array (Ch. 2) in the direction of increased $^{87}Sr/^{86}Sr$ and form good linear correlations. These trends could be interpreted as a consequence of near-surface contamination of basaltic magmas with oceanic sediments intercalated in the oceanic-island crust. However, Hawkesworth et al. (1979b) have argued against this for the Azores (São Miguel) and

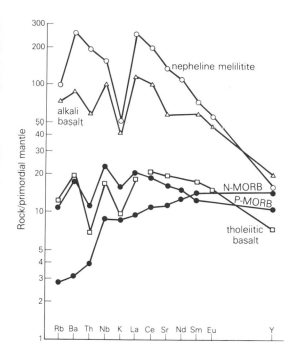

Figure 9.22 Incompatible element abundances, normalized to primordial mantle values (Sun 1980), for typical tholeiitic and alkali basalts and a melilite nephelinite from Hawaii. Shown for comparison are patterns for N- and P-type MORB (Sun 1980). Data from Basaltic Volcanism Study Project (1981) and Clague & Frey (1982).

reservoirs exist within the mantle is provided by isotopic data for ultramafic xenoliths from kimberlites and alkali basalts (Menzies and Murthy 1980, Stosch *et al.* 1980) which have $0 > \epsilon_{Nd} > -26$ (Fig. 9.24). These data display the same broad Sr–Nd isotopic correlation as the mantle array, despite the fact that most are samples of subcontinental lithosphere. This would seem to suggest that the overall Nd–Sr correlation of MORB and OIB is a consequence of the relative enrichment and depletion of mantle sources by migrating light-REE enriched small degree partial melts. Regions of the mantle could develop negative ϵ_{Nd} characteristics if such enrichments are ancient features (>1 Ga). The apparent anomaly of negative ϵ_{Nd} values rather than 0 (bulk Earth) values at the extreme end of the mantle array could thus be fairly simply explained by the involvement of a near-primordial lower mantle end-member, which itself has become isotopically heterogeneous over the course of geological time by the migration of small volume partial melts.

Once Pb isotopic data for MORB and OIB are considered it becomes clear that models of Earth differentiation involving unidirectional transport of material into the crustal reservoir are untenable. Instead, bi-directional transport, involving recycling of continentally derived materials via the subduction of oceanic lithosphere, has to be invoked to explain the complex data arrays (Cohen & O'Nions 1982b, Zindler *et al.* 1982, Zindler & Hart 1986). Pb is highly depleted in the upper mantle (DM), rendering it susceptible to contamination by Pb derived from other sources, particularly continentally derived oceanic sediments and oceanic crust that has been altered by seawater interaction to incorporate U of continental derivation. Indeed, Pb isotopes appear to be the most sensitive tracers of mixing processes in the source of oceanic basalts, suggesting a greater difference in isotopic composition between the asthenosphere and the plume (OIB) component for Pb than for Nd or Sr. Galer & O'Nions (1985) have even suggested that there may be a decoupling of the Pb and Nd–Sr isotope systems related to the short residence time of Pb in the upper mantle (<600 Ma). The oceanic crust is a sink for U, Rb and possibly Th, and will conse-

favour the involvement of recycled subducted oceanic sediments in the source region of the magmas. A further anomaly is provided by Sr–Nd isotopic data for basalts from St Helena (Atlantic) and Tubuaii (Pacific) which plot to the low $^{143}Nd/^{144}Nd$ side of the mantle array (Fig. 9.23), and clearly require an additional source component to explain their distinctive isotopic characteristics.

Partial melting of a primordial mantle reservoir with $\epsilon_{Nd} \sim 0$ will produce residua depleted in Nd with respect to Sm and the subsequent radioactive decay of ^{147}Sm will lead to the generation of depleted mantle components with $\epsilon_{Nd} > 0$. In Figure 9.23a the southern Atlantic oceanic islands of Tristan da Cunha and Gough, along with the aseismic Walvis Ridge, clearly display negative values of ϵ_{Nd}. A fundamental question is how such negative ϵ_{Nd} sources originate. Evidence that such

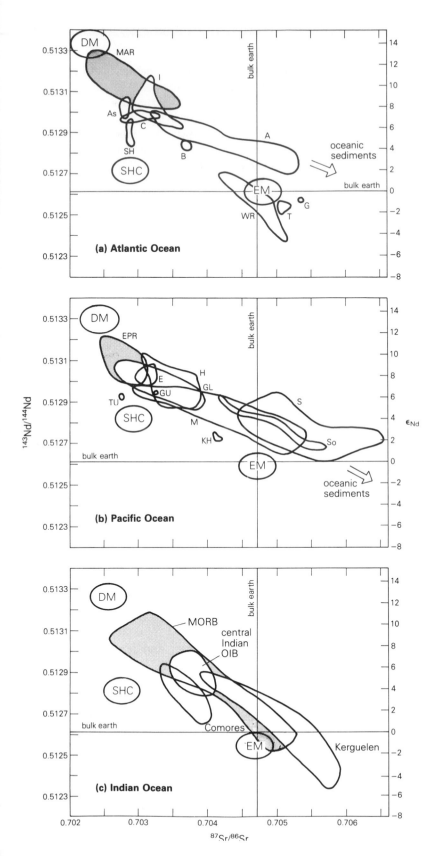

Key

MAR   Mid-Atlantic Ridge
I   Iceland
As   Ascension
C   Canary Islands
SH   St Helena
B   Bouvet
A   Azores
G   Gough
T   Tristan da Cunha
WR   Walvis Ridge
EPR   East Pacific Rise
H   Hawaiian Islands
KH   Koolau, Hawaii
E   Easter Island
GL   Galapagos
GU   Guadaloupe
TU   Tubuaii (Australes)
M   Marquesas
S   Samoa
So   Society Islands

The Central Indian OIB field includes data for the following islands; Marion – Prince Edward, Crozet, Amsterdam, St Paul, Reunion, Rodriguez and Mauritius

Major source components identified

DM   depleted mantle (source of N-type MORB)
EM   enriched mantle
SHC   St Helena component

Figure 9.23 $^{143}Nd/^{144}Nd$ versus $^{87}Sr/^{86}Sr$ for MORB and oceanic-island basalts (after Staudigel *et al.* 1984, and Hamelin *et al.* 1986). Data for the Marquesas and Tubuaii from Vidal *et al.* (1984).

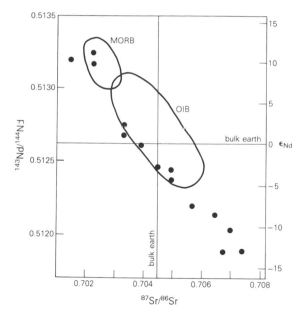

Figure 9.24 $^{143}Nd/^{144}Nd$ versus $^{87}Sr/^{86}Sr$ for lherzolite xenoliths derived from the subcontinental lithosphere. Shown for comparison is the oceanic basalt array, MORB and OIB. Xenolith data from Menzies & Murthy (1980) and Stosch *et al.* (1980).

quently evolve over long periods of time (>1 Ga) to produce more radiogenic Pb isotopic compositions (e.g. high $^{206}Pb/^{204}Pb$) than those of the depleted asthenosphere. Thus, if recycled oceanic crust is indeed involved to a large extent in OIB petrogenesis, this should be revealed in Pb-isotope variation diagrams. The extent to which the isotopic heterogeneities in oceanic basalts reflects recycling of continentally derived material via subduction zones is clearly fundamental to studies of mantle evolution.

Since the early 1960s it has been recognized that the Pb isotopic ratios of OIB have two remarkable characteristics (Chase 1981). One is that the $^{206}Pb/^{204}Pb$ ratio is large compared to $^{207}Pb/^{204}Pb$, such that on a plot of $^{207}Pb/^{204}Pb$ versus $^{206}Pb/^{204}Pb$ (Fig.9.25) the oceanic-island data plot to the right of a primary isochron of zero age through Canyon Diablo troilite Pb − the geochron (Tatsumoto *et al.* 1973; see also Ch. 2). The most logical explanation is that the mantle source regions of OIB have undergone enrichment of U relative to Pb at some time during the past few billion years. The other

remarkable characteristic is that most oceanic-island Pb's fall along an almost linear array in plots of $^{207}Pb/^{204}Pb$ versus $^{206}Pb/^{204}Pb$. MORB data lie on the same array but closer to the geochron. This array can be most simply explained in terms of mixing between a radiogenic Pb component with isotopic characteristics similar to St Helena basalts, hereafter termed the St Helena component (SHC), and a less radiogenic Pb component which could be a mixture of depleted mantle (DM) and enriched mantle (EM). In addition to this overall linear trend of oceanic basalt data, individual islands or groups of islands (e.g. Bouvet and the Canary Islands) also display linear arrays (Sun 1980), the slopes and positions of which are close to but not identical with the average trend of the main data array. Thus, although binary mixing of sources may be the explanation for all of these trends, the end-members, though isotopically similar, are not homogeneous on a global scale. The SHC component has isotopic characteristics consistent with an origin as recycled subducted oceanic crust which has been stored in the mantle for longer than a billion years without rehomogenizing or re-equilibrating isotopically. Thus the evolution of the oceanic-island Pb isotopes may be a consequence of ancient seafloor spreading and subduction processes.

In Figure 9.25 data for the Atlantic oceanic-islands of Discovery, Gough, Tristan da Cunha and the aseismic Walvis Ridge plot closer to the geochron than islands from the North Atlantic (Bouvet, Canaries, Azores, Ascension and Cape Verde), which could reflect the dominance of a more enriched plume component (EM) in their source region. Pb isotopic data for the Hawaiian Islands could also reflect involvement of a similar component.

Figure 9.26 is a plot of $^{208}Pb/^{204}Pb$ versus $^{206}Pb/^{204}Pb$ which also displays a series of subparallel positive correlations for the oceanic basalt data arrays. As for the $^{207}Pb/^{204}Pb$ versus $^{206}Pb/^{204}Pb$ correlations, these may also be explicable in terms of binary mixing processes. Figures 9.25 and 9.26 display good colinearity for MORB and OIB from the Northern Hemisphere, which have been used by Hart (1984) to define Northern Hemisphere

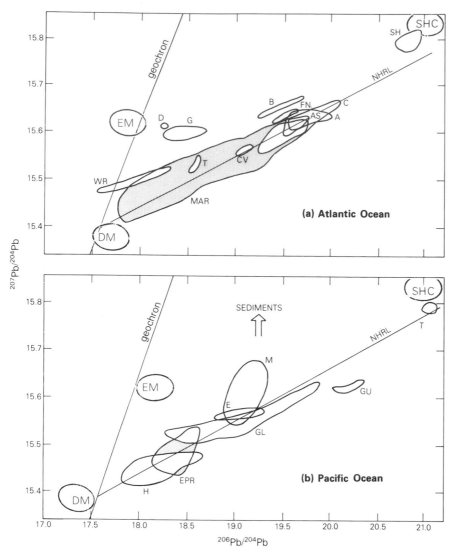

Key

D	Discovery seamount
CV	Cape Verde Islands
FN	Fernando de Noronha
NHRL	Northern Hemisphere Reference Line (Hart 1984)

Figure 9.25 ^{207}Pb/^{204}Pb versus ^{206}Pb/^{204}Pb for MORB and oceanic island basalts from (a) the Atlantic Ocean and (b) the Pacific Ocean. Abbreviations and data sources as in Fig. 9.23.

Reference Lines (NHRL) passing through MAR and EPR MORB, Hawaii, Iceland, Azores, Canaries and Cape Verde data. The equations of these reference lines are:

$$^{207}Pb/^{204}Pb = 0.1084(^{206}Pb/^{204}Pb) + 13.491$$

$$^{208}Pb/^{204}Pb = 1.209(^{206}Pb/^{204}Pb) + 15.627$$

and their positions are shown on the diagrams. The deviation of any data set (DS) from these reference lines is expressed as the vertical deviation in ^{207}Pb/^{204}Pb or ^{208}Pb/^{204}Pb from the line, and is referred to as a Dupal anomaly (Dupré & Allègre 1983). The Dupal anomalies are calculated using the following equations:

$$\Delta 7/4 = [(^{207}Pb/^{204}Pb)_{DS} - (^{207}Pb/^{204}Pb)_{NHRL}] \times 100$$

$$\Delta 8/4 = [(^{208}Pb/^{204}Pb)_{DS} - (^{208}Pb/^{204}Pb)_{NHRL}] \times 100$$

The ^{208}Pb/^{204}Pb versus ^{206}Pb/^{204}Pb plot (Fig. 9.26) clearly shows a grouping of oceanic-island data sets into those which broadly follow the NHRL, and a Dupal group characterized by higher ^{208}Pb/^{204}Pb

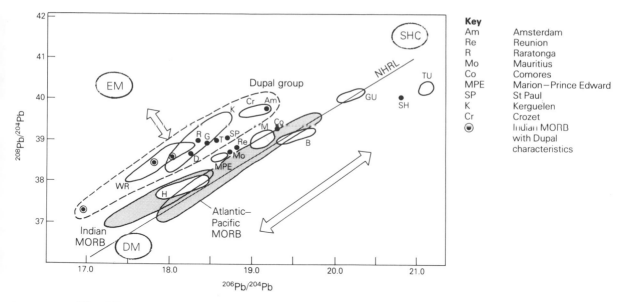

Figure 9.26 ^{208}Pb/^{204}Pb versus ^{206}Pb/^{204}Pb for MORB and oceanic island basalts from the Atlantic, Pacific and Indian oceans. Abbreviations as in Figures 9.23 & 25. The dashed line separates the field of oceanic islands with a distinct Dupal signature (after Hamelin & Allègre (1985) with additional data from Hart (1984) and Vidal *et al.* (1984)). Open arrows indicate two component mixing vectors.

ratios and thus higher values of Δ 8/4 (> 80). Figure 9.27 is a world map contoured for Δ 7/4 and Δ 8/4, which indicates that the Dupal anomaly is a striking feature, occupying a band centred on 30–40°S. The strongest maximum in the anomaly stretches from the southern MAR to the central Indian Ocean, with a second maximum in the central Pacific. This diagram is based on a very variable data set and thus may be subject to considerable modification as more data become available. There is a good correlation between the isotopic anomaly pattern and an equatorial bulge of the geoid, which has been related to a zone of deep-mantle upwelling (Busse 1983). This may suggest that the Dupal signature is a lower mantle characteristic.

Figure 9.28 is a detailed map of the southern Atlantic and Indian oceans showing all the presently available ^{208}Pb/^{204}Pb data, expressed as Δ 8/4, for basalts from both mid-ocean ridge segments and oceanic islands. Data for the Indian Ocean (Hamelin *et al.* 1986) show that the Dupal anomaly is present in both OIB and MORB, which should be anticipated if the spread of MORB isotopic com-

positions can indeed be accounted for in terms of binary mixing between depleted asthenosphere and blobs of OIB-source mantle. A similar Dupal signature is detected in South Atlantic MORB close to the hotspot islands of Tristan da Cunha and Gough.

In the plot of ^{208}Pb/^{204}Pb versus ^{206}Pb/^{204}Pb (Fig. 9.26), MORB from the Indian Ocean clearly define a different trend to the EPR–MAR array (Hamelin & Allègre 1985). Combined with data for the Indian Ocean islands of Marion–Prince Edward, Mauritius, Reunion and the Comores, the trend can be explained in terms of mixing between a SHC component and an asthenospheric component which is more enriched than the normal N–type MORB source (DM). Those samples of Indian ridge MORB displaying strong Dupal characteristics may be extensive partial melts of such an enriched component and lie on a mixing array with an SHC component, including the islands of Kerguelen, Crozet, Amsterdam and St Paul. On the basis of this Indian MORB data, Hamelin & Allègre (1985) conclude that the depleted asthenospheric mantle must be divided into different

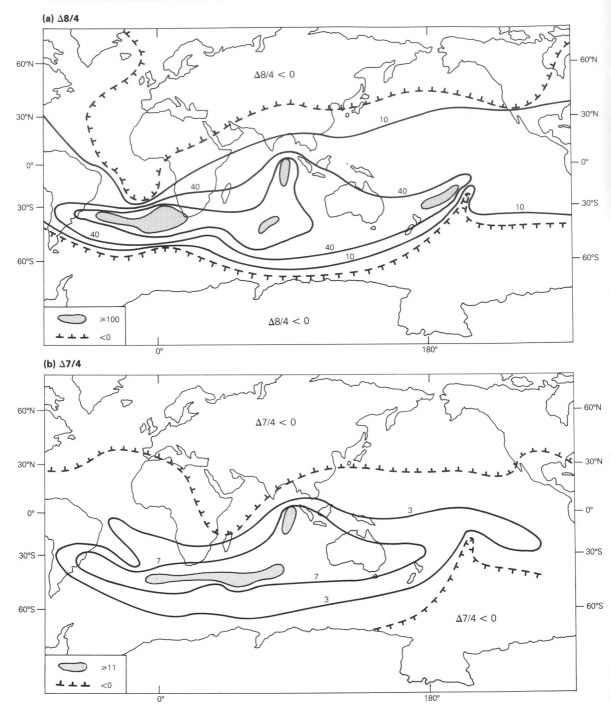

Figure 9.27 World maps (Miller cylindrical projection) showing the distribution of global lead isotopic anomalies for MORB and OIB. The contours shown are subjective, as there are many areas for which isotopic data are lacking. Data for the Ninety East Rise, an aseismic ridge in the Indian Ocean, are included iin the contours, despite being samples of older age which should actually plot in a more southerly position at their time of origin (after Hart 1984, Fig, 2).

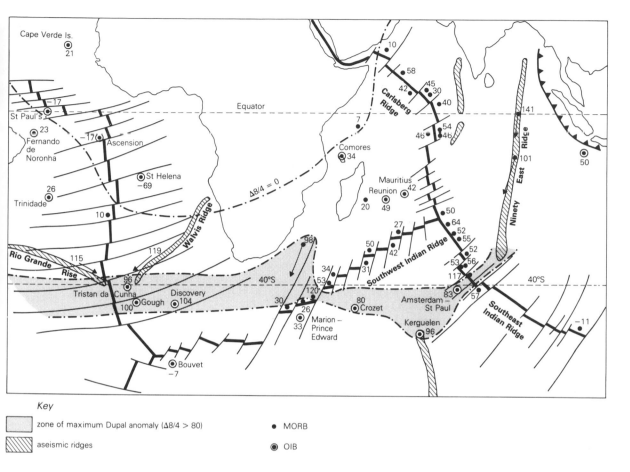

Figure 9.28 Detailed map of the Δ8/4 anomaly for MORB and OIB from the South Atlantic and Indian oceans. Older samples have arrows leading from them pointing in the direction of their original location. Data from Dupré & Allègre (1983), Hart (1984), Hamelin & Allègre (1985) and Hamelin *et al.* (1986).

provinces related to different convective units within the upper mantle.

The coherent linear arrays in $^{208}Pb/^{204}Pb$ versus $^{206}Pb/^{204}Pb$ and $^{207}Pb/^{204}Pb$ versus $^{206}Pb/^{204}Pb$ (Figs 9.25 & 26) could be explained in terms of binary mixing of end-members which show a degree of internal heterogeneity. Alternatively, a third component can be considered in which case the mixing end-members are:

(a) depleted asthenosphere; the source of N-type MORB — DM;

(b) undepleted, variably enriched, lower mantle — EM;

(c) a St Helena component (SHC), which may potentially have the characteristics of recycled oceanic crust.

If this is correct then the same pattern of three component mixing should be apparent in all of the Sr–Nd–Pb isotopic variation diagrams. However, comparison of the plots of $^{143}Nd/^{144}Nd$ versus $^{87}Sr/^{86}Sr$ (Fig. 9.23) with the plot of $^{207}Pb/^{204}Pb$ versus $^{206}Pb/^{204}Pb$ reveals an apparent discrepancy. The Nd–Sr plot appears to be dominated by binary mixing between DM and EM components, whereas the Pb–Pb plot indicates three-component mixing between DM, EM and SHC. Fortunately, this is easily reconciled in terms of the lack of sensitivity of the Nd−Sr diagram to mixtures involving the SHC component, as ancient recycled subducted lithosphere is not likely to develop particularly extreme isotopic compositions of Sr and Nd. More importantly, it is unlikely to contain the high concentrations of Nd necessary to strongly influence the isotopic composition of mixtures.

On the basis of the above, plots of $^{87}Sr/^{86}Sr$ versus $^{206}Pb/^{204}Pb$ (Fig. 9.29), and $^{143}Nd/^{144}Nd$ versus $^{206}Pb/^{204}Pb$ (Fig. 9.30), should be the most sensitive indicators of three-component mixing. The MAR and EPR MORB arrays clearly reflect mixing between a DM and an SHC component, possibly diluted to a small extent by an EM component, whereas Indian Ocean MORB are more complex three-component mixtures. The oceanic islands of Tristan da Cunha, Gough, Kerguelen and the aseismic Walvis Ridge could be samples of a near-primordial mantle component which had become internally heterogeneous on a small scale. However, Hawkesworth *et al.* (1986) have also argued that the Dupal component could reside within the continental lithosphere (Ch. 10). Whether the SHC component is actually close to the St Helena data or more extreme is a matter for speculation at present.

Data arrays for the Azores (Atlantic) and Marquesas (Pacific) indicate the importance of a fourth mixing component which has the isotopic characteristics of terrigenous oceanic sediments. This could be a source component (recycled subducted oceanic sediments) or a near-surface contaminant. Insufficient data are presently available to differentiate between these two models. Tholeiitic and alkali basalts from the Hawaiian islands appear to represent mixtures of EM and DM sources. Their isotopic characteristics have been explained by Staudigel *et al.* (1984) in terms of contamination of magmas with a strong EM signature by the oceanic lithosphere, which should have similar isotopic characteristics to EPR MORB (Section 9.8)

Helium isotopes

Sr, Nd and Pb isotopic variations in oceanic basalts have provided important constraints for models of oceanic mantle structure. In the preceding sections

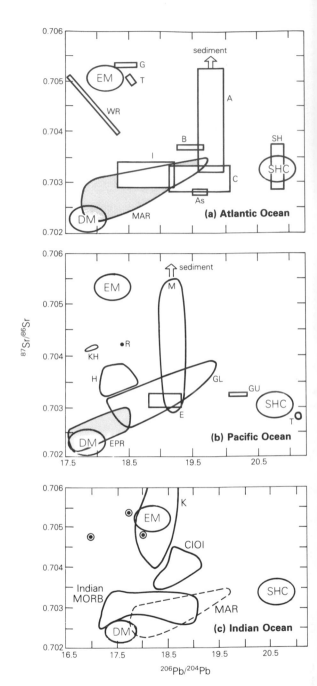

Figure 9.29 $^{87}Sr/^{86}Sr$ versus $^{206}Pb/^{204}Pb$ for MORB and oceanic island basalts form the (a) Atlantic (b) Pacific and (c) Indian oceans (after Staudigel *et al.* (1984). Abbreviations as in previous figures. CIOI are the Central Indian Oceanic Islands. Additional data from Hart (1984), Vidal *et al.* (1984), Hamelin & Allègre (1985) and Hamelin *et al.* (1986).

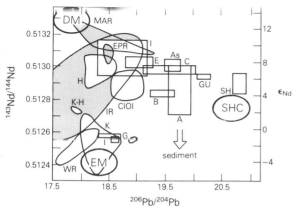

Figure 9.30 $^{143}Nd/^{144}Nd$ versus $^{206}Pb/^{204}Pb$ for MORB and oceanic-island basalts from the Atlantic, Pacific and Indian oceans (after Staudigel *et al.* 1984). IR is the field of Indian MORB, drawn to enclose the rare samples with a strong Dupal signature. Data for Indian Ocean basalts from Hamelin & Allègre (1985) and Hamelin *et al.* (1986).

two reservoir models have been shown to be insufficient to explain the complex data arrays and mixing of at least three independent mantle components is required; depleted asthenospheric mantle, enriched mantle and recycled subducted oceanic lithosphere (basaltic oceanic crust plus sediments).

Studies of the helium isotopic composition of mantle gases may further constrain these mantle mixing models. ^3He in mantle gases is mostly primordial, whereas ^4He is primarily radiogenic, produced by the decay of ^{238}U, ^{235}U and ^{232}Th. Thus high $^3He/^4He$ ratios in young oceanic-island volcanic rocks may be indicative of primordial volatiles, and thus the dominance of a primordial mantle source component in their petrogenesis. From the Nd, Sr and Pb data arrays considered previously, basalts from oceanic islands such as Tristan da Cunha, Gough and Hawaii would be predicted to show the strongest signature of a primordial mantle component.

Figure 9.31 shows the available $^3He/^4He$ data (normalized to atmospheric ratios) for oceanic basalts (Kurz *et al.* 1982). Data for Hawaii and Iceland can be explained in terms of mixing between a primordial component and the MORB source depleted asthenosphere. However, data from Tristan da Cunha and Gough display low $^3He/^4He$ characteristics which would not have been

predicted on the basis of their Sr, Nd and Pb isotopic ratios. To explain these data it is necessary to invoke mixing with a component that has been enriched in Th and U relative to ^3He for time periods long enough to lower the $^3He/^4He$ ratio. It is possible that the mantle source of Tristan da Cunha and Gough basalts has become metasomatically enriched by the migration of small-volume partial melts, enriched in U and Th, over the course of geological time. This has already been suggested to explain their negative ϵ_{Nd} characteristics. These baasalts could have acquired their unusually low $^3He/^4He$ signature if the partial melting event which generated them preferentially sampled the metasomatically enriched fractions of the plume source. Alternatively, the data could be explained by mixing of recycled oceanic sediments into the plume source, although this should have consequences for the other radiogenic isotope systems.

The study of helium isotopes obviously has potential for more detailed understanding of OIB mantle sources. However, considerably more data are required to characterize the range of variation and to understand the processes which may modify the $^3He/^4He$ ratio (Zindler & Hart 1986).

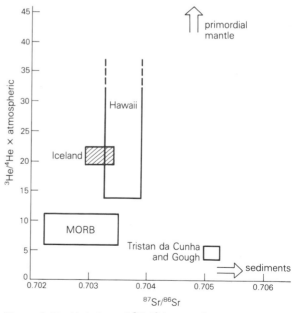

Figure 9.31 Variation of $^3He/^4He$ ratio (normalized to the atmospheric ratio) with $^{87}Sr/^{86}Sr$ for oceanic basalts (MORB and OIB) (after Kurz *et al.* 1982, Fig. 1).

9.8   Detailed petrogenetic model

The geochemical and petrological diversity of OIB compared to MORB has led to the development of petrogenetic models involving the diapiric upwelling of plume components from a deep-seated heterogeneous mantle reservoir (Section 9.2), which is geochemically distinct from the MORB source reservoir. Highly variable Sr, Nd and Pb isotopic ratios recorded from different oceanic islands (Section 9.7.4) indicate that this OIB (plume) source reservoir must involve both enriched mantle and recycled subducted oceanic lithosphere components. Allègre *et al.* (1984) consider that the isotopic heterogeneity of MORB result from mixing of this OIB source reservoir with the depleted asthenosphere.

Although the plume model is conceptually the simplest way in which to explain the characteristic features of oceanic intra-plate volcanism, other models may be equally viable. For example, Zindler *et al.* (1984) proposed that the asthenosphere is heterogeneous on a small scale (> 10 m) and that, while mid-ocean ridge magmatism homogenizes these heterogeneities, smaller degrees of partial melting away from the ridge may preferentially sample pods of more enriched mantle, thus generating the characteristic geochemical signature of OIB and seamount basalts. Geochemically, these two models have similar consequences and are difficult to choose between until we know more about the convective motion of the upper mantle. Nevertheless, it seems inevitable that deeper-mantle upwelling must occur in order to generate significant degrees of mantle partial melting by adiabatic decompression. Thus, despite the fact that we cannot study them directly, mantle plume flows are not simply figments of the geochemists' imagination.

Figure 9.32 depicts the processes involved in OIB petrogenesis. Diapiric upwelling of the mantle causes partial melting of both plume components and depleted asthenospheric (MORB source) mantle caught in the upward flow. These partial melts will mix, and the resultant magma, which segregates at depths of perhaps 50–60 Km (Section 9.3), will carry the isotopic and trace element signature

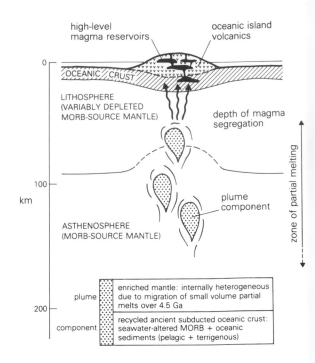

Figure 9.32   Detailed model for OIB petrogenesis.

of both source components, depending on their relative proportions in the ascending plume flow. These primary magmas must then rise through the cold oceanic lithosphere, including the down-warped oceanic crustal layer, before reaching the network of high-level (< 15 Km) storage reservoirs beneath the volcanic edifice. During magma storage, fractional crystallization and magma mixing are important processes which diversify the compositions of erupted magmas. Additionally, crustal contamination by seawater-altered oceanic crust, oceanic sediments and seawater-altered oceanic-island volcanics may further modify the geochemical characteristics of the magmas, particularly their Sr and O isotopic ratios (Sheppard & Harris 1985).

The extent to which isotopic heterogeneities in oceanic basalts reflect recycling of continentally derived material via subduction zones remains uncertain, but is of fundamental importance to studies of mantle evolution. Subducted oceanic crust will have incorporated continentally derived constituents during low- and high-temperature interactions with sea water. Additionally, some pelagic and clastic sedimentary material may also be subducted. Island-arc magmatism recycles some of these constituents back into the crustal reservoir on a comparatively short timescale, but inevitably some will persist in the subducted lithosphere as it travels deeper into the mantle. The times and scales over which isotopic and elemental abundance heterogeneities will survive in recycled oceanic crust will depend upon the efficiency of convective mixing processes in the mantle.

Recognition of recycled slab components in mantle-derived basalts generated at sites far removed from contemporary subduction zones requires that the characteristic isotopic tracer signatures imparted by the subduction process are not reduced by dilution with well mixed mantle material to the extent that they are no longer detectable. Provided that the subducted lithospheric slabs can maintain their geochemical identity in the mantle for significant periods of time ($> 10^8$ yr) then decay of parent isotopes of Rb, U and Th, which are concentrated in the subducted crust, should enhance the isotopic signature of the reservoir, causing, for example, the evolution of high ^{206}Pb/^{204}Pb ratios. Unfortunately, at present it is not possible to make quantitive predictions about the isotopic evolution of recycled oceanic crustal reservoirs in the mantle.

In terms of the Sr, Nd and Pb isotopic data presented in Section 9.7.4, basalts from the Atlantic oceanic island of St Helena and the Pacific island of Tubuaii have characteristics suggestive of a strong subducted slab signature, e.g. high ^{206}Pb/^{204}Pb ratios. Consequently, detailed geochemical studies of volcanic rocks from these islands should provide a fundamental insight into crustal recycling and OIB petrogenesis. At the other extreme, basalts from the southern Atlantic islands of Gough and Tristan da Cunha, and to a lesser extent the Pacific island of Hawaii, have isotopic characteristics indicating the influence of an enriched (?near primordial) mantle plume component. ^3He/^4He data for Hawaiian basalts substantiate the involvement of such a component but appear to negate it for Tristan da Cunha and Gough. However, until more is known about helium isotope systematics in oceanic basalts such data must remain equivocal.

Studies of the Hawaiian Islands have played a fundamental role in the development of models for the origins of linear island chains and the evolution of oceanic-island central volcanoes. All of the Hawaiian volcanoes display a characteristic evolutionary sequence, from a voluminous tholeiitic shield-building phase to a late-stage alkali basaltic phase, ultimately culminating in a post-erosional nephelinitic stage. Similar patterns have been observed in other oceanic islands, e.g. Samoa, Comores and the Canary Islands.

In general, nearly all oceanic-island volcanoes situated on mature oceanic crust evolve from an early less alkalic voluminous shield-building phase to later more alkalic volcanic phases. Traditionally, such evolutionary sequences have been explained in terms of models in which the early eruptives represent moderately large degrees of partial melting at relatively shallow depths, while the late-stage alkali basalts and nephelinites represent smaller degrees of partial melting at greater depths. However, major and trace element and isotope geochemical differences between Hawaiian shield-forming tholeiites and post-erosional alkalic lavas preclude their derivation from a compositionally and isotopically homogeneous source.

Figure 9.33 shows Nd and Sr isotopic data for the older (extinct) Hawaiian islands of Oahu and Maui (Stille et al. 1983, Chen & Frey 1985, Roden et al. 1984). The trend displayed by the Hawaiian basalts is subparallel to the MORB-OIB array, with post-erosional alkalic basalts slightly overlapping the MORB field, and shield tholeiites extending the array towards bulk Earth estimates. The nearly continuous variation in isotopic ratios is consistent with a model involving binary mixing of compositionally distinct source components.

The simplest explanation for the Hawaiian data is that partial melts from the rising plume interact with partial melts of the wall rock, presumed to be oceanic lithosphere. The isotopic characteristics of the erupted volcanics thus reflect the relative contributions of these two isotopically distinct source components. The voluminous shield-building tholeiites have isotopic characteristics close to bulk Earth (primordial mantle) and thus may represent relatively uncontaminated partial

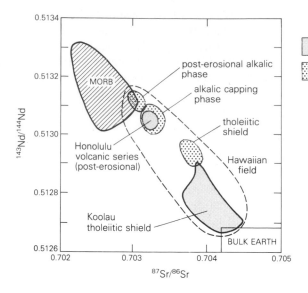

Volcanic rocks

- from Oahu
- from Maui (Haleakala)

Figre 9.33 $^{143}Nd/^{144}Nd$ versus $^{87}Sr/^{86}Sr$ for Hawaiian basalts to show the change in isotopic composition between the tholeiitic shield-building stage and the post-erosional alkalic stage. On Oahu, volcanics of the Koolau tholeiitic shield are overlain by the post-erosional Honolulu Volcanic Series. Data from Chen & Frey (1985) and Stille *et al.* (1983).

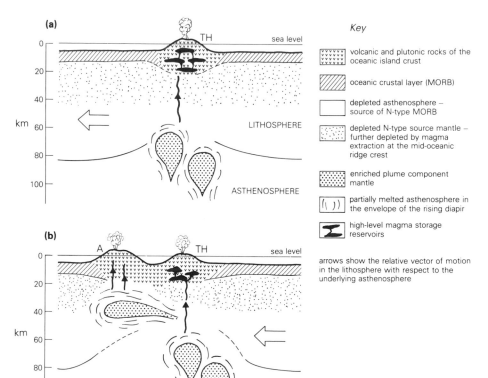

Key

- volcanic and plutonic rocks of the oceanic island crust
- oceanic crustal layer (MORB)
- depleted asthenosphere – source of N-type MORB
- depleted N-type source mantle – further depleted by magma extraction at the mid-oceanic ridge crest
- enriched plume component mantle
- partially melted asthenosphere in the envelope of the rising diapir
- high-level magma storage reservoirs

arrows show the relative vector of motion in the lithosphere with respect to the underlying asthenosphere

Figure 9.34   Models showing the evolution of a Hawaiian volcano from (a) the tholeiitic shield-building stage to (b) the late-stage post-erosional alkalic stage.

melts of the plume source (Fig. 9.34). After the shield-building stage the volcano gradually moves away from the hot spot and the supply of plume material decreases. At this stage the eruption frequency, the degree of partial melting of the plume component and its wall rocks and the relative contribution of the plume component all decrease. During the terminal post-erosional stages of activity nephelinites, basanites and melilitites are generated dominantly by partial melting of the lithosphere (Chen & Frey 1985).

Further Reading

Burke, K. C. & J. T. Wilson 1976. Hotspots on the Earth's surface. In *Volcanoes and the Earth's interior*, R. Decker and B. E. Decker (eds) 1982, 31–42. New York: W. H. Freeman.

Carmichael, I. S. E., F. J. Turner & J. Verhoogen 1974. *Igneous petrology*. New York: McGraw-Hill.

Hart, S. R. 1984. The Dupal anomaly: a large-scale isotope anomaly in the Southern Hemisphere mantle. *Nature* 309, 753–57.

Hofmann, A. W. & W. M. White 1982. Mantle plumes from ancient oceanic crust. *Earth Planet. Sci. Lett.* 57, 421–36.

Hughes, C. J. 1982. *Igneous petrology*. Amsterdam: Elsevier 551 pp.

McKenzie, D. & R. K. O'Nions 1983. Mantle reservoirs and ocean island basalts. *Nature* 301, 229–31.

Morgan, W. J. 1983. Hotspot tracks and the early rifting of the Atlantic. *Tectonophysics* 94, 123–39.

Schilling, J.-G., G. Thompson, R. Kingsley & S. Humphris 1985. Hotspot – migrating ridge interaction in the south Atlantic. *Nature* 313, 187–91.

White, W. M. & A. W. Hofmann 1982. Sr and Nd isotope geochemistry of oceanic basalts and mantle evolution. *Nature* 296, 821–5.

CHAPTER TEN

Continental tholeiitic flood basalt provinces

10.1 Introduction

Large areas of the continents appear to have been covered by vast thicknesses of laterally extensive basaltic lava flows at various stages during the past 1000 Ma, apparently fed from fissures rather than central vent volcanoes. These are referred to as continental flood basalt provinces or CFBs. Originally they were considered to be characterized by the eruption of chemically uniform Fe-rich tholeiitic basalts (Kuno 1969). However, as we shall see in Section 10.6, more recent studies have revealed significant chemical diversity within individual provinces. Nevertheless, most are dominated by eruptions of relatively evolved tholeiitic basalts which show some similarity in terms of their mineralogy and major element chemistry to MORB (Ch. 5), although their trace element characteristics are more akin to those of enriched MORB and oceanic-island tholeiitites (Ch. 9). Next to MORB, they form the largest volcanic features on Earth and like MORB are also generated in extensional tectonic environments.

Table 10.1 lists some of the major CFB provinces, ranging in age from Precambrian to Recent, with an estimate of their maximum thickness and areal extent. These grade downwards into numerous smaller scale but chemically and tectonically similar examples, which clearly cannot be considered here. Of those listed, the Keweenawan and Siberian platform provinces appear to have been associated with intracontinental plate rifts, whereas the younger Karoo, Paraná, Etendeka, Antarctic and North Atlantic provinces are related to sites of successful continental fragmentation and the generation of new ocean basins (Cox 1978).

Table 10.1 Ages and dimensions of the major continental flood-basalt provinces.

Province	Age (Ma)	Maximum thickness (m)	Present area (km²)
Keweenawan (Lake Superior)	Late Precambrian 1100−1200	12 000	>1000 000
Siberian Platform	Permo-Trias 248−216	3 500	>1 500 000
Karoo (Southern Africa)	Jurassic 206−166	9 000	140 000
Kirkpatrick Basalts, Ferrar Dolerites (Antarctica)	Jurassic 179± 7	900	7 800
Paraná (Brazil) }	Late Jurassic − Early Cretaceous 140−110	1 800	1 200 000
Etendeka (Namibia)		900	78 000
North Atlantic Igneous Province	Late Cretaceous − Eocene 65−50	2 000	1 000 000
Deccan Traps (India)	Cretaceous−Tertiary boundary	> 2 000	> 500 000
Columbia River (north-west USA)	Miocene 17−6	> 1 500	200 000
Snake River Plain (north-west USA)	Quaternary 17−0	−	50 000

Data sources: Basaltic Volcanism Study Project (1981), Erlank (1984), Siders & Elliot (1985), Courtillot *et al.* (1986) and Fodor (1987).

In the Karoo, Paraná, Etendeka and Antarctic provinces, rifting and magmatism is associated with the break-up of the Gondwanaland supercontinent during the Jurassic and Cretaceous (Fig. 10.1). Each of these provinces consists of an extrusive phase together with a voluminous suite of doleritic sills which penetrate the underlying platform sediments. The basalts and dolerites of the Karoo and Antarctica (Ferrar dolerites and Kirkpatrick basalts) are mostly of early Jurassic age and were probably erupted shortly before the Indian Ocean was established by the separation of Africa and Antarctica, which occurred sometime between 145 and 170 Ma (Marsh 1987). The Paraná and Etendeka basalts are substantially younger (late Jurassic − early Cretaceous) and are precursive to the opening of the South Atlantic. Several authors (Elliot 1975, Cox 1978, Froidevaux & Nataf 1981) have pointed out that the distribution of these Mesozoic CFB provinces forms a band which parallels the Pacific margin of Gondwanaland (Fig. 10.1), along which subduction had occurred since at least the Devonian and up to the time when the supercontinent was disrupted. Consequently, magma generation in these provinces may involve subduction-modified mantle sources (see Ch. 6), which could explain some of the distinctive trace element characteristics of the flood basalts (Duncan 1987).

Petrogenic interpretation of the Paraná flood basalts has considerable implications for our understanding of the separation of the African and South American continents, and the opening of the South Atlantic Ocean (Bellieni *et al.* 1984, Fodor *et al.* 1985, Hawkesworth *et al.* 1986, Fodor 1987). Flood basalt fields in the Paraná basin (Brazil) and

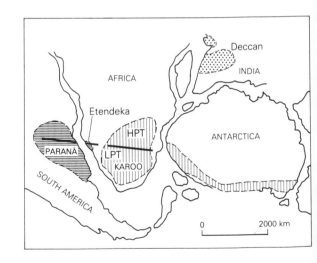

Figure 10.1 Early Mesozoic reconstruction of the supercontinent Gondwanaland and the distribution of major continental flood basalt provinces (after Cox 1978, Fig. 1, p. 47). The solid line divides provinces characterized by eruption of high-phosphorus and titanium basalts (HPT) from low-phosphorus and titanium basalts (LPT).

Etendeka (Namibia, southern Africa) were joined 120 Ma ago (Figs 10.1 & 2), and have subsequently been separated by the opening of the South Atlantic Ocean at the latitude of the Tristan da Cunha hot spot (Ch. 9). Similarly, the extensive Tertiary flood basalts of the North Atlantic igneous province were the precursor to the opening of the North Atlantic Ocean during the early Tertiary.

The magmatism of the Deccan Trap volcanic province of India postdates the fragmentation of Gondwanaland and appears to have occurred in a very short time period, possibly less than 3 Ma, spanning the Cretaceous/Tertiary boundary (Courtillot *et al.* 1986). This was an apparently catastrophic magmatic event, which may be associated with the other anomalous events which apparently took place at this time, such as mass extinctions of certain groups of fauna. Alvarez *et al.* (1980) suggested that such events at the Cretaceous/Tertiary boundary might be caused by the impact of an asteroid. If the Deccan marked the site of such an impact (Courtillot *et al.* op. cit.), it might be considered to represent the terrestrial equivalent of a lunar mare. However, tectonic disturbances related to a major impact event have not been

described in India around the basalt outcrops and an alternative hypothesis, involving the passage of India over an extremely vigorous mantle plume or hot spot (Officer & Drake 1985), seems more appropriate. Plate-tectonic reconstructions suggest that the hot spot, which is presently located beneath Reunion island, stood below the Deccan at the time of major basalt eruption. Thus the present Reunion hot spot could be the weak remainder of the mantle plume which triggered the flood basalt eruptions.

The youngest example of tholeiitic continental flood-basalt volcanism listed in Table 10.1 is that of the Columbia River Plateau, northwestern USA (Fig. 10.3), which was active from 17 to 6 Ma ago (Basaltic Volcanism Study Project 1981). This volcanic province is linked geographically with more recent activity in the Snake River Plain – Yellowstone Park region, which is of a bimodal basalt–rhyolite nature (Christiansen 1984) and appears to be related to the passage of the North American plate over a mantle hot spot,

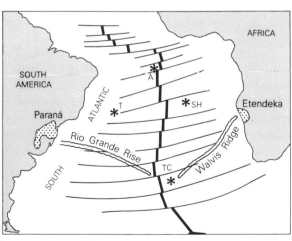

Other hot spots within the South Atlantic

A Ascension   SH St Helena   T Trinidade

Figure 10.2 Location of the Paraná volcanic province of Brazil and the contemporaneous Etendeka province of Namibia (dotted ornament) in relation to the location of the Tristan da Cunha hot spot (TC) and its associated aseismic ridges (Walvis Ridge – Rio Grande Rise).

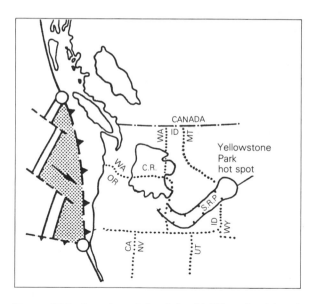

Figure 10.3 Location of the Columbia River flood basalt province (CR) and the associated Snake River Plain (SRP) and Yellowstone provinces. The volcanism is a response to thermal disturbances in the mantle above the subducting Farallon plate (shaded). The volcanism of the SRP and Yellowstone Park traces the locus of a hot spot currently located beneath Yellowstone (after Menzies *et al.* 1984, Fig. 1, p. 645).

currently located beneath Yellowstone (Menzies *et al.* 1984). The basaltic volcanism of the Columbia River province itself appears to be a consequence of thermal anomalies in the upper mantle above a subduction zone. It may be regarded as an example of magmatism within an ensialic back-arc basin (Ch. 8; and see Prestvik & Goles 1985).

All of the examples of continental flood basalt volcanism listed in Table 10.1 are clearly associated with extensional tectonics, which appear to be the consequence of lithospheric stretching associated with the upwelling of deeper hotter mantle material. Such upwelling must be initiated at sublithospheric depths, with its effects propagating upwards by thermal erosion of the base of the lithosphere. It may reflect some sort of back-arc spreading, as in the Columbia River province, or the existence of a mantle hot spot analogous to those proposed in Chapter 9 to explain oceanic intra-plate volcanism. The scale of the phenomenon makes the term 'hot spot' seem somewhat inappropriate, although some large intracontinental plate rifts, such as the East African Rift (Ch. 11), may reflect the surface expression of a linear array of hot spots or a 'hot-line'. In this respect it is interesting to note the apparent association of the Paraná and Etendeka CFB provinces with the Tristan da Cunha hot spot, which was a ridge-centred hot spot during the opening of the South Atlantic (Ch. 9). In the following sections, attention will be focused on the Columbia River, Paraná—Etendeka and Deccan provinces as examples of CFBs formed in different tectonic settings. Erlank (1984) presents a detailed discussion of the Karoo province, and the reader is referred to this work for further details. All three provinces are characterized by their great areal extent, large volumes, immense sizes of individual eruptive units, sub-aerial eruption onto continental basement and dominantly tholeiitic (sub-alkaline) chemistry (Basaltic Volcanism Study Project 1981). In general, they are also commonly associated with high-level basic intrusives, including dyke swarms and extensive dolerite sill complexes, and with continental sediments. Pyroclastic inactivity is insignificant and central volcanic structures rare.

Despite the predominance of tholeiitic basalts, many CFB provinces also contain small ($<$ 10%)

but significant volumes of acidic eruptives, usually in the upper part of the sequence. Intermediate-composition lavas are frequently lacking and the volcanism may be described as being of a bimodal basalt—rhyolite type. In the Karoo, Deccan and Paraná provinces, the acidic volcanics appear to be confined towards the continental margin and have been attributed to partial melting of underplated CFB magmas at the base of the crust, during lithospheric thinning immediately preceding the major rifting events which split the Gondwanaland supercontinent (Cleverly *et al.* 1984, Bellieni *et al.* 1986, Lightfoot *et al.* 1987). Additionally, in the Deccan province small volumes of alkalic volcanics and high-level intrusives occur at scattered locations (Mahoney *et al.* 1985). Such alkaline magmatism is a minor feature of most of the CFB provinces listed in Table 10.1. However, several large flood basalt provinces, such as Ethiopia and the North Atlantic, are substantially alkalic, grading into totally alkalic examples such as the East African Rift. For simplicity, attention is focused in this chapter on the dominantly tholeiitic provinces, leaving consideration of the alkalic examples to Chapter 11. However, the reader should realize that this is a somewhat artificial subdivision and that, in reality, there is a continuum of tectonic and magmatic processes linking the two.

One of the greatest problems in interpreting the petrogenesis of CFB magmas is the lack of recent examples. In previous chapters we have relied heavily on geophysical data to provide information about depths of magma generation, magma transportation and sites of high-level magma storage. This is only possible for the young volcanic zones of the Snake River Plain — Yellowstone Park region of northwestern USA which, although geographically linked with the Columbia River flood basalt province, are not true flood basalt provinces in their own right — nor are they necessarily related to the same tectonic environment.

The origin of CFB magmas thus remains a subject for considerable debate. Kuno (1969) pointed out that, despite their large volumes, relatively uniform chemistries and fissure modes of eruption, most CFB lavas appear to be considerably evolved with respect to our ideas of what primary magma compositions should be (Ch. 2). As a consequence, some authors have considered that they are unusual primary magmas derived by partial melting of iron-rich mantle sources (Wilkinson & Binns 1977), whereas others (e.g. Cox 1980) consider that they are the products of fractional crystallization of picritic basalts near the base of the crust. Thompson *et al.* (1983) concluded that very few CFB magmas could have reached the surface without pausing and re-equilibrating during their ascent, thereby explaining the rarity of picritic eruptives in CFB provinces. The depth of such equilibration will obviously vary from province to province. If flood basalts are indeed produced by extensive low-pressure crystal fractionation, then extensive piles of complementary cumulates must exist within the crust. Cox (1980) suggests that these occur in sill-like bodies near the crust/mantle boundary, which acted as feeder systems for the surface eruptives.

The immense areal extent of some CFB provinces makes them potential sources of information about the nature of the subcontinental upper mantle. Additionally, detailed stratigraphic studies allow elucidation of any temporal changes in the magma generation process (Swanson *et al.* 1979, Reidel 1983, Cox & Hawkesworth 1985, Beane *et al.* 1986). Unfortunately, characterization of CFB mantle source regions is hampered by uncertainties regarding the extent of modification of the magmas by interaction with the continental crust, and also by our lack of detailed knowledge about the chemical and isotopic heterogeneity of the sub-continental mantle. This has led to considerable controversy concerning the interpretation of their trace element and radiogenic isotope geochemistry (Section 10.6), some authors favouring trace element enriched mantle sources (Hawkesworth *et al.* 1983, Menzies *et al.* 1983), whereas others prefer crustal contamination models (Carlson *et al.* 1981, Mahoney *et al.* 1982). In reality, both are probably variably important in different provinces, and the problem then becomes one of trying to resolve their different effects on the geochemical characteristics of the magmas (Section 10.6). It seems likely that magmas erupted during the early stages of volcanism in an extensional continental tectonic setting

should be strongly susceptible to crustal contamination as they establish pathways to the surface. However, once conduits have become established, subsequent batches of magma may reach the surface having experienced little crustal contamination en route. Thus, potentially, the most contaminated magma compositions might be predicted to occur in the lower parts of the flood basalt sequence.

Associated with several CFB provinces are large basic layered intrusions which may have acted as storage reservoirs for the magmas during their ascent to the surface. Examples are the Duluth complex, associated with the late Precambrian Keweenawan flood basalts, the Dufek intrusion, associated with Jurassic flood basalts in Antarctica, the Muskox intrusion, associated with the late Precambrian Coppermine basalts of the North-West Territories, Canada, and the Skaergaard intrusion, emplaced penecontemporaneously with the eruption of flood basalts on the east coast of Greenland during early Tertiary extension of the North Atlantic Ocean. Within the British Tertiary Volcanic Province, central volcanic complexes developed after the cessation of major flood basalt activity and their eroded root zones now include layered plutons such as the Rhum intrusion, which were clearly open system magma chambers, continuously refluxed by basaltic magma (Dunham & Wadsworth 1978). The characteristics of such high-level intrusions have already been discussed in Chapter 4 and thus will not be discussed further here.

10.2 Simplified petrogenetic model

Continental flood basalt provinces are highly complex regions of terrestrial basaltic magmatism and thus models to explain their petrogenesis must be equally complex. Two fundamental problems which need to be addressed are the nature and source characteristics of the primary magmas, and the extent to which such magmas have interacted with rocks of the continental crust en route to the surface.

CFBs characteristically have low Mg numbers (Section 10.6), which means that they cannot have equilibrated with normal mantle lherzolite miner-alogies under water-undersaturated conditions (Cox 1980, Prestvik & Goles 1985). The most obvious explanation for this is that the primary magmas were not basalts but high-MgO picrites, which subsequently underwent low-pressure frac-tionation to produce the range of eruptive magma unusual compositions. However, the possibility remains that they are unusual primary magma compositions derived by partial melting of source regions more iron-rich than normal mantle lherzo-lite (Wilkinson & Binns 1977) or by partial melting of normal mantle lithologies under water-saturated conditions (Prestvik & Goles op. cit.). As discussed in Chapter 6, more silica and iron-rich primary magma compositions may be generated by partial melting of subduction-modified mantle, and this may be relevant to the petrogenesis of CFBs from the Columbia River province.

Cox (1980) considers that the characteristic plagioclase ± olivine ± clinopyroxene phenocryst mineralogy (Section 10.5) of the majority of CFBs suggests that they have been involved in relatively low-pressure crystal fractionation equilibria, sup-porting their evolution from more MgO-rich pri-mary magmas. Figure 10.4 shows a schematic crustal model in which dense picritic magmas pond within deep sill complexes near the base of the crust (25–40 km depth). Such sills provide an ideal environment for the fractionation of large volumes of magma within a comparatively restricted pres-sure range, and also one in which magmas are highly susceptible to crustal contamination (Patch-ett 1980). If correct, this implies that in areas subject to flood-basalt volcanism there may be significant additions of gabbroic and ultramafic cumulates to the lower crust, the total volume of which may exceed that of the surface eruptives (Cox op. cit.). Magmas from the lower sill complexes may feed directly to the surface via dykes as shown, or they may pond again in higher level sill complexes where further fractional crystallization

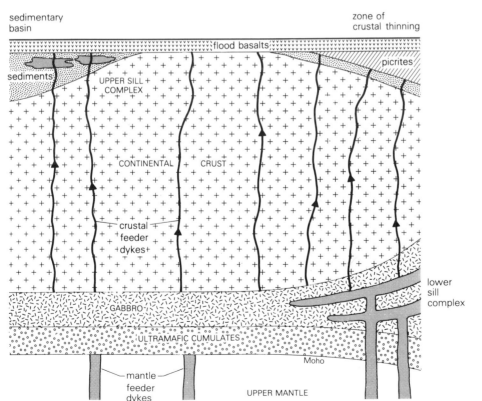

Figure 10.4 Schematic cross section through the continental crust in an area affected by flood basalt volcanism. Picritic basalts only reach the surface in zones of crustal thinning, and high-level sill complexes develop within thick sedimentary successions in subsiding basins (after Cox 1980, Fig. 7, p. 647).

and crustal contamination may occur before eruption. Though geographically rather rare, the occurrence of extensive picritic basalts within parts of the Deccan and Karoo flood basalt sequences lends support for this type of model. Additionally, the conspicuous lack of mantle-derived xenoliths in CFBs provides further evidence for their non-primary nature.

Assuming that primary CFB magmas are more magnesian than the range of eruptive compositions, we must then attempt to elucidate the nature of their mantle source. Figure 10.5 shows a schematic cross section through the continental lithosphere and an adjacent ocean basin, to illustrate the distribution of potential mantle source components. The subcontinental lithosphere may show considerable variations in thickness from 100 to greater than 200 km (Pollack 1986), depending upon the age and tectonomagmatic evolution of the overlying crust. However, unlike the oceanic lithosphere (Ch.5), its base is considerably harder to define seismically and is unlikely to represent a simple thermal boundary layer. Norry & Fitton (1983) consider that the continental lithosphere is lithologically and geochemically complex, recording the same sequence of tectonic, metamorphic and magmatic events as the overlying crust. This is substantiated by the wide range of mantle xenoliths brought to the surface by kimberlites and continental alkali basalts (Harte 1983, Menzies 1983, Nixon 1987). Geochemical studies of such xenoliths (Menzies & Murthy 1980. Erlank *et al.* 1982, Cohen *et al.* 1984) confirm that some portions of the continental lithosphere are both old and relatively enriched in incompatible elements, the more ancient enrichment events being reflected in enhanced isotopic ratios of Sr, Nd and Pb (Section 10.6.3). Thus, in contrast to the relatively refractory nature of the mantle component of the oceanic lithosphere (Ch. 5), that of the continental lithosphere may be significantly fertile due to multiple enrichment events by migrating fluids and partial melts over periods possibly exceeding 1000 Ma. Therefore, as in the active continental margin tectonic setting (Ch. 7), it must be considered as a major potential source component for CFB magmatism.

The convecting asthenosphere lying below the lithosphere must also be considered as a potential source component. This should be relatively homogeneous on a large scale due to convective overturn, and thus cannot preserve the same lithological and chemical heterogeneities as the older portions of the continental lithosphere over the same timescale (1000 Ma). There is no inherent reason to suppose that the convecting mantle beneath the oceans differs from that beneath the continents, although Thompson *et al.* (1984) have

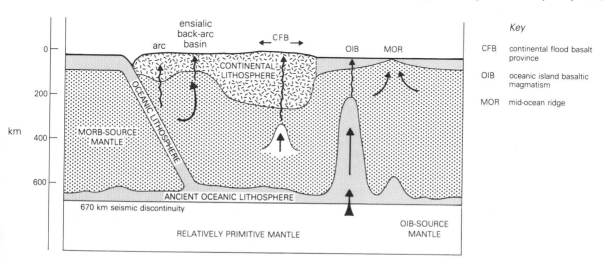

Figure 10.5 Distribution of major mantle source components in relation to sites of continental and oceanic volcanism.

argued that MORB source mantle only underlies the major ocean basins and that OIB source mantle underlies the continental lithosphere. Thompson *et al.* base this argument on the lack of CFBs with MORB-like geochemical characteristics, but this is rather subjective because if the continental lithosphere is variably fertile and, in some instances, greater than 200 km thick it is unlikely that sub-lithosphere sources will be involved extensively in the magma generation process.

In the following sections we will attempt to establish geochemical criteria for distinguishing the relative roles of crustal contamination and enriched mantle sources in the petrogenesis of CFB magmas.

10.3   Crustal structure and magma storage reservoirs

As considered in Section 10.1, a fundamental problem in petrogenetic studies of CFB magmatism is the extent to which rising magmas have undergone crustal contamination and concomitant crystal fractionation (AFC). In order to understand these processes it is necessary to have a detailed knowledge of the crustal structure and, ideally, of the location of high-level magma storage reservoirs. Clearly, the latter is only possible for presently active volcanic provinces, and thus we must rely heavily on data from the Snake River Plain – Yellowstone Park region of northwestern USA in this respect.

Figure 10.6 shows the crust and upper mantle structure beneath the Yellowstone caldera and the adjacent segment of the Snake River Plain (Smith & Braile 1984). There is a marked zone of reduced P-wave velocities (shaded) beneath the caldera, extending to depths of 250 km, which may reflect the location of a zone of partial melting within the lithosphere. The seismic model suggests that although velocities are reduced in the upper crust as a consequence of the magmatism they are not significantly modified in the lower crust. The lack of seismic attenuation in this lower crustal layer may be a consequence of its invasion by a plexus of vertically orientated dykes, transporting magma from deeper levels which are individually too

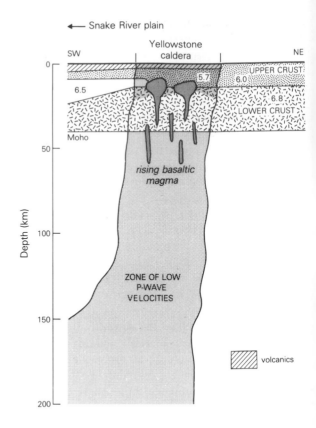

Figure 10.6 Crustal structure and location of major magma storage reservoirs beneath the Yellowstone caldera. The upper crust is characterized by a P-wave velocity of 5.7 km s^{-1}, which is anomalously low compared to that of the adjacent thermally undisturbed crystalline basement (6.0 km s^{-1}). This layer is interpreted to be a hot but relatively solid body, 8–10 km thick, which may represent the storage reservoir for the acid magmas. Basaltic magma chambers are shown shaded black, developing at the upper crust/lower crust boundary. The P-wave velocity of the upper 250km of the mantle is reduced by approximately 5%, suggesting the presence of a partial melt phase (after Smith & Braile 1984, Fig. 7.9, p. 106).

narrow to be resolved seismically. Basaltic magma chambers are shown developing near the upper crust/lower crust boundary, and the 6.5 km^{-1} layer which is well developed beneath the Snake River Plain at a depth of 10–20 km may represent mafic bodies crystallized from such chamber systems.

Gravitational, topographic and seismological studies of recently active intracontinental plate volcanic provinces suggest that considerable lithospheric thinning has taken place beneath these areas (Yuen & Fleitout 1985), which may occur at rates as high as 10 km Ma. Crustal thinning occurs in most cases, concomitant with such lithospheric thinning. Figure 10.7 shows a N–S profile through the Columbia River Plateau, which indicates that the crust has been thinned by as much as 12 km relative to that of adjacent basement terrains.

Thompson *et al.* (1986) have presented a detailed model for the storage and ascent of basaltic magmas beneath the British Tertiary Volcanic Province, based essentially on their geochemical and isotopic characteristics. They suggest that primary picritic magmas ponded at the Moho, where they underwent fractional crystallization until their densities were reduced sufficiently to regain buoyancy and permit further uprise. These more evolved magmas then penetrated the crust through a plexus of dykes, with sill swarms developing at intracrustal density drops such as a mid-crustal boundary

between granulite and amphibolite facies gneisses and an upper crustal boundary between gneisses and pelitic schists. Thompson *et al.* consider that early magma batches within this province may have paused at least twice during their ascent through the crust, becoming contaminated in each instance by a different crustal component. This model agrees rather well with that shown in Figure 10.6 for the crustal structure of the Yellowstone – Snake River Plain province, and thus may potentially have general applicability for CFB magmatism.

10.4 Crustal contamination of magmas

There are several mechanisms for the interaction of mantle-derived magmas with crustal rocks involving both mixing of crustal and mantle partial melts and bulk assimilation. The latter obviously involves a considerable input of heat, much of which may be

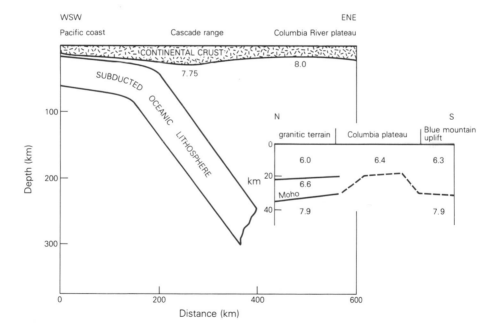

Figure 10.7 Cross section along a WSW–ENE profile across the Cascade range and Columbia River Plateau, to show the present configuration of the subducting Juan de Fuca plate (after Michaelson & Weaver 1986, Fig. 10, p. 2091). The inset shows a N–S profile across the Columbia River Plateau to show the marked crustal thinning which occurs beneath it (after Hill 1972, Fig. 4, p. 1643). Both diagrams show P-wave velocities in km s^{-1}. Higher upper crustal velocities beneath the Columbia River Plateau may be a consequence of the existence of extensive dyke swarms which feed the surface volcanism.

supplied by the latent heat released during simultaneous crystallization (DePaolo 1981). The continental crust contains a complex variety of rock types, ranging from those with low fusion temperatures, such as pegmatites and granites, to those with higher solidus temperatures such as gabbros, mafic granulites and amphibolites. Rising basaltic magmas might therefore be expected to be preferentially contaminated by those components of their wall rocks with the lowest fusion temperatures, with the most favourable site for such contamination being the lower crust (Huppert & Sparks 1985).

A fundamental problem is whether contamination occurs mainly in crustal magma chambers, in the dykes which feed these chambers or in the dykes which vent the chambers to the surface. Magmas stored in sheet-like bodies (dykes or sills) clearly have a greater contact area with their wall rocks and therefore should be the most susceptible to contamination (Patchett 1980, Campbell 1985). The extent of such contamination will vary with the temperature of the magma, the flux of magma through the feeder dykes, the width of the feeder dykes and the composition of the crust through which the magma passes. Where flow in the dyke feeders is fully turbulent, assimilation of wall rocks is most likely (Huppert & Sparks 1985). This appears to develop where the dyke width exceeds 3 m and, under such circumstances, the flowing magma will erode its walls and thus become contaminated by bulk assimilation of crustal material. At low flow rates in narrow dykes, flow is more likely to be laminar and the magma may solidify against the dyke walls, shielding subsequent batches from contamination.

Two contrasting styles of contamination have been recognized in the study of continental flood basalt petrogenesis. In the first assimilation of crustal rocks is accompanied by fractional crystallization (AFC: DePaolo 1981; and see Ch. 4) and the most evolved lavas are also the most contaminated. Lavas with such characteristics are well documented in the Paraná flood basalt province (Section 10.6.3). In the second type the most primitive basalts are the most contaminated and this has been recognized in the plateau lavas of the British Tertiary Volcanic Province (Thirlwall &

Jones 1983, Thompson *et al.* 1986) and in the Deccan Traps (Mahoney *et al.* 1982, Cox & Hawkesworth 1985, Devey & Cox 1987). Combined radiogenic isotope and major trace element geochemical studies clearly have considerable potential for elucidating crustal contamination mechanisms (Ch. 4).

10.5   Petrography of the volcanic rocks

Most flood basalts are variably porphyritic, with total phenocryst contents up to about 25%, although in some provinces (e.g. Columbia River and Paraná) aphyric to sub-aphyric types predominate (Cox 1980). Plagioclase phenocrysts are virtually ubiquitous and these may be accompanied by rare olivine, augite, pigeonite and Ti-magnetite. This assemblage suggests that the magmas have been involved in low-pressure crystal fractionation processes which Cox (op. cit.) considers is likely to reflect a complex series of polybaric events in the pressure range 0–15 kbar.

For descriptive purposes, the petrographic characteristics of the Paraná flood basalts will be taken as typical. The Paraná lavas are mostly aphyric to sub-aphyric, and span the compositional range from basalt to rhyolite. Basalts and basaltic andesites contain phenocrysts of plagioclase (An_{83-50}), augite and pigeonite, accompanied by rare Ti-magnetite and olivine. In the groundmass plagioclase, augite and pigeonite occur, together with abundant Ti-magnetite and ilmenite. It is noteworthy that basaltic andesites with high P_2O_5 and TiO_2 contents (Section 10.6) do not contain pigeonite as a phenocryst phase. In the associated tholeiitic andesites and dacites, similar phenocryst and groundmass mineralogies occur, although phenocryst plagioclase is more sodic (An_{67-50}), olivine is absent and interstitial quartz may occur in the groundmass. In the acid volcanics (rhyodacites and rhyolites) phenocrysts of plagioclase, pyroxene and opaques are typical with plagioclase compositions in the range An_{59-39}, accompanied by occasional alkali feldspar phenocrysts. Quartz and alkali feldspar are now common groundmass phases accompanied by Ca-rich clinopyroxene, pigeonite, Ti—magnetite and ilmenite. Additionally, in the acid rocks large partially resorbed xenocrysts of labradoritic plagioclase are common, which may support models for the origins of the acid rocks by partial melting of lower crustal basic rocks (gabbros or granulites).

The common occurrence of Ca-rich and Ca-poor pyroxenes, the rarity of olivine and the concentration of Fe—Ti oxides in the groundmass of the basic rocks is typical of most tholeiitic volcanic suites. Pyroxenes in rocks ranging from basalt to rhyodacite in composition plot in well defined fields in the pyroxene quadrilateral, similar to the pyroxenes of the tholeiitic Skaergaard intrusion (Fig. 10.8). The TiO_2 contents of the Ca-rich pyroxenes in the basalts and andesites correlate well with that of the host rock.

Picritic basalts are of generally restricted occurrence in CFB provinces, although are locally extensive in the Karoo (Bristow 1984) and Deccan (Krishamurthy & Cox 1977) provinces. Those of the Karoo are enriched in K relative to the Deccan picrites, although even the latter are transitional to mildly alkaline. In general, the Karoo picrites display strongly porphyritic to glomeroporphyritic textures. They typically consist of about 20%

olivine phenocrysts (Fo_{72-92}) set in a groundmass of quench clinopyroxene, quench opaque minerals and glass. Quench plagioclase crystals may be present in the groundmass of some types. Rare varieties may contain clinopyroxene phenocrysts (Cox 1987) and some of the more evolved picrites contain occasional plagioclase phenocrysts (Bristow 1984). Commonly, Karoo picrites contain sparse ($< 1\%$) aggregates of orthopyroxene crystals, mantled by rims of clinopyroxene and subordinate olivine. Additionally, megacrysts of olivine, substantially larger than most of the olivine phenocrysts, have been described from some samples. Together with the orthopyroxene aggregates, these have been conventionally ascribed to the products of high-pressure crystallization (Cox & Jamieson 1974). However, Cox (1987) considers that some of these may actually be fragments of the mantle source region (restite) on the basis of their well developed deformation fabrics.

In contrast, the picrite basalts of the Deccan commonly contain phenocrysts of both olivine and Ca-rich clinopyroxene and may be subdivided into oceanitic (olivine>clinopyroxene) and ankaramitic (clinopyroxene>olivine) types. Olivine is one of the most important constituents, both as phenocrysts (Fo_{90-87}) and in the groundmass (Fo_{88-84}). Some

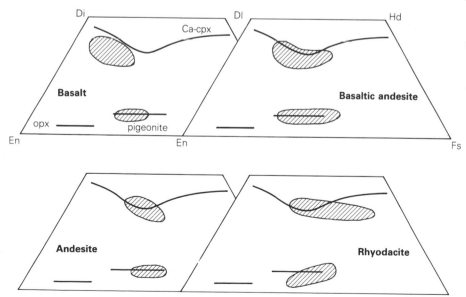

Figure 10.8 Pyroxene compositions in Paraná flood basalts and more evolved rocks (data from Bellieni *et al.* 1984).

of the larger olivine phenocrysts contain inclusions of chromite. Plagioclase does not occur as a phenocryst phase but is common in the groundmass (An_{88-74}) along with Ti-magnetite and ilmenite. The clinopyroxenes are chrome-rich diopsides with up to 1% Cr_2O_3 (Krishnamurthy & Cox 1977).

If we compare the phenocryst assemblage of a typical flood basalt with that of typical N-type MORB (Ch.5) we find that both are dominated by plagioclase. However, in MORB, olivine and Mg−Cr spinel are the common accessory phenocrysts, whereas in CFBs plagioclase is normally accompanied by augite ± pigeonite, with olivine very much rarer. In many continental layered basic intrusions (Ch.4) Ca-poor pyroxene crystallizes before Ca-rich pyroxene, and Campbell (1985) has suggested that this is a consequence of crustal contamination of the basaltic magmas before they entered the high-level magma chamber. The presence of Ca-poor pyroxene, usually pigeonite, in continental flood basalts may therefore reflect the fact that the majority have been variably contaminated by the continental crust en route to the surface.

Figure 10.9 shows the typical petrographic features of a tholeiitic flood basalt from the Deccan province.

10.6 Chemical composition of the erupted magmas

10.6.1 Major elements

SiO_2, TiO_2, Al_2O_3, Fe_2O_3, FeO, MnO, MgO, CaO, Na_2O, K_2O and P_2O_5 can all be considered to be major elements in the description of the geochemistry of continental flood basalt suites. Tables 10.2−4 present major element data for basalts from several different provinces and also for suites of more evolved lavas from the Paraná province, Brazil. In Table 10.3 data are also included for normal and enriched MORB compositions and for a typical oceanic-island tholeiite, for comparative purposes. From these tables it is clear that tholeiitic flood basalts are closely similar to oceanic tholeiites (MORB and OIB) in terms of their major element chemistry.

Figure 10.9 Photomicrograph illustrating the petrographic characteristics of a tholeiitic flood basalt from the Deccan province, India. The rock contains intergrowths of augite, plagioclase and rare olivine (not shown) set in glassy matrix (dark) (×40, ordinary light).

The major element chemistry of CFBs has been the subject of several extensive recent discussions (Cox 1980, Basaltic Volcanism Study Project 1981, Cox & Clifford 1982). Now that a much wider body of data is available it is apparent that the bias towards Mg-poor tholeiitic basalt compositions is much less marked than Kuno (1969) supposed. In terms of a K_2O versus SiO_2 diagram (Fig. 10.10), CFBs span the range from low-K tholeiites comparable to MORB to mildly alkalic basalts, although the majority are sub-alkaline tholeiites with higher K_2O contents than normal MORB. A plot of Alkali Index (A.I.) versus Al_2O_3 (Fig. 10.11) clearly establishes their tholeiitic characteristics.

The marked scatter in variation diagrams such as the K_2O-SiO_2 plot (Fig 10.10) is a natural consequence of the effects of polybaric crystal fractionation combined with source heterogeneity, variable degrees of partial melting and crustal contamination (Cox 1980). Of the three CFB provinces shown (Paraná, Columbia River and Deccan) the Deccan tholeiites show the most restricted range of chemical compositions with distinctively low-K characteristics. However, in parts of the Deccan sequence there is considerable compositional diversity, with the eruption of some of the most alkaline magmas observed in a CFB province (Mahoney *et al.* 1985).

Table 10.2 Major and trace element data for basalts from the Columbia River Province (data from Basaltic Volcanism Study Project 1981, Table 1.2.3.4, p. 82.

	1	2	3	4	5	6	7	8	9	10
%										
SiO_2	48.35	48.14	48.71	53.92	55.40	50.17	52.59	50.84	49.98	47.91
TiO_2	1.57	2.71	0.96	1.83	1.96	3.15	2.80	1.55	3.49	3.26
Al_2O_3	15.49	16.19	16.91	14.23	13.67	13.23	13.09	14.71	12.64	13.23
Fe_2O_3	3.26	3.06	2.44	2.27	4.98	3.52	7.25	1.60	2.30	1.84
FeO	8.05	9.56	7.77	9.10	7.16	10.71	5.88	8.75	12.67	12.49
MnO	0.17	0.18	0.20	0.20	0.23	0.22	0.23	0.19	0.22	0.24
MgO	7.03	5.39	7.86	4.25	3.34	4.41	2.67	6.99	4.27	5.80
CaO	9.92	8.66	11.06	8.42	6.95	8.20	6.01	10.48	8.35	9.99
Na_2O	2.76	3.30	2.48	2.92	2.41	2.85	3.18	2.30	2.46	2.38
K_2O	0.51	0.90	0.28	1.35	1.81	1.26	2.52	0.66	1.33	0.90
P_2O_5	0.24	0.36	0.18	0.31	0.32	0.67	0.85	0.25	0.56	0.78
H_2O	1.52	0.98	0.40	0.83	1.33	0.85	2.56	0.53	0.67	0.50
CO_2	0.05	0.11	0.04	0.04	0.05	0.05	0.08	0.06	0.04	0.08
ppm										
La	7.03	16.7	6.77	21.9	22.2	26.6	42.7	16.2	32.8	36.2
Ce	17.0	43.9	16.0	48.0	45.6	60.1	90.9	35.3	70.0	82.6
Sm	3.72	7.14	2.72	6.00	5.76	8.58	10.0	4.64	9.36	10.7
Eu	1.37	2.27	1.03	1.86	1.63	2.55	4.10	1.54	2.74	3.23
Tb	0.72	1.40	0.59	1.07	1.01	1.44	1.63	0.88	1.68	1.98
Yb	2.60	3.15	2.20	3.45	2.90	4.05	4.26	2.70	4.55	5.05
Lu	0.38	0.47	0.33	0.51	0.43	0.62	0.65	0.40	0.70	0.74
Y	25	34	20	34	34	41	45	26	49	52
Rb	5.9	20	2.3	31	49	33	47	13	33	13
Cs	—	0.34	—	0.84	1.2	0.4	0.53	—	0.52	—
Sr	274	383	338	316	324	301	263	234	242	250
Ba	280	330	180	360	550	600	3000	260	535	560
Hf	2.35	5.79	1.67	4.90	4.42	5.37	11.4	3.64	7.00	6.99
Th	0.57	2.02	0.36	4.25	5.29	4.18	6.73	2.64	5.77	1.94
Pb	5	5	4	10	13	9	5	8	2	12
Ga	19	26	18	22	22	24	21	20	26	24
Sc	40.1	27.0	37.0	34.4	24.6	36.7	26.5	35.0	31.2	41.0
Cr	188	88.7	156	40.3	5	34.0	2.0	118	17.9	206
Ni	77	150	—	—	—	—	—	—	—	—

As the range of SiO_2 contents is frequently restricted, MgO should be used instead as an index of differentiation. Figure 10.12 shows the variation of CaO versus MgO for basalts from the Paraná, Columbia River, Deccan and Karoo provinces. The Columbia River and Paraná data define good positive correlations suggestive of crystal–liquid control by Ca-rich clinopyroxene and plagioclase either during partial melting or crystal fractionation (Cox 1980). The Deccan and Karoo data show considerably greater scatter, although following broadly similar trends.

Figure 10.13 is a schematic illustration of the variation of SiO_2, Al_2O_3, FeO, CaO and MgO during fractionation of olivine–clinopyroxene –plagioclase assemblages from a primary picritic magma (Cox 1980). Fractionation of this common phenocryst assemblage has a very marked buffering effect on SiO_2 and Al_2O_3 and strongly reduces the rate at which MgO is depleted. Only FeO shows an accelerated increase at the onset of clinopyroxene plus plagioclase fractionation (the characteristic

Table 10.3 Major and trace element data for continental flood basalts, compared to a typical oceanic-island tholeiite and normal and enriched MORB.

	Snake River Plain		Deccan	Paraná	West Greenland picrite	Oceanic-island tholeiite	Normal MORB	Enriched MORB
	1	2	3	4	5	6	7	8
%								
SiO_2	46.18	45.89	50.56	50.75	44.20	50.36	50.40	51.18
TiO_2	2.06	3.33	2.57	3.95	0.88	3.62	1.36	1.69
Al_2O_3	14.47	14.63	13.83	13.51	8.34	13.41	15.19	16.01
Fe_2O_3	13.52	16.46	13.79	14.24	12.85	13.63	10.01[a]	9.40
MnO	0.19	0.21	0.17	0.19	0.18	0.18	0.18	0.16
MgO	9.99	6.46	5.12	4.21	25.36	5.52	8.96	6.90
CaO	9.68	9.37	9.62	8.45	8.30	9.60	11.43	11.49
Na_2O	2.63	2.84	2.65	2.80	0.43	2.80	2.30	2.74
K_2O	0.61	0.65	0.93	1.58	0.36	0.77	0.09	0.43
P_2O_5	0.44	0.69	0.22	0.66	0.07	0.42	0.14	0.15
ppm								
Ba	298	464	239	653	31	191	<20	86
Be	0.7	1.1	0.7	1.6	0.1	1.1	—	—
Cr	256	107	44	20	970	81	346	225
Cu	59	50	202	74	60	98	—	—
Ga	20	24	24	25	10	22	—	—
Hf	3.87	7.18	4.49	7.22	1.16	5.95	—	—
Nb	15.1	25.5	15.9	37	1.5	21.5	2.1	8.6
Ni	193	44	44	43	1090	78	177	132
Pb	5	3	—	6	2	~2	—	—
Rb	13.7	10.4	15	44	12.3	15.4	2.3	10.3
Sr	285	370	219	732	76	395	98	155
Ta	0.93	1.68	1.39	1.88	—	1.5	—	—
Th	1.78	1.62	2.12	5.33	0.58	1.64	—	—
Y	31	48	50	42	12.5	42	37	39
Zn	97	131	149	127	64	119	—	—
Zr	167	295	203	398	39	227	97	121
La	18.3	32.3	19.3	46	5.55	24	2.95	6.92
Ce	41.2	74	43.0	100	—	53	12.0	17.8
Pr	4.8	8.4	5.2	11.0	—	6.0	—	—
Nd	23	39	27.6	51	6.55	35.1	9.9	13.6
Sm	5.6	9.4	7.6	10.9	2.1	8.9	3.91	4.64
Eu	1.94	3.35	2.47	3.5	0.73	2.98	1.41	1.55
Gd	5.7	9.6	8.4	9.9	2.3	9.1	6.4	6.0
Dy	5.26	8.28	8.18	7.61	2.32	7.58	5.6	—
Ho	1.05	1.60	1.55	1.42	0.46	1.41	—	—
Er	3.08	4.44	4.20	3.79	1.29	3.64	—	—
Yb	2.78	4.06	3.63	3.07	1.15	3.04	3.61	3.46
Lu	0.42	0.61	0.53	0.45	0.17	0.42	0.50	0.46

[a] All Fe as FeO. *Data sources*: 1–5, Thompson *et al.* (1983); 6, Thompson *et al.* (1984); 7, 8, Humphris *et al.* (1985).

Table 10.4   Major and trace element data for the (a) HPT and (b) LPT series volcanics from the Paraná province (data from Bellieni *et al.* (1986).

(a)

%								
SiO_2	50.82	52.00	54.00	59.17	65.97	68.58	67.46	70.45
TiO_2	2.79	3.29	2.93	1.88	1.35	1.16	1.52	1.31
Al_2O_3	14.15	13.78	13.89	12.82	13.59	13.10	12.60	12.04
FeO_T	13.19	13.67	12.67	11.87	6.35	5.64	6.61	5.79
MnO	0.20	0.20	0.19	0.21	0.14	0.12	0.15	0.13
MgO	4.81	3.69	2.96	1.53	1.35	0.97	1.09	0.65
CaO	9.40	8.40	7.60	5.46	3.02	2.42	1.88	1.19
Na_2O	2.70	2.86	3.05	3.34	3.53	3.34	3.36	3.14
K_2O	1.21	1.52	1.97	2.95	4.25	4.30	4.81	4.87
P_2O_5	0.43	0.59	0.74	0.87	0.45	0.37	0.52	0.43
ppm								
Cr	96	38	30	8	7	3	5	3
Ni	58	28	21	3	5	6	5	6
Ba	496	624	719	910	1096	1178	1193	1283
Rb	27	33	44	86	114	139	130	158
Sr	404	472	453	406	344	318	302	279
La	42	41	47	84	88	91	99	103
Ce	70	92	106	163	180	178	203	201
Zr	205	268	312	596	642	573	716	639
Y	32	37	41	66	69	75	77	84

(b)

%								
SiO_2	52.50	53.10	54.20	54.94	56.73	58.82	61.01	67.04	71.48
TiO_2	0.96	1.27	1.53	1.60	1.70	1.70	1.43	1.03	0.74
Al_2O_3	15.43	15.26	14.74	14.49	14.02	13.29	14.01	13.57	12.54
FeO_T	9.33	11.03	11.70	12.24	12.17	11.78	9.57	5.99	4.78
MnO	0.16	0.17	0.17	0.19	0.18	0.17	0.13	0.11	0.09
MgO	7.63	5.67	4.54	3.68	2.89	2.57	2.29	1.57	0.86
CaO	10.87	9.46	8.77	8.21	6.90	6.46	5.42	3.33	2.05
Na_2O	2.14	2.55	2.71	2.88	2.98	2.93	2.92	3.25	2.73
K_2O	0.82	1.27	1.41	1.54	2.13	2.01	2.93	3.84	4.53
P_2O_5	0.16	0.22	0.23	0.23	0.30	0.27	0.29	0.27	0.20
ppm									
Cr	314	90	51	32	27	31	29	10	6
Ni	123	65	46	37	27	19	22	8	6
Ba	280	413	417	404	470	594	579	652	687
Rb	27	38	47	56	82	102	125	169	200
Sr	232	239	240	209	185	202	170	143	100
La	14	23	26	26	31	38	41	48	61
Ce	35	53	59	59	68	78	79	97	121
Zr	94	144	156	162	181	192	260	267	317
Y	19	26	30	33	36	35	41	45	75

Recalculated to 100% anhydrous.

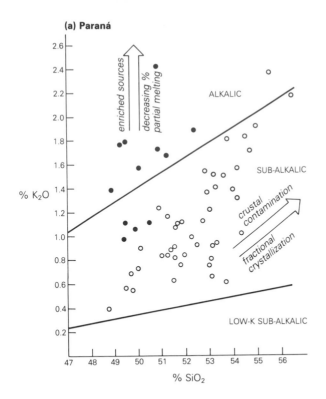

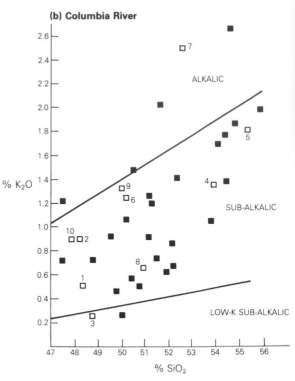

Figure 10.10 Plot of wt.% K_2O versus wt.% SiO_2 for basalts from the (a) Paraná, (b) Columbia River and (c) Deccan CFB provinces. Field boundaries are from Middlemost (1975). Data sources: Paraná, Mantovani et al. (1985) (●, HPT) and Fodor et al. (1985) (○, LPT); Columbia River, Basaltic Volcanism Study Project (1981); Deccan, Cox & Hawkesworth (1985). Numbered Columbia River samples refer to those in Table 10.2.

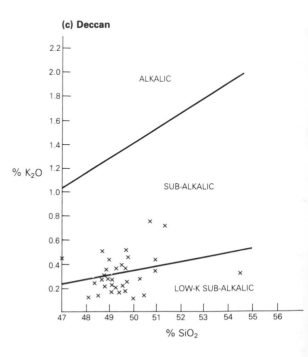

tholeiitic differentiation trend; Ch.1), but even this decreases as soon as Fe–Ti oxides start to crystallize.

Many flood basalts have relatively low Mg' values [Mg/(Mg + ΣFe) <0.7] and therefore if they are primary they must have been generated from mantle more Fe-rich than normal MORB–OIB source mantle. However, their typically low Ni contents combined with the observed CaO–MgO correlations would tend to argue in favour of low-pressure fractional crystallization, dominated by plagioclase and clinopyroxene, in producing the range of erupted magma compositions (Cox 1980). The rare picrites erupted in the Deccan and Karoo

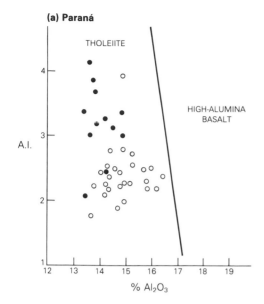

(a) Paraná

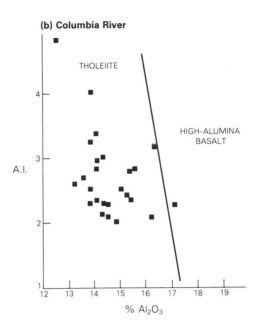

(b) Columbia River

provinces may be appropriate candidates for true primary magma compositions which initially evolve by olivine fractionation alone.

Recent geochemical studies of the Paraná flood-basalt province (Bellieni *et al.* 1984,1986; Mantovani *et al.*1985; Fodor 1987) have revealed the existence of two types of basalt; a low-P_2O_5–TiO_2 (LPT) basalt, dominant in the southern part of the province, and a high-P_2O_5–TiO_2 (HPT) basalt, dominant in the north (Table 10.4). Both types appear to have undergone low-pressure fractional crystallization, in some instances combined with crustal contamination, to produce more SiO_2-rich magmas. For similar MgO values the HPT basalts are characterized by higher contents of FeO_T, P_2O_5, Sr, La, Ba, Ce and Zr than the LPT types. Mantovani *et al.* (1985) have related these different basalt compositions to different mantle sources, although when Sr isotopic data are considered (Section 10.6.3) it is apparent that some lavas of the LPT series may have undergone appreciable crustal contamination (Petrini *et al.* 1987).

Cox *et al.* (1967) had previously established a similar provinciality in Karoo flood basalt types, with basalts from the northern area being enriched in K, P, Ti and incompatible elements relative to those from the southern parts of the province (Fig. 10.1). Similarly, in the Etendeka province of

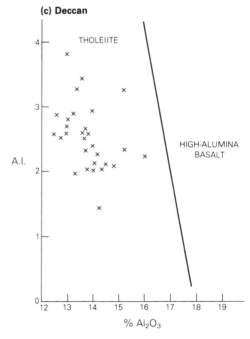

(c) Deccan

Figure 10.11   Plot of Alkali Index (A.I.) versus wt.% Al_2O_3 to illustrate the tholeiitic characteristics of flood basalts from the (a) Paraná, (b) Columbia River and (c) Deccan CFB provinces: Tholeiite/high-alumina basalt field boundary from Middlemost (1975). Data sources as in Figure 10.10. A.I. = $[Na_2O + K_2O]/[SiO_2 - 43] \times 0.17]$.

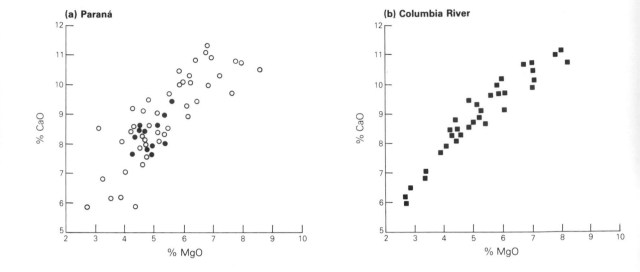

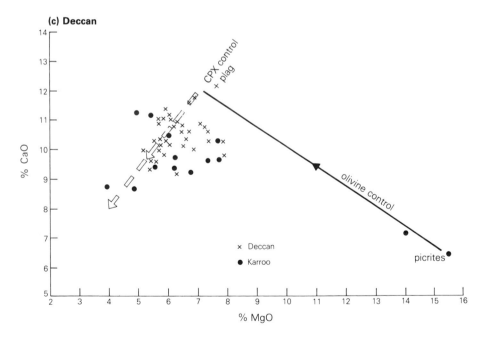

Figure 10.12   Plot of wt.% CaO versus wt.% MgO for basaltic rocks from the (a) Paraná, (b) Columbia River and (c) Deccan CFB provinces: (c) also shows data for Karoo basalts (Basaltic Volcanism Study Project 1981) and a dashed line representing the fractionation trend of Karoo dolerite dykes (Cox 1980). Data sources as in Fig. 10.10.

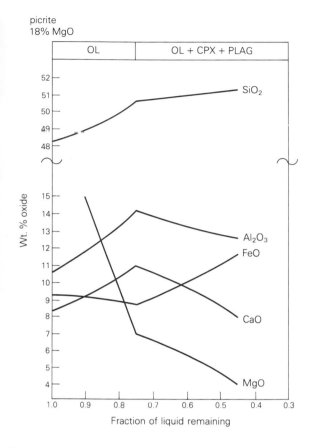

Figure 10.13 Evolution of various oxides in residual liquids during fractional crystallization of olivine + clinopyroxene + plagioclase assemblages from a primary picritic magma (after Cox 1980, Fig. 5, p. 640).

Namibia, high-TiO$_2$ basalts occur to the north of latitude 19.5°S and low-TiO$_2$ basalts to the south (Duncan 1987, Marsh 1987). The Kirkpatrick basalts of Antarctica also contain high-TiO$_2$ and low-TiO$_2$ types, with the high-TiO$_2$ lavas capping the sequence at all localities (Siders & Elliot 1985). In the Karoo province the boundary between the two basalt types typically occurs over a zone some 50–100 km wide in which both magma types are present as dykes and flows. This is also apparent in the Paraná province where there is a broad transitional central zone between the low-TiO$_2$ and high-TiO$_2$ provinces in which both types occur (Petrini *et al.* 1987). The well defined north–south divide across the entire Gondwanaland province

(Fig. 10.1) strongly suggests that the high-TiO$_2$ and low-TiO$_2$ characteristics reflect different mantle source compositions (e.g. Cox 1983, Bellieni *et al.* 1984, Mantovani *et al.* 1985). However, Fodor (1987) favours a model for the Paraná in which low-TiO$_2$ characteristics are the consequence of larger degrees of partial melting of a relatively homogeneous mantle source across the entire province, coupled with crustal contamination.

Figure 10.14a is a plot of TiO$_2$ versus SiO$_2$ for the Paraná province, which clearly differentiates between high-TiO$_2$ and low-TiO$_2$ types at the basaltic end of the spectrum. However, for the acidic rocks (rhyodacites and rhyolites) the distinction is less marked. Figures 10.14b & c are comparable plots for basalts from the Columbia River and Deccan provinces, both of which show a continuous range of TiO$_2$ concentrations rather than the bimodal grouping of the Paraná basalts. The extremely wide range of TiO$_2$ contents (1–4%) at constant SiO$_2$ must reflect the partial melting of heterogeneous mantle sources combined with the effects of crustal contamination.

For the Paraná province, fractional crystallization of LPT and HPT basalts can account for the occurrence of rhyolitic lavas with low- and high-TiO$_2$ characteristics (Fig. 10.14a). However, because of the marked silica gap between 54 and 63% SiO$_2$, Bellieni *et al.* (1986) favour an origin of the acid magmas by partial melting of underplated basic rocks in the base of the crust during the lithospheric thinning associated with the opening of the South Atlantic Ocean. A similar model was proposed by Cleverly *et al.* (1984) for rhyolitic lavas associated with the Karoo flood basalts of southern Africa.

As considered in Section 10.1, the role of crustal contamination in the petrogenesis of continental flood basalts is a fundamental problem. However, the effects of such contamination are rarely visible in terms of the major element geochemistry of the lavas. Nevertheless, for granitic composition contaminants, contaminated basalts are likely to be enriched in SiO$_2$ and K$_2$O and depleted in FeO$_T$ and TiO$_2$ relative to uncontaminated basalts at a similar stage of differentiation (Cox & Hawkesworth 1984, 1985).

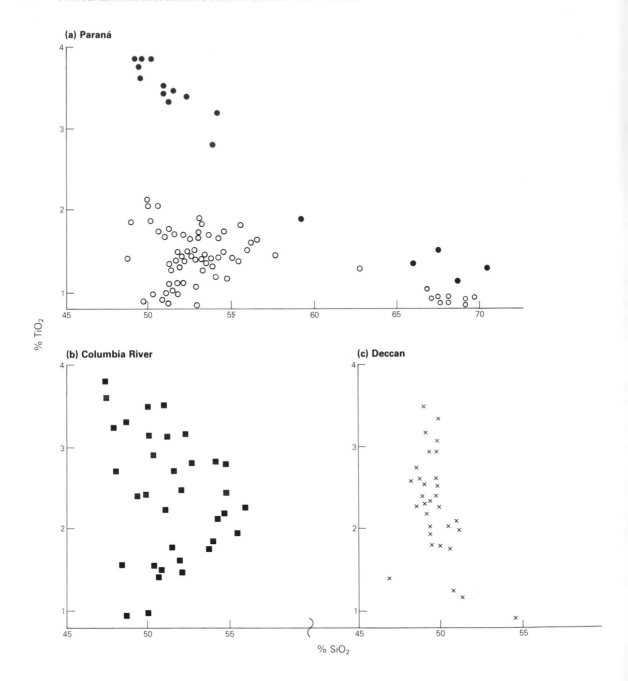

Figure 10.14   Plot of wt.% TiO$_2$ versus wt.% SiO$_2$ for basalts from the (a) Paraná, (b) Columbia River and (c) Deccan CFB provinces. Data sources as in Figure 10.10, with additional Paraná data from Bellieni *et al.* (1984, 1986).

10.6.2   Trace elements

Trace element data for continental flood basalts from different provinces and also for more evolved lavas from the Paraná basin, Brazil, are presented in Tables 10.2–4. Data are also included for normal and enriched MORB compositions and for a typical oceanic island tholeiite for comparative purposes.

Studies of the trace element geochemistry of CFBs have tended to focus on the incompatible elements (Thompson *et al.* 1983, 1984), i.e. those elements which are strongly partitioned into the liquid phase during partial melting of a four-phase lherzolite or during fractional crystallization of basaltic magma. In general, most flood basalts are characterized by low concentrations of compatible trace elements such as Ni, which supports the contention that they are not primary magmas (Ch.2) but have undergone olivine fractionation (at least) en route to the surface. Such fractionation will tend to increase the incompatible element concentrations in the basaltic magmas relative to those of the more MgO-rich primary magmas.

Figure 10.15 is a mantle-normalized trace element variation diagram (spiderdiagram) similar to that used by Sun (1980), in which the elements are plotted in order of decreasing incompatibility from the left to right in a four-phase lherzolite undergoing partial fusion. The data are normalized to chondritic abundances, except for Rb, K and P which are normalized to primitive terrestrial mantle values (Thompson *et al.* 1984). Figure 10.16 is a comparable plot for normal and enriched MORB compositions and a typical ocean-island tholeiite for comparison with the CFB data. It is clear that tholeiitic flood basalts from the Snake River Plain, Paraná and Deccan provinces are enriched in the whole spectrum of incompatible elements relative to normal MORB. However, they show close similarities to enriched MORB and oceanic-island tholeiites. Thompson *et al.* (1983), in a detailed study of CFB spiderdiagram patterns, concluded that many display a marked trough at Nb–Ta similar to, but not as strong as, that displayed by subduction-related magmas (Ch.6). This is most clearly evident in the pattern for the West Green-

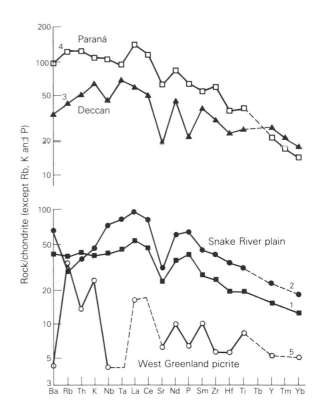

Figure 10.15 Mantle-normalized trace element diagrams for continental flood basalts. Numbered samples are those from Table 10.3, and dashed lines connect points where intermediate data points are missing. Normalization factors from Thompson (1982).

land picrite sample (5 in Table 10.3 and Fig. 10.15). Such a Nb–Ta trough is in marked contrast to the characteristic spiderdiagram pattern of continental and oceanic alkali basalts (Chs. 9 & 11), which normally display a peak at Nb–Ta. Its occurrence in many CFB spiderdiagram patterns could reflect the existence of a residual Nb–Ta-bearing phase during the partial melting process (cf. Ch. 6). However, it is much more likely to be the consequence of crustal contamination (Cox & Hawkesworth 1985). All of the spiderdiagram patterns shown in Figure 10.15 show a distinctive trough at Sr, which is probably a consequence of low-pressure plagioclase fractionation.

Figure 10.17 is a MORB-normalized trace element variation diagram similar to that used in Ch. 6

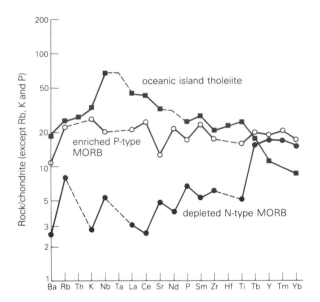

Figure 10.16 Mantle-normalized trace element diagram for depleted and enriched MORB compositions and a typical oceanic-island tholeiite. Dashed lines connect points where intermediate data points are missing. MORB data from Humphris *et al.* (1985); oceanic-island tholeiite data from Basaltic Volcanism Study Project (1981); normalization factors from Thompson (1982).

(Fig. 6.38) to highlight the nature of subduction-zone components in the petrogenesis of island-arc basalts (Pearce 1983). The elements plotted all behave incompatibly during most partial melting and fractional crystallization events, with the exception of Sr which may be concentrated in plagioclase, Y and Yb in garnet and Ti in magnetite. The elements are ordered in the diagram in terms of their mobility in an aqueous fluid phase and their relative incompatibility. Sr, K, Rb and Ba are classed as mobile and plot to the left of the pattern, whereas the group Th−Yb are generally immobile. The elements are arranged so that the incompatibilities of both mobile and immobile elements increase from the outside to the centre of the pattern. In terms of Figure 10.17 it is clear that CFBs are enriched in the whole range of incompatible elements, other than Y and Yb, relative to N-type MORB. Y appears to be little varied by the effects of source heterogeneity and crustal contamination,

and is therefore useful as a fractionation index. This is important because in the study of CFB volcanic suites it is essential to be able to see through the effects of crystal fractionation so that the relative roles of source heterogeneity and crustal contamination in the petrogenesis of the magmas may be evaluated. Assuming that all tholeiitic basalts, both continental and oceanic, are generated by similar degrees of partial melting, then in terms of Figure 10.17 CFBs cannot be derived from normal MORB source mantle. Instead, their spider-diagram patterns are very similar to those of oceanic-island tholeiites and alkali basalts (Ch.9), and Thompson *et al.* (1983, 1984) contend that all CFB tholeiites are simply magmas geochemically similar to oceanic-island tholeiites variably contaminated by the continental crust.

Figure 10.18 is a similar diagram to Figure

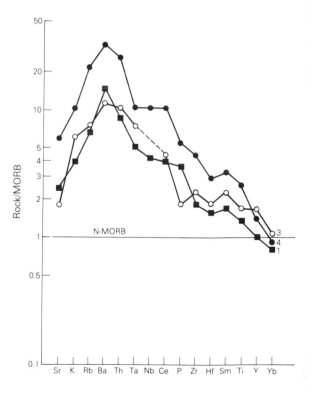

Figure 10.17 MORB-normalized trace element variation diagram (Pearce 1983) for continental flood basalts. Numbered samples are those from Table 10.3, and the dashed line connects points where the intermediate data point is missing.

10.17, in which the flood-basalt data are normalized to the trace element abundances in a typical ocean-island tholeiite (Table 10.3) instead of MORB. This diagram emphasizes the similarity between CFBs and ocean-island tholeiites but also displays a distinctive spiked pattern which may be a consequence of crustal contamination. Fusible crustal rock types are generally much richer than OIB in Ba, Rb, Th, K and light REE, but have similar or lower contents of Nb, Ta, P, Zr, Hf, Y and middle REE (Thompson *et al.* 1984). Thus crustal contamination of a magma with the trace element characteristics of an ocean-island tholeiite could produce the spiked pattern shown.

Continental tholeiites tend to have similar abundances of the high field strength elements (Zr, Hf, P and Ti) to MORB and OIT but, in general, have much higher light-REE contents. CFBs with flat or light-REE depleted REE patterns, comparable to MORB, are very rare. Figure 10.19 shows chondrite-normalized REE patterns for basalts

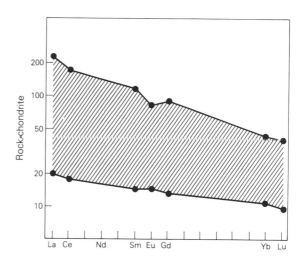

Figure 10.19 Range of chondrite-normalized REE patterns for Columbia River basalts (after Basaltic Volcanism Study Project 1981, Fig. 1.2.3.4, p.86).

from the Columbia River province, all of which are variably light-REE enriched, sometimes with small negative Eu anomalies suggestive of plagioclase fractionation.

The differences between CFBs and oceanic basalts (MORB + OIB) are most evident in the abundances of mobile incompatible elements (K, Rb, Sr, Ba and Th; see Figs 10.17 & 18) which, combined with their strong light-REE enrichment, may either be a consequence of crustal contamination or their derivation from enriched subcontinental mantle sources. Thompson *et al.* (1984) have suggested that the ratio La/Nb might be a useful index of crustal contamination in magmas. OIB, continental alkali basalts and kimberlites all have La/Nb <1, whereas CFBs range from 0.5 to 7, suggesting variable degrees of contamination. On a plot of Th/Yb versus Ta/Yb (Fig. 10.20), which was used to identify the role of subduction-related fluid components in island-arc magma genesis in Chapter 6, Snake River Plain tholeiites plot in the enriched mantle source region, suggesting the involvement of subcontinental lithosphere or OIB source mantle in their petrogenesis.

Figure 10.21 is a plot of Y/Nb versus Zr/Nb similar to that used in Chapter 5 to consider the influence of OIB plumes on MORB geochemistry.

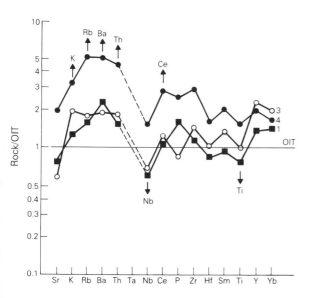

Figure 10.18 Oceanic island tholeiite normalized trace element variation diagram for continental flood basalts. Numbered samples are those from Table 10.3, and dashed lines connect data points where intermediate points are missing. Normalization factors: Sr, 368; K, 0.48%; Rb, 8.4; Ba, 125; Th, 1.14; Ta, ND; Nb, 24; Ce, 36.5; P, 0.27%; Zr, 141; Hf, 4.47; Sm, 5.55; Ti, 2.61%; Y, 22; Yb, 1.90.

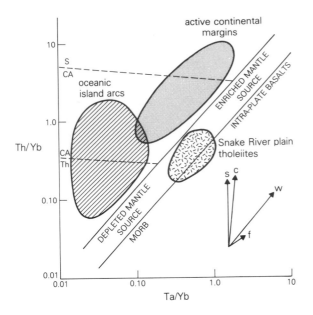

Figure 10.20   Plot of Th/Yb versus Ta/Yb, which shows that the Snake River Plain flood tholeiites plot in the enriched mantle source field and do not show high Th/Yb ratios which might be indicative of crustal contamination. For detailed explanation of this diagram see the caption to Figure 7.26.

Basalts from the Paraná basin plot on a mixing line between a Tristan da Cunha OIB source component and a depleted MORB source component, plotting close to the field of enriched South Atlantic MORB and Tristan da Cunha alkali basalts. This suggests that the Tristan da Cunha mantle plume or hot spot may have been supplying source components to the melting zone beneath the Paraná basin before continental fragmentation began (Fodor 1987).

When trace element data for basalts from the Karoo province are plotted on many of the geochemical discriminant diagrams which have been used to infer palaeotectonic setting (Ch.2), many from the southern province plot within the field of subduction-related basalts (Duncan 1987). In contrast, data from the northern province plot largely in the within-plate fields. This may support the involvement of subduction-modified source components in the petrogenesis of the low-TiO$_2$ flood basalts from the Karoo, and possibly also the other Gondwanaland flood-basalt provinces (Paraná, Etendeka and Antarctica). However, we must be cautious in taking this interpretation of the discri-

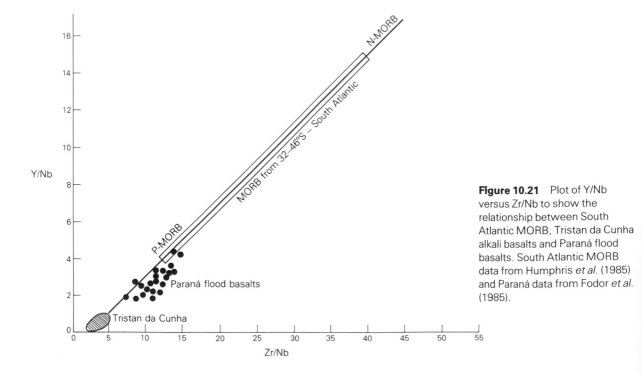

Figure 10.21   Plot of Y/Nb versus Zr/Nb to show the relationship between South Atlantic MORB, Tristan da Cunha alkali basalts and Paraná flood basalts. South Atlantic MORB data from Humphris *et al.* (1985) and Paraná data from Fodor *et al.* (1985).

ɯɯɯɯɯ diagrams too far, as crustal contamination effects may invalidate their usefulness.

In summary, the trace element geochemistry of CFBs appears to suggest that they are derived from enriched mantle sources, but that they have also undergone variable degrees of contamination by rocks of the continental crust.

10.6.3   Radiogenic isotopes

Nd–Sr data

Variations in the isotopic composition of oceanic basalts (MORB and OIB, Chs 5 & 9) clearly indicate that the upper mantle is isotopically heterogeneous. Such variations reflect its time-integrated evolution over the past 4.5 Ga in response to depletion and enrichment events involving the migration of trace element enriched fluids and partial melts, and also the recycling of crustal materials at subduction zones (Ch. 6).

In Section 10.2 we considered that the subcontinental lithosphere may provide a significant source component for CFB magmatism. While the isotopic and chemical composition of this reservoir still remains a subject for considerable speculation, it should potentially exhibit the same range of variation as oceanic basalts. Random samples of the subcontinental upper mantle are rapidly transported to the surface in volcanic diatremes and kimberlites (Ch. 12), and these confirm the existence of extensive isotopic heterogeneities, which have developed in response to perturbations in the Rb/Sr and Sm/Nd systems, exceeding the combined range of MORB and OIB (Allègre *et al.* 1982, Cohen *et al.* 1984; Kramers *et al.* 1983, see also Fig. 10.22a).

On a Nd–Sr isotope diagram (Figs 10.22b–d), some continental flood basalts plot within the mantle array as defined by uncontaminated oceanic basalts, whereas others plot outside this field. This has led to conflicting viewpoints about their origins, particularly for volcanic suites with radiogenic Sr and non-radiogenic Nd which plot to the right of the MORB–OIB array. Many authors have invoked bulk crustal assimilation to explain the high $^{87}Sr/^{86}Sr$ ratios (e.g. Thompson *et al.* 1984), but

Menzies (1983) and Hawkesworth *et al.* (1983, 1984) have shown that such ratios could develop in a comparatively short time in high-Rb/Sr mantle xenoliths and, therefore, that such ratios could be a characteristic of the mantle source of the magmas. A fundamental problem in petrogenetic studies of CFB magmas is thus to find ways of distinguishing between the relatives roles of crustal contamination and enriched mantle sources in the production of their observed geochemical characteristics.

As more data become available for crustal rocks it is apparent that the isotopic compositions of mantle and crustal rocks overlap significantly (Leeman & Hawkesworth 1986). For example, Rb-depleted Archaean granulites may have present-day $^{87}Sr/^{86}Sr$ ratios less than 0.703, comparable to MORB, whereas some garnet peridotites have measured values greater than 0.75, in excess of average upper crustal ratios. However, most crustal rocks have high Rb/Sr and Nd/Sm and thus are characterized, on average, by elevated ε_{Sr} and low ε_{Nd} relative to bulk earth values, and diverge progressively from bulk earth as they age (Ch. 2).

Figure 10.22b presents Nd-Sr isotopic data for tholeiitic flood basalts from the Deccan Traps province of India, in the Mahabaleshwar area. Cox & Hawkesworth (1985) have defined four formations within the basalts of this area which, from the base upwards, are Bushe, Poladpur, Ambenali and Mahabaleshwar. Intraformational boundaries are generally sharp and defined by breaks in initial $^{87}Sr/^{86}Sr$. The Nd-Sr data for these formations define two distinct trends in Figure 10.22b. Data for the Bushe formation are not plotted but would extend the Poladpur field to even higher values of ε_{Sr}. The Ambenali and Mahabaleshwar data plot within the MORB-OIB array and could therefore simply represent uncontaminated magmas derived from isotopically heterogeneous sources, with the Mahabaleshwar lavas with the lowest ε_{Nd} values representing partial melts of the most enriched source component. In contrast, the Poladpur samples are displaced to the high-ε_{Sr} side of the mantle array. Both the Mahabaleshwar and Poladpur fields appear to extrapolate back to that of the Ambenali lavas, and Cox & Hawkesworth (1985) consider that these linear trends reflect some form of mixing process. The Poladpur trend could represent contamination of mantle-derived magmas with Ambenali-type isotopic characteristics with an ancient upper crustal component (Mahoney *et al.* 1982, Cox & Hawkesworth 1985), while the Mahabaleshwar trend could reflect partial melting of a heterogeneous and variably enriched mantle source. However, it might also reflect contamination of an Ambenali-type magma with a different, lower crustal, component from that inferred for the Poladpur trend.

Although the Deccan Traps are a dominantly tholeiitic flood-basalt province, significant volumes of alkalic igneous rocks occur in the upper part of the sequence in the northern part of the province (Krishnamurthy & Cox 1980). Isotopic compositions for the alkalic and tholeiitic volcanics overlap almost completely (Mahoney *et al.* 1985) and span the same range defined by the Ambenali and Poladpur rocks. Additionally, there is no apparent correlation between isotopic composition and degree of fractionation. Therefore, both alkalic and tholeiitic parent magmas must have originated by partial melting of isotopically similar source regions and then become variably contaminated by the low−ε_{Nd} end−member. Mahoney *et al.* (1985) suggest that the mantle source of these Deccan magmas may have recently been metasomatically enriched with the formation of phlogopite-rich veins. Upon partial melting, unenriched regions of the source produced tholeiitic magmas, whereas the phlogopite-rich regions produced the alkalic magmas.

Data for the Columbia River flood basalt province (Fig. 10.22c), define an apparently well constrained mixing array between a component with MORB-source characteristics and an upper crustal component (Basaltic Volcanism Study Project 1981). Also plotted on this diagram are data for tholeiitic basalts from the Snake River Plain which lie on the same trend and could therefore be interpreted as being quite strongly contaminated. However, caution must be exercised in the interpretation of these data in this way because of the tectonic setting of the Columbia River − Snake River Plain province; an ensialic back-arc basin in which the magmas could have a component of

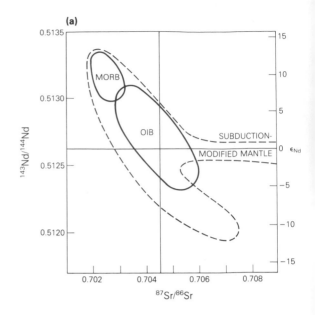

subduction-modified mantle in their source. In this context it is useful to compare Nd−Sr data from the Lesser Antilles (Ch. 6, Fig. 6.40) and the Andes (Ch. 7, Fig. 7.27).

Figure 10.22d shows Nd−Sr isotopic data for lavas from the Paraná province of Brazil and its counterpart in southern Africa, the Etendeka. The HPT and LPT series identified in Section 10.6.1 clearly have different isotopic characteristics, with the HPT lavas plotting at the extreme end of the MORB-OIB array and thus potentially representing uncontaminated partial melts of an enriched mantle source. Volcanic rocks from the LPT series and the Etendeka define an elongate trend, displaced from the mantle array at relatively constant ε_{Nd} in the direction of increasing ε_{Sr}. These trends may indicate the involvement of crustal components in the petrogenesis of the magmas. Figure 10.23 is a plot of $^{87}Sr/^{86}Sr$ initial versus wt.% SiO_2 for the LPT and HPT series of the Paraná. The very strong positive correlation for the LPT lavas provides strong support for crustal contamination models.

Petrini *et al.* (1987) have shown that while low-TiO_2 basalts are the dominant type in the southern part of the Paraná province they also occur infrequently in the central and northern parts. However, in these latter areas they are

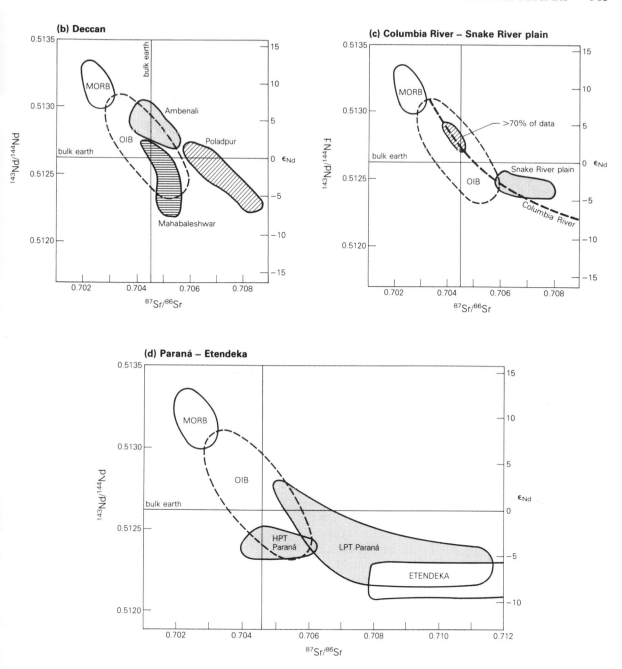

Figure 10.22 Nd—Sr isotope diagrams to show the range of variation displayed by: (a) lherzolite xenoliths from the sub-continental lithosphere (Menzies *et al.* 1984); (b) flood basalts from the Deccan Traps province of India (Cox & Hawkesworth 1985); (c) flood basalts from the Columbia River — Snake River Plain province of the western USA (Basaltic Volcanism Study Project 1981, Menzies *et al.* 1984); (d) HPT and LPT flood basalts form the Paraná province of Brazil and basalts from the Etendeka province of Namibia (Hawkesworth *et al.* 1986, Petrini *et al.* 1987).

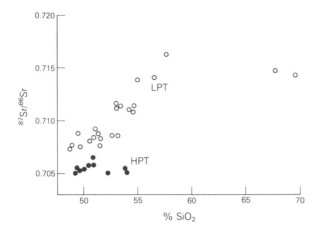

Figure 10.23 Variation of $^{87}Sr/^{86}Sr$ initial versus wt.% SiO_2 for lavas from the LPT series (open circles) and HPT series (closed circles) of the Paraná flood-basalt province of Brazil (data from Mantovani *et al.* 1985).

incompatible element enriched similar to the HPT types and have overlapping Nd–Sr isotopic characteristics. These authors suggest that low-TiO_2 basalts with $^{87}Sr/^{86}Sr$ initial ratios less than 0.706 may be considered to be essentially uncontaminated. If we assume that this is also true for the HPT type (cf. Fodor 1987), then Figure 10.22d suggests that the primary HPT and LPT magmas are derived from isotopically distinct sources (cf. Bellieni *et al.* 1984, Mantovani *et al.* 1985). Additionally, these data suggest that the mantle source in the northern part of the Paraná province may be capable of yielding both HPT and LPT magmas with similar isotopic characteristics. This could be achieved by variable degrees of partial melting, as suggested by Fodor (1987), with the LPT basalts representing greater degrees of melting than the HPT types.

Figures 10.24 and 10.25 are plots of Ba/Yb and Ba/Y respectively versus initial $^{87}Sr/^{86}Sr$ for lavas from the Paraná and Deccan provinces. These may be considered as the trace element analogues of the Nd–Sr isotope diagram and define closely similar trends. The HPT Paraná basalts appear to have strong affinities with those of the Deccan Mahabaleshwar formation and may similarly represent partial melts of an enriched mantle source, possibly

the subcontinental lithosphere. In contrast, the LPT series is similar to the contaminated Poladpur formation.

From the above it is apparent that the TiO_2 content of CFB magmas may be quite a sensitive indicator of crustal contamination. Crustal rocks and their partial melts in general are characterized by very low TiO_2 contents and thus low-TiO_2 magmas may carry an imprint of crustal contamination. However, caution must be exercised, as crystal fractionation of Fe–Ti oxides can rapidly deplete a magma in TiO_2. The corollary of this is that high TiO_2 abundances in basic magmas are most unlikely to result from any form of crustal contamination, and must therefore reflect derivation of the magmas from enriched mantle sources.

Figure 10.26 is a plot of ε_{Nd} versus Ti/Yb ratio for the Deccan and Paraná flood basalts, which shows a striking decrease in both parameters for the Poladpur and Bushe rocks consistent with a crustal contamination model, as also inferred from the Nd–Sr isotope data. In contrast, the Mahabaleshwar basalts are characterized by high Ti/Yb ratios and, with decreasing ε_{Nd}, show progressive enrichment of K, Ti, Zr, Nb, Rb, Ba, P and Sr (Leeman & Hawkesworth 1986). The Paraná HPT lavas have high Ti/Yb ratios compared to the LPT

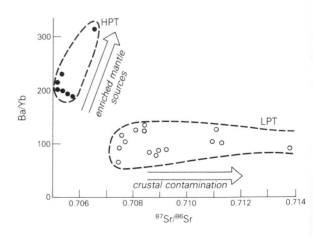

Figure 10.24 Variation of Ba/Yb ratio versus $^{87}Sr/^{86}Sr$ initial for basalts from the HPT and LPT series of the Paraná province, Brazil; symbols as in Figure 10.23 (data from Mantovani *et al.* 1985). Note the analogy with the Ba/Y versus $^{87}Sr/^{86}Sr$ plot for the Deccan province (Fig. 10.25).

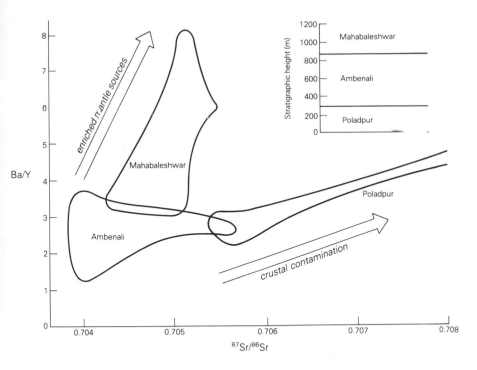

Figure 10.25 Variation of Ba/Y ratio versus ^{87}Sr/^{86}Sr initial for flood basalts from the Ambenali, Mahabaleshwar and Poladpur formations of the Deccan flood basalt province, India (data from Cox & Hawkesworth 1985).

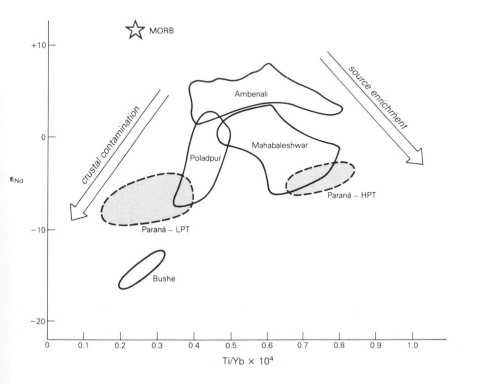

Figure 10.26 Plot of ε_{Nd} versus Ti/Yb ratio for basaltic rocks from the Deccan flood basalt province, compared to LPT and HPT basalts from the Paraná province. Data from Mantovani *et al.* (1985), Hawkesworth *et al.* (1986) and Leeman & Hawkesworth (1986).

types, confirming the contaminated nature of the latter.

Pb isotopes

Figure 10.27 shows the variation of $^{208}Pb/^{204}Pb$ versus $^{206}Pb/^{204}Pb$ for flood basalts from the Paraná (a), Deccan (b) and Columbia River − Snake River Plain − Yellowstone (c) provinces compared to fields for MORB (shaded) and oceanic-island basalts (OIB). In Chapter 9 we considered the existence of a large-scale isotopic anomaly in the upper mantle, the Dupal anomaly, which provides a source component for the basalts of certain oceanic islands (e.g. Tristan da Cunha) at low latitudes south of the Equator. In the South Atlantic the anomaly appears to be restricted to the area between the CFB provinces of Paraná and Etendeka, which are associated with the earliest stages of continental rifting. Dupal OIB have been distinguished as a separate field in Figure 10.27 and, rather interestingly, most CFB data show substantial overlap with it, plotting to the high $^{208}Pb/^{204}Pb$ side of the Northern Hemisphere Reference Line (Hart 1984). Also shown for comparison are the Pb isotopic compositions of separated clinopyroxenes from lherzolite xenoliths derived from the African subcontinental lithosphere (Cohen *et al.* 1984), which also plot within the OIB field.

Thus far we have tended to favour a model for CFB magmatism in which the primary magmas are derived by partial melting of enriched subcontinental lithosphere and then subsequently modified by interaction with the continental crust en route to the surface. How do we then reconcile such a model with the Pb isotopic similarity of CFB basalts to the

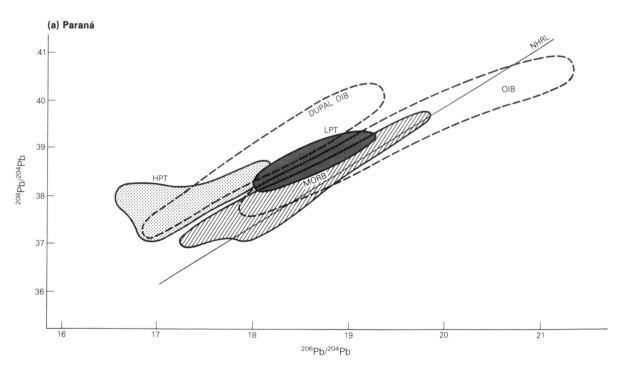

Figure 10.27 Variation of $^{208}Pb/^{204}Pb$ versus $^{206}Pb/^{204}Pb$ for tholeiitic flood basalts from: (a) the Paraná province of Brazil (Hawkesworth *et al.* 1986); (b) the Deccan province of India (Allègre *et al.* 1982); (c) the Columbia River − Snake River Plain − Yellowstone province of the western USA (Church 1985, Doe *et al.* 1982). Shown for comparison are fields for MORB (shaded) and oceanic-island basalts (OIB) from Figure 9.26. NHRL is the Northern Hemisphere Reference Line of Hart (1984). Also shown in (c) are data for separated clinopyroxenes from lherzolite xenoliths (open circles) from the African sub-continental lithosphere (Cohen *et al.* 1984).

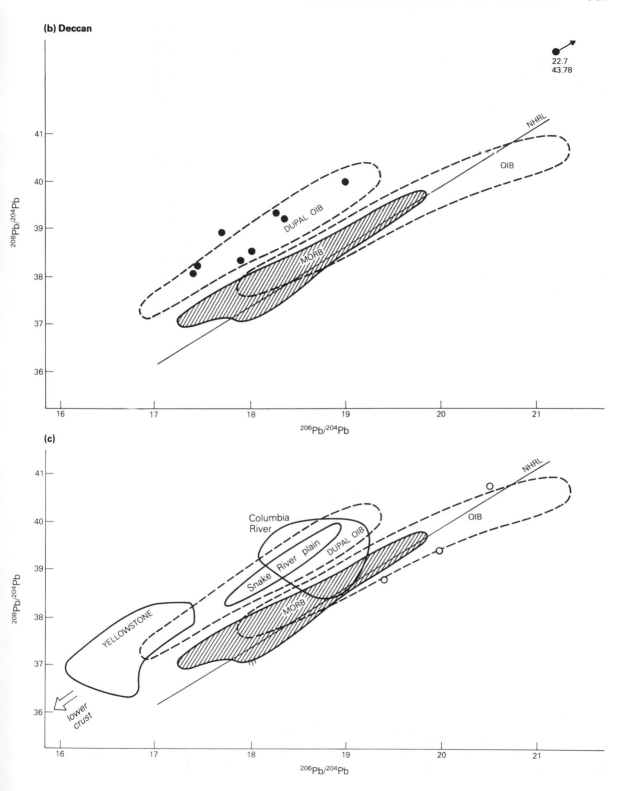

Dupal group of OIB? Does this mean that the Dupal anomaly is actually a shallow level feature of the upper mantle rather than a deep-seated one, or is the similarity between the two groups of basalts merely fortuitous?

Figure 10.28 shows the variation of $^{207}Pb/^{204}Pb$ versus $^{206}Pb/^{204}Pb$ for basalts from the same provinces plotted in Figure 10.27. Again, the majority of flood basalts are displaced to the high side of the NHRL, similar to Dupal OIB. It is interesting to compare this diagram with similar plots in Chapters 6 and 7 (Figs 6.43 & 7.29) for subduction-related magmas to which CFBs also show considerable similarity in terms of their Pb isotopes. In terms of this comparison it is possible that subduction-modified mantle could also provide a source component for CFB magmatism. This may have particular relevance to the Columbia River province, an ensialic back-arc basin.

The presence of Dupal Pb isotope features in both OIB and CFBs and the obvious tectonic link in the South Atlantic between Tristan da Cunha and the Paraná and Etendeka flood basalt provinces suggests that there may actually be a strong link between continental and oceanic magmatism. The model proposed in Chapter 9 was that the Dupal anomaly was caused by a deep-seated upwelling that initiated continental break-up and persisted as an OIB plume throughout the opening of the South Atlantic. However, Hawkesworth et al. (1986) have suggested an alternative model in which, at least for the South Atlantic, the Dupal signature is a relatively shallow level phenomenon which evolved in the continental lithosphere. In this model the continental lithosphere is heated and remobilized during extension and detached from the continental plates, enabling it to contribute to oceanic volcanism as the continents move apart and the ocean basin develops.

In Figures 10.27 and 10.28 the Paraná HPT and

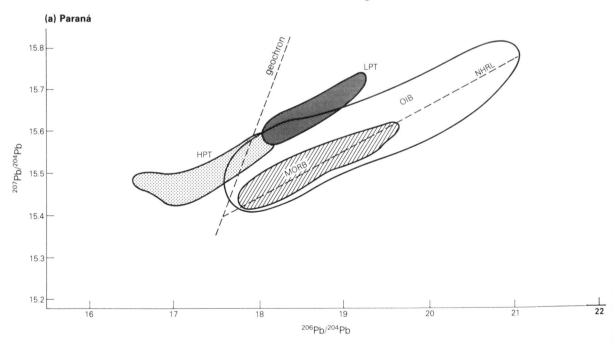

Figure 10.28 Variation of $^{207}Pb/^{204}Pb$ versus $^{206}Pb/^{204}Pb$ for tholeiitic flood basalts from: (a) the Paraná province of Brazil (Hawkesworth et al. 1986); (b) the Deccan province of India (Allègre et al. 1982; (c) the Columbia River – Snake River Plain – Yellowstone province of the western USA (Church 1985, Doe et al. 1982). Shown for comparison are fields for MORB (shaded) and oceanic-island basalts (OIB) from Figure 9.25. NHRL is the Northern Hemisphere Reference Line of Hart (1984). The Pb isotopic composition of lherzolite clinopyroxenes (open circles) are plotted in (c) (see Fig. 10.27 for details).

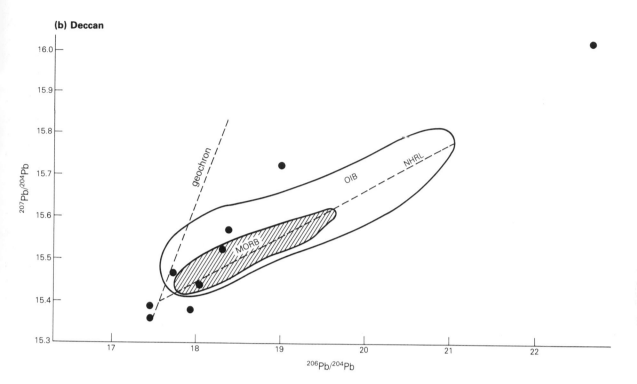

(b) Deccan

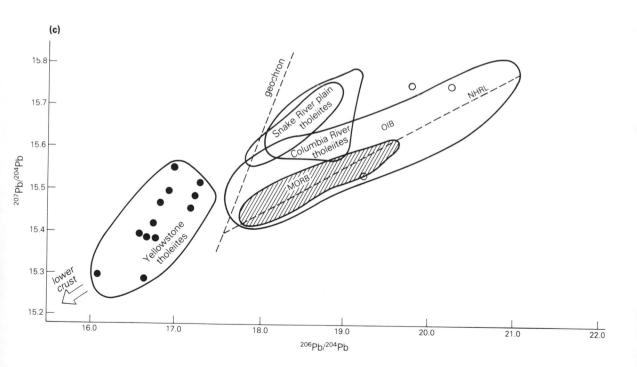

(c)

LPT series define subparallel arrays to the MORB field, displaced to higher $^{208}Pb/^{204}Pb$ and $^{207}Pb/^{204}Pb$ respectively. The HPT basalts could clearly have been derived from sources similar to those which fed the magmatism on Gough, Tristan da Cunha and the Walvis Ridge (Ch. 9). However, as we have already suggested, the LPT basalts owe much of their characteristics to crustal contamination. This is clearly shown in Figure 10.29, a plot of $^{87}Sr/^{86}Sr$ initial versus $^{206}Pb/^{204}Pb$ in which LPT basalts with $^{206}Pb/^{204}Pb$ ratios between 18 and 19 show a wide range of $^{87}Sr/^{86}Sr$ ratios between 0.707 and 0.716. Thus, in terms of Figures 10.27 and 10.28, vectors subparallel to the MORB array from HPT to LPT fields represent a trend of increasing crustal contamination. Data for the Deccan flood basalts are less well constrained but show a similar pattern.

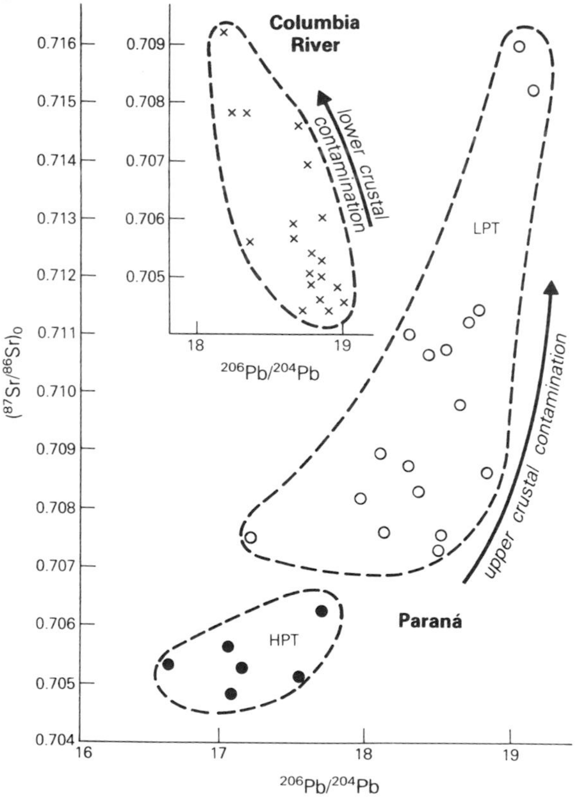

However, when we consider data for the Columbia River − Snake River Plain − Yellowstone province, a completely different pattern emerges. Columbia River and Snake River Plain tholeiites have a restricted range of $^{206}Pb/^{204}Pb$ ratios between 18 and 19, similar to the Paraná basalts. In contrast, tholeiitic basalts from Yellowstone plot in a completely different field with very low $^{206}Pb/^{204}Pb$ ratios between 16 and 17. The Columbia River basalts are plotted as an inset on Figure 10.29, the plot of $^{206}Pb/^{204}Pb$ versus initial $^{87}Sr/^{86}Sr$, where they clearly define a negative correlation in contrast to the positive one displayed by the Paraná basalts. This suggests that they have been contaminated by a lower-crustal component, whereas the Paraná basalts have experienced contamination in the upper crust. On this basis the Yellowstone tholeiites would appear to have experienced the greatest degrees of lower-crustal contamination. However, this is at variance with the conclusions of Doe et al. (1982), based on the same set of Pb isotopic data, who consider that the Yellowstone basalts have experienced little crustal contamination and that their isotopic compositions therefore reflect those of their mantle source.

From the above it is clear that combined Sr and Pb isotopic studies can provide strong evidence for contrasting styles of crustal contamination in CFB suites. However, although the isotopic variability of upper crustal rocks is fairly well characterized much less is known about the lower crust, which often makes constraining specific contaminants more difficult. In general, lower crustal contaminants are invoked to explain distinctive isotopic characteristics such as low $^{87}Sr/^{86}Sr$, low $^{143}Nd/^{144}Nd$ and unradiogenic Pb. Many of the basalts erupted within the British Tertiary Volcanic Province display such characteristics and appear to have been contaminated by high-grade Lewisian metamorphic rocks in the base of the crust (Thirlwall & Jones 1983, Dickin et al. 1984, Thompson et al. 1986).

Figure 10.29 Variation of initial $^{87}Sr/^{86}Sr$ versus $^{206}Pb/^{204}Pb$ for HPT and LPT basalts from the Paraná (Hawkesworth et al. 1986), compared with similar data for the Columbia River flood basalts (Church 1985).

10.6.4   Stable isotopes: oxygen

Interaction between crustal materials and mantle-derived magmas may involve either bulk assimilation of crustal rocks, with resulting elemental mixing of the two reservoirs, or selective elemental or isotopic exchange. The upper continental crust constitutes an ^{18}O-enriched reservoir relative to the mantle, and thus mantle-derived magmas which have been contaminated by upper crustal rocks should have distinctive ^{18}O isotopic signatures (James 1981). In contrast, granulite facies components of the lower crust may be relatively depleted in ^{18}O and thus may not significantly modify the oxygen isotopic characteristics of contaminated magmas.

For a suite of petrogenetically related volcanic rocks the clearest indication of post-partial melting magma contamination is the correlation of stable and radiogenic isotope variations with variations in chemical composition. For example, a mantle-derived partial melt which has been affected by upper crustal contamination due to AFC processes (Ch. 4) involving plagioclase fractionation should exhibit increases in $^{18}O/^{16}O$ and $^{87}Sr/^{86}Sr$ which correlate with increases in SiO_2 and decreases in Sr content. The implications of such correlations for the petrogenesis of subduction-zone magmas have been considered in Chapter 6.

The $\delta^{18}O$ values of continental tholeiites are generally near 5.5, but can range as high as 7 (Kyser *et al.* 1982; and see Fig. 10.30). These overlap with the range of values for oceanic basalts (MORB + OIB) and mantle-derived ultramafic xenoliths. Low values of $\delta^{18}O$ (<5) may be the result of interaction of the rocks with meteoric waters. Tholeiitic basalts from the Columbia River province have $\delta^{18}O$ in the range 5.8−6.8 and show no correlation of oxygen isotopic composition with $^{87}Sr/^{86}Sr$ or $^{143}Nd/^{144}Nd$ (Carlson *et al.* 1981). The limited range of $\delta^{18}O$ clearly precludes contamination with young upper crustal materials which would have $\delta^{18}O > 10$, but provides no evidence for or against lower crustal contamination. In contrast, Mahoney *et al.* (1985), in a study of tholeiitic flood basalts from the northern Deccan Traps province, have shown that $\delta^{18}O$ ranges as high as 8.3‰ and correlates with Nd and Sr isotopic ratios, thereby supporting upper crustal contamination models. For both provinces the oxygen isotopic data strongly support previous

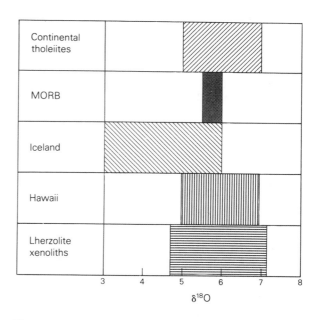

Figure 10.30 Comparison of the oxygen isotopic composition of continental flood basalts with those of MORB, oceanic tholeiites from Iceland and Hawaii, and spinel and garnet lherzolites form the sub-continental lithosphere (data from Kyser *et al.* 1982).

conclusions about the nature of crustal contaminants based on radiogenic isotopic data alone.

Unfortuantely, there is very little published oxygen isotope data available for continental flood basalts. However, such studies clearly have tremendous potential for elucidating the nature of crustal contamination processes.

10.7   Detailed petrogenetic model

Fundamental chemical and isotopic differences exist between continental and oceanic tholeiitic basaltic magmas, which can be expressed in terms of abundances and ratios of incompatible elements such as K, Rb, Ba, Ti, P and the light REE, and also in terms of their Rb−Sr, Sm−Nd and U−Th−Pb isotopic systematics (Dupay & Dostal 1984). In general, the continental basalts display greater elemental and isotopic diversity, which has been attributed to a variety of processes including:

(a) crustal contamination (Carlson et al. 1981, Dickin 1981, Thompson et al. 1986);

(b) melting of enriched subcontinental mantle (Menzies et al. 1983, 1984);

(c) mixing between depleted and enriched mantle sources (Hart 1985);

(d) combinations of melting of enriched mantle and crustal contamination (Mahoney et al. 1982, Carlson 1984, Cox & Hawkesworth 1985).

The isotopic and chemical composition of the subcontinental upper mantle and, indeed, the thickness of the continental lithosphere, still remain subjects for considerable speculation. Ultramafic xenoliths brought to the surface by continental alkaline and kimberlite magmas (Chs 11 & 12) are chemically and isotopically heterogeneous, recording distinct syles of trace element enrichment that may have developed recently or, in some cases, as much as 1000 Ma ago (Menzies et al. 1984). In some instances, marked isotopic heterogeneities have developed in these xenoliths in response to ancient perturbations in the Rb$-$Sr, Sm$-$Nd and U$-$Th$-$Pb systems (Kramers 1977, Allègre et al. 1982, Cohen et al. 1984) which exceed the combined range of MORB$-$OIB data (Fig. 10.22a). Thus those flood basalts with similar isotopic characteristics could be derived by partial melting of lithospheric mantle sources, without recourse to crustal contamination models.

However, many tholeiitic flood basalts have isotopic compositions plotting to the right of the Nd$-$Sr mantle array, sometimes with markedly radiogenic Sr isotope compositions. Many authors have invoked crustal contamination to explain these characteristics but, as shown in Figure 10.22a, such ratios can also develop in high-Rb$-$Sr mantle xenoliths, which may represent samples of upper mantle metasomatized by subduction-zone fluids. Thus Sr$-$Nd isotopic data alone do not provide unambiguous evidence for crustal contamination.

Studies of mantle-normalized trace element variation diagrams (spiderdiagrams; Section 10.6.2) confirm that enriched MORB, oceanic-island tholeiites and alkali basalts and continental alkali basalts (Chs 5, 9 & 11) could all be derived by

variable degrees of partial melting of an enriched mantle source. However, it is equally clear that many continental tholeiites could not be derived from the same source (Section 10.6.2) unless their characteristic depletion in Nb and Ta reflects the removal of these elements in a fractionating or residual phase during magma genesis (Dupuy & Dostal 1984). The distinctive Nb$-$Ta trough characteristic of their spiderdiagram patterns must either reflect unusual characteristics of regions of the subcontinental upper mantle not sampled by continental alkaline magmatism or crustal contamination. Since not all CFB spiderdiagram patterns show a Nb$-$Ta trough, the latter seems more likely.

Correlations between trace element ratios and isotopic ratios seem to provide good evidence for crustal contamination and, in particular, Sr$-$Pb isotopic variation diagrams appear to have considerable potential in resolving the effects of upper versus lower crustal contaminants. Correlations between Sr isotopic composition and SiO_2 content provide strong evidence for the operation of high-level AFC processes (Ch. 4), whereas a lack of correlation between isotopic ratios and major and trace element abundances may favour either partial melting of heterogeneous mantle sources (Menzies et al. 1983) or mixing of mantle-derived magmas and crustal melts at lower crustal levels.

In general, there appears to be good evidence to support both the involvement of enriched mantle sources and crustal contamination in producing the diversity of geochemical characteristics displayed by continental flood basalts. Mechanisms for crustal contamination may vary considerably from province to province, ranging from mixing between mantle melts and lower crustal partial melts to high-level AFC processes. In some provinces, magmas may experience contamination at both lower and upper crustal levels, making interpretation of their geochemical characteristics even more difficult (Thompson et al. 1986). Thompson et al. (1984) have pointed out that there are no recorded examples of CFBs with MORB-like trace element and isotopic characteristics and, on this basis, have concluded that all continental tholeiites are ultimately derived from the OIB source reservoir, experiencing variable degrees of crustal contamination en route to the surface. However, such conclusions must be regarded as speculative given the considerable overlap in the geochemical characteristics of OIB source reservoirs and the subcontinental lithosphere.

Crystal fractionation processes, in some instances combined with assimilation of crustal rocks, have clearly been responsible for the production of more siliceous magmas in many CFB provinces. However, in nearly all those provinces in which highly acid magmas (rhyolites) occur, they are associated with basalts in a bimodal association with a distinct lack of intermediate compositions. This has led many authors (Doe *et al.* 1982, Cleverly *et al.* 1984) to propose that the acid magmas are generated by partial melting of underplated basic igneous rocks at the base of the continental crust.

The close association between theoleiitic flood basalt provinces and continental fragmentation means that their petrogenetic interpretation is fundamental to our understanding of the formation of new ocean basins (Ch. 5). Unfortunately, the lack of recent examples makes combined geophysical–geochemical studies effectively impossible. The youngest example of the type associated with continental break-up we have for direct study is the 60–65 Ma British Tertiary Volcanic Province, which is just too old for geophysical modelling of lithospheric thinning as its age is the same order of magnitude as the thermal relaxation time of the lithosphere. Thus in the development of models for continental fragmentation we have to rely heavily on data from active intra-plate rifts,

which need not necessarily be the precursor to ocean-basin formation (Ch. 11).

Further Reading

Basaltic Volcanism Study Project 1981. *Basaltic volcanism on the terrestrial planets.* 30–107. New York: Pergamon Press

Cox, K. G. 1980. A model for flood basalt volcanism. *J. Petrol.* 21, 629–50.

Cox, K. G. & C. J. Hawkesworth 1985. Geochemical stratigraphy of the Deccan Traps at Mahabaleshwar, Western Ghats, India, with implications for open system magmatic processes. *J. Petrol.* 26, 355–77.

Dupuy, C & J. Dostal 1984. Trace element geochemistry of some continental tholeiites. *Earth Planet. Sci. Lett.* 67, 61–9.

Erlank, A. J. (ed.) 1984. *Petrogenesis of the volcanic rocks of the Karoo province.* Geol. Soc. South Africa Spec. Publ. 13, 395 pp.

Huppert, H. E & R. S. J. Sparks 1985. Cooling and contamination of mafic and ultramafic magmas during ascent through the continental crust. *Earth Planet. Sci. Lett.* 74, 371–86.

Leeman, W. P. & C. J. Hawkesworth 1986. Open magma systems: Trace element and isotopic constraints. *J. Geophys. Res.* 91, 5901–12.

Thompson, R. N., M. A. Morrison, A. P. Dickin & G. L. Hendry 1983. Continental flood basalts... arachnids rule OK? In *Continental basalts and mantle xenoliths*, C. J. Hawkesworth & M. J. Norry (eds), Nantwich: Shiva. 158–85.

CHAPTER ELEVEN

Continental rift zone magmatism

11.1 Introduction

Basaltic magmatism appears to be a common manifestation of extensional tectonics within continental plates. In Chapter 10 we focused our attention on the dominantly fissure-fed sub-alkaline (tholeiitic) flood basalt provinces, whereas in this chapter we shall concentrate on those more alkaline provinces, closely associated with rift/graben structures, fed from both fissure and central vent activity. As we have previously stated, this is a totally artificial subdivision, and in reality there is a complete spectrum of intracontinental plate magmatic activity, from extensive sub-alkaline flood basalt provinces to largely amagmatic graben structures with localized centres of alkaline volcanism.

Figure 11.1 shows the distribution of the major presently and recently (Tertiary–Recent) active intracontinental plate rifts and also the location of several classic postulated ancient rift structures, such as the Precambrian Gardar province of Greenland and the Permian Oslo Graben of Norway. A comprehensive review of all aspects of the igneous activity of each of these continental rifts is clearly beyond the scope of a single chapter. Thus we shall focus our attention on the East African Rift system (Fig. 11.2), because of its scale and the diversity of its tectonic and magmatic activity, making comparisons with other rift systems where appropriate. It is undoubtedly the largest presently active intracontinental plate rift, with significantly greater volumes of volcanic products than other recently active rifts. For example, Mohr (1982) estimates the volumes of volcanic rocks in the East African Rift in Kenya and Ethiopia alone to be 500 000 km^3, compared with 12 000 km³, in the Rio Grande Rift of the western USA and 5000 km³ in the Baikal Rift.

Continental rift zones are areas of localized lithospheric extension characterized by a central depression, uplifted flanks and a thinning of the underlying crust. High heat flow, broad zones of regional uplift and magmatism are often associated with such structures. In general, they are a few tens of kilometres wide and tens to a few hundred kilometres in length. However, in the extreme case, extension and associated magmatism may be distributed over broad zones hundreds of kilometres in each direction, as in the Basin and Range province

of the western USA (Fig. 11.3).

Traditionally, there has been a tendency to think of intracontinental plate volcanism in terms of models of incipient continental fragmentation, as exemplified by the East African Rift system. This forms part of the much larger Afro-Arabian rift system, extending some 6500 km from Turkey to Mozambique, including the Dead Sea (Levantine Rift), Red Sea, Gulf of Aden and East African rifts (Fig. 11.2). In the Red Sea and Gulf of Aden continental separation has occurred and new ocean basins are forming. Many authors have suggested that the eastern and western branches of the East African Rift represent an early stage in such a

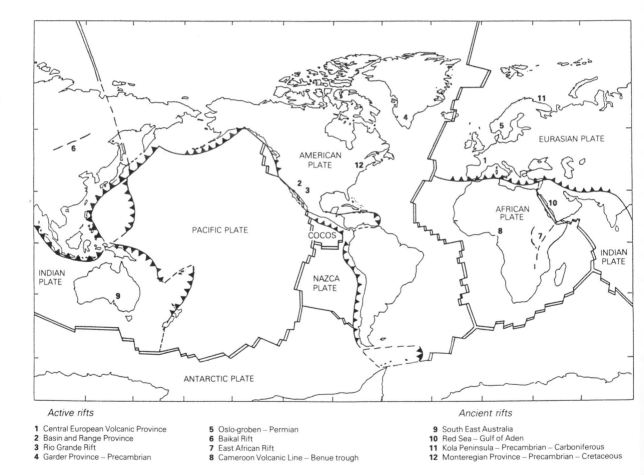

Active rifts

1 Central European Volcanic Province
2 Basin and Range Province
3 Rio Grande Rift
4 Garder Province – Precambrian

5 Oslo-groben – Permian
6 Baikal Rift
7 East African Rift
8 Cameroon Volcanic Line – Benue trough

Ancient rifts

9 South East Australia
10 Red Sea – Gulf of Aden
11 Kola Peninsula – Precambrian – Carboniferous
12 Monteregian Province – Precambrian – Cretaceous

Figure 11.1 Global distribution of the major active and recently (Tertiary–Recent) active intracontinental plates rifts. Also shown are the locations of some postulated ancient rift structures, now largely represented by alkaline plutonic complexes. Major plate boundaries are from Figure 1.1.

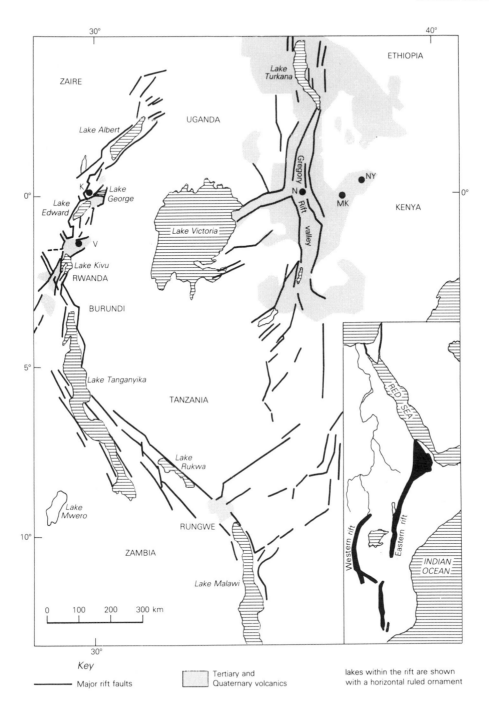

Figure 11.2 Sketch map of the southern termination of the East African rift system and its relation to the Red Sea and Gulf of Aden oceanic rifts (inset). Volcanic fields specifically referred to in the text include the following: K, Katwe–Kikorongo; V, Virunga (including Bufumbira and Nyiragongo); N, Naivasha; MK, Mt Kenya; NY, Nyambeni (after Barberi *et al.* 1982, Fig. 6, p. 234).

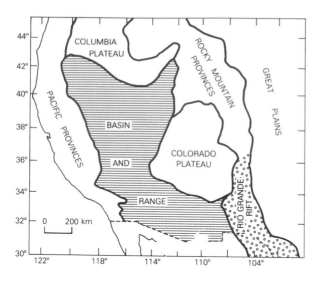

Figure 11.3 Simplified tectonic map of the western USA to show the location of the Basin and Range province and the Rio Grande Rift.

process. However, extensional tectonics and associated magmatism are not restricted to such environments. Volcanically active structural basins may develop within continents as a consequence of continent–continent collision (e.g. the Central European Volcanic Province, including the Rhinegraben) or as subduction related back-arc basins (e.g. the Columbia River province, Ch. 10). Nevertheless, despite the diversity of plate-tectonic settings which may give rise to continental rift zones (CRZs), they show many similarities in terms of their structural development and volcanicity. This must ultimately reflect a similarity in partial melting processes in the underlying mantle, regardless of the tectonic environment.

Processes of continental rifting must precede the formation of new ocean basins, and it is obviously of great importance to be able to understand how the transition from a continental to an oceanic rift takes place. The Red Sea provides an excellent natural laboratory in which to study such a transition. In the southern sector new oceanic crust has been generated by seafloor spreading processes related to a linear array of asthenospheric hot spots, spaced approximately 50 km apart, superimposed

upon a broader zone of mantle upwelling (Bonatti 1985). The growing rift appears to propagate from each of these hot spots and discontinuities develop where the tips of rift segments propagating from adjacent hot spots join; these may subsequently become transform faults in the newly opened ocean basin (Ch. 5). Bonatti (op. cit.) considers that this mechanism of opening may also be applicable to the Atlantic Ocean, where individual ridge segments are also approximately 50 km long (Schouten et al. 1985). A regular spacing of around 40 km has also been noted for Pleistocene and active volcanoes in the northern Kenya and Ethiopian rifts, which may provide a link between continental and oceanic rift processes in terms of mantle upwelling. Away from the active axial zone in the southern Red Sea and across the northern sector, true oceanic crust is absent and the Red Sea is instead floored by thinned continental crust injected by basic dykes. This would appear to be compatible with a model for the transition from continental to oceanic crust in which the continental crust is stretched and progressively invaded by a plexus of mafic dykes (Nicolas 1985).

Rift valleys, both active and ancient, are found over much of the area of the continents. Some may eventually evolve into new ocean basins but many abort after only a few kilometres of horizontal extension, forming so-called 'failed rifts'. Along many passive continental margins (e.g. the Atlantic ocean) there are examples of rift-rift-rift triple junctions in which two of the three rifts evolved into a new ocean basin while the third 'failed rift' extends into the interior of the continent forming an *aulacogen* (Burke & Wilson 1976). Such failed rifts are very important economically as their sediment infill may contain oil-bearing strata (Reeves et al. 1987).

Studies of the evolution of continental rift systems and rifted continental margins have led to the development of many hypotheses for the driving forces which create them (McKenzie 1978b, Illies 1981, Bott 1982, Mohr 1982, Keen 1985). An element of controversy exists as to whether rift zones are produced by upwelling mantle splitting the continent along pre-weakened zones (active models) or whether the mantle is forced to rise as the continents are pulled apart during lithospheric stretching (passive models) (Fig. 11.4) (Spohn & Schubert 1982; see also Section 11.2). Clearly, a spectrum of processes is likely to exist in nature between these two extremes. However, regardless of the mechanism of inception, available geophysical data demonstrate that asthenospheric upwelling is a fundamental feature of all presently and recently active continental rift zones (Basaltic Volcanism Study Project 1981; see also Section 11.3).

Theoretically, it should be possible to distinguish between active and passive rift models on the basis of the relative timing of rifting and volcanism. In the active case upwelling asthenospheric mantle is responsible for updoming and cracking of the lithosphere, and the observed tectonomagmatic sequence should be doming−volcanism−rifting. In contrast, in the passive case, cracking of the lithosphere due to differential stresses within the plate induces asthenospheric diapirism and partial melting. In this case the predictive sequence would be rifting−doming−volcanism. Unfortunately,

such simplistic models, when tested against observed field relationships in well studied rift segments, rarely appear uniformly applicable and the processes of rift development must actually be much more complex.

There is still some dispute about the classification of modern rifts as active or passive (Spohn & Schubert 1982). For example, Sengor & Burke (1978) and Buck (1986) consider the Rhinegraben and Baikal rifts are passive, produced by horizontal stresses induced by the Alpine orogeny, whereas Illies (1981) considers that they are both active. However, there appears to be a general consensus that the East African Rift has to be an active rift (Buck 1986) and that lithospheric stretching is necessary for the creation of rifted sedimentary basins and rifted continental margins (McKenzie 1978, Jarvis & McKenzie 1980, England 1983).

All of the modern rifts cited above are characterized by anomalously high heat flow and anomalous seismic structures, which suggests that they are underlain by zones of thinned lithosphere (Spohn & Schubert op. cit.). These anomalies are often of considerable lateral extent, far exceeding the dimensions of the surface graben. For example in East Africa, beneath the Kenya rift, the lithosphere is thinned by about 20% of its thickness beneath the adjacent cratons and the anomaly has a wavelength of the order of 1000 km (Fairhead & Reeves 1977). Additionally, most are also associated with broad zones of residual negative gravity anomalies (e.g. Fig. 11.5; see also Fairhead 1979), which may be explained in terms of upwelling of low-density, possibly partially molten, low-velocity asthenospheric mantle material. Many also have a narrow zone of residual positive gravity anomaly, sited over the axial graben, which may be related to the intrusion of basaltic magmas into the sialic crust (Searle 1970, Cordell 1978, Ramberg et al. 1978).

The Afro-Arabian rift system represents the most extensive currently active zone of continental rifting on Earth (Shudofsky 1985) and, as such, it has been the focus for an extensive range of geological and geophysical studies aimed at understanding the fundamental processes involved in continental rifting. As a consequence, much of the discussion in the following sections will be based upon this

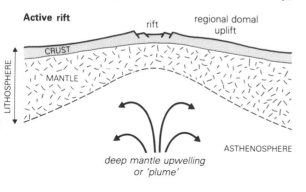

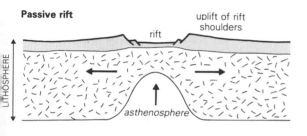

Figure 11.4 Active versus passive rifting models (after Keen 1985, Fig. 18, p. 116).

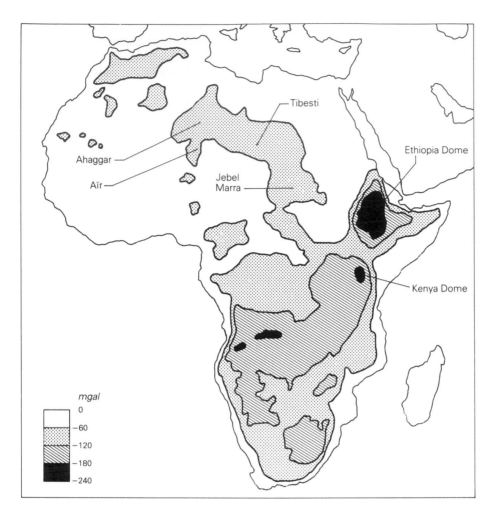

Tibesti

Ethiopia Dome

Ahaggar

Aïr

Jebel
Marra

Kenya Dome

mgal
0
−60
−120
−180
−240

Figure 11.5 Bouger
anomaly map of Africa
(after Fairhead 1979, Fig. 1,
p. 110).

database. It provides an invaluable natural laboratory in which to study continental rift zone magmatic processes because nearly all stages in the rifting
process have been recognized, from incipient rifting in southern Africa to the development of new
oceanic crust in the Red Sea/Gulf of Aden. Within
the African sector the rift divides into eastern and
western branches (Fig. 11.2), separated by the
Tanganyika shield. These two branches are characterized by volcanism of quite different character,
composition and intensity (Williams 1982).

The Afro-Arabian rift system has been evolving
over the past 45 Ma, but evidence suggests that its
development has been episodic rather than continuous (Girdler 1983). The eastern rift in Kenya is
thought to have evolved in three major pulses (late

Eocene (44−38 Ma), middle Miocene (16−11 Ma)
and Plio−Pleistocene (5−0 Ma); see Baker *et al.*
1972, with phases of uplift being broadly synchro-
nous in both Ethiopia and Kenya. Uplift has
produced two major domal structures (Savage &
Long 1985); the Afro-Arabian dome, trisected by
the Red Sea, Gulf of Aden and Ethiopian rifts, and
the Kenya dome which is bisected by the Kenya, or
Gregory, Rift. Almond (1986) has synthesized the
structural and volcanic history of the former and
concludes that there is no simple link between
doming, rifting, oceanic spreading and volcanism.
The upper limits for crustal extension are estimated
to be 30 km in Ethiopia, 10 km in Kenya and 2−3
km in Tanzania (Shudofsky 1985). This gives a
maximum spreading rate of 1 mm yr^{-1} since the

Miocene, which is one to two orders of magnitude less than for oceanic spreading centres (Ch. 5).

A fundamental problem in tectonic studies of continental rift zones is the extent to which ancient basement structures control the orientation and development of the presently active rifts. In Ethiopia and Kenya any direct connection between Cenozoic rifting and Precambrian basement structure is obscured by the volcanic cover. Nevertheless, the general NNE−SSW trend of the rift zone is parallel to that of the 600 Ma Mozambique fold belt which runs along the adjacent coast. However, in detail, there are many cross-cutting relationships which would appear to negate a strong basement control (Shudofsky op. cit.). While the eastern branch of the rift appears to have a basement of relatively young mobile belt rocks, the western branch is founded on Precambrian rocks along the western boundary of the Tanganyika shield. The western rift has significantly less magmatic activity than the eastern branch, and it is interesting to speculate whether the nature of the basement may exert a fundamental control on the upward passage of magmas.

In addtion to the East African rift system there are several other major centres of Tertiary−Recent volcanic activity within the African plate, all of which are associated with crustal doming. These include Tibesti, Ahaggar and Jebel Marra (Fig. 11.5). The simplest model to explain their location appears to be the impingement of localized mantle upwellings (plumes) on the base of the African lithosphere.

Recently, the Rio Grande Rift of the western USA (Fig. 11.3) has become a focus for extensive geochemical and geophysical studies which provide a useful database for comparison with the East African Rift system (Aldrich et al. 1986, Dungan et al. 1986, McMillan & Dungan 1986, Sinno et al. 1986). This occurs at the eastern margin of a broad zone of late Cenozoic uplift, extensional tectonics and basaltic volcanism that characterizes much of the western USA (Basaltic Volcanism Study Project 1981). Prior to the onset of extensional tectonics in the Oligocene, subduction-related magmatism occurred over the same area and thus, in its early developmental stages, the rift may be considered as

an example of extension behind a waning magmatic arc (Lipman 1980). Some of the earlier rift-related basalts clearly have subduction-related characteristics, whereas more recent eruptives are more typical of intra-plate rifts.

A wide spectrum of magma compostions characterizes CRZ magmatism, in contrast to the rather uniform sub-alkaline basalts which dominate the continental flood basalt provinces (Ch. 10). Basalts range from transitional sub-alkaline types through alkali basalts to silica-undersaturated basanites and nephelinites, and in some instances ultrapotassic magmas such as leucitites (Ch. 12). In some rifts, carbonatite magmatism may be associated with the more silica-undersaturated volcanic centres (Le Bas 1987). Unlike the CFB provinces described in the previous chapter, the volcanism is not predominantly basaltic but involves large volumes of felsic eruptives (trachytes, phonolites and rhyolites). At certain periods within the evolution of the East African Rift, trachytic and phonolitic magmas have been erupted on a regional scale (Baker et al. 1972, Williams 1982).

Basaltic magmas erupted within most of the major continental rift zones indicated on Figure 11.1 are predominantly alkalic. However, within zones of greatest crustal extension, such as Ethiopia, transitional basalts dominate, perhaps suggesting a correlation between the rate of extension and the degree of partial melting of upwelling mantle. In general, CRZ volcanism is highly explosive and pyroclastic rocks may locally dominate the volcanic sequence. This suggests an enrichment of volatiles in the magma source region (Bailey 1983, 1985).

Within CRZs there are many recorded instances of basalts of contrasted chemical compositions being erupted in close spatial and temporal association, which clearly has important implications in terms of petrogenetic modelling. Additionally, the volcanic products may vary transversely across rifts, with flank lavas tending to be more alkaline than those erupted in the axial graben. Many of the better documented continental rifts also show a distinct periodicity to their magmatic activity (e.g. the eastern branch of the East African Rift; Baker et al. 1972, Baker 1987) which may be related to changing patterns of tectonic activity. This may reflect fundamental changes in the state of the mantle beneath the rift, although once asthenospheric upwelling has been initiated episodic faulting and volcanism are likely to be second-order phenomena (Basaltic Volcanism Study Project 1981). As a consequence of such periodicity there may be distinct changes in the chemistry of the predominant magma type, in the relative proportions of basic and acid magmas, in rates of magma production and transmission and in modes of eruption (e.g. pyroclastic versus lava emission, fissure versus central vent activity).

A major problem in petrogenetic studies of continental rift zone volcanic suites involves the origin of magmas more silica-rich than basalt. In some rifts basalts and more evolved lavas (trachytes, phonolites and rhyolites) can be clearly related by processes of fractional crystallization. However, in others the volume, timing and isotopic and trace element geochemical characteristics of the salic magmas suggests the involvement of crustal rocks in their petrogenesis. This is to be expected, particularly in the more active rifts, because the high heat flows, generated as a consequence of the emplacement of large volumes of high-temperature basaltic magma into the crust, may locally elevate crustal temperatures above the solidus, facilitating crustal contamination and AFC processes (Ch. 4).

Detailed geophysical studies of the crustal structure within the Rio Grande and Kenya rifts have provided important constraints on the location of crustal magma chambers (Section 11.3). Additionally, for many of the older rifts located on Figure 11.1, such as the Precambrian Gardar province of Greenland and the Permian Oslo graben of Norway, extensive erosion has revealed the plutonic root zones of former trachytic/phonolitic stratvolcanoes, now exposed as syenite/nepheline syenite/alkali granite plutonic complexes (Sørensen 1974). The geology and geochemistry of such complexes potentially provide important information about subvolcanic magmatic processes. However, their detailed description is clearly beyond the scope of this work. Within the southern part of the East African Rift, Mesozoic alkaline plutonic complexes provide evidence for the exist-ence of trachyte/phonolite stratovolcanoes similar to those of the presently active Kenya Rift, e.g. the Chilwa alkaline province of southern Malawi (Bailey 1974).

11.2   Simplified petrogenetic model

It is now generally accepted that the formation of sedimentary basins and intracontinental plate rifts is connected with a stretching deformation of the crust and mantle components of the lithosphere. Simple models (e.g. McKenzie 1978b) involve stretching of the two layers by an equal amount, whereas more realistic models assume that the mantle component of the lithosphere is thinned more efficiently than the crust (Fleitout *et al.* 1986). As a consequence of such thinning heat is transferred upwards from the asthenosphere, and the resultant thermal anomaly may promote domal uplift of the overlying crustal rocks.

A fundamental problem in understanding the processes of development of continental rift zones centres on mechanisms for inducing the observed upward migration of the lithosphere/asthenosphere boundary. Two limiting cases may be considered (Sengor & Burke 1978), as shown schematically in Figure 11.4:

(a) *Active rifting*. Asthenospheric upwelling thins and causes uplift of the lithosphere and controls rift formation. The upwelling could be of the two-dimensional type, associated with mid-ocean ridges (Ch. 5) or an axisymmetric mantle plume. In such an environment volcanism and doming should precede rifting.

(b) *Passive rifting*. This is caused by differential stresses in the lithosphere (McKenzie 1978b). In this case the rift forms first, and uplift of the rift flanks may follow due to the development of small-scale convection cells beneath them (Buck 1986). Passive rifting has been considered previously in Chapter 9 as a mechanism to explain the origin of some linear oceanic-island chains.

While both active and passive rifting may produce uplift, in passive models it is confined to the stretched and rifted near-surface region (Keen 1985, Buck 1986). In contrast, in active rifting uplift may extend hundreds of kilometres beyond the rift. Similarly, lithospheric thinning is laterally confined to the rift zone in passive rifting, while in the active case the zone of thinning is several times the width of the zone of mantle upwelling (Keen op. cit.). Crustal thinning may not always be a good discriminator between active and passive models. In the active model there is likely to be little crustal thinning until the base of the lithosphere reaches the Moho. However, the crust may eventually be thinned over the width of the surface rift zone in a manner similar to that produced by passive rifting (Keen op. cit.).

In order to decide whether a rift is active or passive, it is necessary to have good constraints on the deep structure of the lithosphere, and to be able to correlate the horizontal extent of uplift with the near-surface geology. On this basis most modern continental rifts would be classified as active, with the possible exception of the Basin and Range province. However, in practical terms it is probably almost impossible to distinguish active from passive types, as an initial phase of passive rifting could easily trigger a more dynamic asthenospheric upwelling. Additionally, W. J. Morgan (1983) has suggested that hotspot tracks within continental plates (equivalent to aseismic ridges within the ocean basins; see Ch. 5) may pre-weaken the lithosphere and could then become the sites of later continental rifting.

Within the African plate there is generally a very good correlation between areas of domal uplift and Cenozoic alkali basalt—trachyte volcanism (Fairhead 1979). Some of the best examples are Aïr, Ahaggar, Tibesti, Jebel Marra, and Ethiopia and Kenya within the East African Rift system. All of these domal uplifts are associated with negative Bouger anomalies (Fig. 11.5) implying mass deficiencies at depth. This suggest that these are sites of marked lithospheric thinning and upwelling of hotter asthenospheric mantle material. Burke & Whiteman (1973) noted that several of these areas of domal uplift are not rifted and, on this basis,

proposed a model in which doming preceded rifting, consistent with an active origin. However, for the Ethiopian and Kenyan rifts, Baker et al. (1972) have clearly demonstrated that rifting and subsidence preceded the major phases of domal uplift, suggesting that here at least differential stresses within the lithosphere may have activated the initial stages of rifting.

Any model for CRZ magmatism must be capable of explaining the remarkable compositional diversity of the erupted magmas from silica-poor melilitites, basanites and nephelinites through carbonatites and ultrapotassic magmas to a range of mildly alkalic and transitional basalts and their differentiates. The general characteristics of CRZ magmas are their alkaline nature, enrichment in volatiles (particularly halogens and CO_2) and enrichment in LILE, suggesting derivation from enriched mantle sources (Bailey 1983). A fundamental question is why the magmas consistently appear to come from enriched rather than depleted asthenospheric (MORB-source) mantle. It is possible that beneath most continental rifts the enriched source is simply the old subcontinental lithosphere, and that MORB-source asthenospheric mantle only becomes extensively involved in the most actively extending rift segments. The occurrence of sub-alkaline basalts in the Ethiopian rift which appear to have a strong MORB-source component (Section 11.5.2) appears to support such a model. Detailed trace element and isotope geochemical studies of rift basalts are required to test this hypothesis in detail. In general, we might expect to see a temporal progression in an actively extending rift from early lithosphere-dominated magma sources to later asthenosphere-dominated sources. Fundamental to this question is the role of deep-mantle plumes in the petrogenesis of rift magmas. If a deep-mantle upwelling (plume or hot spot) is responsible for the initiation of a particular rift (active model), then the magmas may be generated from OIB-source mantle components within the plume (Ch. 9) and contain little contribution from the asthenospheric MORB-source component in the edges of the diapir. Unfortunately, one of the major problems in petrogenetic modelling of CRZ magmas lies in the difficulty in distinguishing between enriched mantle source components derived from mantle plumes as opposed to those which undoubtedly exist within the subcontinental lithosphere.

Bailey (1983) considers that CRZs are characterized by two distinct magmatic associations. The first involves a range of parental basalts from transitional or mildly alkaline types to more strongly alkaline basalts and basanites, each of which may fractionate to produce a spectrum of more evolved hawaiites, mugearites, benmoreites, trachytes and alkali rhyolites. The second involves more highly alkaline and silica-undersaturated primary magmas, including nephelinites, melilitites and leucitites, which fractionate towards phonolitic residua. This will be considered further in Section 11.5.

In the African plate the strongest rift developments are around the margins of the ancient cratonic blocks, with the main rifts following the broad structural trends of the younger fold belts (Rosendahl 1987). Geothermal gradients are likely to be steeper in these mobile belts, possibly accounting for the greater volume of magmatism and its more sodic alkaline characteristics, compared to the more sporadic and highly potassic volcanism of the western branch of the East African Rift, which is situated on a craton.

Barberi *et al.* (1982) have classified intracontinental plate rifts into high-volcanicity and low-volcanicity types, on the basis of their relative volumes of volcanic products. Examples of low-volcanicty rifts (LVRs) include the western branch of the East African Rift, the Rhinegraben and the Baikal Rift. These are characterized by relatively small volumes of eruptive products, low rates of crustal extension, discontinuous volcanic activity along the rift, a wide spectrum of basaltic magma compositions and generally small volumes of salic differentiates (Fig. 11.6). Strongly alkalicundersaturated magmas predominate (nephelinites, basanites and leucitites), with transitional types becoming more abundant as the volume of eruptives increases. In such environments Bailey (1983) has suggested that deep lithospheric fractures could permit volatile fluxing from the asthenosphere, causing metasomatism of the lithosphere which may subsequently undergo partial melting to provide a major source component for the magmatism.

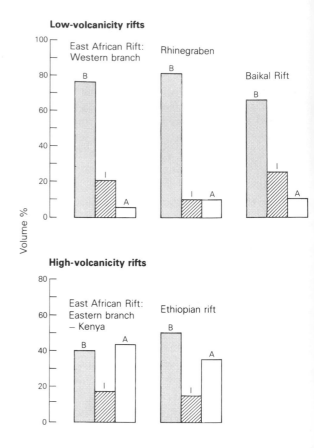

Figure 11.6 Comparison of the relative volumes of basic (B), intermediate (I) and acid (A) magmas erupted in high- and low-volcanicity rifts (after Barberi *et al.* 1982, Fig. 13, p. 248).

In contrast, high-volcanicity rifts (HVRs) are characterized by more voluminous magmatic activity, higher rates of crustal extension, predominantly mildly alkalic basalts and a bimodal distribution of basic and acid magma types (Fig. 11.6). Examples of HVRs include the Kenyan and Ethiopian sectors of the East African Rift and some sectors of the Rio Grande Rift. In general, there is a close relationship between the chemistry of the basic and acid magmas erupted in the same rift sector at the same time. Thus phonolites occur in association with nephelinites and basanites, trachytes with alkali basalts, peralkaline rhyolites with mildly alkaline basalts and sub-alkaline rhyolites with sub-alkaline basalts. In most cases, the acid magmas could originate by fractional crystallization of the associated basalts, with the apparent bimodality of the suites simply being a function of the magmatic plumbing system. However, detailed isotopic studies are required to confirm that apparently simple fractionation trends on Harker variation diagrams (Section 11.5.1) represent true liquid lines of descent.

In Kenya there is an apparent decrease in the alkalinity of the erupted magmas with time, and also spatially towards the rift axis (Baker 1987). This may be explained in terms of increasing degrees of partial melting in the ascending asthenospheric mantle (Baker et al. 1972). On this basis it would appear that the Ethiopian sector of the East African Rift was subjected to higher rates of crustal extension and mantle upwelling, producing greater degrees of partial melting and the generation of sub-alkaline basaltic magmas. In the Ethiopian rift, such sub-alkaline basalts appear to have a stronger MORB-source signature than spatially associated alkalic basalts, suggesting a greater asthenospheric contribution to the partial melts (Section 11.5).

In studies of the geochemistry of CRZ magmas, the role of continental crustal contamination must be considered, as indeed it must in all intra-continental plate magmatic environments. The anomalously high heat flow characteristic of active rift zones (P. Morgan 1983) indicates elevated crustal temperatures which may permit the generation of some acid magmas by crustal melting or mixing of crustal and mantle partial melts. For example, it is possible that older sequences of volcanic rocks deeply buried within rifts could become remelted, as could intrusions of basic magma within the lower crust. In detail, such models can only be tested by systematic trace element and Sr−Nd−Pb−O isotopic studies (Section 11.5).

A fundamental question in understanding the magmatic activity of continental rifts is whether there is any predictable sequence in time and space between crustal uplift, rifting and magmatism. Early authors (Le Bas 1971, Burke & Whiteman 1973) emphasized a sequence of regional domal uplift, alkaline volcanism upon the dome and then crestal rifting across the dome in the evolution of the East African Rift system. However, more recent studies (Mohr 1982, Almond 1986) do not support such a simplistic model. For example, for the Kenyan sector of the rift the first tectonic phenomenon was crustal downwarping, forming a proto-rift depression (Baker et al. 1972). This appears to be true for many rifts, suggesting that lithospheric stretching is the dominant factor in the initial stages of rift development. As a consequence of such stretching asthenospheric upwelling occurs, causing heating of the lithosphere and crustal doming. Thus, in general, it may be that many rifts evolve from a passive phase to an active phase.

11.3 Crust and upper mantle structure

It is clearly beyond the scope of this chapter to investigate the structure of the crust and upper mantle beneath each of the active continental rift zones shown in Figure 11.1 in detail. Consequently, we shall use the East African and Rio Grande rifts as examples to illustrate some of the characteristic features.

All of the young continental rifts considered thus far are characterized by seismic structures suggesting anomalously thin lithosphere (Mohr 1982). For example, regional seismic studies have indicated the presence of anomalous mantle with low P-wave velocities beneath the East African Rift in Ethiopia and Kenya (Savage & Long 1985). This is typically a very narrow feature (Fig. 11.7) and, within 50 km of the rift axis on the adjoining plateau, shield-type crust approximately 40 km thick is underlain by normal velocity mantle ($V_p = 8.1$ km s^{-1}).

Nolet & Mueller (1982) have compared the crust and upper mantle structure beneath the eastern and western branches of the East African Rift in Kenya and have shown that they are markedly different. In the eastern rift (Fig. 11.7), a 40 km thick crust overlies a low-velocity zone, persisting to some 200 km depth. In contrast, in the western rift there is a 35 km thick crust underlain by a high-velocity layer. This overlies a low-velocity zone, bottoming at approximately 140 km, below which high-velocity mantle reappears. Nolet & Mueller (op. cit.) suggest that a diapir of partially molten mantle material has almost reached the surface in the eastern rift, but has stalled at about 55 km beneath the western rift. Shudofsky (1985) explains this in terms of a model whereby a sublithospheric thermal anomaly, originally located beneath the eastern rift, is now beneath the western rift, due to the northeasterly migration of the African lithosphere over the past 40 Ma. The continuation of rifting and magmatism within the eastern rift to the present day may be related to tectonic stresses associated with seafloor spreading in the Red Sea/Gulf of Aden. During this 40 Ma period, the East African plateau between the eastern and western rifts must also have overridden the thermal anomaly, but because it is an ancient cratonic block it did not rift but underwent uplift.

Figure 11.8 shows an E−W cross section through the southern part of the Rio Grand Rift (Sinno *et al.* 1986) for comparison with that of the Kenya Rift (Fig. 11.7). Here there is also a marked upwarp of low-density/low-velocity mantle material beneath the rift axis. Similar upwarps have also been observed beneath the central and northern sectors of the rift (Olsen *et al.* 1979). The Baikal rift

(Fig. 11.1) is also underlain by low-velocity mantle, which reaches up to the crust beneath the graben (Zorin 1981).

In regions of continental extension earthquakes, in general, have maximum focal depths of 12−15 km (Chen & Molnar 1983). These occur in the cold upper crust, which responds by brittle failure. In contrast, in the lower crust, where temperatures are higher than normal due to the thermal effects of lithospheric thinning, crustal rocks are weaker, deforming by ductile flow, and therefore earthquakes should be rare. However, in the East African Rift earthquakes do occur in the 20−30 km depth range suggesting that the lower crust may be unusually cold (Shudofsky op. cit.). This appears to be at variance with the voluminous magmatic activity associated with the rift until the location of these deep earthquake epicentres is considered. Most occur within the least magmatically active segments of the rift (i.e. the western branch of the rift and the southern part of the Gregory Rift). As shown in Figure 11.7, lithospheric thinning is less extensive in such rift segments and consequently

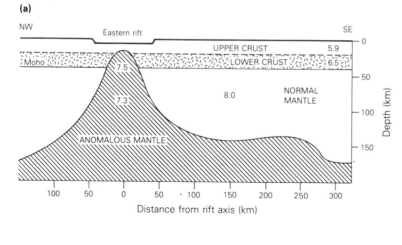

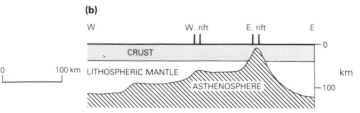

Figure 11.7 (a) Schematic cross section across the Kenya rift, showing the upwelling of low-velocity mantle beneath the rift axis. P-wave velocities of crust and mantle layers are in km s^{-1} (after Savage & Long 1985, Fig. 6, p. 470). (b) Schematic E−W profile across the eastern and western branches of the East African Rift, to show the differing levels of asthenospheric penetration of the lithosphere (after Girdler 1983, Fig. 2, p. 245).

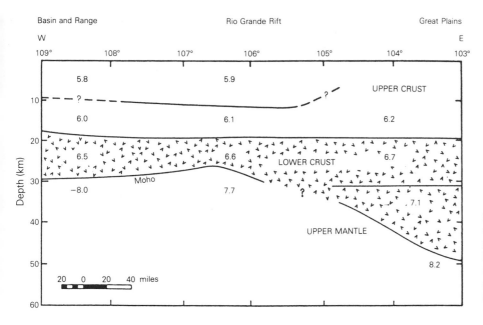

Figure 11.8 East–West cross section across the southern Rio Grande Rift in New Mexico; numbers are P-wave velocities in km s^{-1} (after Sinno *et al.* 1986, Fig. 18, p. 6154).

geothermal gradients in the overlying crust will correspond more closely to normal continental geotherms.

Determination of earthquake focal mechanisms can provide important constraints on the orientation of the present-day stress field within a plate. Shudofsky (1985) has shown that there is a remarkable consistency in the axes of least compressive stress for earthquakes distributed widely across East Africa and at varying focal depths, with tensional axes trending perpendicular to the trends of surface rifting. Significantly, earthquakes in regions that have not experienced recent rift faulting have tensional axes oriented in a similar direction to those in rifted areas. These data may indicate that rifting within East Africa is initiated passively by lithospheric stretching.

As considered in Ch. 10, a fundamental problem in petrogenetic studies of all intracontinental plate magmatic provinces is the extent to which rising basaltic magmas have interacted with crustal rocks. to understand such processes it is necessary to have a detailed knowledge of the crustal structure and, ideally, of the location of high-level magma-storage reservoirs.

Almost all continental rifts have one or more geophysical characteristics that can be interpreted in terms of the presence of magma bodies at different levels in the crust and upper mantle. These include observations of low-resistivity, low-seismic-velocity layers in the crust, lower than normal upper mantle velocities and higher than normal heat flows (Sanford & Einarsson 1982). However, the detection of significant accumulations of magma beneath or within the crust of continental rifts is rare. Possibly the best example occurs within the Rio Grande Rift of central New Mexico, where a thin (0.5–1 km) almost horizontal lens of magma with a minimum areal extent of 1700 km^2 has been seismically detected at mid-crustal depths (~20 km) (Sanford & Einarsson op. cit.). Additionally, on the basis of P- and S-wave attenuation studies, small pockets of magma may exist at very shallow levels (5–10 km) in the crust near Socorro in the Rio Grande Rift.

Knowledge of the variation of electrical conductivity within the crust and upper mantle can provide important constraints for the location of magma reservoirs beneath active rifts. However, because of the large number of factors influencing electrical conductivity, the data cannot always be interpreted unequivocally as indicating the presence of a melt phase without supporting seismic data. Hermance (1982) has compared the results of deep electrical studies of the Baikal, East African and Rio Grande rifts and used the data to infer the location of melt zones within the crust (Fig. 11.9). The variation of resistivity with depth for each of these rifts is shown schematically, and represents a synthesis of a wide variety of data. Each is clearly underlain by a conducting layer some 20–30 km thick, located either within the lower crust (Rio Grande Rift) or spanning the crust/mantle boundary (Baikal, East African and Rhinegraben rifts),

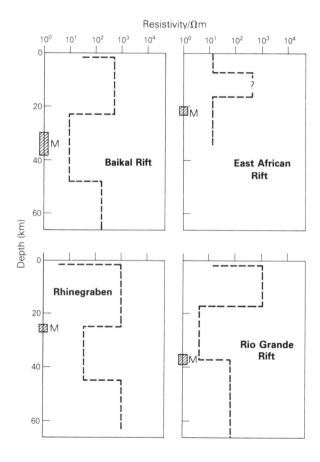

Figure 11.9 Schematic variation of electrical resistivity with depth for Baikal, East African, Rhinegraben and Rio Grande Rifts (data from Hermance 1982); M marks the approximate position of the Moho. The high-resistivity layer within the lower crust of the East African Rift is not well proven, and low resistivity could be a feature of the entire crustal profile. In each of these rifts, the marked low-resistivity channel at lower crustal–mantle depths may represent a major zone of magma accumulation.

which may represent the site of major magma accumulation.

It is well known that centres of silicic volcanism exist along some of the major active rift zones, such as the East African and Rio Grande rifts, and it has been suggested by many authors that these have developed by partial melting of crustal material. Mantle-derived basaltic magmas rising beneath the rift axis may tend to pond at hydrostatically controlled levels in the crust, leading to the development of lensoid bodies of magma. Conduction of heat from such bodies may locally raise the temperature of surrounding basement rocks above their solidus and cause partial melting. The resultant low-density silicic melts will rise rapidly and may thus feed silicic volcanism at the surface. Figure 11.10 shows a schematic cross section of the crust beneath the Valles caldera, which was the site of a cataclysmic eruption of silicic tephra on the western flank of the Rio Grande Rift 1.1 Ma ago (Olsen et al. 1986). This is amongst the largest of the world's giant calderas, and may still be underlain by a residual crustal magma body located within Precambrian basement rocks (Ankeny et al. 1986).

Most continental rifts are strongly asymmetric in cross sections normal to their long axes and display great diversity in their internal fault patterns, with both planar fault arrays and curved listric fault systems. Recent geological and geophysical studies have indicated that most of the extension is accommodated by displacements on low-angled (listric) normal faults (Bosworth 1985). As listric faults curve in plan view as well as in cross section, this effectively limits the horizontal extent of a single rift bounding fault (or system of parallel faults) to a few tens of kilometres. As a consequence, most rifts are actually made up of a series of sub-basins separated by 'accommodation zones' (Bosworth op. cit.). In some instances rift asymmetry reverses at sub-basin boundaries, suggesting that the underlying detachment system also changes polarity. Young rifts may thus be considered as large-scale half-graben, the sense of asymmetry of which alternates more or less regularly along the axis of the rift (Fig. 11.11). Deep-crustal fault systems may provide important path-

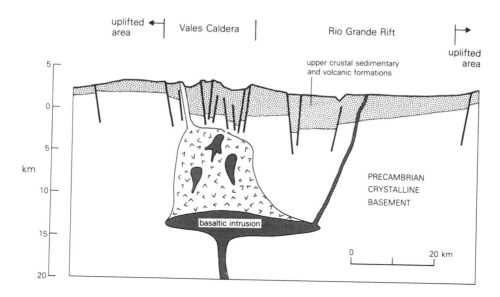

Figure 11.10   Schematic cross section across the Rio Grande rift in the vicinity of the Valles caldera, showing the location of a crustal magma chamber in Precambrian basement rocks. An intrusion of basaltic magma at about 15 km depth produces a broad zone of melting in the overlying crustal rocks. Acidic melts from this zone feed volcanic activity in the overlying caldera, while the basaltic layer is occasionally tapped to feed surface flows (after Ankeny *et al.* 1986, Fig, 9, p. 6196).

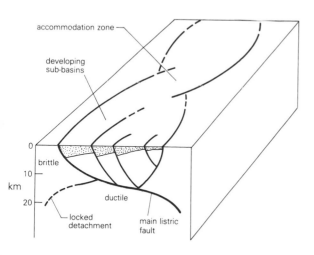

Figure 11.11   Model for the propagation of continental rifts (after Bosworth 1985, Fig. 4, p. 627).

ways for magmas rising towards the surface, and thus adjacent rift basins may be characterized by different styles of magmatic activity.

11.4   Petrography of the volcanic rocks

Given the compositional diversity of CRZ volcanic suites it is clearly impossible to document the entire range of petrographic variation. Consequently, three volcanic suites of differing alkalinity have been selected to illustrate the changes in mineralogy which occur with increasing SiO_2 content of the magma. On the basis of their major and trace element geochemistry (Section 11.5) these suites appear to represent coherent fractionation sequences which are not obviously modified by the effects of crustal contamination. They are:

(a)   a basanite−phonolite suite from the Nyambeni Range of eastern Kenya (Brotzu *et al.* (1983);

(b)   an alkali basalt−trachyte suite from the southern part of the Gregory (Kenya) Rift (Baker *et al.* 1977);

(c)   a transitional basalt−rhyolite suite from the Boina centre, Ethiopia (Barberi *et al.* 1975).

The alkalinity (% Na_2O + K_2O) of these suites decreases from (a) to (c). Generally, there is a broad spectrum of textural variation within these suites from aphyric to strongly porphyritic rock types. Figure 11.12 shows the range of phenocryst mineralogies as a function of silica content. In general, groundmass mineralogies are similar to the observed phenocryst assemblages. Those minerals restricted to the groundmass are indicated by the symbol 'g' in Figure 11.12.

Olivine appears to crystallize throughout the compositional range in each of these suites, ranging from Fo_{80} in the basic end-members to almost pure fayalite in the trachytes and rhyolites. However, it is apparently less common in rocks of phonolitic composition and does not show such extreme iron enrichment.

Ca-rich clinopyroxene also exhibits a wide range of crystallization, ranging from pale brown augite in the basic rocks through pale green hedenbergite in the trachytes and rhyolites to green−yellow pleochroic aegirine−augite in the phonolites. The degree of Na substitution in the clinopyroxenes increases with increasing SiO_2 content of the magma and is a function of the alkalinity of the suite as a whole. Thus the most alkaline suite (basanite−phonolite) also shows the most extreme Na enrichment in the pyroxenes of the salic rocks.

Fe−Ti oxides crystallize throughout the compositional range in the basanite-phonolite and alkali basalt−trachyte suites. In contrast, their appearance is suppressed to an advanced stage of fractionation in the transitional basalt−rhyolite suite. The Boina volcanics display typical tholeiitic characteristics, with marked iron enrichment in the early stages of fractionation due to the non-crystallization of Fe−Ti oxides.

Hydrous minerals (amphibole and biotite) are essentially restricted to the more evolved members of the basanite−phonolite suite. They occur sporadically as phenocrysts in the Nyambeni phonolitic tephrites and phonolites and are often extremely altered, indicating disequilibrium. Amphibole has also been recorded as a rare groundmass phase in trachytes from the southern Gregory Rift. The amphiboles are alkali amphiboles, generally pleochroic in shades of greenish brown, and show a

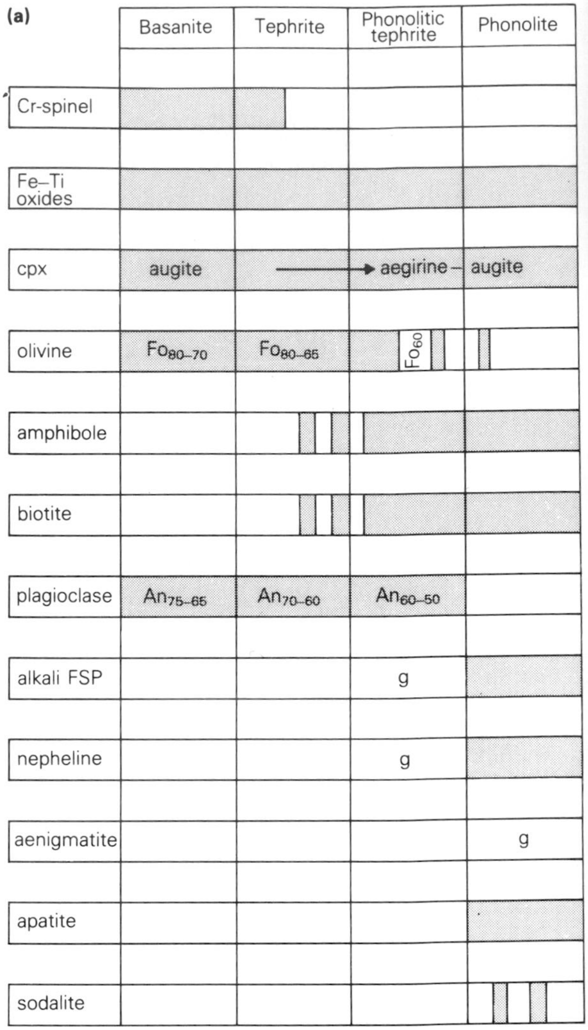

(a)

	Basanite	Tephrite	Phonolitic tephrite	Phonolite
Cr-spinel				
Fe−Ti oxides				
cpx	augite		aegirine −	augite
olivine	Fo_{80-70}	Fo_{80-65}	Fo_{60}	
amphibole				
biotite				
plagioclase	An_{75-65}	An_{70-60}	An_{60-50}	
alkali FSP			g	
nepheline			g	
aenigmatite				g
apatite				
sodalite				

Figure 11.12 Phenocryst mineralogy of continental rift zone volcanic suites of varying alkalinity: (a) basanite−phonolite suite from the Nyambeni Range of eastern Kenya (Brotzu et al. 1983); (b) alkali basalt−trachyte suite from the southern part of the Gregory (Kenya) Rift (Baker et al. 1977); (c) transitional basalt−rhyolite suite from the Boina centre, Ethiopia (Barberi et al. 1975). The alkalinity (% Na_2O + K_2O) of these suites decreases from (a) to (c). Minerals occuring only as groundmass phases are indicated by the symbol 'g'.

(b)

	Alkali basalt	Hawaiite	Mugearite	Benmoreite	Trachyte
opaques					
cpx	g	g	g	augite	aegirine–augite
olivine	Fo_{65}				Fo_{10}
amphibole					g
biotite					
plagioclase	An_{85}			An_{30}	
alkali FSP	g	g	g	anortho-clase	sanidine
apatite	g	g	g	g	g
quartz					⌒

(c)

	Transitional basalt	Trachy-andesite	Trachyte	Rhyolite
opaques	g	g		
cpx	augite			hedenbergite
olivine	Fo_{80}			Fo_{2}
amphibole				
biotite				
plagioclase	An_{75}			An_{17}
alkali FSP				anorthoclase
apatite				
quartz				

wide range of compositional variation.

Plagioclase is a common phenocryst phase in rocks from all three suites, ranging from bytownite–labradorite (An_{85-75}) in the basic end-members to oligoclase (An_{17}) in the acid rocks. In general, it appears to cease crystallizing in the most fractionated members of the three suites, generally coinciding with the appearance of alkali feldspar (anorthoclase or sanidine) as a major phenocryst phase.

Nepheline only crystallizes in the most evolved members of the basanite–phonolite suite, where it may be joined by sodalite. In contrast, the alkali basalt–trachyte and transitional basalt–rhyolite suites fractionate towards oversaturated residua with the development of quartz in the most acid rocks.

Apatite is a common accessory mineral in all three suites, occurring as both a phenocryst and a groundmass phase. The Na–Fe–Ti silicate aenigmatite ($Na_2Fe_5TiSi_6O_{20}$) is a distinctive groundmass constituent in many phonolites, recognized by its blood red to black pleochroism.

The ultrabasic ultrapotassic magmas (e.g. leucitites and leucite basanites) characteristic of some low-volcanicity rift segments (e.g. the western branch of the East African Rift) tend to have distinctive mineralogies as a consequence of their high K_2O contents. This has led to a profusion of exotic rock names (e.g. ugandite, mafurite and katungite) in the literature. In general, if the wt.% Na_2O in the rock is approximately equal to the per cent K_2O then nepheline is present, whereas if the per cent K_2O is greater than the per cent Na_2O the potassium feldspathoids leucite and kalsilite may occur. Many of these ultrapotassic rocks are feldspar free. Ugandites are characterized by the phenocryst assemblage olivine + augite + leucite, mafurites by olivine + augite + kalsilite, and katungites by olivine + melilite. These assemblages may be accompanied by Ti-rich accessory minerals such as magnetite, ilmenite, perovskite and melanite garnet.

Figure 11.13 shows some of the characteristic petrographic features of CRZ volcanic rocks.

11.5 Chemical composition of the erupted magmas

Any review of the geochemical characteristics of CRZ volcanic suites is complicated by the large number of rift-related magmatic settings and their extreme variability in terms of age, tectonic setting and intensity of magmatism. Additionally, the available geochemical data are extremely variable in quality and trace element and Sr−Nd−Pb isotopic data are lacking for many suites, making general comparisons difficult. As a consequence, in the following sections we have tended to focus on the East African rift system as its magmas span the complete range of compositional variation observed within continental rift zones. However, even for this well studied rift, trace element and Sr−Nd−Pb

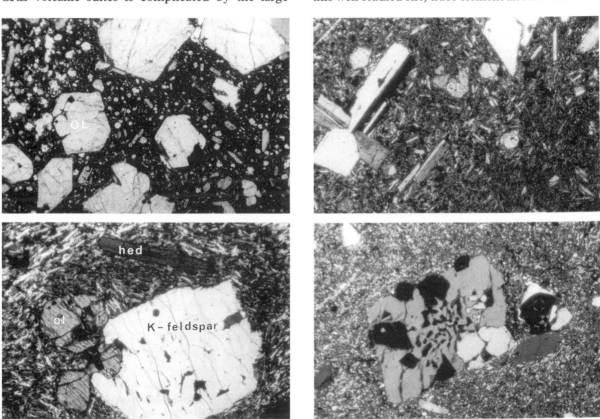

Figure 11.13 Photomicrographs illustrating the petrographic characteristics of CRZ volcanic rocks. (a) Ugandite from the western branch of the East African Rift, containing abundant phenocrysts of olivine set in a fine-grained groundmass containing rounded crystals of leucite (colourless) and clinopyroxene (augite) (×40, ordinary light). (b) Olivine basalt from the Naivasha complex (Kenya) in the eastern branch of the East African Rift, containing phenocrysts of olivine and plagioclase set in a fine-grained matrix of olivine, augite, opaques and plagioclase (×40, crossed polars). (c) Trachyte from the Naivasha complex, containing phenocrysts of clinopyroxene (hedenbergite). olivine (fayalite) and K-feldspar set in a matrix of aligned feldspar laths (×40, crossed polars). (d) Rhyolite from the Naivasha complex, containing a complex intergrowth of quartz and K-feldspar set in a fine-grained quartzofeldspathic matrix (×40, crossed polars).

isotopic data are rather limited, and these are supplemented with data from other CRZ provinces where appropriate.

As we have considered in Section 11.2, it is possible to divide continental rift zones into high-volcanicity (HV) and low-volcanicity (LV) types (Barberi *et al.* 1982). The former include the eastern branch of the East African Rift in Kenya, the Ethiopian Rift and possibly some sectors of the Rio Grande Rift, whereas the latter include the western branch of the East African Rift, the Rhinegraben and the Baikal Rift. HV and LV rifts differ in the spectrum of eruptive magma conpositions, and also in the degree of alkalinity of the parent magmas. In general, HV rifts are characterized by the eruption of mildly alkaline to sub-alkaline basalts and their differentiates, whereas LV rifts typically erupt more highly alkaline silica-undersaturated basic and ultrabasic magmas. In HV rifts there is apparently a close relationship between the chemical characteristics of basic and salic magmas erupted in the same rift sector at the same time. Thus phonolites are associated with nephelinites and basanites, trachytes with alkali basalts, peralkaline rhyolites with mildly alkaline basalts and sub-alkaline rhyolites with sub-alkaline basalts. This suggests that the acid magmas could be produced by fractional crystallization of the associated basalts. Such a model seems most appropriate for those central vent volcanoes which show a complete spectrum in their eruptive products from basic to acid. However, as we shall show in the following sections, in some instances it may also explain some of the more strongly bimodal (basic—acid) volcanic sequences, in which the rarity of intermediate composition lavas has in the past been used to argue against crystal fractionation models (e.g. Cox *et al.* 1969). Nevertheless, for some of these bimodal suites there is strong geochemical evidence (trace element and radiogenic isotopes) to indicate that crustal rocks are involved in the petrogenesis of the more silica-rich members.

In the following sections we shall evaluate the use of major and trace element and Sr—Nd—Pb isotopic data in distinguishing between the relative roles of fractional crystallization and crustal contamination in the petrogenesis of CRZ volcanic suites.

Additionally, we shall attempt to characterize the various source components involved in the petrogenesis of the primary basaltic magma spectrum.

11.5.1 Major elements

SiO_2, TiO_2, Al_2O_3, Fe_2O_3, FeO, MnO, MgO, CaO, Na_2O, K_2O and P_2O_5 can all be considered major elements in the description of the geochemistry of continental rift zone volcanic suites. Tables 11.1−5 present major element data for basalts and more evolved lavas from several suites from the East African and Ethiopian rift systems, to illustrate the spectrum of parental magma compositions and their characteristic fractionation sequences. These data may be considered to be representative of the range of CRZ magma compositions in general.

The lavas of these volcanic suites can be most easily classified using the total alkalis (Na_2O + K_2O) versus silica diagram of Cox *et al.* (1979) (Fig. 11.14). This is similar to the K_2O versus SiO_2 diagram that we have used more frequently in previous chapters, in that both Na_2O and K_2O essentially behave incompatibly until advanced stages of fractionation. The basaltic end-members clearly span the complete range from low-K transitional basalts in the Ethiopian Rift through to a range of mildly to strongly alkalic basalts in the East African Rift. In general, for these basic lavas wt.% Na_2O is greater than wt.% K_2O although for the ultrapotassic lavas of the western branch of the East African Rift (Fig. 11.14b) wt.% K_2O is greater than wt.% Na_2O. The transitional basalts from the Ethiopian Rift are geochemically comparable with the continental flood basalt suites we have discussed previously in Chapter 10.

In terms of Figure 11.14, some of the selected volcanic suites display a spectrum of compositions from basic to acid, whereas others are strongly bimodal. For those suites showing a continuous range of variation, it may be reasonable to suppose that the more acid magmas are produced by fractional crystallization of the associated basalts. However, for the bimodal suites the genetic relationships between basic and acid magmas are less

Table 11.1   Major and trace element analyses of ultrapotassic lavas from the western branch of the East African Rift.

	a	a	b
%			
SiO_2	43.22	37.69	43.15
TiO_2	4.53	5.64	3.71
Al_2O_3	9.98	7.37	12.24
Fe_2O_3	12.55	13.72	13.08
MnO	0.17	0.20	0.20
MgO	8.80	8.61	8.54
CaO	11.56	16.84	11.83
Na_2O	2.69	1.16	2.03
K_2O	4.77	3.50	3.40
P_2O_5	0.55	1.01	0.62
ppm			
Cr	464	173	369
Ni	124	60	74
V	350	449	—
Rb	115	69	127
Sr	1411	2252	1005
Y	15	14	29
Zr	280	326	306
Nb	209	269	108
Ba	1376	1496	1119
La	141.8	171.0	81
Ce	277.1	327.8	164
Nd	105.6	128.5	64
Sm	14.0	17.3	10.7
Eu	3.5	4.5	3.0
Gd	8.1	8.0	8.1
Dy	4.9	4.9	—
Er	1.6	1.6	—
Yb	1.2	1.1	2.1
Th	—	—	11.3
Ta	—	—	9.1
Hf	—	—	10.4
U	—	—	2.3

Data sources: [a] Davies & Lloyd (1988); [b] Thompson *et al.* (1984).

obvious. In general, the scatter in variation diagrams such as Figure 11.14 is a natural consequence of the combined effects of polybaric fractional crystallization, source heterogeneity, variable degrees of partial melting and crustal contamination. However, despite the apparent coherence of some of these trends, caution must be exercised in interpreting the data as representing a true liquid line of descent for, as we have shown in previous chapters, linear trends can be preserved in suites undergoing AFC. Only detailed trace element and radiogenic isotope studies can prove a true liquid line of descent.

Figure 11.15a shows the variation of wt.% K_2O versus SiO_2 for the Ethiopian Rift basalts plotted in Figure 11.14e. K_2O correlates positively with silica and the trend is similar to those displayed by typical continental flood basalt suites (Ch. 10). The bulk of

Table 11.2 Major and trace element analyses of members of the basanite−phonolite suite from Mt Kenya (Data from Price *et al.* 1985).

	Basanite	Mugearite	Benmoreite	Trachyte	Phonolite
%					
SiO_2	41.43	50.07	58.28	65.04	55.74
TiO_2	3.64	2.23	0.81	0.22	0.85
Al_2O_3	11.87	16.37	15.99	15.77	18.26
Fe_2O_3	2.74	1.63	1.69	0.90	1.51
FeO	11.55	9.32	7.11	2.84	4.63
MnO	0.23	0.25	0.31	0.17	0.25
MgO	10.52	2.93	0.74	0.15	1.01
CaO	11.10	5.96	2.79	0.99	2.57
Na_2O	2.33	5.71	6.69	6.98	8.53
K_2O	1.48	2.74	4.16	5.57	4.82
P_2O_5	0.94	1.18	0.51	0.10	0.41
H_2O	0.87	0.28	0.27	0.49	0.36
CO_2	0.07	0.89	0.01	0.07	0.02
S	0.01	0.05	0.05	0.04	0.04
F	—	—	0.12	0.36	0.15
ppm					
Rb	52	56	88	168	94
Ba	622	1028	1337	236	1324
Sr	1230	1375	481	26	881
Pb	4	8	11	29	12
Th	5	5	13	39	15
U	<1	1	2	2	<1
La	73	81	96	228	107
Ce	100	149	161	314	167
Y	26	29	44	88	33
Zr	197	283	545	1119	545
Nb	59	86	127	301	145
Sc	22	9	7	2	3
V	350	113	14	<1	23
Ni	137	2	<1	<1	<1
Cu	84	38	14	7	11
Zn	100	113	111	157	107
Ga	16	18	22	30	21
F	—	—	1164	3564	1455

the basalts plot in the sub-alkaline field, despite plotting above the alkalic−sub-alkalic boundary in Figure 11.14. Consequently, they should really be classified as transitional basalts. The coherent variation of K_2O with silica could reflect variable degrees of partial melting of a common source, or it could represent a crustal contamination trend, with the most K_2O-rich basalts being the most contaminated. Figure 11.15b shows a similar trend for transitional basalts from the Taos Plateau Volcanic Field of the Rio Grande Rift, which can be readily related to the effects of crustal contamination on the basis of Sr−Nd−Pb isotopic studies (Dungan *et al.* 1986; see also Section 11.5.3).

In the continental rift zone tectonic setting, the more salic lavas may be the products of fractional crystallization of the spatially and temporally associated basaltic magmas, possibly combined

Table 11.3 Major and trace element analyses of an alkali basalt–trachyte suite from the southern Gregory (Kenya) Rift. (Data from Baker *et al.* 1977).

	Basalt	Ferrobasalt	Benmoreite	Trachyte
%				
SiO_2	47.93	47.48	58.48	63.65
TiO_2	2.11	3.09	1.57	0.94
Al_2O_3	15.01	14.31	16.16	14.12
Fe_2O_3	2.99	3.40	1.59	2.01
FeO	8.96	10.21	4.78	6.03
MnO	0.20	0.25	0.21	0.27
MgO	6.94	5.43	2.14	0.04
CaO	12.05	10.83	4.61	1.31
Na_2O	2.69	3.07	5.53	6.34
K_2O	0.80	1.29	4.04	5.22
P_2O_5	0.32	0.64	0.39	0.07
ppm				
Sc	37	30	14	5.6
Cr	83	97	20	—
Co	47	44	14	1.1
Ni	76	67	—	—
Rb	15	30	67	115
Sr	428	415	337	10
Ba	300	510	1040	160
Y	24	45	44	93
La	24.7	45.2	76.7	152
Ce	—	—	128	185
Nd	—	51	57	91
Sm	5.48	9.42	10.83	17.9
Eu	1.86	2.69	3.39	3.17
Tb	0.90	1.28	1.32	2.4
Yb	2.6	3.4	3.6	8.9
Lu	0.41	0.61	0.58	1.64
Zr	112	183	259	764
Hf	2.9	4.7	9.0	17.7
Nb	35	55	113	207
Ta	1.4	2.6	5.2	11.1
Th	2.5	5.0	10.4	19.4

with variable degrees of crustal contamination. Alternatively, they could be the products of partial melting of sources either related to or independent of the associated basalts, possibly under volatile-rich conditions. Such a model has previously been suggested in Chapter 6 to explain the petrogenesis of magmas more SiO_2-rich than basalt by the partial melting of mantle lherzolite modified by hydrous subduction-zone fluids. In general, these various hypotheses cannot be distinguished on the basis of major element data alone.

For those volcanic suites showing continuous chemical variation from basic to acid end-members, major element variation diagrams can provide powerful constraints on the nature of fractional crystallization processes. From Figure 11.14 we have selected two such suites, from the Boina centre, Ethiopia (Fig. 11.14e), and the Nyambeni Range of the eastern branch of the East African Rift in Kenya (Fig. 11.14c) for further discussion. These suites show evolutionary trends from transitional basalt to rhyolite and basanite to phonolite

Table 11.4 Major and trace element analyses of a transitional basalt–rhyolite suite from the Boina centre, Afar Rift, Ethiopia (Data from Barberi *et al.* 1975).

	Basalt	Trachyandesites		Rhyolite
%				
SiO_2	46.75	56.81	65.02	72.11
TiO_2	2.30	1.76	0.36	0.38
Al_2O_3	13.03	13.88	14.88	9.35
Fe_2O_3	3.20	0.70	1.75	2.30
FeO	8.08	9.37	3.48	3.80
MnO	0.19	0.29	0.13	0.21
MgO	9.75	2.13	0.04	<0.01
CaO	10.08	5.04	1.34	0.34
Na_2O	2.70	5.00	5.90	5.74
K_2O	0.80	2.15	4.30	4.40
P_2O_5	0.35	0.72	0.04	0.01
ppm				
Cr	370	6	56	77
Ni	75	6	2	8
Rb	18	49	106	147
Sr	382	360	60	3
Zr	121	405	902	1170
Ba	250	408	736	<10
La	19.6	60.0	111.5	159.5
Ce	35.9	114.5	193.8	287.4
Nd	21.8	70.7	93.0	129.2
Sm	4.5	16.6	17.4	27.0
Eu	1.7	5.3	3.3	3.3
Gd	4.3	17.4	13.8	25.0
Dy	4.1	13.9	15.4	26.0
Yb	1.9	7.0	10.3	14.1
Th	—	4.7	—	14.9
Hf	4	14	27	32
U	0.65	1.65	—	3.5

respectively. In terms of Figure 11.14c, the Nyambeni suite actually displays two apparently fractionation-controlled lineages; a dominant basanite to phonolite sequence and a subordinate alkali basalt to mugearite sequence.

Figure 11.16 shows the variation of MgO and Al_2O_3 versus SiO_2 for the Boina suite. The strongly segmented trends can be interpreted in terms of crystal fractionation dominated by olivine and plagioclase in the early stages. Clinopyroxene does not appear to be a dominant phase in the fractionation sequence, but may crystallize along with plagioclase to produce the observed trends. Figure 11.17 shows the variation of MgO and Al_2O_3 with SiO_2 for the Nyambeni suite for comparison. These diagrams show that, at least in terms of the major elements, the alkali basalt–mugearite series could be derived from a similar basanitic parent magma to the basanite–phonolite series, the only difference being the point at which a clinopyroxene-plagioclase (in this case cpx dominant) assemblage began to fractionate. This could reflect differences in the depth of crystallization of the magmas or differences in parameters such as oxidation state or volatile content. Clearly, isotopic data are required to prove such a contention. It is significant that, in the basanite–phonolite suite, clinopyroxene dominates the fractionation sequence, whereas in the transitional basalt–rhyolite suite plagioclase is apparently the most important phase. This is most

Table 11.5 Major and trace element analyses of a bimodal transitional basalt–rhyolite suite from the Naivasha volcanic field, Kenya Rift. Data from Davies & Macdonald (1987) and Macdonald *et al.* (1987).

	Basalts		Rhyolite
%			
SiO_2	47.2	47.8	75.2
TiO_2	1.95	1.95	0.17
Al_2O_3	15.83	15.43	12.11
Fe_2O_3	1.60	—	0.83
FeO	9.61	12.09	1.06
MnO	0.20	0.21	0.04
MgO	7.34	6.55	0.07
CaO	12.27	11.12	0.44
Na_2O	2.62	2.26	4.59
K_2O	0.48	0.81	4.73
P_2O_5	0.24	0.42	—
H_2O	0.12	0.46	0.06
ppm			
Cr	—	67	—
Ni	—	20	—
V	—	274	—
Rb	9	20	290
Sr	382	448	6.15
Y	19	25	108
Zr	71	108	439
Nb	23	24	200
Ba	254	461	13
La	16.9	23.1	74
Ce	36.3	48.7	142
Nd	18.1	24.4	53
Sm	4.11	5.09	12.3
Eu	1.64	1.80	0.15
Gd	4.50	5.05	15.6
Dy	4.05	4.56	—
Er	2.25	2.49	—
Yb	2.13	2.14	10.1
Th	—	—	38.8
Ta	—	—	14.4
Hf	—	—	14.9
U	—	—	7.1

clearly demonstrated in the Al_2O_3 versus SiO_2 diagrams for the two suites (Figs. 11.16 & 17). Unfortunately, the relative order of crystallization of clinopyroxene and plagioclase from basaltic magmas is a function of a variety of parameters including the chemistry of the magma, the depth of crystallization, the volatile content and the oxida-

tion state of the magma (Cox *et al.* 1979). In the absence of appropriate experimental data it is impossible to decide which is the most important factor in this case.

11.5.2 Trace elements

Tables 11.1–5 include trace element data for basalts and more evolved lavas from several volcanic suites from the East African and Ethiopian rifts, which may be regarded as broadly representative of the range of CRZ magmas. As in previous chapters, we will tend to focus our discussion on the incompatible trace elements, i.e. those which are strongly partitioned into the melt phase during

Figure 11.14 Total alkalis (% $Na_2O + K_2O$) versus silica diagrams for volcanic suites from the East African and Ethiopian rifts. (a) Nomenclature of non-potassic volcanic rocks (after Cox *et al.* 1979, Fig. 2.2, p. 14). For potassic suites in which $K_2O > Na_2O$ the following equivalent rock names are used:

Rock name	Potassic equivalent
basanite	leucite basanite
nephelinite	leucitite
tephrite	leucite tephrite
phonolite	leucitophyre
alkali basalt	potassic alkali basalt

The boundary between the alkalic and sub-alkalic fields is extrapolated from Macdonald & Katsura (1964). (b) Ultrapotassic lavas from the western branch of the East African rift system. Data from Sahama (1973), Ferguson & Cundari (1975) and Davies & Lloyd (1988). Average compositions of ugandite (U), mafurite (M) and katungite (K) from Higazy (1954). (c) Basanite–phonolite fractionation series from the eastern branch of the East African Rift in Kenya. (d) Bimodal alkali basalt–trachyte suite from the southern part of the Gregory (Kenya) Rift in Kenya (Baker *et al.* 1977). (e) Transitional – mildly alkaline plateau basalts from Ethiopia (solid circles) and a transitional basalt–rhyolite fractionation sequence from the Boina centre (crosses). Data from Barberi *et al.* (1975) and Jones (1976). (f) Bimodal transitional basalt–rhyolite suite from the Naivasha volcanic complex in the eastern branch of the East African Rift in Kenya. Data from Davies & Macdonald (1987) and Macdonald *et al.* (1987).

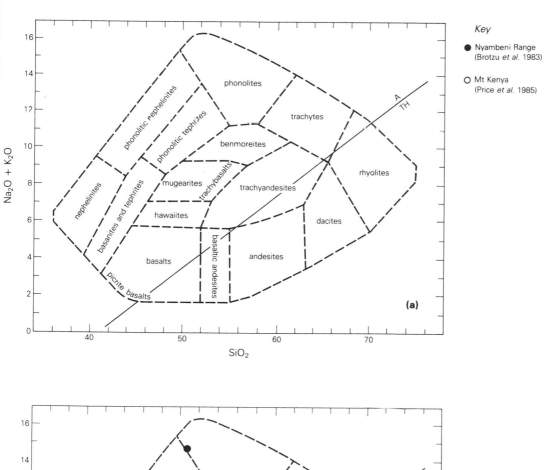

(a)

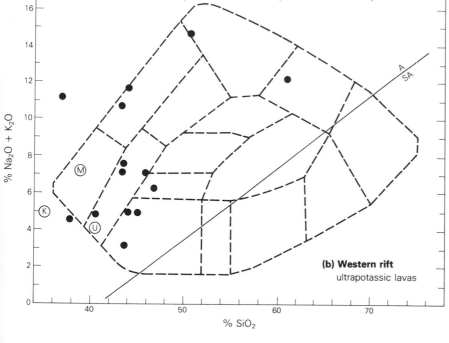

(b) **Western rift**
ultrapotassic lavas

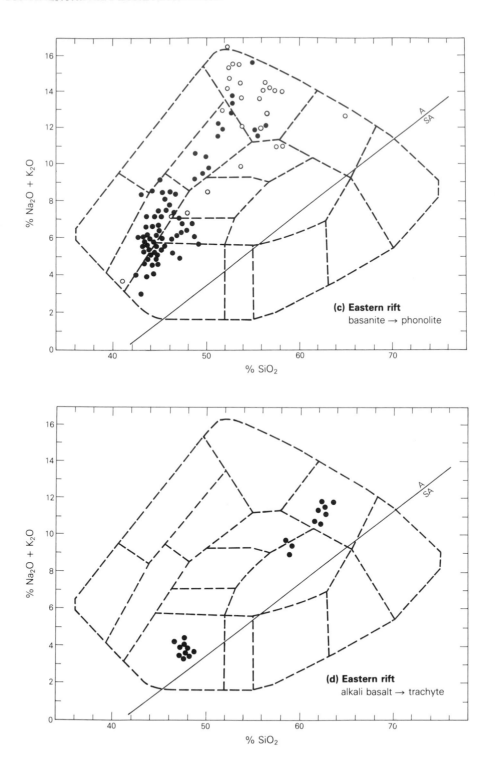

(c) **Eastern rift**
basanite → phonolite

(d) **Eastern rift**
alkali basalt → trachyte

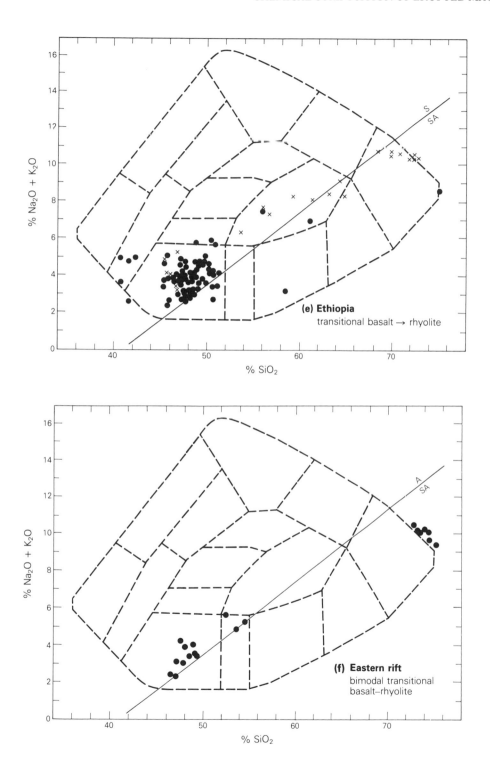

(e) **Ethiopia**
transitional basalt → rhyolite

(f) **Eastern rift**
bimodal transitional
basalt–rhyolite

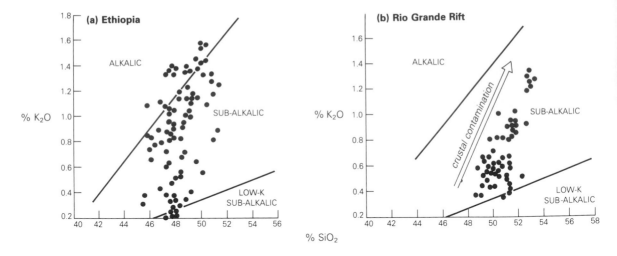

Figure 11.15 Variation of wt.% K_2O versus wt.% SiO_2 for (a) plateau basalts from Ethiopia (Jones 1976), and (b) transitional basalts from the Taos Plateau Volcanic Field of the Rio Grande Rift (Dungan *et al.* 1986). Boundaries between the alkalic, sub-alkalic and low-K subalkalic fields are from Middlemost (1975).

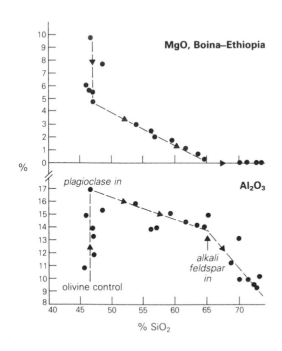

Figure 11.16 Variation of wt.% MgO and Al_2O_3 versus wt.% SiO_2 for the Boina suite, Ethiopia (data from Barberi *et al.* 1975).

partial melting and fractional crystallization processes. In general, the most basic end-members of these suites have rather low concentrations of compatible trace elements, such as Ni, suggesting that they have experienced some degree of olivine fractionation en route to the surface. Such fractionation will tend to increase the incompatible trace element concentrations in the basaltic magmas, relative to those of the more MgO-rich primary magmas, but will not produce any marked inter-element fractionations.

A characteristic feature of many East African CRZ volcanic suites is their relatively constant ratios of certain incompatible trace elements (e.g. Nb/Zr, Ce/Zr, La/Zr and Rb/Zr) in rocks of widely varying silica content (Weaver *et al.* 1972, Lippard 1973). This provides a useful test for the fractionation-controlled origin of the more SiO_2-rich magmas, because if trace element ratios are constant throughout a suite then significant crustal contamination is unlikely to have occurred. Indeed, in the absence of radiogenic isotope data to confirm a liquid line of descent, constancy of incompatible element ratios provides strong evidence that fractional crystallization has been the dominant process in the evolution of a particular suite. Even for the more strongly bimodal suites, constancy of such

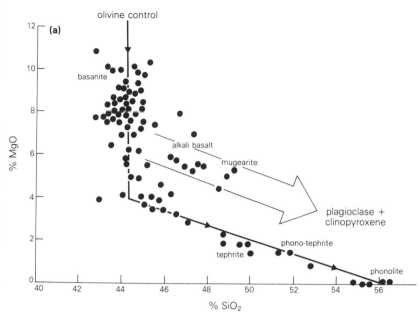

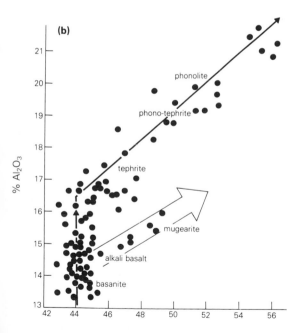

Figure 11.17 Variation of wt.% MgO (a) and Al$_2$O$_3$ (b) versus wt.% SiO$_2$ for the Nyambeni volcanic suite (data from Brotzu *et al.* 1983).

ratios would tend to negate petrogenetic models involving partial melting of unrelated sources to generate the basic and acid magmas.

Figure 11.18 shows the variation of Nb versus Zr (ppm) for a basanite−phonolite suite from Mount Kenya (Price *et al.* 1985) and an alkali basalt−trachyte suite from the southern part of the Gregory (Kenya) Rift (Baker *et al.* 1977). Both suites display a remarkably constant Zr/Nb ratio of ~3.5, suggesting that the phonolitic and trachytic eruptives may represent the products of fractional crystallization of the associated basic magmas. Additionally, these data show that the basanites and alkali basalts could be derived from similar mantle sources, at least with respect to their Zr/Nb ratio. Barberi *et al.* (1975) have shown similar good correlations between ratios of other incompatible elements, such as Hf/La and Zr/Ce, for the transitional basalt−rhyolite suite from the Boina centre, Ethiopia (Fig. 11.19), which thus may also be regarded as dominantly fractionation-controlled. Coherence in trace element plots such as these suggests that the continuous trends in the major element Harker variation diagrams for these suites (Section 11.5.1) do indeed reflect liquid lines of descent. Only fractional crystallization is likely to preserve unchanged the concentration ratio of two incompatible elements. Any other process such as crustal contamination would tend to change them.

In contrast to the above suites, Figure 11.20 shows the variation of Nb versus Zr for the bimodal basalt−rhyolite suite from the Naivasha complex in the Kenya Rift (Davies & Macdonald 1987, Macdonald *et al.* 1987). The acid

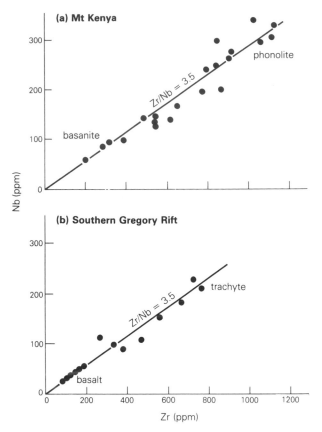

Figure 11.18 Variation of Nb versus Zr (ppm) for (a) a basanite−phonolite suite from Mt Kenya (Data from Price *et al.* 1985), and (b) an alkali basalt−trachyte suite from the southern Gregory (Kenya) Rift (data from Baker *et al.* 1977).

rocks clearly have different Zr/Nb ratios from the associated basalts, suggesting that other source components are involved in their petrogenesis. Davies & Macdonald (op. cit.) consider that many of the acid magmas are in fact generated by volatile-induced partial melting of the crust, including both crystalline basement rocks and their cover of Miocene−Recent volcanics.

While basalts from the eastern branch of the East African Rift have rather constant Zr/Nb ratios of ~3.5, transitional basalts from Ethiopia (Jones 1976) and the Taos Plateau Volcanic Field of the Rio Grande Rift (Dungan *et al.* 1986) display wide ranges. Figure 11.21 shows the variation of Y/Nb

versus Zr/Nb for these two basaltic suites. A similar diagram was used in Chapter 5 to illustrate the influence of OIB plumes on MORB geochemistry. The Ethiopian basalts plot on an apparent mixing trend between an enriched component, characterized by magmas from the Kenya Rift, and a depleted MORB-source component. This provides strong evidence for the role of asthenospheric or MORB-source mantle in the petrogenesis of transitional basalts in actively extending rift segments. Basalts from the Taos Plateau volcanic field show a similar correlation, although on a very much reduced scale, which can be related to increasing degrees of crustal contamination, which decreases both Zr/Nb and Y/Nb ratios. Thus the very high ratios of some of the Ethiopian basalts are an original source characteristic, not produced by crustal contamination. Detailed Sr−Nd−Pb isotopic data are clearly required to confirm the involvement of MORB-source mantle in their petrogenesis.

Spiderdiagrams

Having established the usefulness of incompatible trace element ratios in deducing a fractionation-controlled origin for at least some of the salic CRZ volcanic rocks, we now turn our attention to the basic end-members of these suites and the information they can provide about the nature of their mantle source. Figure 11.22 shows a series of mantle-normalized trace element variation diagrams (spiderdiagrams) for basalts from the East African and Rio Grande rifts, in which the elements are plotted in order of decreasing incompatibility from left to right in a four-phase lherzolite undergoing partial fusion. The data are normalized to chondritic abundances except for Rb, K and P, which are normalized to primitive terrestrial mantle values (Thompson *et al.* 1984). Shown for comparison in Figures 11.22c & d are typical patterns for mid-ocean ridge basalts (MORB), oceanic-island tholeiites (OIT) and oceanic-island alkali basalts. Basalts from the East African Rift are clearly enriched in the whole spectrum of incompatible elements relative to MORB, with the most strongly alkaline ultrapotassic lavas from the western rift

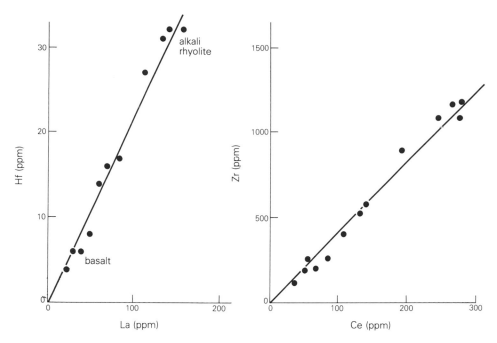

Figure 11.19 Variation of Hf versus La and Zr versus Ce (ppm) for a transitional basalt–rhyolite suite from the Boina centre, Ethiopia (data from Berberi *et al.* 1975).

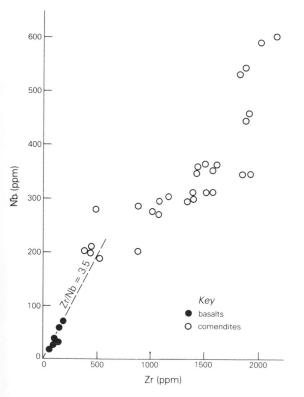

showing the greatest enrichments. Unfortunately, there are insufficient trace element data for basalts from the Ethiopian rift to plot spiderdiagram patterns. Instead, data are shown in Figure 11.22c for chemically similar transitional basalts from the Rio Grande Rift. These basalts are clearly similar to enriched MORB and oceanic-island tholeiites in their trace element geochemistry.

Thompson (1985) has compared typical spiderdiagram patterns for strongly silica-undersaturated ultrapotassic lavas (leucitites) from the western branch of the East African Rift (Fig. 11.22a) with those of potassic alkali basalts from the oceanic island of Tristan da Cunha (Fig. 11.22d). Despite differences in their major element chemistry, the spiderdiagram patterns are remarkably similar and, on this basis, Thompson has suggested that there is no necessity for involving enriched

Figure 11.20 Variation of Nb versus Zr (ppm) for a bimodal basalt–rhyolite suite from the Naivasha complex, Kenya Rift. Data from Davies & Macdonald (1987) and Macdonald *et al.* (1987).

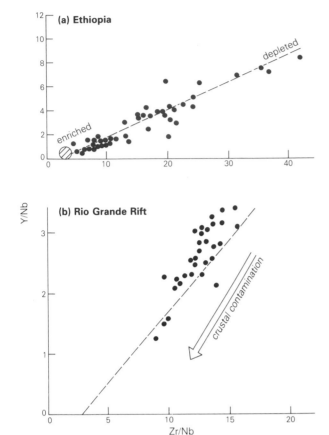

Figure 11.21 Variation of Y/Nb versus Zr/Nb for transitional basalts. (a) From the Ethiopian Rift (Jones 1976). The dashed line represents a mixing trend between an enriched source component and a depleted MORB-source component; the shaded field is the average composition of Kenya Rift basalts. (b) From the Taos Plateau Volcanic Field of the Rio Grande Rift (Dungan *et al.* 1986). The arrow represents increasing degrees of crustal contamination.

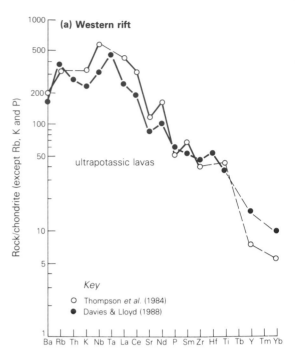

Figure 11.22 Mantle-normalized trace element variation diagrams (spiderdiagrams) for CRZ basalts from the East African and Rio Grande rifts. The data are normalized to chondritic abundances, except for Rb, K and P, which are normalized to primitive terrestrial mantle values (Thompson *et al.* 1984). (a) Ultrapotassic lavas (leucitites) from the western branch of the East African Rift. (b) Basalts from the eastern branch of the East African Rift. (c) Transitional basalts from the Taos Plateau Volcanic Field, Rio Grande Rift (Dungan *et al.* 1986): also shown for comparison are typical patterns for N- and P-type MORB and a typical oceanic-island tholeiite (OIT) from Figure 10.16. (d) Typical patterns for oceanic-island alkali basalts. (e) Basalts from the oceanic and continental sectors of the Cameroon volcanic line (Fitton & Hughes 1985).

subcontinental lithosphere mantle sources in the petrogenesis of potassic CRZ volcanic suites. A similar argument can be made for CRZ basalts in general, based upon the characteristic spider-diagram patterns of basalts from the oceanic and continental sectors of the Cameroon volcanic line, which straddles the African continental margin (Fig. 11.1). Basalts from the two sectors show identical patterns (Fig. 11.22e), suggesting the non-involvement of sub-continental lithosphere

mantle sources in the petrogenesis of basalts erupted within the continental sector (Fitton & Dunlop 1985).

The spiderdiagram patterns of all the East African Rift basalts are similar to those of oceanic and continental alkali basalts in general, having a marked peak at Nb–Ta. They must therefore be similarly derived from enriched mantle sources. In contrast, the Rio Grand Rift (Taos Plateau Volcanic Field) transitional basalts display positive spikes at

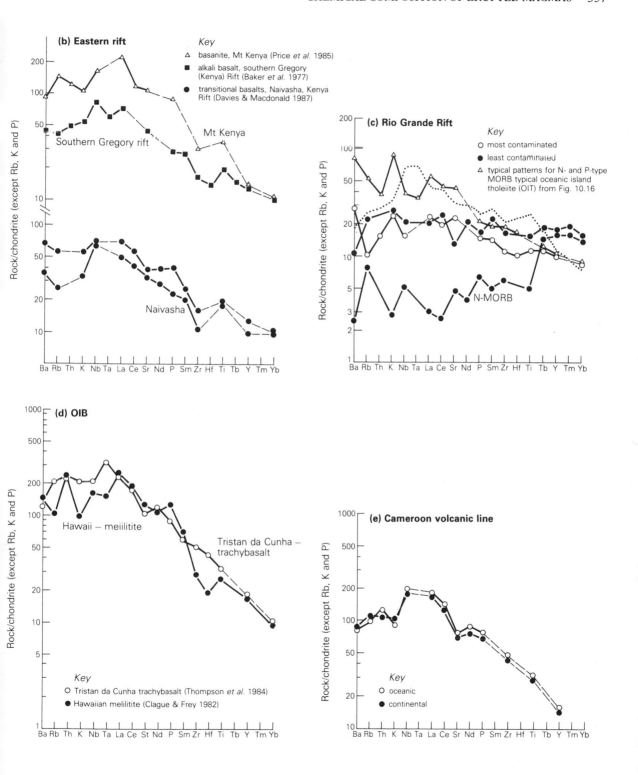

Ba, K and Sr and a Nb—Ta trough. The latter could reflect the existence of a residual Nb—Ta-bearing phase in the source during the partial melting process (cf. Ch. 6) or the effects of crustal contamination (Cox & Hawkesworth 1985). Nb and Ta depletion is typical of magmas erupted in subduction-related tectonic settings (Ch. 6). As calc-alkaline magmatism was widespread in the region of the Rio Grand Rift in Oligocene times it is possible that there is a subduction-modified mantle source component involved in the petrogenesis of the magmas. However, on the basis of Sr—Nd—Pb isotopic data (Section 11.5.3) there appears to be strong evidence in favour of crustal contamination. Intuitively, we should expect the greatest degrees of crustal contamination in regions with high rates of crustal extension, as these are associated with higher magma production rates, which will induce higher geothermal gradients in the crust.

Figure 11.23 and 11.24 show the trace element concentrations in transitional basalts from the Rio Grande and Ethiopian rifts normalized to those in N-type MORB and a typical oceanic island tholeiite

respectively (Table 10.3). Figure 11.24 clearly shows the similarity between these basalts and oceanic-island tholeiites. Nevertheless, they display the same distinctive spiked pattern characteristic of many continental flood basalts (Fig. 10.18), which appears to be a consequence of crustal contamination. Relative to oceanic-island tholeiites, the Rio Grande Rift basalts are depleted in Rb and Nb and enriched in Ba and K. Comparing these patterns with spiderdiagram patterns for amphibolite and granulite facies crustal rocks (Fig. 11.25) suggests that these basalts have been contaminated by granulite facies lower crust en route to the surface. In contrast, the Ethiopian Rift basalts show a relative enrichment in Rb, and these may have been contaminated by amphibolite facies upper crust.

REE

Figure 11.26 shows chondrite-normalized REE patterns for ultrapotassic lavas from the western branch of the East African Rift, transitional to mildly alkalic basalts from the eastern rift and

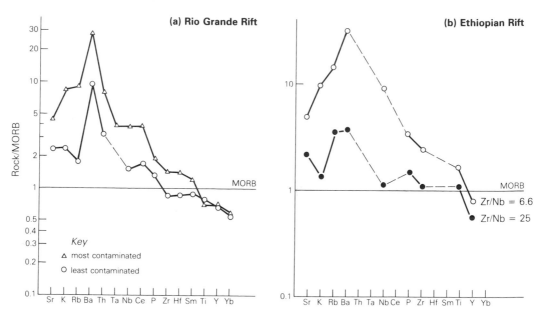

Figure 11.23 Trace element abundances in transitional basalts from the (a) Rio Grande and (b) Ethiopian rifts, normalized to N-type MORB. The elements are ordered according to Pearce (1983), and the normalization factors are from Table 10.3. The two basalts plotted for the Ethiopian rift have high and low Zr/Nb ratios respectively. Note that the highest ratio has the most MORB-like trace element pattern, consistent with its derivation from a depleted asthenospheric mantle source.

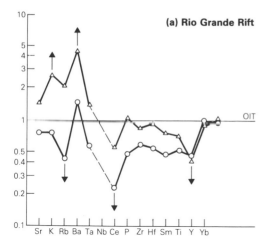

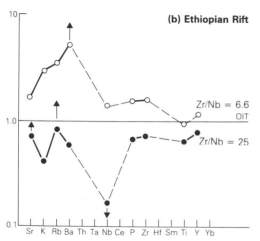

Figure 11.24 Trace element abundances in transitional basalts from the (a) Rio Grande and (b) Ethiopian rifts, normalized to typical oceanic-island tholeiite (OIT) values. The elements are ordered as in Figure 11.23, and the normalization factors are from Table 10.3. Symbols are as Figure 11.23: arrows indicate those elements which are added (↑) or subtracted (↓) during crustal contamination.

transitional to sub-alkalic basalts from Ethiopia and the Rio Grande Rift. All of these patterns are light-REE enriched, with the ultrapotassic lavas showing the most extreme LREE enrichment and heavy- to light-REE fractionation. Heavy-REE abundances are only 2−5 times chondritic for the ultrapotassic lavas, suggesting the presence of residual garnet in the source.

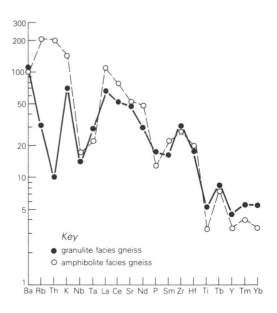

Figure 11.25 Mantle-normalized trace element variation diagrams for granulite and amphibolite facies crustal rocks. Normalization factors from Thompson et al. (1984); data from Weaver & Tarney (1981).

11.5.3   Radiogenic isotopes

Nd−Sr data

We have already established in previous chapters (Ch. 5 & 9) that the upper mantle can show considerable isotopic variability, reflecting its time-integrated evolution over the past 4.5 Ga. Additionally, in Chapters 7 and 10 we have shown that the subcontinental lithospheric mantle should, in theory, preserve the most extreme isotopic heterogeneities and may represent an important magma-source component in all regions of intra-continental plate magmatic activity. While the isotopic and chemical composition of this reservoir remains a matter for speculation, the available data for ultramafic xenoliths contained in kimberlites and continental alkali basalts confirms the existence of extensive heterogeneities, far exceeding the combined range of MORB and OIB (Allegrè et al. 1982, Kramers et al. 1983, Cohen et al. 1984).

Figure 11.27 shows the variation of $^{143}Nd/^{144}Nd$ versus $^{87}Sr/^{86}Sr$ for a range of CRZ volcanic suites.

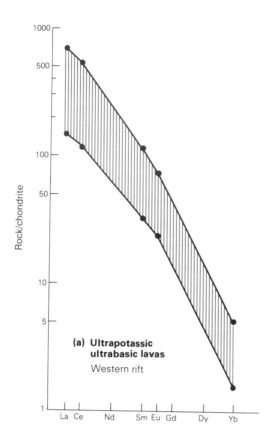

(a) Ultrapotassic ultrabasic lavas
Western rift

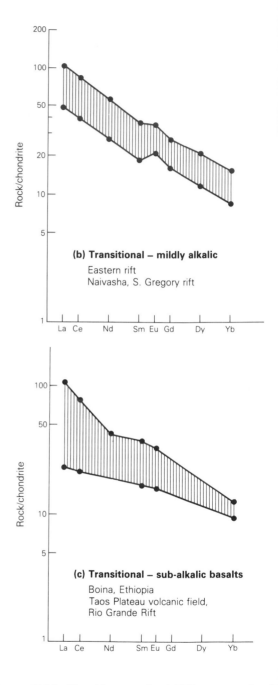

(b) Transitional – mildly alkalic
Eastern rift
Naivasha, S. Gregory rift

(c) Transitional – sub-alkalic basalts
Boina, Ethiopia
Taos Plateau volcanic field,
Rio Grande Rift

Many CRZ basalts plot within the mantle array as defined by uncontaminated oceanic basalts (MORB + OIB), whereas others plot outside this field. This has led to conflicting views concerning the origins of the latter, specifically as to whether they have undergone crustal contamination or are simply partial melts of enriched subcontinental lithosphere. In general, as we have established previously for continental flood basalts in Chapter 10, it is likely that both crustal contaminants and enriched mantle sources may be involved to different extents in different CRZ provinces. The problem then becomes one of recognizing their different trace element and isotopic signatures. The range of isotopic variability displayed by lherzolite xenoliths derived from the subcontinental lithosphere is outlined by the dashed field in Figure 11.27. Clearly, the entire range of isotopic variability of CRZ basalts could be explained in terms of their derivation from this mantle reservoir.

Figure 11.26 Chondrite-normalized REE patterns for (a) Ultrapotassic lavas (leucitites) from the western branch of the East African Rift, (b) transitional – mildly alkalic basalts from the Eastern branch of the East African Rift, and (c) transitional–sub-alkaline basalts from Ethiopia and the Rio Grande rifts. Normalization constants from Nakamura (1974). Data sources as in previous diagrams.

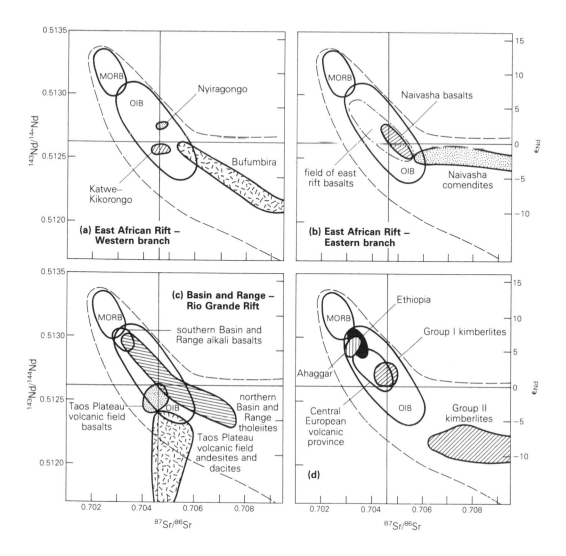

Figure 11.27 Variation of $^{143}Nd/^{144}Nd$ versus $^{87}Sr/^{86}Sr$ for volcanic rocks from continental rift zones. (a) The western branch of the East African Rift. Katwe–Kilkorongo data from Davies and Lloyd (1988); Virunga province (Nyiragongo and Bufumbira) data from Vollmer and Norry (1983a,b). The dashed line outlines the field for subcontinental lithosphere based on data for mantle-derived xenoliths (Zindler & Hart 1986). b) The eastern branch of the East African Rift. Data for field of east rift basalts from Norry *et al.* (1980)); Naivasha data from Davies & Macdonald (1987). (c) The Basin and Range and Rio Grande rift provinces of the western USA. Taos Plateau Volcanic Field data from Dungan *et al.* (1986); Basin and Range tholeiites data from Hart (1985); Basin and Range alkali basalt data from Menzies *et al.* (1983). (d) The Central European Volcanic Province (Downes 1984, Wörner *et al.* 1986), the Ahaggar volcanic complex, north Africa (Allègre *et al.* 1981) and Afar, Ethiopia (Betton & Civetta 1984). Shown for comparison are fields for Group I and Group II kimberlites from Chapter 12.

Considering first the western branch of the East African rift system; basalts from the volcano Nyiragongo in the Virunga province (Vollmer & Norry 1983a,b) and the Katwe−Kikorongo field (Davies & Lloyd 1988) plot close to bulk Earth, similar to Dupal OIB (Ch. 9); and see Fig. 11.27a. In contrast, a series of basalts and more silica-rich volcanics from the Bufumbira area of the Virunga province (Vollmer & Norry 1983a) plot in an array extending away from the MORB−OIB field to higher values of $^{87}Sr/^{86}Sr$. This trend may be attributable to the effects of high-level crustal contamination (AFC) as $^{87}Sr/^{86}Sr$ is well correlated with 1/Sr and SiO_2.

Within the eastern branch of the East African Rift, the available Nd-Sr isotopic data for basalts (Norry et al. 1980, Davies & Macdonald 1987) all plot within the OIB field (Fig. 11.27b). For the bimodal basalt−rhyolite suite from the Naivasha volcanic complex (Davies & Macdonald op. cit.) the basalts plot close to bulk Earth, whereas the acid rocks form an almost horizontal array extending to higher values of $^{87}Sr/^{86}Sr$. As the rhyolite glasses have very low Sr contents (<8 ppm), these acid magmas would have been highly susceptible to the effects of crustal contamination and therefore, in terms of the Nd-Sr data, they could represent AFC-controlled residua of the associated basalts. In Figure 11.28 $^{87}Sr/^{86}Sr$ is plotted as a function of wt.% SiO_2 for the Naivasha basalts. The diagram shows that while some have clearly undergone AFC processes, others have fractionated without concomitant crustal contamination. In Section 11.5.2 we showed, on the basis of incompatible trace element ratios, that the bulk of the Naivasha rhyolites must be derived from a different source to the associated basalts, and this is confirmed by the Nd−Sr data. In terms of such trace element ratios we also showed that the Mt Kenya basanite−phonolite suite (Price et al. 1985) could be explained in terms of simple crystal fractionation models without crustal contamination. This is confirmed by Figure 11.29, a plot of $^{87}Sr/^{86}Sr$ versus SiO_2, which shows that within error there is no Sr isotopic variation within the series.

In the western USA Nd−Sr isotopic data are available for basalts and more evolved lavas from

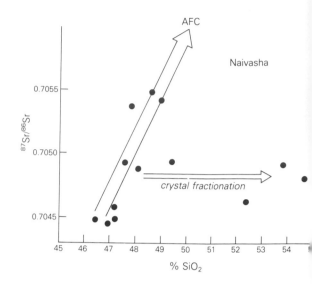

Figure 11.28 Variation of $^{87}Sr/^{86}Sr$ versus wt.% SiO_2 for basalts from the Naivasha volcanic complex, Kenya Rift (data from Davies & Macdonald 1987). AFC is assimilation and fractional crystallization trend.

the Taos Plateau Volcanic Field of the Rio Grande Rift (TPVF) and also from the adjacent Basin and Range province (Fig. 11.27c). Basalts from the TPVF (Dungan et al. 1986) plot near bulk Earth, whereas associated andesites and dacites define a large spread in $^{143}Nd/^{144}Nd$ values at approximately constant $^{87}Sr/^{86}Sr$ (Williams & Murthy 1979). This may be interpreted as an AFC trend between a mantle-derived basaltic component and an ancient continental crustal component with low Rb/Sr and high Nd/Sm. Oxygen isotope data support this, with the basalts having $\delta^{18}O$ values of 6‰, increasing up to 9.5‰ in the associated dacites (Dungan et al. op. cit.).

Menzies et al. (1983) have shown that alkali basalts from the southern part of the Basin and Range province have a very restricted range of Nd−Sr isotopic compositions similar to P-type MORB. However, these basalts contain mantle xenoliths derived from the subcontinental lithosphere, with ε Nd values ranging from + 13 to −1. This suggests that while the lithosphere is isotopically heterogeneous, the magmas may be derived from a more homogeneous asthenospheric mantle source (Perry *et al.* 1987). In contrast, tholeiitic basalts from the northern part of the province (Hart 1985) show an array of Nd−Sr isotopic compositions extending from the field of southern Basin and Range alkali basalts to low values of ^{143}Nd/^{144}Nd and more radiogenic ^{87}Sr/^{86}Sr. Hart (op. cit.) has explained this array in terms of mixing of source components, a depleted MORB-source component and an old isotopically enriched subcontinental lithosphere component. Nevertheless, a similar trend could also be produced by crustal contamination of primary mantle-derived magmas, initially plotting within the P-MORB field, by a relatively young upper crustal component. However, oxygen isotope data appear to argue against crustal contamination models as δ^{18}O only varies between 5.4 and 5.9‰.

Figure 11.27d shows Nd−Sr isotopic data for volcanic rocks from the Central European volcanic province (Downes 1984, Worner *et al.* 1986), the Ahaggar volcanic complex in north Africa (Allègre *et al.* 1981) and the Afar region of Ethiopia (Betton & Civetta 1984). The Ahaggar volcanics, like those from the southern Basin and Range province and Ethiopia, plot close to the MORB field, and may therefore be derived from mantle sources with a dominant asthenospheric (MORB-source) compo-

nent. On the basis of trace element data we would predict that those Ethiopian transitional basalts with the highest Zr/Nb ratios (Section 11.5.2) should have the most MORB-like isotopic signatures. Unfortunately, data are not available to confirm this. In general, in terms of Figure 11.27, it appears that MORB-source mantle is not a major magma source component in most CRZ volcanic provinces. Instead, we must consider the possibility that the bulk of the magmas are either derived from OIB-source mantle plumes (active rifts) or from within the subcontinental lithosphere (passive rifts).

Pb isotopes

Figure 11.30 shows the variation of ^{208}Pb/^{204}Pb versus ^{206}Pb/^{204}Pb for volcanic rocks from the East African Rift (a,b), the Basin and Range − Rio Grande Rift province of the western USA (c) and the Central European volcanic province and Ahaggar (d), compared to fields for MORB and oceanic-island basalts (OIB). In Chapter 9 we considered the existence of a large-scale isotopic anomaly in the upper mantle, the Dupal anomaly, which provides a source component for the basalts of certain oceanic islands at low latitudes south of the Equator, e.g. Tristan da Cunha. This has been distinguished as a separate field in Figure 11.30. Also shown for comparison is the location of the Northern Hemisphere Reference Line (NHRL) (Section 9.7.4) of Hart (1984).

In general, basalts from these various CRZ provinces have Pb isotopic compositions which lie to the high-^{208}Pb/^{204}Pb side of the NHRL, in some instances overlapping with the Dupal OIB field. Shown for comparison in Figure 11.30e is the Pb isotopic composition of clinopyroxene and amphibole separated from lherzolite xenoliths derived from the African subcontinental lithosphere and of a suite of mafic granulites derived from around the crust/mantle boundary (Kramers 1977, Kramers *et al.* 1983, Cohen *et al.* 1984). These data may be considered to represent the range of isotopic variation within the African subcontinental lithosphere, and may thus be used to assess the relative importance of this reservoir in the petrogenesis of East African CRZ volcanics.

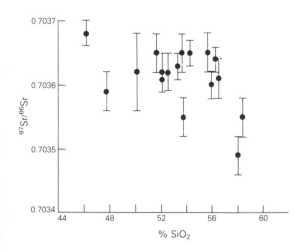

Figure 11.29 Variation of ^{87}Sr/^{86}Sr versus wt.%SiO$_2$ for the basanite−phonolite suite from Mt Kenya (data from Price *et al.* 1985). Error bars are shown for each sample.

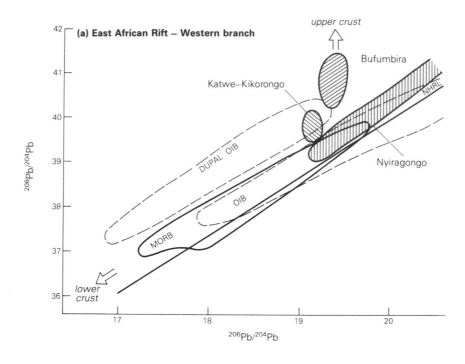

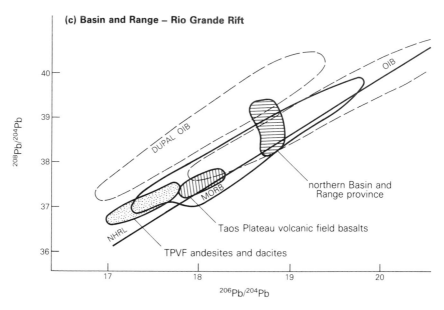

Figure 11.30 Variation of $^{208}Pb/^{204}Pb$ versus $^{206}Pb/^{204}Pb$ for CRZ volcanics from (a) The western branch of the East African Rift; Katwe–Kikorongo data from Davies & Lloyd (1988); Bufumbira and Nyiragongo data from Vollmer & Norry (1983a,b). (b) The eastern branch of the East African Rift. Eastern rift basalt data from Norry *et al.* (1980); Naivasha data from Davies & Macdonald (1987) and Macdonald *et al.* (1987). (c) The Basin and Range province and the Rio Grande Rift of the western USA. Northern Basin and Range data from Hart (1985); Taos Plateau Volcanic Field (TPVF) (Rio Grande Rift) data from Dungan *et al.* (1986).

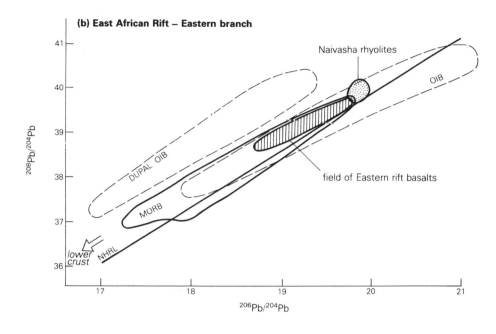

(b) East African Rift — Eastern branch

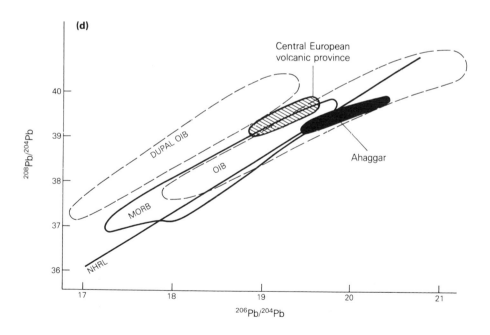

(d)

(d) Central European Volcanic Province (Wörner *et al.* 1986) and Ahaggar (Allègre *et al.* 1981). (e) The Pb isotopic composition of materials derived from the African sub-continental lithosphere. Data from Kramers (1977), Kramers *et al.* (1983) and Cohen *et al.* (1985); fields of Group I and II kimberlites from Fraser *et al.* (1985). Shown for comparison are fields for MORB and oceanic-island basalts (OIB) from Figure 9.26. NHRL is the Northern Hemisphere Reference Line of Hart (1984).

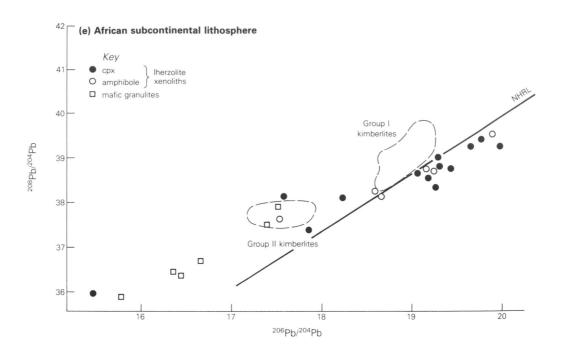

The silica-poor ultrapotassic lavas of the Katwe–Kikorongo field of the western rift (Davies & Lloyd 1988) display limited Pb isotopic variation (Fig. 11.30a), overlapping with the radiogenic end of the Dupal OIB field. These lavas have high Cr and MgO contents and contain mantle-derived xenoliths, which suggests that they have risen rapidly through the crust and not become contaminated. In contrast, the Pb isotopic compositions of lavas from the Bufumbira field of the Virunga province (Vollmer & Norry 1983a) can be accounted for in terms of AFC processes, the contaminant being young upper crustal rocks. Despite having rather uniform Nd and Sr isotopic compositions, nephelinites from Nyiragongo (Vollmer & Norry 1983b) display the largest range of Pb isotopic variation ever observed for young volcanic rocks, with $^{206}Pb/^{204}Pb$ ratios up to 62. In this case some form of unusual mantle metasomatism in the source region must have caused extreme fractionation of U from Pb to produce these extraordinarily high ratios.

Basalts from the eastern branch of the East African Rift plot in a linear trend in Figure 11.30b within the radiogenic end of the MORB field. This suggests that they are derived from an enriched mantle source and may have undergone lower crustal contamination en route to the surface. In contrast, the Naivasha rhyolites plot as a discrete cluster at the high-$^{206}Pb/^{204}Pb$ end of the basalt trend. A plot of $^{87}Sr/^{86}Sr$ versus $^{206}Pb/^{204}Pb$ (Fig. 11.31) shows the isotopic differences between the Naivasha basalts and rhyolites more clearly. The rhyolites define a near-vertical array parallel to the $^{87}Sr/^{86}Sr$ axis, which might be considered to be a crustal contamination effect due to their low Sr contents. In contrast, the basalts have high Sr contents (>400 ppm) in comparison to typical crustal materials, and therefore their Sr isotopic composition is much less susceptible to contamination. However, as their Pb contents are very low (<5 ppm) they are highly susceptible to Pb contamination by crustal materials (20–25 ppm Pb). Thus the shallow trend of decreasing $^{206}Pb/^{204}Pb$ for basaltic rocks shown in Figure 11.31 may be considered to be a consequence of AFC-controlled contamination of the basic magmas by the lower crust. Davies & Macdonald (1987) consider that the Naivasha rhyolites are in fact derived by partially melting a combination of

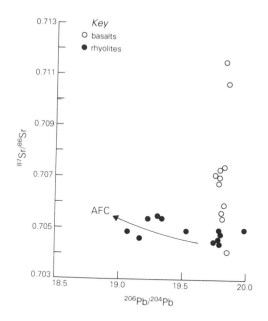

Figure 11.31 Variation of $^{87}Sr/^{86}Sr$ versus $^{206}Pb/^{204}Pb$ for basalts and rhyolites from the Naivasha volcanic complex. Data from Davies & Macdonald (1987) and Macdonald *et al.* (1987).

volcanic cover and crystalline basement rocks in the upper crust.

For the volcanic rocks of the TPVF, Rio Grand Rift (Fig. 11.30c), Pb isotope ratios correlate well with bulk composition and $^{208}Pb/^{204}Pb$, $^{207}Pb/^{204}Pb$ and $^{206}Pb/^{204}Pb$ ratios all become less radiogenic with increasing SiO_2. The Pb isotope arrays provide strong evidence for the mixing of a mantle-derived component, plotting in the least radiogenic portion of the field of MORB glasses, with a Proterozoic crustal component with low Rb/Sr, U/Pb and Th/Pb. Similar isotopic trends have been noted for basalts from the British Tertiary volcanic province (Thompson *et al.* 1983). In terms of the Pb-isotope data, it would seem that those basalts with the highest $^{206}Pb/^{204}Pb$ ratios are uncontaminated. However, their MORB-like Pb-isotopic characteristics appear at variance with their Nd−Sr isotopic compositions which plot near bulk Earth (Fig. 11.27c). Basalts from the northern Basin and Range province define a steep array

parallel to the $^{208}Pb/^{204}Pb$ axis, which could be explained as an upper crustal contamination trend or in terms of mixing of sources with enriched MORB-like and Dupal characteristics.

Basalts from the Central European volcanic province and Ahaggar plot at the high-$^{206}Pb/^{204}Pb$ end of the MORB array in Figure 11.30d. For Ahaggar this is in agreement with the Nd−Sr isotopic data (Fig. 11.27d), which appear to indicate enriched MORB-source characteristics.

Figure 11.32 shows the variation of $^{207}Pb/^{204}Pb$ versus $^{206}Pb/^{204}Pb$ for volcanics from the same CRZ provinces shown in Figure 11.30. Apart from basalts from Ahaggar, all are displaced to the high-$^{207}Pb/^{204}Pb$ side of the NHRL. The Bufumbira trend (Fig. 11.32a), subparallel to the $^{207}Pb/^{204}Pb$ axis, may again be attributable to high-level contamination of the magmas by young upper crustal materials. In the eastern rift the Naivasha basalts have rather high $^{207}Pb/^{204}Pb$ ratios, while the associated rhyolites have the highest $^{207}Pb/^{204}Pb$ and constant $^{206}Pb/^{204}Pb$ ratios. The basalts display a diffuse trend to less radiogenic Pb isotopic compositions, which may be caused by lower crustal contamination. The Taos Plateau volcanic field basalts plot above the MORB field in Figure 11.32c and display an excellent lower crustal contamination trend.

Comparison of the data in Figures 11.32a−d with the available data for xenoliths from the subcontinental lithosphere (Fig. 11.32e), reveals considerable overlap. Thus, in principle, much of the observed Pb isotopic variation amongst these CRZ volcanic suites could reflect their derivation from heterogeneous lithospheric mantle sources. However, for those suites such as the TPVF of the Rio Grande rift which define linear Pb isotopic trends, crustal contamination seems more likely. It is interesting to compare Figure 11.32 with similar plots in Chapters 6 and 7 (Figs. 6.34 & 7.29) for subduction-related magmas. Clearly, it is possible that subduction-modified mantle may provide a source component for subsequent CRZ magmatism in some tectonic settings, particularly in the western USA where active rifting postdates a subduction-related magmatic phase. Such a component is not obviously involved in the petrogenesis of

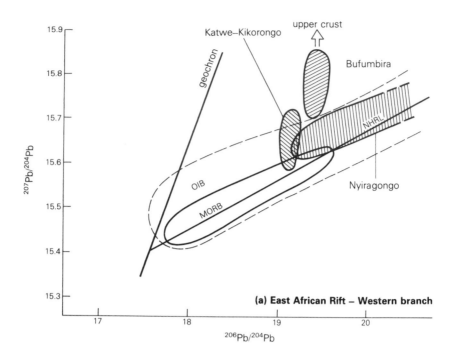

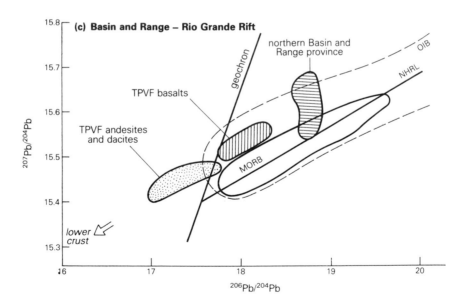

Figure 11.32 Variation of $^{207}Pb/^{204}Pb$ versus $^{206}Pb/^{204}Pb$ for CRZ volcanics from: (a) the western branch of the East African Rift; (b) the eastern branch of the East African Rift; (c) The Basin and Range province and Rio Grande Rift of the western USA; (d) the Central European Volcanic province and Ahaggar. (e) (p.370) The Pb isotopic composition of materials derived from the African subcontinental lithosphere. Data sources as Figure 11.30.

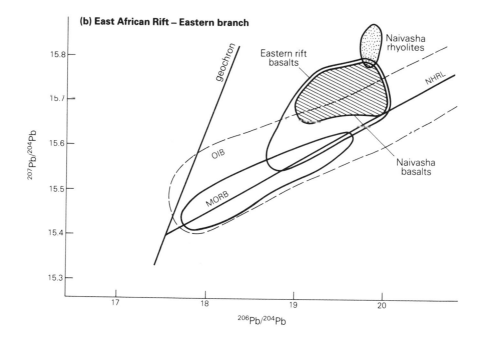

(b) East African Rift – Eastern branch

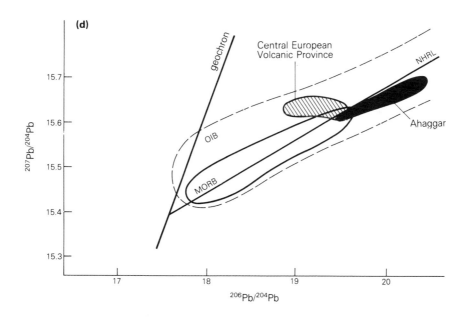

(d)

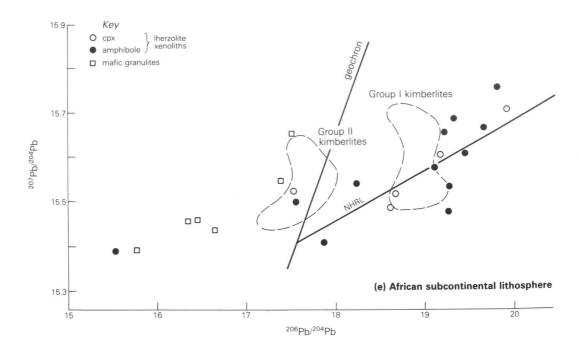

the TPVF basalts of the Rio Grande Rift. However, one could be important in the generation of the northern Basin and Range basaltic magmas, at least in terms of their Pb isotopes.

11.5.4   Stable isotopes

Variations in the isotopic composition of oxygen and hydrogen in volcanic rocks can provide important constraints on magmatic processes, involving the interaction of mantle-derived magmas with crustal rocks. For example, Kyser *et al.* (1982) and Harmon & Hoefs (1984) have shown that, in general, continental intra-plate basalts have higher $^{18}O/^{16}O$ ratios than their oceanic counterparts, which may be attributable to the effects of crustal contamination. However, caution must be exercised in the interpretation of oxygen isotope data simply in terms of crustal contamination models, as Kyser *et al.* (op. cit.) have demonstrated that the range of mantle $\delta^{18}O$ values is actually much larger than the 5−6‰ commonly assumed. Similarly, Boettcher & O'Neil (1980) have shown that *D/H* ratios for mantle-derived phlogopite and amphibole have a much wider range than the δD values of

−60±20 typically assigned to the mantle.

Nevertheless, with the above reservations, stable isotope studies, preferably in conjunction with Sr−Nd−Pb radiogenic isotope data, can be extremely useful in elucidating the role of crustal contamination in the petrogenesis of CRZ volcanic suites. Additionally, they may also provide important information about the mantle source of the primary magmas. Unfortunately, few data are available for those volcanic suites upon which we have based our discussion of the geochemistry of CRZ magmas thus far.

Harmon *et al.* (1987) have made a detailed study of the O, H and S isotope geochemistry of tholeiites, alkali basalts, nephelinites and basanites from the northern Hessian Depression, western Germany, part of the Central European Volcanic Province (Fig. 11.1). Their main objective was to assess any differences in mantle-source characteristics for these different basaltic magma types. Oxygen and hydrogen isotopic compositions show little variation with magma chemistry, $\delta^{18}O$ ranging between +5.3 and +8.0‰, while δD is approximately constant at −90. However, in contrast, sulphur isotopes indicate that the tholeiites and alkali basalts (including nephelinites and basanites) must be derived from different mantle sources. $\delta^{34}S$ varies from -0.6 to $+1.4‰$ (mean $= -0.03$) in the tholeiites and from $+0.9$ to $+8.6‰$ (mean $= +2.5$) in the alkali basalts. Typical MORB-source mantle has a $\delta^{34}S$ value of approximately 0, close to the average for the tholeiitic basalts, which may suggest that they are derived from a depleted asthenospheric mantle source. In contrast, the alkali basalts are enriched in $\delta^{34}S$ relative to MORB-source mantle, and Harmon et al. (op. cit.) consider that they are the products of partial melting of a metasomatically enriched mantle source. On the basis of the stable isotope data they suggest that the metasomatic event was two-phase, involving both CO_2- and H_2O-rich fluids, and that it must have occurred approximately contemporaneously with the melting event.

Wörner et al. (1987) have made a detailed O and H isotopic study of the phonolitic Laacher See volcano in the Quaternary East Eifel volcanic field of the Central European Volcanic Province. They show that the parental basanitic magmas from which the phonolites have evolved were characterized by $\delta^{18}O$ values of $+5.5$ to $+7.0‰$ and δD of -20 to $-10‰$. The oxygen isotope ratios are comparable to those of basalts from the northern Hessian Depression and are typical of continental alkali basalts (Kyser et al. 1982). However, the D/H ratios of these basanites show considerable D enrichment, which Wörner et al. (op. cit.) interpret in terms of metasomatic enrichment of their mantle source by fluids derived from recycled oceanic crust. During the evolution of the Laacher See volcano, $\delta^{18}O$ values changed from $+5.5‰$ in the basanites to $+8‰$ in the phonolites, which may be attributable to the assimilation of amphibolite facies crustal rocks.

11.6 Detailed petrogenetic model

The chemical composition of magmas erupted in intra-continental plate rift zones depends upon a variety of factors including the chemical and mineralogical heterogeneity of the mantle source, the degree of melting, the depth of melting, the rate of magma transfer to the surface and the existence of high-level magma storage reservoirs. Generally, in those provinces characterized by cinder-cone fields, basaltic lavas which appear to have risen relatively rapidly to the surface without undergoing significant fractional crystallization or crustal contamination dominate. Within these provinces there is little geophysical evidence to support the existence of high-level magma chambers. In contrast, in those provinces characterized by the development of large central volcanic structures, crustal magma reservoirs exist in which fractional crystallization can produce a wide spectrum of more silica-saturated magmas, from intermediate compositions to trachytes, phonolites and alkali rhyolites. Clearly then, fractional crystallization, possibly in combination with crustal contamination, is an important process in controlling the geochemical evolution of many CRZ magmas.

A fundamental problem in the study of CRZ magmatism concerns the relative roles of asthenospheric and lithospheric mantle sources in the petrogenesis of the primary basaltic magma spectrum. In Section 11.3 we considered the geophysical evidence for the ascent of asthenospheric mantle diapirs beneath the axes of the Kenya and Rio Grande rifts. Extensive thinning of the lithosphere in this manner should produce a significant contribution to the developing melt zone beneath the rift axis from asthenospheric mantle sources. In general, it seems reasonable to suppose that those rifts with the greatest degrees of crustal extension and mantle upwelling should also contain the greatest asthenospheric contribution to the partial melts. Additionally, we might predict that during the evolution of a rift there should be a progressive changeover from lithosphere- to asthenosphere-dominated sources. Detailed trace element and Sr−Nd−Pb isotopic studies of extensive stratigraphic sections would be required to test such a hypothesis, but unfortunately such data are presently unavailable. Part of the problem for modern rifts is that the older eruptive sequences tend to be deeply buried beneath younger rift volcanics and therefore cannot be easily sampled.

The role of subcontinental lithospheric mantle in the petrogenesis of CRZ basalts remains a controversial one. Random samples of this reservoir, transported to the surface in continental alkali basalts and kimberlites, display marked heterogeneities in their Sr−Nd−Pb isotopic characteristics (Kramers *et al.* 1983, Menzies *et al.* 1983, Cohen *et al.* 1984), preserving a unique record of magmatic, metasomatic and metamorphic events within the lithosphere, possibly spanning more than 2000 Ma of Earth history. In terms of their radiogenic isotope compositions it is clearly possible to generate virtually the entire spectrum of CRZ basalts by partial melting of such a lithospheric source. However, as we have already shown in Section 11.5.3, many of the distinctive isotopic trends displayed by CRZ volcanic suites can equally well be explained in terms of crustal contamination models.

Geochemical studies of the Cameroon volcanic line (Fitton & Dunlop 1985) have provided a unique opportunity to assess the relative roles of subcontinental and sub-oceanic lithosphere in the petrogenesis of intra-plate alkali basalts. This is a line of Tertiary−Recent volcanoes stretching for 1600 km from the Atlantic oceanic island of Pagalu across the continental shelf and into the African continent. It is not a hotspot trace but an incipient rift associated with the trend of the Cretaceous Benue trough (Fig. 11.1). The trace element (Fig. 11.22e) and isotopic characteristics of basalts from the continental and oceanic sectors are identical, suggesting that significant involvement of lithospheric mantle sources can be ruled out. However, caution must be exercised in applying such a conclusion too liberally to all examples of CRZ magmatic activity.

Barberi *et al.* (1982) have subdivided CRZs into high- and low-volcanicity types, on the basis of the relative volume of their eruptive products. The different geodynamics of these two rift types must clearly influence the magma generation process. In low-volcanicity rifts the rates of crustal extension are lower, corresponding to smaller degrees of lithosphere attenuation and hence lesser asthenospheric penetration into the lithosphere. Such rifts might therefore be expected to contain the greatest contribution from subcontinental lithosphere mantle sources to the partial melts. In contrast, in high-volcanicity rifts it seems inevitable that the basic magmas should be derived by partial melting of asthenospheric mantle rising beneath the rift axis, possibly mixing with lithospheric partial melts en route to the surface.

Generally, there appears to be a good relationship between the volume of erupted basic lavas and their chemical composition, transitional to mildly alkaline basalts being characteristic of increased magma production rates. The volume of magma is obviously an expression of the intensity of the melting process in the ascending mantle diapir beneath the rift axis. Ethiopia is affected by a greater rate of crustal extension and a correspondingly faster rate of asthenospheric upwelling than many other CRZs. As a consequence, the magmatic activity is dominated by transitional basalts, which presumably represent the greatest degrees of partial melting of the mantle source. In contrast, the rate of extension is very much lower in the Kenya Rift and the magmas are consequently more alkaline, being generated by smaller degrees of partial melting, possibly at greater depths.

To add to the complexity of the model we must accept that the upwelling asthenosphere beneath the rift axis is unlikely to be homogeneous MORB-source mantle. In Section 11.2 we considered that there was a spectrum of rift types, from those driven by deep-mantle upwellings (active rifts) to those generated passively as a consequence of lithospheric stretching (passive rifts). In active rifts the asthenosphere rising beneath the axis may be dominated by an OIB-source mantle plume, whereas in passive rifts it may be MORB-source mantle. Consequently, we should predict lithosphere/MORB-source mantle sources for basalts generated in passive rifts and lithosphere/OIB-source mantle sources for active rifts.

In any intracontinental plate magmatic environment it seems inevitable that at least some of the magmas should become contaminated by the crustal rocks through which they rise. The imprint of such contamination will depend both upon the chemistry of the primary magmas and the crustal rocks themselves. Specifically, contamination within the upper or lower crust should produce recognizable isotopic fingerprints. However, given the extreme isotopic heterogeneity of the mantle part of the subcontinental lithosphere it is often extremely difficult to prove unequivocally that crustal contamination has occurred.

The petrogenesis of evolved alkaline magmas (trachytes, phonolites and alkali rhyolites) within continental rift zones remains a subject for considerable speculation. In some instances trace element and radiogenic isotope data support an origin by low-pressure fractional crystallization of basaltic magma (e.g. Barberi *et al.* 1975, Baker & McBirney 1985, Price *et al.* 1985). In others the relative volumes of basic and salic magmas and the lack of intermediate compositions appears to argue against such a simple model (Davies & Macdonald 1987, Macdonald *et al.* 1987). Instead, partial melting of crustal rocks or even the upper mantle under anomalous volatile-rich conditions (Bailey 1980) have been proposed to explain the origin of the salic magmas. For the Naivasha volcanic complex, in the eastern branch of the East African Rift in Kenya, Davies & Macdonald (op. cit.) have shown that crustal components were extensively involved in the petrogenesis of the rhyolitic end-members of a bimodal basalt–rhyolite suite. Their model involves intrusion of considerable volumes of transitional basaltic magma into the crust, which raises the geothermal gradient promoting the development of hydrothermal systems. The latter then induce partial melting of the upper crust and its volcanic cover at depths of only few kilometres to produce the rhyolites.

Volcanic suites with unusually large volumes of salic rocks, Daly gaps and a scarcity of associated basalts appear to be peculiar to the continental rift zone tectonic setting. The Kenya Rift contains the largest accumulation of peralkaline salic lavas in the world (Lippard 1973), and the great volumes of these lavas has led to a continuing debate about their petrogenesis. Despite this there have been no systematic isotopic and trace element studies of the flood trachytes and phonolites, necessary to resolve the various arguments. Obviously, to produce such large volumes of alkaline salic magmas by fractional crystallization requires even larger volumes of alkali basaltic parent magmas, which must pond in high-level crustal magma reservoirs. In such an

environment it seems inevitable that AFC processes will be important in the evolution of the magmas.

McKenzie (1984) has suggested that if convective upwellings or mantle plumes are initiated from the 670 km seismic discontinuity, then they and their partial melts should rise indiscriminately beneath both continental and oceanic regions. This suggests that there may be some similarities between active intracontinental plate rifts and oceanic islands. Figure 11.33 shows simplified models of these two environments, with an estimate of the Sr–Nd-isotopic compositions of the various mantle reservoirs which may become involved in the partial melting process. The isotopic composition of the plume component in each case is based on the observed range in OIB (Ch. 9), whereas that of the

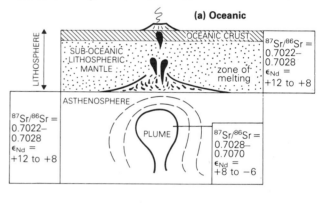

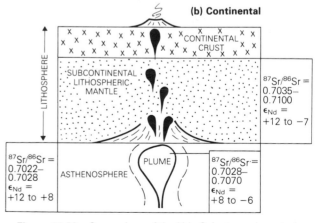

Figure 11.33 Comparison of the Nd–Sr isotope characteristics of the different mantle reservoirs involved in the petrogenesis of (a) oceanic-island and (b) active continental rift basalts. (After McDonough *et al.* 1985, Fig. 7, p. 2063).

sub-oceanic lithosphere and the depleted astheno-spheric upper mantle is taken to be the same as normal (N-type) MORB (Ch. 5). The range for the subcontinental lithospheric mantle is based on available data for garnet and spinel lherzolite xenoliths in kimberlites and alkali basalts. This is far greater than that of any of the other mantle reservoirs, which is to be expected given the variety of tectonic and magmatic processes involved in the evolution of this part of the mantle. These models are clearly a gross oversimplification, as the depleted upper mantle may itself be internally heterogeneous on a small scale (Cohen & O'Nions 1982a, Zindler *et al.* 1984).

In terms of Figure 11.33, in both the oceanic and continental cases, intrusion of a mantle plume into the base of the lithosphere will raise the temperature and may initiate partial melting of the lithosphere. Lithospheric mantle melts would then mix with plume-derived melts, generating a spectrum of chemical and isotopic compositions. Additionally, we must also consider the possibility of incorporating partial melts from the depleted asthenospheric mantle reservoir, forming the envelope to the rising plume.

The chemical and isotopic composition of tholeiitic basalts from oceanic islands has been shown in Chapter 9 to provide the most direct information about the nature of certain plume components. In the Hawaiian Islands there is a consistent trend in which the tholeiitic shield-building lavas are characterized by higher $^{87}Sr/^{86}Sr$ and lower ϵNd than post-caldera collapse alkalic lavas, the isotopic compositions of which overlap with the MORB field (Chen & Frey 1983, Feigenson 1984). It has been suggested that the tholeiitic lavas are dominated by the plume component, while the late-stage alkalic lavas are derived from the sub-oceanic lithosphere. For continental intra-plate basalts it is unfortunately much more difficult to characterize the plume component because of the potentially extreme isotopic heterogeneity of the subcontinental lithosphere.

In terms of their major element chemistry and mineralogy, there seems no reason to suppose that the mantle sources of the majority of transitional to mildly alkalic CRZ basalts are anything other than normal four-phase spinel or garnet lherzolites. However, those ultrapotassic magmas generated within some low-volcanicity rifts have such high K/Na ratios that it seems unlikely that they could be partial melts of similar materials. Experimental studies (Wendlandt 1984, Foley *et al.* 1986) suggest instead that small degrees of partial melting of a phlogopite-bearing peridotite at depths below the level of amphibole stability might produce a high-K_2O partial melt. This modification of the mantle source of ultrapotassic magmas has been attributed to metasomatism associated with the early stages of continental rifting (Bailey 1985). This will be considered further in the following chapter (Ch. 12).

Further reading

Bailey, D.K. 1983. The chemical and thermal evolution of rifts. *Tectonophysics* **94**, 585−97.

Basaltic Volcanism Study Project 1981. *Basaltic volcanism on the terrestrial planets.* New York: Pergamon Press 108−31.

Bosworth, W. 1985. Geometry of propagating continental rifts. *Nature* **316**, 625−7.

Girdler, R.W. 1983. Processes of planetary rifting as seen in the rifting and breakup of Africa. *Tectonophysics* **94**, 241−52.

Keen, C.E. 1985. The dynamics of rifting: deformation of the lithosphere by active and passive driving forces. *Geophys. J.R. Astron. Soc.* **80**, 95−120.

Palmason, G. 1982. *Continental and oceanic rifts.* Washington, DC: American Geophysical Union.

CHAPTER TWELVE

Potassic magmatism within continental plates

12.1 Introduction

Most petrogenetic models assume that primary basic and ultrabasic magmas are generated by varying degrees of partial melting of fertile lherzolite within the upper mantle (Ch. 3). The degree of partial melting and the depth of segregation of the magmas are considered to be the main variables in controlling the composition of the melt. Additionally, variations in the volatile content and mineralogy of the source mantle, and the extent of subsequent fractional crystallization and crustal contamination are invoked to explain the wide range of terrestrial basaltic magma compositions (Chs. 3 & 4). In the majority of tectonic settings which we have discussed thus far, an important characteristic of the primary basaltic magmas is that they contain significantly greater concentrations of Na_2O than

K_2O on a weight per cent basis. Exceptions to this are the more potassic members of the subduction-related magmatic series (Chs. 6 & 7) and certain potassic oceanic-island suites (e.g. Tristan da Cunha; Ch. 9). However, even within these relatively potassic suites Na_2O is still greater than K_2O in the basaltic end-members. Rather more rarely, and almost totally restricted to within-continental plate-tectonic settings, basic and ultrabasic magmas are generated in which the content of K_2O exceeds that of Na_2O, often significantly so. Kimberlites are included within this category in addition to a range of highly potassic igneous rocks with exotic names and frequently equally exotic mineralogies, generated in a wide variety of tectonic settings.

Foley *et al.* (1987) have suggested that such potassium-rich igneous rocks should be termed 'ultrapotassic' if they contain high contents of K_2O (>3 wt.%) and have high K_2O/Na_2O ratios (>3 on a wt. % basis). An additional requirement is that they have high Mg numbers (100 Mg/(Mg + Fe)), MgO contents greater than 3 wt. % and high Ni and Cr contents. These criteria are necessary to restrict the definition to relatively primitive compositions, as the fractionation of plagioclase from basic magmas can significantly enhance the K_2O/Na_2O ratio of evolved liquids. This definition excludes most kimberlites which, despite having high K_2O/Na_2O ratios, have only low total alkali contents. Additionally, it also excludes most lamprophyres apart from some minettes (Rock 1986, 1987; Bergman 1987).

Clearly, this separation of ultrapotassic rocks from other alkaline rocks, based on K_2O/Na_2O ratio and K_2O content, is a rather arbitrary one. In some provinces basic and ultrabasic magmas with widely varying K_2O/Na_2O ratios and K_2O contents occur in close association, suggesting that ultrapotassic magmas should be considered as one end-member of a continuum.

Three geodynamic situations appear to favour the production of potassic to ultrapotassic mafic magmas (Thompson & Fowler 1986):

(a)   They are the rare products of magmatism above active subduction zones, where they are intimately associated with members of the calc-alkaline suite (Chs. 6 & 7). In this respect they may be considered to include members of the shoshonite series in which $Na_2O \approx K_2O$ (Morrison 1980). Examples include the Sunda arc of Indonesia (Wheller et al. 1987), Fiji and New Guinea (Gill 1981). Considerable diversity of opinion exists as to the origin and tectonic significance of highly potassic volcanic rocks in island arcs and active continental margins (Cundari 1980, Edgar 1980). A fundamental question is whether K-rich magma production is directly related to the subduction of oceanic lithosphere or to subsequent crustal uplift and rifting comparable to that in continental rift zones (Ch. 11).

(b)   They frequently occur during or after continental collision following ocean-basin clo-

sure. The post-collisional phase of potassic magmatism may continue for tens of millions of years before grading into extension-related alkalic intracontinental plate volcanism (Ch. 11). Much of the Neogene–Quaternary potassic volcanism of the Mediterranean region, including the Roman province of Italy, is related to such a tectonic setting (Keller 1983).

(c)   Very rare examples of ultrapotassic magmatism occur in regions of extensional intracontinental plate magmatic activity. Examples include the south-west Ugandan segment of the western branch of the East African Rift (Thompson 1985) and the West Kimberley (Varne 1985) and New South Wales (Nelson et al. 1986) provinces of Australia. This tectonic setting has been discussed previously in Chapter 11.

Kimberlites are the products of continental intra-plate magmatism, confined to regions of the crust underlain by ancient cratons. No occurrences have been described from oceanic environments or young fold belts (Mitchell 1986). The tectonic environment in which they are emplaced continues to be the subject of considerable debate (Crough et al. 1980, Wyllie 1980, Mitchell 1986). Particular regions of the lithosphere appear to have acted as foci for repeated cycles of kimberlite magmatism, and the reactivation of major zones of weakness in the lithosphere plays a significant role in their emplacement. Kimberlite magmatism has been related to the rise of sublithospheric mantle plumes beneath ancient cratonic nucleii by several workers (England & Houseman 1984, Le Roex 1986). As we have considered in previous chapters (Chs. 9–11) such plumes or hot spots appear to be the driving force for many intracontinental plate-tectonic phenomena, causing both lithospheric thinning and rifting. However, there is no evidence to suggest that kimberlites occur in association with major continental rift zones (Mitchell 1986), which tend to form along zones of weakness in mobile belts and rarely traverse the old cores of cratons. Crough et al. (1980) and Crough (1981) have shown, on the basis of reconstructions of plate motions, that many post-Jurassic kimberlites apparently formed within

5 degrees of a known mantle hot spot. They suggest that kimberlites in North America, South America and Africa are related to the palaeotracks of presently active Atlantic hot spots. England & Houseman (1984) consider that kimberlite magmatism only occurs when such hot spots impinge on the base of a slowly moving continental plate.

Foley et al. (1987) and Bergman (1987) have presented useful reviews of the major occurrences of potassic and ultrapotassic volcanism, while Mitchell (1986) provides a comprehensive description of kimberlite magmatism. On the basis of these data, Figures 12.1 and 12.2 indicate the major localities of kimberlite and ultrapotassic magmatism on a global scale.

In discussing the petrogenesis of potassic and ultrapotassic magmas we are clearly dealing with magmas generated in a diversity of tectonic settings. Both groups are characterized by extremely high concentrations of incompatible elements (Section 12.5.2), which appears to preclude an origin by partial melting of normal garnet or spinel lherzolite mantle, as extremely small degrees of partial melting (<1%) would be required. Consequently, most authors agree that they must be derived from metasomatized or enriched mantle sources. Thus, despite the obvious differences in tectonic setting, it is useful to consider this rare and remarkably diverse group of igneous rocks together, as they are linked both in their potassic character and in being the products of small degrees of partial melting of anomalous mantle material. As we have already established in previous chapters, there are a number of ways in which metasomatism of the mantle may occur, ranging from the migration of small-volume partial melts or fluids over the course of geological time to the input of H_2O-rich fluids from subduction zones (Ch. 6). All of these kinds of

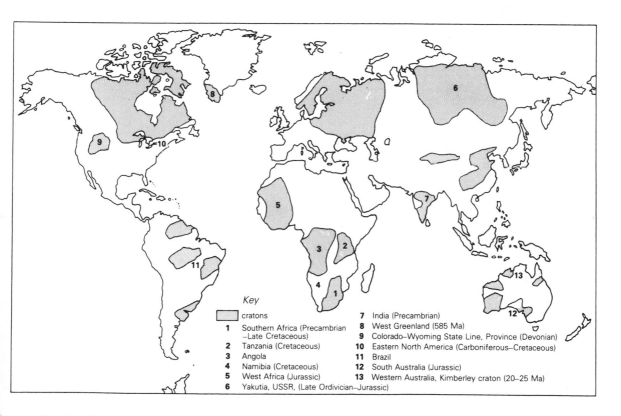

Key

cratons

1	Southern Africa (Precambrian–Late Cretaceous)
2	Tanzania (Cretaceous)
3	Angola
4	Namibia (Cretaceous)
5	West Africa (Jurassic)
6	Yakutia, USSR, (Late Ordivician–Jurassic)
7	India (Precambrian)
8	West Greenland (585 Ma)
9	Colorado–Wyoming State Line, Province (Devonian)
10	Eastern North America (Carboniferous–Cretaceous)
11	Brazil
12	South Australia (Jurassic)
13	Western Australia, Kimberley craton (20–25 Ma)

Figure 12.1 Distribution of major kimberlite provinces in relation to the positions of cratonic nuclei (shaded) (data from Mitchell 1986).

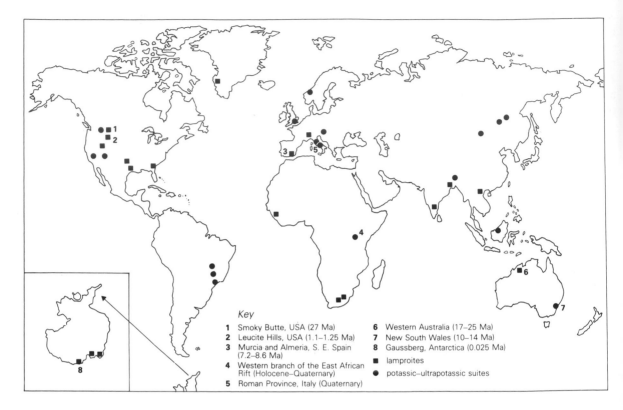

Figure 12.2   Distribution of the major lamproite occurrences and of other potassic−ultrapotassic suites. Data from Bergman (1987) and Foley *et al.* (1987) Specific localities referred to in the text are numbered.

mantle metasomatism are potentially capable of providing suitable source components for potassic and ultrapotassic magmas, but may imprint very different trace element characteristics.

Potassic and ultrapotassic magmas display considerable mineralogical diversity (Section 12.4), which has led to considerable confusion over their nomenclature, resulting in multiple names for rocks which may be very similar chemically. A number of recent publications have proposed useful simplifications but these still rely partly or wholly on modal mineralogy (Jaques *et al.* 1984, Scott-Smith & Skinner 1984b, Mitchell 1985, Bergman 1987). Foley *et al.* (1987) have subdivided the ultrapotassic rocks into three subgroups based upon their major element chemistry (Section 12.5.1), as follows.

Group I: lamproites (orenditic class of Sahama 1974)

Lamproites are ultrapotassic magnesian igneous rocks, typically with $K_2O/Na_2O > 5$ (wt. % basis) and SiO_2 in the range 45−55% (Bergman 1987). They range in age from Proterozoic to Quaternary and occur in every possible igneous form, most commonly in dykes and flows but also in diatremes or pipes. In general, they are the products of post-orogenic magmatic phenomena in regions that have experienced continental collision, with underlying fossil Benioff zones, several tens to hundreds of millions of years previously (Bergman op. cit.). In contrast with the general restriction of kimberlites to craton interiors, lamproites generally occur closer to craton margins. However, like kimberlites, they are commonly associated with lineaments or fault trends which may reflect deep zones of weakness within the lithosphere.

Generally, lamproites overlap in major element composition with both kimberlites and lamprophyres, and Rock (1987) considers that both kimberlites and lamproites should be classified as subgroups of the lamprophyre clan. However, this remains a subject for debate. Dawson (1987) has demonstrated considerable geochemical similarities between olivine-bearing lamproites and some types of micaceous kimberlite, which may indicate their derivation from similar mantle sources.

In the past a variety of rock names were assigned to members of the lamproite group, mainly on the basis of mineralogical differences. These include madupite, orendite, wyomingite, katungite, mafurite, fortunite, jumillite and verite (Section 12.4). Such names are essentially redundant within a modern petrogenetic framework and, for the most part, have little applicability outside their type locality.

Some of the classic lamproite occurrences are located on Figure 12.2. These include: Leucite Hills, Wyoming, USA; Gaussberg, Antarctica; Smoky Butte, Montana, USA; Murcia and Almeria, southern Spain; Western Australia. Of these only the Spanish lamproites can be directly related to a recent subduction-related magmatic event (Venturelli et al. 1984), although Nixon et al. (1984) consider that their petrogenesis is more likely to be related to Pliocene post-nappe block faulting. Gaussberg lies at the intersection of the Kerguelen-Gaussberg aseismic ridge with the Antarctic continent and may therefore be hot spot related. From Figure 12.2 we can see that all these major lamproite localities are Cenozoic in age, in contrast to the vast majority of kimberlites which are pre-Cenozoic (Fig. 12.1).

Group II: ultrapotassic rocks of continental rift zones (kamafugitic class of Sahama 1974)

Included within this group are leucitites, olivine melilitites, some ultrabasic lamprophyres and some micaceous kimberlites. The type locality is the Toro—Ankole province of the western branch of the East African Rift. This has been considered

previously in Chapter 11, and will not be considered further here.

Group III: ultrapotassic rocks of active orogenic zones

The occurrence of highly potassic leucite-bearing lavas is a typical feature of many Neogene to Quaternary volcanic provinces in the Mediterranean region, for example the Roman Province of central Italy and the active Aeolian and Hellenic arcs. Similar examples of potassic magmatism also occur in Anatolia, Turkey and Iran (Keller 1983). Such volcanism is dominantly related to the complex plate convergence of Africa and Europe.

In all of these provinces, ultrapotassic magmas are intimately associated with less potassic types. For example, in the Roman Province of Italy four groups of potassic igneous rocks occur (Peccerillo & Manetti 1985):

(a)   a potassic (K) series ranging from trachybasalts to trachytes;
(b)   a highly potassic (high-K) series ranging from leucite tephrite to leucite phonolite;
(c)   ultrapotassic leucite- and kalsilite-bearing melilitites;
(d)   minettes.

Experimental studies (Wendlandt & Eggler 1980a,b) suggest that the parental magmas of each of these groups were generated at different depths in a K-rich phlogopite-bearing mantle source. These primary magmas then underwent a complex process of evolution in the crust involving AFC (assimilation and fractional crystallization; Ch. 4) and magma mixing.

The Roman Province is one of the classic occurrences of potassic alkaline magmatism and, as such, has been a focus for extensive geochemical studies (Cox et al. 1976, Thompson 1977, Hawkesworth & Vollmer 1979, Civetta et al. 1981, Holm et al. 1982, Peccerillo et al. 1984, Peccerillo & Manetti 1985). High levels of enrichment of incompatible elements and high $^{87}Sr/^{86}Sr$ ratios in the most primitive lavas have been interpreted in terms of their derivation from enriched mantle sources with subordinate amounts of crustal contamination. At least some of this mantle enrichment must be induced by subduction-zone fluids, as the mafic potassic lavas display clear geochemical affinities with subduction-related volcanic rocks (Section 12.5).

Kimberlites

Kimberlites are volatile(CO_2 + H_2O)-rich potassic ultrabasic igneous rocks with low Na_2O contents, high K_2O/Na_2O ratios and high concentrations of incompatible elements (Smith *et al.* 1985, Mitchell 1986). They occur as small volcanic diatremes and pipes and also as dykes and sills in clusters, fields and provinces. Kimberlite provinces may consist of a single field, several fields of similar age or several fields of differing age and petrological character. A kimberlite field may consist of up to 100 individual intrusions. Provinces exhibiting multiple episodes of kimberlite magmatism (e.g. Southern Africa and Yakutia, USSR) are clearly important with respect to the elucidation of the tectonic controls on the emplacement of kimberlite magmas and the location of their mantle sources. Kimberlites commonly contain inclusions of upper mantle derived ultramafic rocks (Ch. 3) and variable quantities of crustal xenoliths and xenocrysts. Occasionally they may contain xenocrysts of diamond. Economically important diamond-bearing kimberlites are found in regions underlain by continental crust older than 2.4 Ga, while those occurring within younger accreted mobile belt terranes tend to be barren (Mitchell 1986). Kimberlites are thought to originate at depths of 100−200 km by partial melting of mantle peridotite in the presence of H_2O and CO_2 and to ascend rapidly to the surface within at most a few hours to a few years (England & Houseman 1984). In cases where diatremes are formed they break explosively through the upper few kilometres of the crust. The magmatic episodes are of relatively short duration and involve small volumes of magma.

Kimberlite magmatism is a rare essentially pre-Cenozoic phenomenon (Fig. 12.1). In North America there have been four major periods of activity since the beginning of the Devonian (early−mid Devonian, mid-Permian, late Jurassic − early Cretaceous and post-Eocene). In contrast, in Africa post-early Palaeozoic kimberlite activity occurs in an apparently continuous episode from the early Jurassic to the late Cretaceous, with peaks in the mid-Jurassic and mid−late Cretaceous (Mitchell 1986).

The petrography of kimberlites is complex because, in addition to exhibiting modal mineralogical variations arising from magmatic differentiation processes, they are also hybrid rocks containing crystals from fragmented crustal and mantle xenoliths. They have been divided into two petrographic types, non-micaceous (Group I) and micaceous (Group II) (Smith 1983). Both types may contain diamond, and within southern Africa display a broad age correlation with Group I (80−114 Ma) being generally younger than Group II (114−200 Ma). These two groups also have significantly different Nd−Sr−Pb isotopic characteristics (Section 12.5.3), which must reflect fundamental differences in their mantle source (Smith op. cit.). Group I kimberlites have Nd−Sr isotopic compositions which suggest that they are derived from asthenospheric-mantle sources, whereas Group II may be derived from enriched mantle sources within the subcontinental lithosphere.

In west Africa, Angola and Namibia, kimberlites were intruded subsequent to the opening of the South Atlantic ocean. In many cases they appear to have been emplaced along reactivated Precambrian shear zones in the basement (Mitchell 1986). Similarly, the emplacement of kimberlites in eastern North America is believed to be related to stresses generated during the opening of the North Atlantic which have reactivated old basement fractures (Taylor 1984).

12.2 Simplified petrogenetic model

Many of the petrogenetic problems posed by ultrapotassic igneous rocks are similar to those of other alkaline rock types such as kimberlites, melilitites and lamprophyres. All are characterized by extreme enrichment in incompatible elements, which makes them difficult to derive by partial melting of normal garnet or spinel lherzolite mantle sources. To achieve the observed incompatible element concentrations would require very small degrees of partial melting (<1%) for conventional melting models (Ch. 3), and there are clearly considerable problems in separating such small-volume melts from the mantle. A variety of hypotheses have been suggested to explain this extreme incompatible element enrichment (Foley *et al.* 1987) including high degrees of high-pressure eclogite fractionation from more normal basaltic melts, assimilation of crustal material rich in incompatible elements, zone refining of a large vertical column of mantle and partial melting of a pre-enriched (metasomatized) phlogopite-bearing mantle source. While each of these processes may be important in a particular tectonic setting, most authors favour some sort of source enrichment process.

Sr−Nd−Pb isotopic studies (Section 12.5.3) have provided substantial evidence for the involvement of enriched mantle sources in the petrogenesis of potassic intracontinental plate magmas. In some instances the trace element geochemistry of the lavas indicates enrichment of Nd/Sm and Rb/Sr in the source which is not reflected in their Nd−Sr isotopic compositions, suggesting a very recent (<200 Ma) metasomatic event. However, the isotopic compositions of the majority of ultrapotassic lavas and Group II kimberlites indicate much older events (>1 Ga). Opinions differ as to whether the enriching agents are small degree volatile-rich partial melts or H_2O-CO_2 fluids (Fraser *et al.* 1985, Hawkesworth *et al.* 1985, Foley *et al.* 1987, Menzies 1987).

In order to elucidate the nature of the anomalous mantle source of these magmas it is important to constrain the composition of the primary partial melts, i.e. those which have been least modified by the effects of fractional crystallization and crustal contamination. A major problem in this respect is that the usual criteria established for partial melts in equilibrium with typical upper-mantle mineralogies (olivine + orthopyroxene + clinopyroxene + garnet or spinel), high Mg number (>70), high Ni (>500 ppm), high Cr (>1000 ppm) and SiO_2 not exceeding 50%, may not be applicable. This is because the metasomatism of the source may be so extreme that harzburgite (olivine + orthopyroxene) is no longer the residue from partial melting and thus these phases will no longer buffer the Mg number, Ni and Cr contents of the partial melts. In the absence of the above criteria the presence of mantle-derived ultramafic xenoliths can be used to infer relatively primary characteristics for the host lava. These would be expected to settle out if significant fractional crystallization had occurred.

Given the range of tectonic settings in which ultrapotassic magmas occur it is quite clear that a number of different source components are likely to be involved in their petrogenesis. In those cases in which there is an obvious link between the magmatism and a precursor phase of subduction-related magmatism, subducted sediments and/or subduction-zone fluids, released from the descending lithospheric slab, may provide a source of potassium and incompatible elements. Such components may be important in the petrogenesis of Group I and Group III ultrapotassic magmas. In contrast, in those provinces situated on thick stable Precambrian basement, an enriched source component situated within the continental lithosphere seems more likely. As we have considered in previous chapters (Chs. 10 & 11) the stable non-convecting mantle beneath the cratons preserves a complex record of depletion and enrichment events, perhaps extending over periods greater than 1 Ga. Migrating silicate melts and aqueous fluids from the underlying asthenosphere may have significantly modified the trace element characteristics of portions of the subcontinental lithosphere. Enhanced trace element ratios will eventually register as modified isotopic ratios if left undisturbed for considerable periods of time. Extremely heterogeneous mantle may be produced by such processes and may extend to depths greater than 200 km beneath some stable continental regions (Richardson *et al.* 1984). In Chapter 9 we considered the role of recycled subducted oceanic lithosphere (on a timescale >1 Ga) as a potential source component for oceanic-island volcanism. Smith *et al.* (1985) and Le Roex (1986) have suggested that this may also be involved in the petrogenesis of hot spot related kimberlites.

Any magma generated within the asthenosphere must first rupture the lithospheric mantle and crust and establish by magma fracturing a network of conduits in order to ascend to the surface (Ch. 3). The lithospheric mantle thus provides a potential source of 'mantle contamination' for asthenospheric melts (Menzies 1987). Mixing of asthenospheric and lithospheric melts may actually be a more significant process than crustal contamination in determining the isotopic and trace element characteristics of some ultrapotassic magmas.

Hawkesworth *et al.* (1985) consider that lamproites and kimberlites are among the most extreme products of mantle enrichment processes. A fundamental question is whether their generation occurs in response to the introduction of fluids or partial melts into fairly normal mantle or whether they only occur when old, volatile-rich, anomalous mantle is remobilized. On the basis of the available Nd−Sr−Pb isotopic data (Section 12.4.3) the latter would seem more likely in the majority of occurrences.

One of the main problems in developing petrogenetic models for kimberlites and ultrapotassic magmas involves the depths at which the magmas separate from their mantle source. The depth at which the upper mantle may melt to form kimberlite can be estimated from the constraints imposed by the presence of xenocrystal diamond and coesite, and from the thermodynamic calculation of equilibration pressures for the entrained suites of mantle-derived ultramafic xenoliths (Ch. 3). Stishovite has never been reported from kimberlites, and therefore the maximum depth for kimberlite generation should be limited by the intersection of the stishovite−coesite inversion curve with a typical shield geotherm (Fig. 12.3; and see Mitchell 1986). However, it is always possible that pre-existing stishovite could have inverted to coesite during slow upwelling of material from greater depths. The diamond−graphite inversion curve indicates the minimum depths required for diamond stability, approximately 135 km for the geotherm shown. All diamond-bearing kimberlites should have originated below this depth. Calculated equilibration pressures for garnet lherzolite xenoliths should give

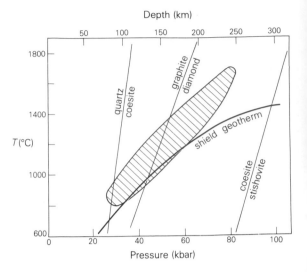

Figure 12.3 Constraints on the depth of origin of kimberlite magmas based on the pressures and temperatures of equilibration of mantle xenoliths (shaded field) and the relative stabilities of quartz, coesite, stishovite and diamond (after Mitchell 1986, Fig. 9.1, p. 373).

another estimate of the minimum depth of kimberlite generation. Unfortunately, there is still considerable disagreement between the results of the various $P-T$ methods used (Mitchell 1986). The field shown in Figure 12.3 encompasses the results of all the commonly used methods, and suggests that the minimum depth of kimberlite generation is of the order 200−250 km. This confirms that kimberlites are derived from deeper levels in the mantle than any other magma type.

Diamondiferous lamproites must also originate at depths within the diamond stability field, suggesting minimum depths of ∼ 135 km for typical shield geotherms. The depth of origin of non-diamond bearing lamproites and Group II and III ultrapotassic magmas is rather more difficult to constrain. Nixon et al. (1984) consider that the presence of spinel lherzolite xenoliths in some of the Spanish lamproites suggests their derivation from mantle depths shallower than those at which garnet is stable, perhaps less than 60−70 km. Experimental data relevant to the petrogenesis of the Roman Province (Group III) ultrapotassic lavas (Wendlandt and Eggler (1980 a,b)) suggest magma generation in the pressure range 14−30 kbars, 45−100 km depth. Additionally Arima and Edgar

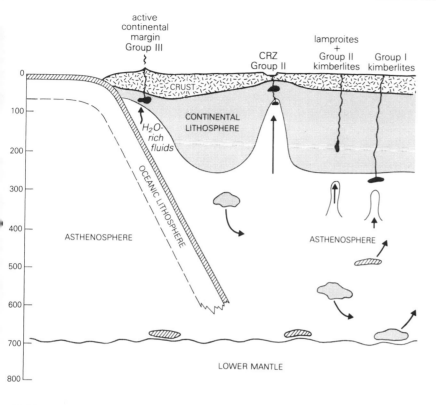

Figure 12.4 Schematic illustration of the range of tectonic environments and source components involved in the petrogenesis of potassic intracontinental plate magmas.

(1983a) have shown that it is highly unlikely that Group II ultrapotassic magmas from the East African Rift could have equilibrated with a garnet or spinel lherzolite mantle source, containing H_2O and CO_2, at depths up to 40 kbars. This is inferred from the absence of olivine + orthopyroxene + clinopyroxene + garnet or spinel near the liquidus of these lavas up to these pressures. Instead they conclude that a more reasonable source is a phlogopite clinopyroxenite at pressures greater than 20 kbars. In contrast appropriate experimental data relevant to the petrogenesis of kimberlites suggest that they might be derived from a phlogopite-magnesite garnet lherzolite at pressures of 40–50 kbars (Mitchell (1986), Wyllie (1980)).

Group II kimberlites and lamproites have Nd–Sr isotopic ratios (Section 12.4.3) distinctly different from those of oceanic basalts and more akin to those of continental crustal rocks or ultramafic xenoliths derived from the sub-continental lithosphere. As these isotopic ratios are unlikely to have been modifed by crustal contamination, due to the high ascent rates and high

Sr, Nd and Pb contents of the magmas, we could argue that the source of these potassium-rich magmas resides totally within the subcontinental lithosphere. However, it is likely that in many cases an asthenospheric component (e.g. a mantle plume) is required to provide the heat and possibly volatiles necessary to trigger partial melting (Menzies 1987). If the ultrapotassic magmatism is generated passively as a consequence of intra-plate stresses reactivating deep lithospheric fractures, then partial melting of the lithospheric mantle could simply be a response to adiabatic decompression.

Figure 12.4 shows the range of tectonic environments in which potassium rich magmas may be generated within continental plates. The source of Group I kimberlites is shown to be within the asthenospheric upper mantle, on the basis of their Nd–Sr isotopic compositions, whereas Group II kimberlites originate at somewhat shallower depths within the subcontinental lithosphere. The chemical characteristics of ultrapotassic magmas in orogenic areas can be attributed to the involvement of both asthenospheric and lithospheric mantle components, the partial melting of which is triggered by H_2O-rich subduction-zone fluids. Within the majority of these ultrapotassic provinces the range of magma compositions observed is primarily a consequence of partial melting processes. However, within Type III provinces (e.g. the Roman Province) large subvolcanic magma chambers may develop in which significant low-pressure fractional crystallization and crustal contamination may occur. Even in kimberlites some near-surface differentiation may occur (Mitchell 1986).

According to the model shown in Figure 12.4, some kimberlites are formed as a consequence of adiabatic decompression in mantle diapirs rising from the asthenosphere. Segregation of the melt may occur within the lithosphere or at its base, followed by rapid ascent of the kimberlite magma into the crust. Ascent within the lower lithosphere may be controlled by pressure-induced fracture propagation (Spera 1984). At higher levels the magma may be channelled into pre-existing fracture zones that have been reactivated by intra-plate stresses. In contrast, Bailey (1980, 1982, 1985) favours a more passive mechanism which he terms 'volatile fluxing'. In his model fractures, developed or reactivated by tectonic processes external to the craton, penetrate to the base of the lithosphere and provide channels for the escape of volatiles from the asthenosphere. Regions adjacent to these channels eventually become metasomatized and may ultimately melt to provide a source component for kimberlite magmatism. This model may also have some relevance to the generation of ultrapotassic magmas within continental rift zones.

12.3 Partial melting processes in the upper mantle

A variety of factors may influence the compositions of mantle-derived partial melts, including the mineralogy of the source, variations in the composition and abundance of volatiles, and variations in the pressure and temperature of partial melting. As we have shown in Chapter 3, varying the degree and depth of partial melting, under essentially anhydrous conditions, can generate a spectrum of basaltic magma compositions from a garnet or spinel lherzolite source. Additionally, a range of experimental studies on the melting of peridotite in the presence of excess volatiles (Kushiro 1972, Boettcher et al. 1975, Mysen & Boettcher 1975) have shown that a wide spectrum of primary magma compositions could also be generated by varying the H_2O/CO_2 ratio of the source. However, large quantities of volatiles are unlikely to be present in the mantle, and therefore these experiments are probably unrealistic in terms of interpreting the petrogenesis of magmas.

The mineralogy of the upper mantle is in part dependent upon the abundance and proportions of H_2O and CO_2 present. These volatile species may be fixed in hydrous phases such as amphibole or biotite and carbonate minerals (calcite, magnesite and dolomite). Alternatively, under appropriate conditions they may exist as a separate fluid phase. Modal mantle metasomatism, involving the introduction of hydrous phases and a range of accessory minerals, has been proposed as a critical precursor to alkaline magmatism (Boettcher & O'Neil 1980, Menzies & Murthy 1980, Wass & Rogers 1980, Bailey 1982, Wilkinson & Le Maitre 1987) and seems to be essential in the generation of ultrapotassic magmas.

Mica and amphibole (K-richterite) have long been recognized as important repositories in the mantle for K, Ba and Rb as well as volatiles. Amphibole is probably the major source of Ti and K for low K/Na and medium K/Na magma types, whereas mica is more likely to be the major K-bearing phase in the source regions of high-K/Na magmas (Wilkinson & Le Maitre op. cit.). However, the observed enrichments in incompatible elements in kimberlites and ultrapotassic magmas require sites for a much greater number of elements than those contained in mica and amphibole. A variety of minor phases have been recorded in metasomatized mantle xenoliths, including crichtonite group minerals, wadeite, priderite, rutile, perovskite and ilmenite (Haggerty 1983). While these minerals could clearly be appropriate sites for the storage of incompatible elements in the upper mantle, their existence in the source regions of ultrapotassic magmas remains somewhat equivocal. These accessory phases could crystallize from a number of sources, including fluids ascending from a subducted lithospheric slab, those emanating from ascending mantle plumes and silicate melts of a variety of origins.

Schneider & Eggler (1986) consider that any C–O–H fluids which exist in the upper mantle are mixtures of H_2O and CO_2 and that CH_4 does not play any significant role in mantle metasomatism, as the mantle is for the most part too oxidized. Separate fluid phases can only exist in the colder lithospheric mantle which is subsolidus. In the asthenosphere any fluid present would immediately cause partial melting and become incorporated into the melt phase due to the higher temperatures.

The association of Group II ultrapotassic rocks from the East African Rift with carbonatites argues for a CO_2-rich magma-generation environment (Bailey 1980), and this may be true for all Group II rocks (Foley *et al.* 1987). Bergman (1987) considers that both kimberlite and alkali basalt mantle sources are also characterized by low $H_2O/(H_2O + CO_2)$ ratios, whereas lamproite sources have high H_2O/CO_2. This is reflected in the volatile contents of lamproites, which are H_2O dominated, whereas kimberlites have sub-equal amounts of H_2O and CO_2. Additionally, Foley *et al.* (1986) have shown that fluorine is important in the petrogenesis of lamproites.

In order to understand the petrogenesis of kimberlites and ultrapotassic magmas it is useful to consider the results of three different types of melting experiments:

(1) melting of 'normal' peridotite in the presence of H_2O and CO_2;
(2) melting of metasomatized peridotite in the presence of H_2O and CO_2;
(3) studies of the liquidus behaviour of ultrapotassic magmas themselves in order to constrain their $P–T$ of formation.

12.3.1   Melting of 'normal' peridotite in the presence of H_2O and CO_2

The addition of H_2O to lherzolite compositions lowers the solidus and causes the production of more SiO_2-rich partial melts than under anhydrous conditions. In contrast, CO_2 has the opposite effect and causes the production of SiO_2-poor melts at temperatures only slightly lower than the volatile-free solidus (Ch. 3). 'Normal' mantle can only store small amounts of H_2O; about 0.06% and 0.4% (by weight) are required to make the maximum amounts of phlogopite or amphibole respectively (Wendlandt and Eggler 1980b). Addition of H_2O to an originally anhydrous bulk composition will result in the formation of amphibole at lower pressures and phlogopite at higher pressures.

CO_2 reacts with mantle minerals to form carbonates in a series of carbonation reactions (Wyllie & Huang 1976). The two most important reactions are:

$$olivine + clinopyroxene + CO_2 \rightleftharpoons orthopyroxene + dolomite$$

and

$$olivine + dolomite + CO_2 \rightleftharpoons orthopyroxene + magnesite$$

Approximately 5% CO_2 is required to make the first of these reactions go to completion in a lherzolite containing 10–15% clinopyroxene.

As both H_2O and CO_2 are likely to be important in the source region of kimberlites and ultrapotassic magmas, we clearly need a range of experiments in the presence of H_2O–CO_2 fluids of fixed composition. Unfortunately, there is some conflict concerning phase relationships in the system peridotite–H_2O–CO_2 (Brey *et al.* 1983, Olafsson & Eggler 1983, Wyllie 1987) and the compositions of the partial melts have been inadequately characterized. However, there is a general consensus that partial melts of phologopite peridotite will be K_2O- and MgO-rich, while those of carbonated peridotite will be carbonate-rich. The MgO/CaO ratio of the latter will depend upon whether dolomite or magnesite is stable at the solidus.

During partial melting of a volatile-enriched peridotite, hydrous phases, such as amphibole and phlogopite, and carbonates are rapidly consumed over a small temperature interval close to the solidus (Brey *et al.* op. cit., Olafsson & Eggler op. cit.). Thus near-solidus partial melts will have distinctive compositions reflecting the nature of the minerals stable at the solidus, since melt compositions are determined by element partitioning between the liquid and the residual crystals. These chemical characteristics will be progressively diluted at greater degrees of partial melting by increased consumption of olivine and orthopyroxene into the melt.

Figure 12.5 shows the subsolidus phase relations of peridotite in the presence of small amounts of H_2O (0.3%) and CO_2 (0.7%) (after Olafsson & Eggler op. cit.) Brey et al. (op. cit.) have investigated a similar system with 0.3% H_2O and 5% CO_2, and although the configuration of their diagram is similar there is considerable disagreement between the two experiments over the pressure at which the various subsolidus phase boundaries occur (Wyllie 1987). For the purpose of this discussion we will ignore this discrepancy and use the diagram of Olafsson & Eggler, as it is the principles involved, rather than the absolute pressures, which are more important.

The solidus for amphibole peridotite in the presence of H_2O-CO_2 vapour (V) extends from low pressures to point A where dolomite becomes stable. Between A and B amphibole and dolomite coexist at the solidus, with olivine + orthopyroxene + clinopyroxene + garnet + spinel in the absence of a free vapour phase. At pressures higher than B the solidus is for dolomite peridotite plus vapour, with phlogopite present in rocks containing sufficient K_2O. Above about 27 kbar dolomite is replaced by magnesite at the solidus.

When magnesite, dolomite or amphibole exist at the solidus the composition of any coexisting vapour phase is buffered by reactions involving these hydrated or carbonated phases. Thus when amphibole is present at the solidus, the coexisting vapour, at pressures below A, is enriched in CO_2, whereas when carbonate is stable at pressures above B the vapour becomes increasingly H_2O-rich. Schneider & Eggler (1986) consider that such buffering reactions will separate the subsolidus lithosphere into two regions. At depths greater than 70 km will be a region of carbonate + phlogopite peridotite in which fluids are H_2O-rich. Above this is a region of amphibole peridotite in which fluids are CO_2-rich. Schneider & Eggler consider that H_2O-rich fluids will migrate out of and leach the carbonate region, enriching the overlying amphibole region. If the upward-migrating hydrous fluids are solute-undersaturated, considerable leaching of K and LIL elements may occur from phlogopite-enriched regions. These fluids may then metasomatically enrich the shallower levels of the mantle (<70 km) by amphibolitization, producing an appropriate source for alkaline magmatism.

Figure 12.6 shows the Olafsson & Eggler solidus in relation to the positions of the major carbonation reactions of Brey et al. (1983). Magnesite is the stable carbonate phase present at depths greater than about 140 km, whereas dolomite is stable

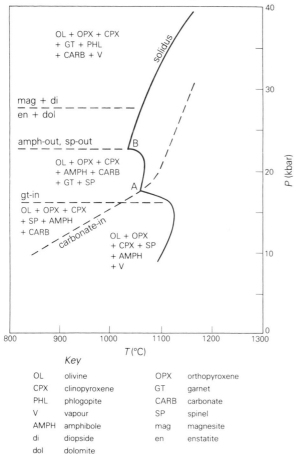

	Key		
OL	olivine	OPX	orthopyroxene
CPX	clinopyroxene	GT	garnet
PHL	phlogopite	CARB	carbonate
V	vapour	SP	spinel
AMPH	amphibole	mag	magnesite
di	diopside	en	enstatite
dol	dolomite		

Figure 12.5 Near-solidus phase relations of peridotite in the presence of small amounts of H_2O and CO_2 (0.3% H_2O and 0.7% CO_2) (after Olafsson & Eggler 1983, Fig. 3, p. 309).

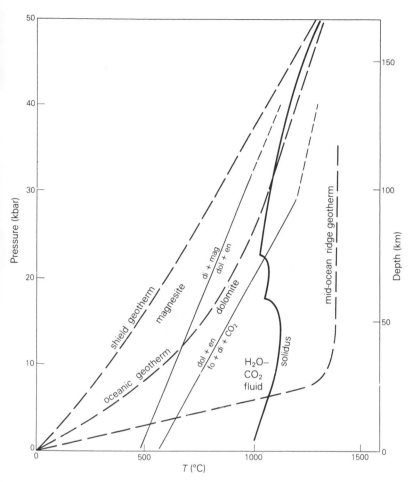

Figure 12.6 The peridotite solidus in the presence of small amounts of H_2O and CO_2 (Olafsson & Eggler 1983) in relation to the major mantle carbonation reactions (Brey *et al.* 1983). Dashed lines represent typical shield, oceanic and mid-ocean ridge geotherms.

between 75 and 140 km. In any given tectonic setting it is the geothermal gradient as well as the mantle composition which will determine whether or not partial melting occurs. Typical shield, oceanic and mid-ocean ridge geotherms are shown in Figure 12.6 for comparative purposes. The shield geotherm grazes the solidus around 150 km depth within the magnesite stability field. Brey *et al.* (op. cit.) suggest that near-solidus liquids at this depth will be highly magnesian and enriched in incompatible elements but relatively low in CaO and Al_2O_3, similar to kimberlites. Along the oceanic geotherm, the solidus is crossed in the dolomite stability field and the resultant near-solidus partial melts are enriched in CaO, similar to olivine melilitites. In many tectonic environments further degrees of partial melting above the solidus are associated with adiabatic decompression of the

source in ascending mantle diapirs (Ch. 3). However, for alkaline ultramafic magmatism it has been suggested that volatile-induced partial melting, where the geotherm intersects a vapour-present solidus, is more common (Wyllie 1980, Bailey 1985).

12.3.2 Melting of metasomatized peridotite in the presence of H_2O and CO_2

Within the western branch of the East African Rift in south-west Uganda, explosive Quaternary–Recent volcanism has brought to the surface nodules of phlogopite clinopyroxenite, which appear to be of mantle origin. Lloyd & Bailey (1975) and Lloyd (1981) consider that these represent the products of reaction between K-rich fluids and mantle rocks, orthopyroxene and olivine hav-

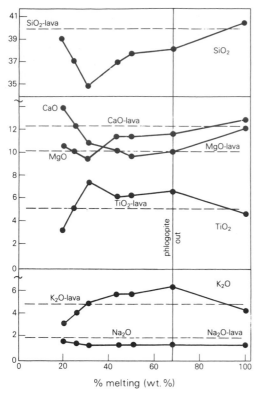

Figure 12.7 Chemical composition of partial melts of a Ugandan phlogopite–clinopyroxenite nodule, at 30 kbar pressure, as a function of the degree of melting (data from Lloyd *et al.* 1985). Dashed horizontal lines represent the composition of an average Ugandan ultrapotassic lava.

ing been replaced by mica and clinopyroxene. Wass (1980) considers that such pyroxenites are common in the source regions of many alkaline magmas. If these pyroxenite xenoliths are actually representative of a large portion of the sub-rift mantle in south-west Uganda they may provide an important source component in the petrogenesis of the highly potassic magmas characteristic of this segment of the rift.

Figure 12.7 shows the results of melting experiments (Lloyd *et al.* 1985) at 30 kbar on one of the Ugandan pyroxenite nodules, in which the composition of the partial melt is plotted as a function of the degree of melting. Shown for comparison is the composition of an average Ugandan ultrapotassic lava. Interpretation of these data is unfortunately somewhat equivocal, because the bulk composition of the xenoliths is quite close to the average lava

composition, and therefore partial melting is bound to generate liquids of a broadly similar composition given the phases involved (i.e. biotite and clinopyroxene). Of significance is the observation that the highest K_2O content and K_2O/Na_2O ratio in the melt occurs at the point at which biotite completely melts out. This may also be the case for partial melting of less strongly metasomatized source rocks. Ideally, to understand the partial melting behaviour of metasomatized mantle source rocks we require melting experiments on a spectrum of compositions, from peridotite–H_2O–CO_2 to pyroxenite–H_2O–CO_2. Unfortunately such data are not presently available.

Wendlandt & Eggler (1980b) have used the synthetic system $KA1SiO_4$–MgO–SiO_2–CO_2–H_2O to model the melting behaviour of phlogopite peridotite and its role in the genesis of K-rich magmas, kimberlites and carbonatites. In this system at pressures up to 30 kbar partial melting of a phlogopite peridotite with small amounts of H_2O and CO_2 produces increasingly potassic liquids with increasing pressure. Above 30 kbar magnesite becomes stable and the partial melts are carbonatitic. Phlogopite ceases to be a solidus phase at 50 kbar.

12.3.3   Melting relations of ultrapotassic lavas

Edgar (1987) has summarized the available experimental data relevant to the melting behaviour of ultrapotassic lavas. High-pressure experiments in the presence of H_2O and CO_2 have been carried out for a range of compositions from south-west Uganda (Edgar *et al.* 1976, 1980; Arima & Edgar (1983a), from Leucite Hills, Wyoming (Barton & Hamilton 1979, 1982) and from the West Kimberley area, Western Australia (Arima & Edgar 1983b). Barton & Hamilton (op. cit.) studied orendites, wyomingites and madupites (Section 12.4) and found that only the orendites and wyomingites could equilibrate with a garnet lherzolite assemblage at pressures greater than 26 kbar and $PH_2O = P_{total}$. They suggest that the madupites probably represent partial melts of a phlogopite pyroxenite or a phlogopite–olivine pyroxenite. The association of low-SiO_2 madupites with high-

SiO$_2$ orendites at Leucite Hills thus appears to result from variations in the mineralogy of the source, possibly combined with variations in its H$_2$O/CO$_2$ ratio. Ultrapotassic lavas from south-west Uganda could never have equilibrated with an orthopyroxene- or garnet-bearing assemblage in the presence of H$_2$O and CO$_2$ at pressures up to 40 kbar, as only diopside, olivine, ilmenite and phlogopite occurred on the liquidus. However, they could be generated by partial melting of a clinopyroxenite or phlogopite peridotite source.

12.4 Petrographic characteristics of kimberlites and ultrapotassic lavas

Mitchell (1985) and Bergman (1987) have presented excellent reviews of the mineralogy of lamproites, and Mitchell (1986) of the mineralogy of kimberlites. Consequently, only a brief summary of the petrographic characteristics of these rocks will be presented here; the reader is referred to the above works for more detailed information. The mineralogy of Group II ultrapotassic rocks from the western branch of the East African Rift was considered in Chapter 11, and therefore will not be discussed further in this section. Brief descriptions of the petrographic characteristics of the Roman Province ultrapotassic lavas are given as an example of Group III ultrapotassic rocks.

12.4.1 Kimberlites

Kimberlites exhibit extensive mineralogical variation coupled with considerable textural diversity. They have distinctive inequigranular textures resulting from the presence of macrocrysts set in a finer-grained matrix (Clement *et al.* 1984). This matrix may contain, as primary phenocryst or groundmass constituents, olivine (Fo90−87), phlogopite, carbonate (commonly calcite), serpentine, clinopyroxene (commonly diopside), monticellite, apatite, spinel, perovskite and ilmenite. The macrocrysts are anhedral grains mostly derived from the fragmentation of entrained mantle xenoliths and include olivine, phlogopite, picroilmenite, chrome spinel, magnesian garnet, clinopyroxene (commonly chrome diopside), and orthopyroxene

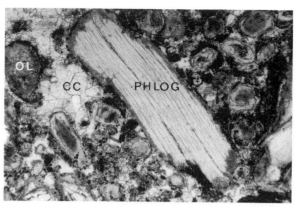

Figure 12.8 Photomicrograph to illustrate the petrographic characteristics of a Group I kimberlite from South Africa. The field of view shows a large phlogopite megacryst set in a pelletal matrix which includes altered olivine, carbonate (CC) and opaques (×40, ordinary light).

(commonly enstatite). Olivine is extremely abundant relative to the other macrocrysts, all of which are not necessarily present. The macrocrysts and relatively early formed matrix minerals are commonly altered by deuteric processes to serpentine and carbonate. Additionally, kimberlite may contain xenocrysts of diamond, but only as a very rare constituent.

Kimberlites can be subdivided into those having a highly micaceous matrix and those having little or no mica. These have been termed Group II and I respectively by Smith (1983). Micaceous kimberlites tend to contain considerable amounts of matrix calcite and apatite (Dawson 1987) and also groundmass diopside, which is quite rare in non-micaceous types. Macrocrystal picroilmenite is common in non-micaceous kimberlites but rare in micaceous varieties.

Figure 12.8 shows the petrographic characteristics of a Group I kimberlite.

12.4.2 Lamproites

Lamproites are characterized by widely varying modal proportions of Ti-rich Al-poor phlogopite, Ti−tetraferriphlogopite, K−Ti richterite, forsteritic olivine (Fo94−87), diopside, sanidine and leucite as the major phases. Minor phases include enstatite, priderite ((K,Ba)(Ti,Fe)$_8$O$_{16}$), wadeite

Table 12.1   Mineralogical classification of lamproites (Mitchell 1985).

New classification	Old classification
diopside−leucite−phlogopite lamproite	wyomingite
diopside−sanidine−phlogopite lamproite	orendite
olivine−phlogopite lamproite	verite
enstatite−phlogopite lamproite	fortunite
leucite−phlogopite lamproite	fitzroyite
leucite−diopside lamproite	cedricite
leucite−richterite lamproite	mamillite
diopside−leucite−richterite madupitic lamproite	wolgidite
olivine−richterite madupitic lamproite	jumillite
diopside madupitic lamproite	madupite

$(Zr_2K_4Si_6O_{18})$, apatite, magnesiochromite, ilmenite, shcherbakovite $(NaK(Ba,K)Ti_2Si_4O_{14})$, armalcolite $((Fe,Mg)Ti_2O_5)$, perovskite and jeppeite $(K_2Ti_6O_{13}−Ba_3Ti_5O_{13}ss)$. Diamond is an important accessory mineral in some lamproites (Jaques *et al.* 1984, Scott-Smith & Skinner 1984a) and may be regarded as a mantle derived xenocryst (cf. diamond in kimberlites). Important textural features of petrogenetic significance are the occurrence of phlogopite as either phenocrysts or poikilitic groundmass plates and the late (post-phlogopite) crystallization of K−Ti richterite. Ti−phlogopite, although not ubiquitous, is one of the most characteristic minerals of lamproites.

Analcime is common as a secondary mineral replacing leucite and sanidine. Other secondary phases include carbonate, chlorite, zeolites and barite. In most lamproites, the olivines are partially or totally pseudomorphed by serpentine, iddingsite, carbonate or quartz. Similarly, fresh leucite is rare, typically being pseudomorphed by sanidine, analcite, quartz, zeolite or carbonate.

Scott-Smith & Skinner (1984b) and Mitchell (1985) have proposed that lamproites can be divided into subgroups, on the basis of the dominant minerals present, i.e. phlogopite, richterite, olivine, diopside, sanidine and leucite. In this classification (Table 12.1), rocks with phenocrystal phlogopite are termed 'phlogopite lamproites',

whereas those with poikilitic groundmass phlogopite are termed 'madupitic lamproites'.

Dawson (1987) has noted the following mineralogical differences between lamproites and kimberlites:

(1) lamproites contain glass;
(2) lamproite groundmass contains K-richterite;
(3) compared with typical groundmass micas in kimberlites lamproite micas are richer in Ti, Fe and Na but poorer in A1;
(4) groundmass diopsides in lamproites have higher Ti contents than those in micaceous kimberlites;
(5) calcite is virtually absent in lamproites.

However, there are some mineralogical similarities between olivine lamproites and Group II kimberlites. Lamproites typically lack nepheline, melilite, kalsilite, alkali feldspar, plagioclase and Al-rich augite, which are characteristic of Group II and III ultrapotassic rocks.

Of the major minerals occurring in lamproites only phlogopite exhibits extensive compositional variation. Amphiboles and spinel show limited variation, while pyroxenes, leucite, sanidine and olivine are of essentially constant composition. These latter phases are, however, significantly different in their chemistry from similar minerals in

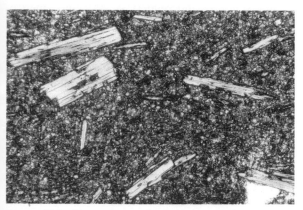

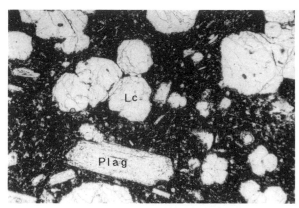

Figure 12.9 Photomicrograph of a lamproite (wyomingite) from the Leucite Hills province, USA, showing flow-aligned biotite phenocrysts set in a fine-grained matrix (×40, ordinary light).

Figure 12.10 Photomicrograph of a leucitite from the Roman Province, Italy, showing large rounded leucite phenocrysts and an elongate plagioclase phenocryst (×40, ordinary light).

other groups of highly potassic rocks. In addition, different lamproite provinces appear to be characterized by phlogopites, amphiboles, sanidines, leucites and priderites of distinctly different compositions. All lamproite micas appear to be poor in Ba, despite their occurrence in Ba-rich rocks, which sets them apart from micas in other potassic rocks which are Ba rich.

Figure 12.9 shows the characteristic petrographic features of a Leucite Hills lamproite.

12.4.3 Group III ultrapotassic rocks

Highly potassic volcanic rocks from the Roman Province of Italy (Peccerillo & Manetti 1985) may be used to illustrate the mineralogical variation of Group III ultrapotassic rocks. Lavas from the high-K series range in SiO_2 content from 38 to 55% and display a range of mineralogies which may be related to fractional crystallization processes. They range from aphyric to strongly porphyritic types. Clinopyroxene, leucite and plagioclase are the most common phenocryst phases, with olivine occurring in addition in the most mafic volcanics, and sanidine in the most evolved. Hauyne occurs in accessory amounts as phenocrysts. Biotite and magnetite occur in the most evolved rocks, along with melanite garnet in some cases. Brown amphibole may occasionally occur as a phenocryst and in the groundmass. The groundmass consists of the

same phases as the phenocrysts with nepheline and sometime melilite as additional components. Apatite is the most common accessory mineral.

Clinopyroxene is a ubiquitous phase in these rocks, ranging in composition from diopside to ferrosalite, and frequently displays complex zoning patterns. Mica compositions differ from those in lamproites in being relatively Fe-rich and Ti-poor with extreme Ba enrichments.

Figure 12.10 shows the characteristic petrographic features of a leucitite from the Roman Province.

12.5 Chemical composition of kimberlites and ultrapotassic rocks

Ultrapotassic rocks are chemically extremely heterogeneous, displaying wide variations in most major element abundances and high concentrations of volatile species (H_2O, CO_2, F, Cl and SO_2). Additionally, they contain the highest concentrations of incompatible elements of all terrestrial magmas. They have been divided into three groups (Foley *et al.* 1987) on the basis of their chemistry and tectonic setting, which should realistically be regarded as end-members between which transitional types can occur. The term 'ultrapotassic' encompasses a variety of magmas generated as a consequence of variable source compositions, depth and degree of melting, volatile compositions and fO_2. In Group I much of the compositional diversity can be attributed to partial melting processes, whereas in Groups II and III low-pressure fractional crystallization, combined with crustal contamination, is important in producing a range of more silica-rich magmas.

Geochemical studies of kimberlites are complicated by problems inherent in the intrusive style of this volatile-rich magma type. Most diatreme facies rocks are hybrids, containing a range of crustal and mantle xenoliths and xenocrysts. As a consequence, many of the older analyses of kimberlite in the literature may be suspect. In contrast, hypabyssal facies kimberlites contain few crustal xenoliths and hence contamination from this source is of lesser importance. However, they may contain greater concentrations of mantle-derived megacrysts and high-pressure phenocrysts, which make it equally difficult to estimate the primary kimberlite magma composition. Mitchell (1986) considers this problem in some detail.

Table 12.2 Major and trace element analyses of some typical Group I ultrapotassic rocks (lamproites) (data from Bergman 1987).

	Gaussberg	Leucite Hills	Smoky Butte	W. Australia
%				
SiO_2	52.2	52.7	53.5	51.3
TiO_2	3.5	2.4	5.6	5.1
Al_2O_3	10.1	10.8	9.8	7.4
FeO	6.1	5.1	5.4	7.1
MnO	0.09	0.09	0.12	0.09
MgO	8.2	8.4	7.4	11.7
CaO	4.7	6.7	6.4	6.0
Na_2O	1.7	1.3	1.5	0.5
K_2O	11.9	10.4	7.4	8.3
P_2O_5	1.5	1.5	1.7	1.1
ppm				
La	230	213	392	282
Ce	420	427	774	416
Nd	150	166	304	133
Sm	19	21	36	16
Eu	4.5	4.5	7.3	3.5
Gd	9.8	10.8	—	8.4
Tb	—	—	1.7	—
Dy	—	4.2	6.9	4.5
Yb	0.45	1.09	2.26	1.16
Lu	—	0.15	0.28	0.15
Y	18	20	31	15
U	2.5	—	3.3	5.4
Th	30	—	6.5	35
Rb	300	253	102	424
Sr	1830	2840	3160	1 180
Ba	5550	6600	9810	11 100
Zr	1000	1440	1660	1 040
Ta	—	—	4.4	7.4
Hf	—	—	64	26
Nb	90	74	113	167
Ni	230	253	344	167
Cr	310	460	501	550
V	110	—	—	700

Table 12.3 Major and trace element analyses of some Group II ultrapotassic rocks from the western branch of the East African Rift (Data from Higazy 1954).

	Katungites		Mafurite	Ugandite
%				
SiO_2	37.93	34.23	39.06	40.47
Al_2O_3	6.59	8.02	8.18	5.38
Fe_2O_3	6.81	6.62	4.01	4.03
FeO	4.37	5.34	4.98	6.47
MnO	0.18	0.22	0.26	0.23
MgO	14.54	9.92	17.66	24.84
CaO	15.23	16.54	10.40	8.06
Na_2O	0.88	1.20	0.18	0.68
K_2O	2.65	3.39	6.98	3.46
H_2O^+	3.38	2.80	1.42	1.11
H_2O^-	1.42	1.72	0.50	0.57
CO_2	0.50	4.02	tr	0.36
TiO_2	4.12	4.56	4.36	3.52
P_2O_5	1.03	0.96	0.61	0.29
Cl	0.01	—	—	0.01
F	0.16	0.14	0.13	0.10
S	—	0.12	0.13	0.04
ppm				
Rb	150	220	450	450
Ba	2600	2000	7500	2000
Sr	3800	7500	7000	1800
Cr	800	650	1300	1200
Ni	140	200	300	900
Zr	1100	850	900	300
La	<30	50	80	35
Y	<30	<30	<30	<30
V	350	260	220	110

In the following sections we will evaluate the usefulness of major and trace element and Sr−Nd−Pb isotopic studies of kimberlites and ultrapotassic rocks in characterizing the nature of their mantle source. This is generally accepted to be variably metasomatized in comparison with normal mantle.

12.5.1   Major elements

SiO_2, TiO_2, Al_2O_3, Fe_2O_3, FeO, MnO, MgO, CaO, Na_2O, K_2O, P_2O_5, H_2O and CO_2 can all be considered major element oxides in the description of the geochemical characteristics of kimberlites and ultrapotassic rocks. Tables 12.2−5 present representative major element analyses of Group I,

II and III ultrapotassic rocks and kimberlites respectively. Table 12.6 compares average analyses of kimberlite, lamproite and lamprophyre.

Following Foley *et al.* (1987) we have used the term 'ultrapotassic' to describe those igneous rocks with high contents of K_2O (>3%) and other incompatible elements, high K_2O/Na_2O ratios (>3 on a wt.% basis) and high Mg numbers and Ni and Cr contents, characteristic of relatively primitive basaltic magmas. Ultrapotassic rocks thus defined have been further subdivided into three groups (I, II and III) on the basis of their major element chemistry (Foley *et al.* op. cit.). Figures 12.11 and 12.12 are plots of wt.% CaO versus Al_2O_3 and SiO_2 respectively, which readily differentiate members of these three groups. Group 1 rocks are lamproites, characterized by their low contents of Al_2O_3, CaO and Na_2O. They have variable SiO_2 contents, between 36 and 60%, and generally have higher Mg numbers than members of the other groups. Group II rocks, by comparison, have consistently low SiO_2 contents (<46%) but high CaO. They also have relatively low Al_2O_3 contents. In terms of Figures 12.11 and 12.12 it can be seen that kimberlites plot within the field of Group II rocks, which may have some significance in terms of petrogenetic models. The principal major element characteristic of Group III rocks is their high Al_2O_3 content. Rocks with extremely low SiO_2 contents (<42%) do not occur in this group, although silica contents less than 50% are common. Mg numbers are generally lower than in Groups I and II and much of the compositional variation, particularly the range in SiO_2 contents, may be attributable to the combined effects of low-pressure fractional crystallization and crustal contamination. Figure 12.13 is a plot of wt.% K_2O versus wt.% SiO_2 for the two major potassic magma series of the Roman Province of Italy, the K-series and the high-K series (Peccerillo & Manetti 1985). This shows that the two series have distinctly different primary magma compositions, both of which evolve along subparallel vectors as a consequence of low-pressure fractional crystallization in subvolcanic magma chambers.

As we have stated previously in Section 12.1, Group I ultrapotassic rocks occur in stabilized

Table 12.4   Major and trace element analyses of Group III ultrapotassic lavas from Vulsini, Roman Province (data from Rogers *et al.* 1985).

	Leucite basanite	Magnesian leucitite	Leucitite
%			
SiO_2	47.12	46.74	47.10
TiO_2	0.64	0.75	0.81
Al_2O_3	11.84	12.30	15.56
Fe_2O_3	3.26	3.30	3.16
FeO	4.16	4.59	4.58
MnO	0.15	0.19	0.15
MgO	12.84	8.96	6.04
CaO	13.88	15.35	12.67
Na_2O	1.42	0.81	1.50
K_2O	3.29	4.59	6.54
P_2O_5	0.30	0.34	0.45
H_2O	0.12	0.22	0.21
ppm			
V	186	212	198
Cr	911	230	22
Ni	261	109	64
Rb	253	396	558
Sr	769	791	1278
Y	24	28	42
Zr	128	182	308
Nb	8	7	16
Ba	620	738	1161
La	40.4	51.8	77.0
Ce	92.0	124.0	163.0
Nd	41.7	55.5	69.0
Sm	8.27	11.0	12.4
Eu	1.79	2.33	2.88
Gd	6.40	8.8	10.5
Tb	1.01	1.27	1.78
Yb	1.63	1.81	2.79
Th	20.5	30.3	46.8
Ta	0.45	0.60	0.88
Hf	3.59	4.98	7.29
U	4.50	4.6	10.3

orogenic areas which have previously experienced a subduction-related magmatic event. Group II rocks occur dominantly in continental rift zone environments (Ch. 11), while Group III rocks occur in active orogenic zones. Within Group I, ultrapotassic rocks generated in areas which have been stabilized for a considerable period of time can be distinguished on the basis of their high TiO_2 contents (2–8%) from those erupted within more recently tectonically active areas (<2%).

The major element characteristics of Group I ultrapotassic rocks (low CaO, Al_2O_3 and Na_2O) suggest that they are derived from an originally depleted mantle source which has subsequently been metasomatically enriched in K_2O and incompatible elements (Foley *et al.* 1987). Group II rocks also have low Al_2O_3 and Na_2O suggesting a depleted source. However, they have high CaO contents, requiring either incomplete elimination of clinopyroxene during the depletion event or subsequent re-introduction of CaO during a later metasomatic event. Group III rocks have high contents

Table 12.5 Average major and trace element analyses of kimberlites (data from Smith *et al.* 1985).

	IA	IB	II
%			
SiO_2	32.1	25.7	36.3
TiO_2	2.0	3.0	1.0
Al_2O_3	2.6	3.1	3.2
Fe_2O_3	9.2	12.7	8.4
MnO	0.2	0.2	0.2
MgO	28.5	23.8	29.7
CaO	8.2	14.1	6.0
Na_2O	0.2	0.2	0.1
K_2O	1.1	0.6	3.2
P_2O_5	1.1	1.1	1.1
H_2O^-	1.1	0.5	0.7
H_2O^+	8.6	7.2	5.3
CO_2	4.3	8.6	3.6
ppm			
Nb	165	210	120
Zr	200	385	290
Y	13	30	16
Sr	825	1020	1140
Rb	50	30	135
Th	18	27	30
Ni	1360	800	1400
Cr	1400	1000	1800
V	75	170	85
Ba	1000	850	3000
Sc	13	20	20
La	90	125	200
Ce	140	220	350
Nd	90	100	145
Gd	4	8	6
U	4	6	5
Pb	7	10	30

of CaO, Al_2O_3 and Na_2O, and hence there is no evidence for a previous depletion event in their source region.

Kimberlites can be subdivided into Group I (poorly micaceous) and Group II (micaceous) types on the basis of their isotopic and major and trace element compositions (Smith *et al.* 1985). Group I kimberlites have been further subdivided into subgroups IA and IB. Group IA occur totally within cratons, whereas IB occur mostly off the cratons. Table 12.5 compares average major element analyses of Groups IA, IB and II. These averages have been plotted in Figures 12.11 and 12.12 for comparison with the three major groups

of ultrapotassic rocks, and clearly lie within the Group II field. In view of the wide range of compositions that exist within the kimberlite group, it is unlikely that such averages have any real geochemical significance. However, they are useful in illustrating the broad range of variation, and for comparison with other groups of potassic igneous rocks. Group IB have lower SiO_2 and higher CaO, Fe_2O_3 total, volatiles and TiO_2 than Group IA. Group II kimberlites have higher abundances of SiO_2, K_2O and MgO, and lower TiO_2, CaO and CO_2 relative to Group I. In general, kimberlites have lower SiO_2, TiO_2 and alkalis and higher MgO than lamproites (Scott-Smith &

Table 12.6 Average major and trace element analyses of kimberlite, lamproite and lamprophyre (data from Bergamn 1987).

	Kimberlite	Lamproite	Lamprophyre
%			
SiO_2	38.4	53.3	46.3
TiO_2	2.6	3.0	2.6
Al_2O_3	4.7	9.1	13.5
FeO	11.3	6.3	11.0
MnO	0.18	0.10	0.21
MgO	28.7	12.1	9.1
CaO	11.3	5.8	10.7
Na_2O	0.5	1.4	3.1
K_2O	1.4	7.2	2.9
P_2O_5	0.9	1.3	0.9
H_2O^+	6.6	2.7	2.6
CO_2	5.6	2.8	2.5
ppm			
La	150	240	105
Ce	200	400	195
Nd	85	207	100
Sm	13	24	22
Eu	3.0	4.8	4.9
Gd	8.0	13	14.3
Tb	1.0	1.4	1.8
Dy	—	6.3	5.7
Yb	1.2	1.7	1.9
Lu	0.16	0.23	0.37
Y	22	27	36
U	3.1	4.9	5.0
Th	16	46	24
Rb	65	272	115
Sr	740	1530	1010
Ba	1000	5120	1345
Zr	250	922	350
Ta	9	4.7	2
Hf	7	39	9
Nb	110	95	83
Ni	1050	420	1553
Cr	1100	580	40

Skinner 1984b).

Figure 12.14 is a plot of wt.% TiO_2 versus K_2O for Group I and II kimberlites. Group I clearly have high Ti/K ratios, whereas Group II have low Ti/K. The extremely good positive correlation between TiO_2 and K_2O for Group II kimberlites is consistent with the storage of Ti in groundmass phlogopite (Smith *et al*. 1985). In contrast, Ti in Group I kimberlites is primarily stored in oxide minerals.

Figure 12.15 is a triangular $K_2O-MgO-Al_2O_3$ (wt.%) diagram, which usefully shows the main compositional differences between kimberlites, lamproites and lamprophyres. Dawson (1987) has shown that Group II kimberlites have closer chemical similarities to olivine lamproites than Group I kimberlites; hence the overlapping fields in Figure 12.15, which may reflect their derivation from similar mantle sources. This similarity is further reflected in their isotopic compositions (Section 12.5.3). Table 12.6 shows average analyses of these three major groups of biotite-rich igneous rocks.

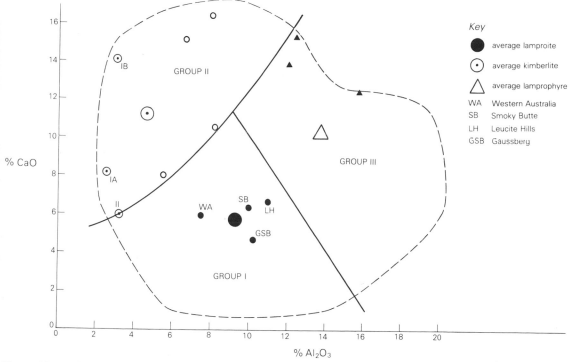

Figure 12.11 Classification of ultrapotassic igneous rocks into Groups I, II and III on the basis of wt.% CaO versus Al_2O_3 (after Foley *et al.* 1987). Average analyses of kimberlites and ultrapotassic rocks are from Tables 12.2−6.

Lamprophyres are a complex polygenetic group of hypabyssal minor intrusions, characteristically containing phenocrysts of amphibole and phlogopite. They can be subdivided into calc-alkaline, alkaline and ultramafic types (Rock 1987). Calc-alkaline lamprophyres may be regarded as the hypabyssal equivalents of potassic subduction-related volcanic rocks, while alkaline lamprophyres may be the equivalents of alkali basalts, basanites and nephelinites, Ultramafic lamprophyres have no recognized volcanic equivalents and include some exceptionally low-SiO_2 rocks (down to 15%) which grade into carbonatites. Of these three lamprophyre subgroups, only calc-alkaline lamprophyres are potassic, but then only mildly so. Rock (1987) considers that kimberlites and lamproites should be considered as subgroups of the lamprophyre clan, as they are also characterized by the presence of biotite. Table 12.7 shows Rock's proposed subdivision of the lamprophyric rocks. Lamprophyres occur in a wide variety of tectonic settings, including continental rift zones, oceanic islands, island

arcs, active continental margins and continental collision zones. The overlap between the lamproite and lamprophyre fields in Figure 12.15 is therefore not surprising, as both groups may be generated in areas which have previously experienced a subduction-related magmatic event.

From the data in Tables 12.2−6 it is clear that H_2O and CO_2 are important major constituents of kimberlites, ultrapotassic rocks and lamprophyres.

12.5.2   Trace elements

Tables 12.2−5 present representative trace element analyses for Groups I, II and III ultrapotassic rocks and kimberlites, while Table 12.6 compares average analyses of kimberlite, lamproite and lamprophyre. All of these rock types are characterized by extreme enrichments in incompatible elements, which must reflect both the chemical composition of their mantle source and the partial melting processes involved in their formation.

The concentrations of the compatible trace ele-

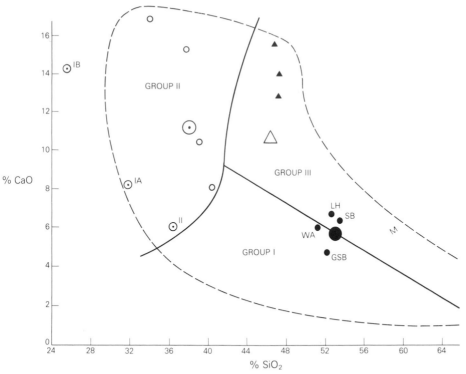

Figure 12.12 Classification of ultrapotassic igneous rocks into Groups I, II and III on the basis of wt.% CaO versus SiO$_2$ (after Foley *et al.* 1987). Data sources and abbreviations as in Figure 12.11.

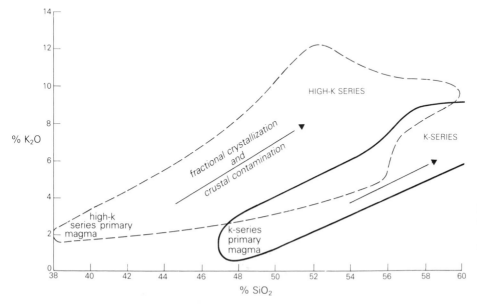

Figure 12.13 Variation of wt.% K$_2$O versus SiO$_2$ for K-series and high-K series volcanic rocks from the Roman Province of Italy (data from Peccerillo & Manetti 1985).

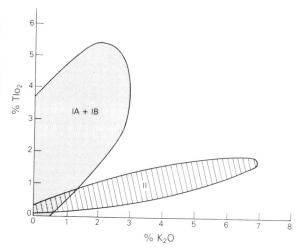

Figure 12.14 Variation of wt.% TiO_2 versus K_2O for Group I (IA and IB) and Group II kimberlites (after Smith *et al.* 1985, Fig. 3, p. 273).

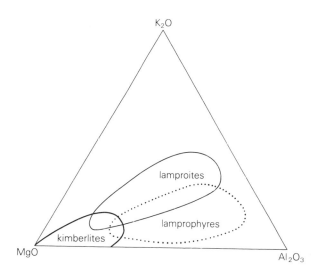

Figure 12.15 Triangular diagram showing the compositions of kimberlites, lamproites and lamprophyres in terms of wt.% K_2O, MgO and Al_2O_3 (after Bergman 1987, Fig. 21, p. 148).

ments Ni, Cr, Sc and V are important in the identification of primary magma compositions. In general, magmas with greater than 500 ppm Ni and 1000 ppm Cr may be considered primary (Ch. 2). However, for these potassic igneous rocks this may not always be a valid assumption because, as we have stated previously, extensive metasomatism of a lherzolite mantle source may completely eliminate olivine and orthopyroxene, in favour of assemblages such as clinopyroxene + phlogopite + garnet. Partial melts of such modified mantle may no longer be constrained by the Mg numbers, Ni and

Cr contents conventionally assumed to characterize primary magmas in equilibrium with a harzburgitic (olivine + orthopyroxene) residuum. Identification of primary magma compositions is necessary in order to demonstrate that Group I, II and III ultrapotassic rocks do indeed stem from distinct parental magmas, rather than being related to each other by processes such as fractional crystallization or crustal contamination.

Table 12.7 Subdivision of the lamprophyre clan of rocks (after Rock 1987)

calc-alkaline lamprophyres		minette, vogesite, kersantite, spessartite, kentallenite, appinite
alkaline lamprophyres		camptonite, monchiquite, sannaite
ultramafic lamprophyres		aillikite, alnoite, bergalite, damkjernite, ouachitite
lamproites	phlogopite lamproite	wyomingite, orendite, verite, fortunite, fitzroyite, cedricite, mamillite
	madupitic lamproite	madupite, jumillite, wolgidite
kimberlites	Group I Group II	

The most useful way of comparing the incompatible element geochemistries of kimberlite and ultrapotassic rocks is by plotting mantle-normalized trace element variation diagrams, i.e. spiderdiagrams. In these diagrams the elements are plotted in order of decreasing incompatibility from left to right in a four-phase lherzolite undergoing partial fusion. The data are normalized to chondritic abundances, except for Rb, K and P, which are normalized to primitive terrestrial mantle values (Thompson 1982). Figure 12.16 shows spiderdiagram patterns for selected Group I (a, b, c), Group II (d) and Group III (e) ultrapotassic rocks and kimberlites (f). All these patterns indicate extreme trace element enrichments, particularly of the most highly incompatible elements.

Group I ultrapotassic rocks in general show the highest overall abundances of incompatible elements. The spiderdiagrams for Western Australian, Smoky Butte, Gaussberg and Leucite Hills lamproites (Figs. 12.16a,b) are broadly similar to those of leucitites from the East African Rift (Group II) (Fig. 12.16d), and to those of continental (Ch. 11) and oceanic (Ch. 9) alkali basalts, although with considerably higher degrees of incompatible element enrichment than the latter. All of these Group I and II rocks show a distinctive but fairly small negative Sr spike. In addition, the Gaussberg and Leucite Hills lamproites also show a negative Nb spike.

The spiderdiagram patterns of primitive Group III rocks (Fig. 12.16e) are markedly different from those of Groups I and II, with overall lower enrichments of most incompatible elements and distinctive negative spikes at Ba, Nb−Ta, Ti and P. However, Rb, Th and K are strongly enriched. The patterns are clearly similar to those of subduction-related volcanic rocks (Chs 6 & 7), reinforcing the

Figure 12.16 Mantle-normalized trace element variation diagrams (spiderdiagrams) for ultrapotassic lavas and kimberlites. The data are normalized to chondritic abundances, except for Rb, K and P which are normalized to primitive terrestrial mantle values (Thompson 1982). (a–c) Group I – lamproites; (d) Group II; (e) Group III; (f) kimberlites. Data sources: (a,b) Bergman (1987); (c) Nixon *et al.* (1984); (d) Thompson *et al.* (1984) and Davies & Lloyd (1987); (e) Rogers *et al.* (1985); (f) Smith *et al.* (1985).

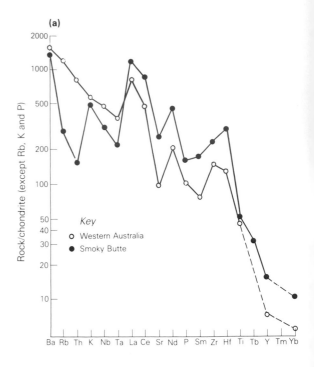

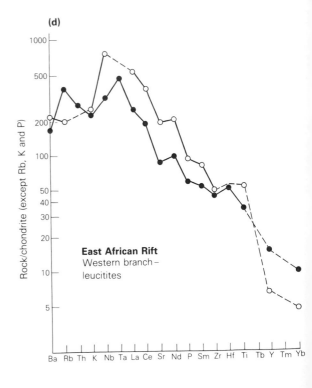

(b)

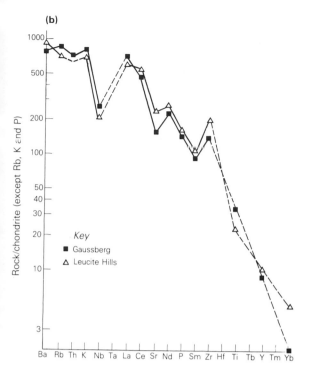

(c)

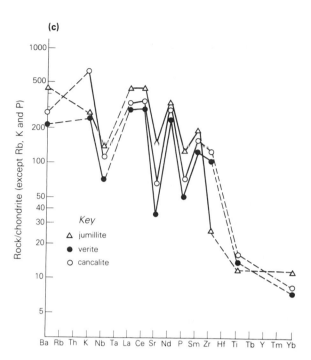

(e)

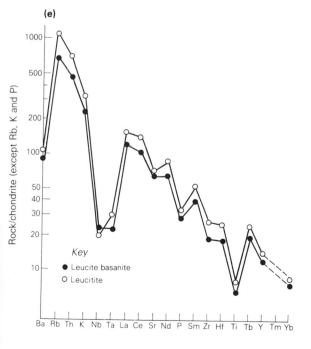

(f)

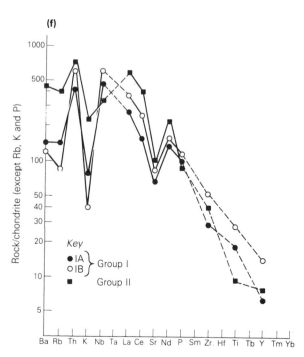

association of Group III ultrapotassic rocks with active or recently active subduction systems. The very slight negative Sr anomaly is most probably a consequence of the low-pressure fractional crystallization of plagioclase. Group I lamproites from south-east Spain (Fig. 12.16c) show similar spiderdiagram patterns to Group III rocks, but with much stronger negative spikes at Sr and P. The negative Sr spike in this instance, however, is much more likely to be a source characteristic, as lamproites do not crystallize plagioclase. The Spanish lamproites clearly postdate a subduction-related magmatic event (Nixon et al. 1984), and therefore it is not surprising that they display similar incompatible element characteristics to subduction-related magmas. The Gaussberg and Leucite Hills lamproites (Fig. 12.16b) also display weak negative Nb spikes. However, while Rowell & Edgar (1983) have suggested that the Leucite Hills magmatism may be related to a fossil subduction zone, Gaussberg magmatism is much more clearly hot spot related. Crustal contamination may also be important in the generation of the negative Nb−Ta, Sr, P and Ti spikes (see Fig. 11.25 for typical spiderdiagram patterns of crustal rocks) in these Group I and III rocks, but the effects cannot be isolated easily from those of subduction-modified mantle sources.

Figure 12.16f shows characteristic spiderdiagram patterns for Group I (IA and IB) and Group II kimberlites, which are remarkably similar to each other in view of the fundamental Nd−Sr−Pb isotopic differences between the two groups (Section 12.5.3). The shapes of these patterns are similar to those of OIB (Ch. 9) and continental alkali basalts (Ch. 11), but with much higher incompatible element enrichments and distinctive negative K and Sr spikes. Group II kimberlites have higher concentration of Ba, Rb and K than Group I. Both groups have Nb concentrations approximately ten times those of OIB, and Group I display a significant positive Nb anomaly. Ti and Nb are higher in concentration in Group I than in Group II, while Zr concentrations are broadly similar. Consequently, Group II kimberlites have higher Zr/Nb ratios than Group I, which is largely a reflection of their relative Nb concentrations. They also have higher concentrations of light REE.

Incompatible elements are generally not considered to be fractionated from each other during partial melting of normal mantle lherzolite, nor to a significant extent during subsequent fractional crystallization processes. As a consequence, ratios of incompatible elements in basalts are frequently postulated to reflect those in their mantle source. Thus the difference in incompatible element ratios (e.g. Zr/Nb) between Group I and Group II kimberlites could be used to argue their derivation from distinct mantle sources. However, this line of reasoning may be invalid for kimberlites and ultrapotassic magmas if, as their incompatible element concentrations imply, they are the products of very small degrees of partial melting of normal (Kramers et al. 1981) or metasomatically enriched mantle (Smith et al. 1985). Fractionation of incompatible elements during partial melting could occur if phases which concentrate them are residual in the source at low degrees of melting. For examples, mica (phlogopite) and amphibole (K-richterite) could preferentially concentrate K, Rb and Ba, while more exotic accessory minerals (e.g. crichtonite group minerals, wadeite, priderite, rutile, perovskite and ilmenite) could concentrate a much wider range of incompatible elements. The distinctive negative K spike in the spiderdiagram patterns of Group I and II kimberlites could thus reflect residual phlogopite in the source, while the negative Sr spike could in part reflect residual clinopyroxene (Smith et al. 1985).

Figure 12.17 shows fields for Group II and III ultrapotassic rocks on a diagram of Th/Yb versus Ta/Yb, which we have used previously (Chs 6 & 7) to characterize the mantle source of subduction-related volcanic rocks from island arcs and active continental margins. Group III ultrapotassic rocks clearly overlap with the field of active continental margin basalts, whereas Group II ultrapotassic lavas from the western branch of the East African Rift plot within the enriched mantle field for intra-plate basalts. This diagram strongly supports the involvement of subduction-modified mantle in the petrogenesis of Group III ultrapotassic rocks. A plot of Nb versus Zr (Fig. 12.18) is equally useful in distinguishing potassic and ultrapotassic rocks related to intra-plate activity from those directly or

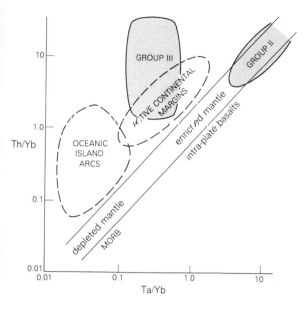

Figure 12.17 Variation of Th/Yb versus Ta/Yb (cf. Fig. 7.26) to show the similarity between Group III ultrapotassic rocks from the Roman Province (Rogers *et al.* 1985) and subduction-related basalts. Group II ultrapotassic lavas from the western branch of the East African Rift (Mitchell & Bell 1976) plot within the field of intra-plate basalts derived from enriched mantle sources.

indirectly related to subduction. The former are characterized by high Nb contents (>80 ppm), whereas the latter contain less than 50 ppm. Low Nb is a characteristic feature of all subduction-related volcanic rocks (Ch. 6 & 7).

Figure 12.19 is a plot of Ba/Nb versus La/Nb for Group I and II kimberlites and oceanic basalts from the South Atlantic ocean (MORB and OIB). Assuming that these ratios are not normally fractionated during partial melting processes, we can use this diagram to elucidate the characteristics of the kimberlite source relative to those of other mantle reservoirs, i.e. MORB and OIB source mantle. In this diagram Dupal OIB (Gough, Tristan da Cunha, Discovery and Walvis Ridge) define a trend away from compositions typical for normal OIB (e.g. Bouvet and Ascension) towards a field of high Ba/Nb and La/Nb ratios characteristic of recycled oceanic lithosphere, plus sediment or continental lithosphere and Group II kimberlites. Group I kimberlites overlap with the field of non-Dupal OIB. Trace element and Nd−Sr−Pb isotopic variations in South Atlantic OIB suggest that the Dupal anomaly may reflect an area where ancient recycled material exists within the asthenosphere (see Chs 9 & 10). Upwelling from this zone has resulted in hotspot volcanism with an apparent lithospheric signature (Ch. 9), whereas upwelling outside this region has led to hotspot volcanism with a more normal OIB signature (Le Roex 1986). In terms of Figure 12.19, Group I and II kimberlites appear to be derived from source regions

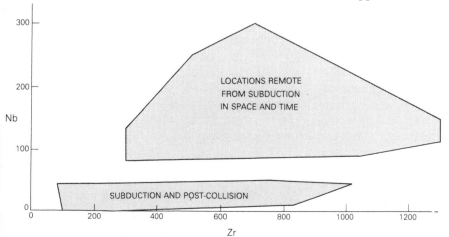

Figure 12.18 Variation of ppm Nb versus Zr for potassic and ultrapotassic volcanic rocks with less than 60% SiO$_2$ (after Thompson & Fowler 1986, Fig. 9, p. 518).

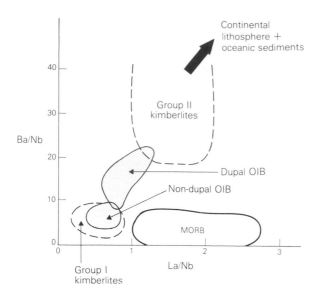

Figure 12.19 Variation of Ba/Nb versus La/Nb for Group I and II kimberlites compared to fields for MORB and South Atlantic OIB (after Le Roex 1986, Fig. 2, p. 244).

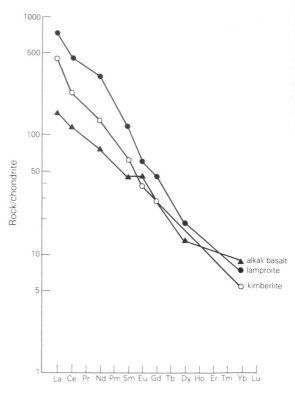

Figure 12.20 Chondrite-normalized REE diagram for a typical kimberlite, lamproite and alkali basalt. Data from Bergman (1987); normalization constants from Nakamura (1974).

similar to the enriched components involved in non-Dupal and Dupal hot spots respectively. Recent plate-tectonic models (Crough *et al.* 1980, Le Roex op. cit.) advocate a correlation between the palaeopositions of South Atlantic hot spots and Southern African kimberlite magmatism. Figure 12.19 clearly indicates that Group I kimberlites are derived from OIB-source asthenospheric mantle rather than MORB-source mantle. However, it does not constrain the source of Group II kimberlites to the continental lithosphere.

Figure 12.20 is a chondrite-normalized REE diagram showing typical REE patterns for lamproites, kimberlites and alkali basalts. The degree of light-REE enrichment increases from alkali basalts through nephelinites and melilitites to kimberlites and lamproites. Lamproites and kimberlites are markedly light-REE enriched and heavy-REE depleted relative to alkali basalts. The similarity in their REE patterns suggests that they are derived from similar upper-mantle sources. However, the clearly different mineralogy and evolution of these different magma types requires

that the major element compositions of their sources are quite different. Addition of a common metasomatic component to a variety of mantle source rocks may be the origin of the similar incompatible element contents (Mitchell 1986). Mitchell (op. cit.) has calculated that 1–8% partial melting of a source containing 2.5% apatite, 2.5% K-richterite, 5% clinopyroxene, 10% garnet, 20% orthopyroxene and 60% olivine could account for the observed REE contents and La/Yb ratios of many kimberlites.

12.5.3 Radiogenic isotopes

Nd–Sr data

A number of recent studies of highly potassic igneous rocks (McCulloch *et al.* 1983, Vollmer &

Norry 1983a,b, Fraser *et al.* 1985, Nelson *et al.* 1986) have documented Nd−Sr isotopic compositions far exceeding the combined range of MORB + OIB, indicative of long histories of high Rb/Sr and Nd/Sm in their mantle source. For example, diamond-bearing lamproites from Western Australia appear to have been derived from mantle sources enriched in Rb/Sr and Sm/Nd for at least 1000 Ma (McCulloch *et al.* op. cit.). Such data strongly support the derivation of these continental volcanic rocks from metasomatized mantle sources.

Figure 12.21 shows the variation of initial $^{87}Sr/^{86}Sr$ versus ε_{Nd} for kimberlites, and for a range of potassic and ultrapotassic rocks. The ε_{Nd} notation has been used because of the age range of the samples involved (Ch. 2). The data define two distinct trends, one shallow and one steep, which indicates that the processes which concentrate incompatible elements in the upper mantle generate low Sm/Nd ratios but both high and low Rb/Sr (Hawkesworth *et al.* 1985, Fraser *et al.* op. cit.). As more data become available it is possible that these

two trends may merge into a continuum, as suggested by the Gaussberg data, which plot in an intermediate position. Hawkesworth *et al.* (1984) have argued that the low-Sm/Nd−low-Rb/Sr style of trace element enrichment reflects the migration of small-volume silicate melts in the upper mantle, and is the process responsible for the high Ti, Nb and Zr abundances of within-plate basalts. However, Menzies & Wass (1983) have argued that such enrichment may also result from the migration of light-REE enriched, low-Rb, CO_2-rich fluids. The shallow $^{87}Sr/^{86}Sr-\varepsilon_N$ trend requires the remobilization of old relatively high Rb/Sr portions of the upper mantle. The existence of K-richterite and phlogopite-bearing garnet-free peridotite xenoliths in kimberlites with similar isotopic compositions confirms that such high Rb/Sr material does indeed exist within the subcontinental lithosphere (Erlank *et al.* 1982, Jones *et al.* 1982). Such xenoliths are clearly H_2O rich relative to normal mantle, and may be the products of mantle metasomatism induced by H_2O-rich fluids. While the origin of

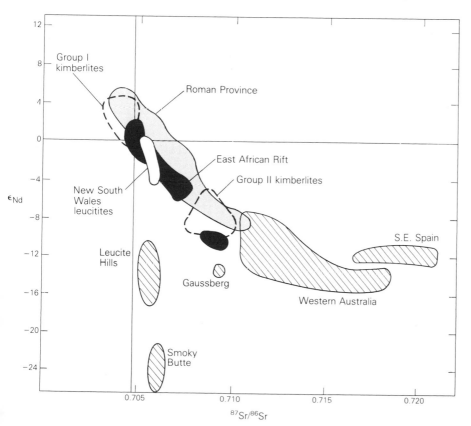

Figure 12.21 Variation of ε_{Nd} versus initial $^{87}Sr/^{86}Sr$ for kimberlites and potassium-rich volcanic rocks (after Nelson *et al.* 1986, Fig. 3, p. 236). Data for Smoky Butte lamproites from Fraser *et al.* (1985).

such fluids remains a matter for speculation, there is clear evidence in some provinces that they may be related to subduction processes.

In terms of Figure 12.21 it is clear that Group I and II kimberlites have fundamentally different Nd-Sr isotopic characteristics, consistent with their derivation from mantle source rocks which have been modified by different types of enrichment processes and with different evolutionary histories (Fraser *et al.* 1985, Smith *et al.*1985). These differences in Nd−Sr isotopic characteristics are correlated with differences in emplacement ages and the presence or absence of certain high-$P−T$ xenolith suites, as well as petrographic characteristics (Smith 1983, Smith *et al.* op. cit.; see also Table 12.8). Group I kimberlites have isotopic signatures consistent with their derivation from time averaged primitive to slightly depleted sources. Thus their enrichment in incompatible elements must have been close enough in time to preclude the development of significant isotopic changes. The slightly depleted Nd−Sr isotopic composition relative to bulk Earth suggests that they are derived from an asthenospheric mantle source, as their isotopic characteristics are similar to those of many oceanic island basalts (Ch. 9). The two subgroups IA and IB are isotopically similar and will not be considered further in this section.

In contrast, the isotopic characteristics of Group II kimberlites suggest that incompatible element enrichment is a time-averaged feature of the source. Their source clearly involves a component which is old and LREE enriched (low Sm/Nd), which most authors would place within the lithosphere, either in the subcontinental mantle or the continental crust. High $^{87}Sr/^{86}Sr$ and low ε_{Nd} in continental volcanic rocks have often been regarded as indicative of crustal contamination (Chs 7,9 & 11). However, the extremely high Nd and Sr contents of kimberlites mean that their isotopic compositions are very difficult to modify by contamination with crustal rocks. In addition, the presence of diamonds and mantle xenoliths in some kimberlites provides strong evidence against the possibility of contamination at crustal levels. Le Roex (1986) has argued that Group II kimberlites are derived from

Table 12.8   Characteristic features of Group I and II southern African kimberlites (Smith *et al.* 1985).

	Group I	Group II
initial isotope ratios:		
$^{87}Sr/^{86}Sr$	0.702−0.705	0.7075−0.710
$^{143}Nd/^{144}Nd$	0.51268−0.51276	0.51206−0.51227
$^{206}Pb/^{204}Pb$	18.3−20.0	17.2−17.7
emplacement age	80−100 Ma	114−127 Ma
mineralogy	phlogopite-poor, perovskite-rich, zircon present, ilmenite present	phlogopite-rich, perovskite-poor, zircon absent, ilmenite absent
xenoliths high $P−T$ sheared lherzolite	present in some	believed absent
source	OIB source	ancient enriched subcontinental lithosphere

source regions similar to the enriched component involved in the petrogenesis of Dupal OIB (Ch. 9), whereas Group I kimberlites are derived from sources similar to those involved in non-Dupal OIB.

Figure 12.22 shows the variation of initial $^{87}Sr/^{86}Sr$ versus Zr/Nb for Group I and II kimberlites compared to fields for MORB and Dupal and non-Dupal OIB. The data appear to define two mixing arrays between a depleted MORB source component and two highly enriched components defined by Group I and II kimberlites. This appears to support the derivation of Group I and II kimberlites from source regions similar to the enriched components involved in the petrogenesis of non-Dupal and Dupal OIB respectively (Le Roex op. cit.). In the South Atlantic, Bouvet, with a high $^3He/^4He$, can be regarded as a hot spot with a significant primordial component in its source region, whereas the Gough, Tristan da Cunha, Discovery and Shona hot spots appear to involve a major recycled source component (Kurz et al. 1982; Allègre & Turcotte 1985; Le Roex 1985, 1986).

Group I ultrapotassic rocks have representatives in both of the trends in Figure 12.21. Lamproites from Leucite Hills and Smoky Butte display some

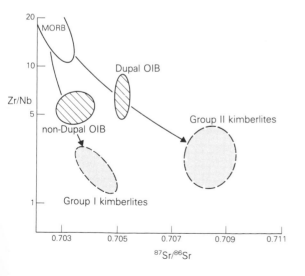

Figure 12.22 Variation of Zr/Nb versus initial $^{87}Sr/^{86}Sr$ for Group I and II kimberlites, compared to fields for MORB and South Atlantic OIB (after Le Roex 1986, Fig. 1. p. 243).

of the least radiogenic Nd—Sr isotopic compositions observed for potassium-rich volcanic rocks. Their negative ε_{Nd} values indicate, as in all the other high-K provinces, that the magmas contain a contribution from an old light-REE enriched source. The extremely negative ε_{Nd} isotopic compositions of these lamproites cannot be a consequence of crustal contamination (see, e.g., Vollmer et al. 1984) as the light-REE concentrations of the lavas far exceed those of average continental crust. Lamproites from Western Australia, south-east Spain and Gaussberg plot on the shallow Nd—Sr trend, along with Group II kimberlites, Group III potassic to ultrapotassic volcanics from the Roman Province of Italy and Group II lavas from the western branch of the East African Rift. It is clear that the sources of lamproite magmas have different styles of trace element enrichment, characterized by low Sm/Nd but both high and low Rb/Sr. Hawkesworth et al. (1985) have estimated that average Rb/Sr ratios in the source range from 0.2 in Western Australia to 0.04 at Smoky Butte. Lamproites from the shallow trend are characterized by high Rb/Sr but low Rb/Ba and K/Ti ratios, suggesting that their source has been metasomatized by the introduction of small-volume partial melts. In contrast, Group II kimberlites from the same trend have high Rb/Sr, Rb/Ba and K/Ti, suggesting that in this case the source metasomatism has been accomplished by the introduction of H_2O-rich fluids (Hawkesworth et al. 1984, 1985). High Rb/Sr and thus, with time, high $^{87}Sr/^{86}Sr$ ratios have therefore developed in mantle sources with both high and low Rb/Ba.

Nd—Sr isotopic studies of volcanic rocks from the Roman Province of Italy (Turi & Taylor 1976; Vollmer 1976, 1977; Hawkesworth & Vollmer 1979; Civetta et al. 1981; Vollmer et al. 1981; Holm & Munksgaard 1982; Holm et al. 1982) have led to suggestions that crustal contamination has played a fundamental role in the petrogenesis of these rocks. There is a clear trend of increasing $^{87}Sr/^{86}Sr$ and decreasing ε_{Nd} from south to north within the province, which has been interpreted in terms of the progressive contamination of the magmas by a crustal component. However, it is equally possible that the mantle source of these magmas is heterogeneous (Peccerillo 1985). Such heterogeneity may have been introduced by fluids rising from a subducted lithospheric slab which was underthrusting the Appennines during the Tertiary (Civetta *et al.* 1981, Peccerillo *et al.* 1984, Peccerillo 1985). Clearly, in the case of these ultrapotassic magmas it is difficult to use Sr isotopes to monitor crustal contamination, as the primary magmas have such high Sr contents that their isotopic compositions are not very susceptible to change as a consequence of interaction with the relatively low-Sr rocks of the continental crust.

Pb data

Figures 12.23 and 12.24 show the variation of $^{207}Pb/^{204}Pb$ and $^{208}Pb/^{204}Pb$ versus $^{206}Pb/^{204}Pb$ respectively for ultrapotassic rocks and Group I and

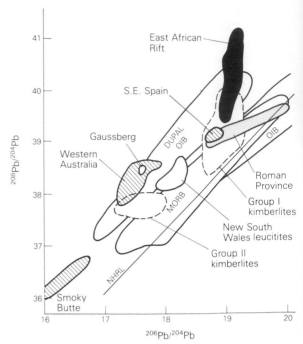

Figure 12.24 Variation of $^{208}Pb/^{204}Pb$ versus $^{206}Pb/^{204}Pb$ for kimberlites and potassic volcanic rocks, compared to fields for MORB, normal OIB and Dupal OIB (Ch. 9). Data sources as in Figure 12.23.

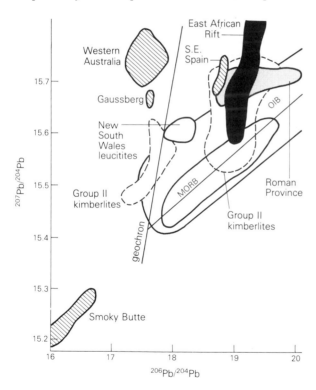

Figure 12.23 Variation of $^{207}Pb/^{204}Pb$ versus $^{206}Pb/^{204}Pb$ for kimberlites and potassic volcanic rocks, compared to fields for MORB and OIB. NHRL is the Northern Hemisphere Reference Line (see Ch. 9 for explanation). Data from Fraser *et al.* (1985), Nelson *et al.* (1986) and Davies & Lloyd (1988).

II kimberlites, compared to fields for MORB and OIB. These diagrams reveal a remarkable diversity of isotopic characteristics, with $^{206}Pb/^{204}Pb$ ratios ranging from 16 to 20.

In terms of Figure 12.23, lamproites from Western Australia, Gaussberg and Smoky Butte and some Group II kimberlites are distinctive in plotting to the left of the geochron. Nelson *et al.* (1986) interpret the Pb isotopic signature of the Western Australian and Gaussberg lamproites in terms of involvement of an extremely ancient source component which had high U/Pb early in its history, followed by a more recent lowering of U/Pb. They estimate a minimum of 2.1 Ga for this differentiation event. The high $^{207}Pb/^{204}Pb$ and low $^{206}Pb/^{204}Pb$ ratios of these rocks contrast with those of oceanic-island basalts (OIB; Ch. 9), which define a linear array of positive slope, extending from the unradiogenic Pb field of MORB to the high

^{207}Pb/^{204}Pb and ^{206}Pb/^{204}Pb OIB of St Helena and Tubuaii. The OIB array lies to the right of the geochron, on which all present-day single-stage Pb should lie, indicating that the mantle sources of OIB have undergone an increase in U/Pb (either progressively or episodically) within the past 2.5 Ga. The position of the MORB−OIB array to the right of the geochron has been attributed to the progressive loss of Pb from the mantle to the core. However, if the Western Australia and Gaussberg lamproites are generated from low-U/Pb source components within the subcontinental lithosphere, the existence of such reservoirs could also compensate for a general increase of U/Pb in OIB source mantle (Nelson *et al.* op. cit.)

In Figure 12.23 it is possible to divide the various ultrapotassic suites into two groups, using a dividing line of ^{206}Pb/^{204}Pb of 18.5. Group II lavas from the East African Rift, subduction-related Group III Roman Province rocks and Group I lamproites from south-east Spain plot to the high-^{206}Pb/^{204}Pb side of this line. The Pb isotopic compositions of the Spanish lamproites and Roman Province rocks closely resemble those of oceanic sediments (Sun 1980), possibly supporting their derivation from asthenospheric mantle sources which have been metasomatized by a subduction-zone fluid, involving a component derived from subducted sediments. Ultrapotassic lavas from the western branch of the East African Rift display similar Pb isotope characteristics, although with a wider range of ^{207}Pb/^{204}Pb ratios, overlapping with the MORB−OIB field, suggesting that they may have experienced some crustal contamination. The field of Group I kimberlites encompasses these three groups of potassic and ultrapotassic rocks.

Lamproites from Smoky Butte have the most extreme unradiogenic Pb isotopic compositions recorded for ultrapotassic rocks. Their isotopic characteristics are similar to those of old granulite facies rocks from the lower crust. However, on the basis of their trace element characteristics (Section 12.5.2) it is extremely unlikely that crustal contamination could have modified the isotopic characteristics of these magmas. Thus their isotopic composition must reflect that of their mantle source.

Xenoliths derived from the subcontinental upper mantle display a wide range of Pb isotopic ratios which are thought to have developed in old mantle material within the continental lithosphere (Kramers *et al.* 1983, Cohen *et al.* 1984, Richardson *et al.* 1984). Figure 12.25 shows the variation of ^{207}Pb/^{204}Pb versus ^{206}Pb/^{204}Pb for such xenoliths from the African subcontinental lithosphere (Ch. 11). The array clearly encompasses the entire field of ultrapotassic rocks and kimberlites, and thus all of them could be derived from the lithosphere on the basis of their Pb isotope compositions.

Model Pb ages for Smoky Butte lamproites are ∼2.5 Ga (Fraser *et al.* 1985), while Western Australian lamproites give a minimum age of 1.8 Ga for their source enrichment, with the added requirement that before 1.8 Ga evolution took place in a higher U/Pb environment. Pb isotopic data for Group II kimberlites are more ambiguous, but are consistent with ages of 1−1.4 Ga for portions of the South African upper mantle, inferred from Karoo picrites (Hawkesworth *et al.* 1983, Smith 1983).

For those ultrapotassic suites for which combined Nd−Pb-isotopic data are available there is a striking positive correlation between ε_{Nd} and ^{206}Pb/^{204}Pb (Fraser *et al.* 1985; and see Fig. 12.26). This contrasts markedly with the Nd−Pb isotopic compositions of oceanic basalts, which either exhibit a negative correlation as in MORB (Ch. 5) or a wide range in ^{206}Pb/^{204}Pb with very little variation in ε_{Nd} as in OIB (Ch. 9). Similar Nd−Pb isotopic compositions have been reported for mantle xenoliths from East Africa (Cohen *et al.* 1984). This suggests that the trend in Figure 12.26 reflects a low-U/Pb−low-Sm/Nd style of trace element enrichment in the subcontinental upper mantle. Whether such characteristics are due to the low U/Pb ratios of the enriching agent or to the stabilization of a low-U/Pb phase such as amphibole is unclear (Nelson *et al.* op. cit.). It is interesting to note that the Spanish and Italian provinces plot in a subparallel array to the main trend, displaced to slightly higher ^{206}Pb/^{204}Pb ratios. However, until further Nd−Sr−Pb isotopic data are available it is difficult to conclude whether or not this is significant in terms of petrogenetic models.

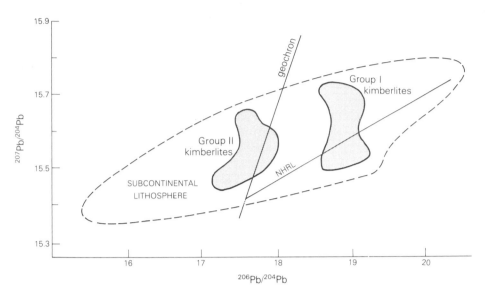

Figure 12.25 Pb isotopic composition of materials derived from the subcontinental lithosphere, compared to fields for Group I and II kimberlites. NHRL is the Northern Hemisphere Reference Line (see Ch.9 for details). Data from Kramers *et al.* (1983), Cohen *et al.* (1984), Richardson *et al.* (1984).

12.5.4   Stable isotopes

As we have considered in previous chapters, variations in the isotopic composition of oxygen in continental volcanic rocks can provide important constraints on the extent of interaction between mantle-derived magmas and the crustal rocks through which they pass. In general, continental intra-plate basalts have higher $^{18}O/^{16}O$ ratios than their oceanic counterparts, which may be attributable to the effects of crustal contamination (Kyser *et al.* 1982). However, Kyser *et al.* have shown that the range of mantle $\delta^{18}O$ values is actually much larger than the 5–6‰ commonly assumed, and therefore caution must be exercised when interpreting oxygen-isotope data in isolation.

In theory, combined Sr–O isotopic studies should provide better constraints on the role of crustal contamination in the petrogenesis of continental volcanic rocks. The reader is referred to Chapter 6 for a detailed discussion of this approach, in which we used combined Sr–O data to differentiate between source contamination and crustal contamination in the petrogenesis of island-arc volcanic rocks. Unfortunately, Sr contents in ultrapotassic magmas are so high that Sr isotopic ratios are not very susceptible to change as a result of interaction with the continental crust.

Taylor *et al.* (1984) have shown that the oxygen isotope compositions of leucite-bearing highly potassic lavas from New South Wales (Australia), Bufumbira (western branch, East African Rift) and Gaussberg (Antarctica) lie in a very narrow range from + 5.8 to + 7.5‰ and, as such, are virtually indistinguishable from the range of mantle-derived magmas worldwide. In marked contrast, leucite melilitites from San Venanzo, in the northern part of the Roman Volcanic Province, show significantly higher $\delta^{18}O$ values between 10.8 and 12‰. Turi & Taylor (1976) have shown that there is a general northward increase in $\delta^{18}O$ within the Roman Province which might be attributable to the effects of crustal contamination.

The Roman Province represents a rather unique geological environment, within the spectrum of K-rich magmatic provinces, in that the magmas were erupted through a thick, tectonically complex, section of ^{18}O-rich sedimentary rocks, including limestones, marls and mudstones with $\delta^{18}O$ up to 20‰, overlying a relatively high-^{18}O pelitic metasedimentary basement (Taylor *et al.* op. cit.). Contamination effects should be marked when there is such a strong isotopic contrast between the magmas and the crustal rocks which they assimilate. Nevertheless, contamination is only pronounced in the northern part of the province,

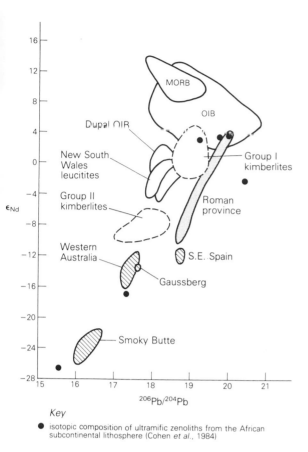

Key

● isotopic composition of ultramafic zenoliths from the African subcontinental lithosphere (Cohen et al., 1984)

Figure 12.26 Variation of ϵ_{Nd} versus $^{206}Pb/^{204}Pb$ for potassic volcanic rocks and Group I and II kimberlites, compared to fields of MORB, normal OIB and Dupal OIB (Ch. 9). Data from Fraser et al. (1985)] and Nelson et al. (1986).

where the crustal rocks had previously undergone anatexis during an earlier magmatic event.

In contrast, in Australia, East Africa and Antarctica the potassic magmas were erupted through relatively old regionally metamorphosed cratonic basement with $\delta^{18}O$ in the range +7 to +10‰. Clearly, with such little isotopic contrast between the primary magmas and their potential crustal contaminant, it is going to be difficult to identify contaminated samples on the basis of oxygen isotope data alone.

Ferrara et al. (1986) have demonstrated the importance of crustal contamination effects in the

northern part of the Roman Province by a combined Sr-O study of potassic volcanics from the Vulsini centre (Fig. 12.27). The lavas erupted here have $\delta^{18}O$ and $^{87}Sr/^{86}Sr$ (0.7097−0.7168) values that are much higher than in any of the other major volcanic centres of the Roman Province, and display a clear mixing trend towards the isotopic composition of local metasedimentary basement rocks. Nevertheless, Ferrara et al. consider that the primary magmas at Vulsini (high-K series) had $^{87}Sr/^{86}Sr$ ratios in the range 0.7101−0.7102 and $\delta^{18}O$ of + 5.5 to 7.5‰, indicating their derivation from an enriched mantle source.

12.6 Detailed petrogenetic model

Despite many recent publications (Fraser et al. 1985, Hawkesworth et al. 1985, Mitchell 1986, Nelson et al. 1986, Bergman 1987, Foley et al. 1987), the origins of K-rich mafic and ultramafic magmas is still not fully understood. Given the diversity of tectonic settings in which they occur, it is clearly impossible for a single petrogenetic model to be devised. However, most workers agree that the ultimate source of this relatively rare group of magmas lies in the upper mantle and is enriched in incompatible elements, particularly K, Rb, Ba, Sr, REE, P and Zr.

In most areas of the mantle, high concentrations of incompatible elements probably reflect the migration of small-volume silicate melts (Hawkesworth et al. 1985), although in some tectonic settings H_2O-rich fluids may be important (Schneider & Eggler 1986). In general, silicate melts are much more effective agents for the transportation of incompatible elements than fluids (Schneider & Eggler op. cit.). Infiltration of such partial melts is not in itself mantle metasomatism, although localized metasomatism may occur in the wall rocks adjacent to the veins. When such veined mantle is re-melted the incompatible element characteristics of the resultant magmas will be dominated by the material in the veins, if the degree of melting is small. Clearly, melting of veined mantle that is heterogeneous in mineralogy and chemistry, in spacing of veins and sometimes in isotopic

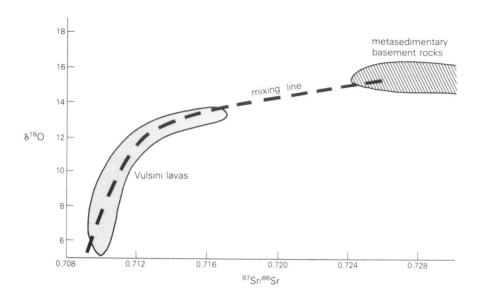

Figure 12.27 Variation of $\delta^{18}O$ versus $^{87}Sr/^{86}Sr$ for potassic-rich volcanics from Vulsini, Roman Volcanic Province. The dashed line shows a mixing relationship between these lavas and metasedimentary basement rocks (after Ferrara *et al.* 1986, Fig. 9, p. 276).

composition (due to disequilibrium between vein and host peridotite) can produce considerable diversity in the geochemical characteristics of the resultant partial melts. Vein-type mantle enrichment of this nature, involving the introduction of hydrous phases (amphibole and phlogopite), apatite and other minerals has been considered by many authors as a critical precursor to alkaline magmatism in general (Boettcher & O'Neil 1980, Menzies & Murthy 1980, Wass & Rogers 1980, Bailey 1982). Wilkinson & Le Maitre (1987) suggest that amphibole is probably the major source of Ti and K for low-K/Na and medium-K/Na magma types, whereas phlogopite is more likely to be the major K-bearing phase in the source regions of high-K/Na magmas. A fundamental question is whether the generation of potassium-rich magmas occurs in response to the introduction of melts/fluids into fairly normal mantle, or whether they only occur when volatile-rich mantle is remobilized, perhaps after billions of years.

Mantle enrichment processes may vary considerably in response to variations in the nature, composition and origin of the enriching agent and in the stabilization of minor phases. Clearly, different enrichment chemistries can be expected where the enriching agent is a silicate melt as opposed to an H_2O- or a CO_2-rich fluid. There is

little evidence to suggest that CO_2-rich fluids play an important role as they can carry very little dissolved solute compared to H_2O-rich fluids (Schneider & Eggler 1986). However, H_2O-rich fluids are clearly important agents of mantle metasomatism in subduction-related magmatic provinces (Wyllie & Sekine 1982; and see Chs. 6 & 7), and thus may be particularly important in the petrogenesis of Group III ultrapotassic rocks. Fraser *et al.* (1985) consider that Group II kimberlites are derived from mantle sources metasomatized by H_2O-rich fluids, on the basis of their high Rb/Sr, Rb/Ba and K/Ti ratios. In contrast, lamproites have high Ta/Yb, low Rb/Ba and low K/Ti, consistent with the enrichment of their mantle source by small-volume silicate melts. As we have shown in Section 12.5.3, the Rb/Sr ratio of the mantle source of lamproites must vary considerably, which may be a feature of the introduced melts.

The chemistry of partial melts of enriched mantle sources will depend upon the nature of residual accessory phases and the relative concentrations of volatile species (H_2O, CO_2, F etc.) in the source, as well as the depth and degree of melting. Different accessory phases may selectively incorporate different incompatible elements, which may explain the negative spikes in the spiderdiagram patterns of some ultrapotassic rocks (Section 12.5.2), provided that the crystal/liquid partition coefficients for the appropriate elements are sufficiently high. For example, the low Ti, Nb and Ba contents of orogenic ultrapotassic rocks may reflect the retention of these elements in accessory phases residual to partial melting. Similarly, the negative K spike in the spiderdiagram patterns of kimberlites may reflect residual phlogopite in the source.

The existence of a range of silica contents in apparently primary lamproite magmas could be explained by partial melting over a range of pressures. Experimental studies have shown that partial melts of peridotite in the presence of small amounts of H_2O and CO_2 become poorer in SiO_2 with increasing pressure. Additionally, for the more silicic lamproite magmas to be primary, H_2O/CO_2 ratios in the source would have to be high, as the presence of CO_2 leads to lower-SiO_2 partial melts (Foley *et al.* 1986). Foley *et al.* consider that the presence of CH_4 and HF will also promote the formation of silica-rich partial melts. Jaques *et al.* (1984) and Foley *et al.* (op. cit.) have suggested that fluorine is important in the petrogenesis of lamproite magmas.

An equally important influence on the chemistry of the partial melts is the composition of the mantle source prior to the enrichment event. For example, Foley *et al.* (op. cit.) consider that Group I and II ultrapotassic rocks are derived from variably depleted mantle sources, whereas Group III sources were originally fairly fertile prior to the enrichment event.

Experimental studies (e.g. Barton & Hamilton 1979, 1982; Arima & Edgar 1983a,b; Nicholls & Whitford 1983; and see Section 12.3) have failed to resolve the precise nature of the mantle source of ultrapotassic magmas. This is hardly surprising given the diverse range of tectonic settings in which this compositionally heterogeneous group of igneous rocks occur. There is probably a complete spectrum of mantle source compostions from phlogopite−garnet lherzolite to phlogopite clinopyroxenite. Available data suggest that kimberlites may be derived by partial melting of a phlogopite-magnesite garnet lherzolite at pressures of 40−50 kbar (Wyllie 1980, Mitchell 1986), whereas some silica-poor lamproites may be derived from a phlogopite pyroxenite source (Barton & Hamilton 1979, Arima & Edgar 1983b). Experimental studies have constrained the stability of amphibole in metasomatized mantle to relatively shallow depths (<70 km), whereas phlogopite may extend to as deep as 180 km (Bailey 1987). One of the major problems of these experimental studies involves the necessity of controlling the total volatile content, its H_2O/CO_2 ratio and oxygen fugacity, parameters which can have a profound effect on the compositons of partial melts. Ideally, we require experiments performed under the complete range of conditions appropriate to the generation of ultrapotassic magmas. Unfortunately such data are not available at present. Using the volatile contents of potassic magmas to constrain their proportions in the mantle source, Bergman (1987) has proposed that the mantle sources of kimberlites and alkali basalts have low H_2O/CO_2 ratios, whereas lamproite sources have high H_2O/CO_2.

In many of the preceding chapters we have associated mantle partial melting with the adiabatic decompression of mantle material in ascending diapirs or plumes. However, Bailey (1985, 1987) and Wyllie (1980) consider that, in the case of alkaline ultramafic volcanism, the trigger is volatile-induced partial melting, where the ambient geotherm intersects a vapour-present solidus. Similar arguments have been invoked for the generation of basaltic magmas in subduction-zone environments (Chs 6 & 7). In the generation of potassic and ultrapotassic magmas within continental plates there is undoubtedly a spectrum of processes operating between these two extremes. For example, the causes of kimberlite magmatism have been variously attributed to source metasomatism (Bailey op. cit.), diapiric upwelling (Le Roex 1986) or to passive tectonic disturbances involving the propagation of deep lithosphere fractures (Haggerty 1982).

In focusing our attention on the role of metasomatized or enriched mantle sources in the petrogenesis of potassic and ultrapotassic magmas we have tended to dismiss older hypotheses for the generation of highly incompatible element enriched alkaline magmas. For example, Harris & Middlemost (1969) suggested that silicate melts, originating at great depth in the mantle, rise towards the surface by a solution stoping process, zone refining (Ch. 4), which continuously enriches incompatible elements in the liquid, while keeping the major element composition relatively constant. This model requires no pre-enrichment of the mantle source in incompatible elements. However, as Sr−Nd−Pb isotopic data provide unequivocal evidence for source enrichment we have not considered zone refining to be an important process. Nevertheless, until we have a greater understanding of the transport mechanism of small-volume silicate melts through the lithosphere we should keep an open mind on this subject.

Ultrapotassic magmas display a remarkably diverse range of Nd−Sr−Pb isotopic compositions (Section 12.5.3). As a consequence of their extremely high concentrations of Sr, Nd and Pb they are comparatively insensitive to the effects of crustal contamination and thus, in general, their isotopic compositions may be considered to reflect those of their mantle source. Most petrogenetic models favour their generation as a consequence of the invasion of the subcontinental mantle by isotopically distinct incompatible element enriched fluids or partial melts. In many cases these added components possess high $^{207}Pb/^{204}Pb$, radiogenic Sr and unradiogenic Nd isotopic compositions relative to MORB, indicating that their sources had long and complex histories within the upper mantle (Nelson et al. 1986). The trace element and isotopic characteristics of partial melts of such enriched mantle sources are strongly influenced by those of the added component, resulting in isotope−isotope and isotope − trace element relations indicative of mixing. Those potassic-ultrapotassic rocks having Nd-Sr isotopic compositions close to bulk Earth (e.g. New South Wales leucitites and Group II kimberlites) may have sources in the asthenospheric upper mantle. However, most examples have higher $^{87}Sr/^{86}Sr$ and lower ε_{Nd} than bulk Earth and clearly contain a contribution from old light-REE enriched (low Sm/Nd) material. Most authors would place such a component within the mantle part of the subcontinental lithosphere (Hawkesworth et al. 1983, Richardson et al. 1984). Com-

bined Nd−Sr−Pb isotopic studies indicate that most lamproites and Group II kimberlites are derived from upper mantle source regions in the age range 1−2.5 Ga (McCulloch et al. 1983, Smith 1983, Fraser et al. 1985, Hawkesworth et al. 1985).

The depth to which the continental lithosphere extends beneath ancient cratons is not well constrained geophysically. However, Richardson et al. (op. cit.) speculate that it could extend into the diamond stability field. Thus even diamond-bearing Western Australian lamproites and Group II kimberlites could be derived from lithospheric sources. The Western Australian, Gaussberg, Leucite Hills and East African Rift provinces are all situated on thick stable Precambrian basement, supporting the contention that ancient enriched subcontinental lithosphere may be the major magma source. In contrast, the Spanish lamproites and Roman Province volcanic rocks are emplaced within much younger basement terranes, and therefore their mantle sources may be somewhat different.

Figure 12.28 is a schematic illustration of the solidus of essentially unmodified peridotite in the presence of small amounts of H_2O and CO_2. Depending upon the geothermal gradient, partial melting may occur at a variety of depths. Figure 12.29 shows vertical mantle sections, corresponding to the three geotherms shown in Figure 12.28, indicating the changing mineralogy of the mantle source with depth, the beginning of partial melting (the solidus) and the field of diamond stability. In each of these environments, possible compositions of near-solidus partial melts are indicated (Bailey 1985). If the source of potassic and ultrapotassic magmas really is unmodified mantle lherzolite in the presence of H_2O and CO_2, then from Figure 12.29 we would conclude that the sources of Group II and III ultrapotassic rocks were located at shallower depths in the mantle than those of Group I ultrapotassic rocks and kimberlites. Even for highly metasomatized mantle sources this will still be true, as the diamond-graphite phase transition is independent of the bulk composition of the source, and diamonds have never been recorded from Group II and III ultrapotassic rocks. Geothermal gradients in both these provinces are likely to be

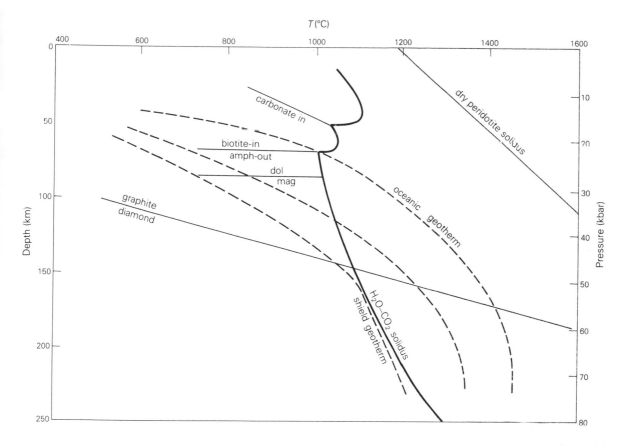

Figure 12.28 Schematic relationship between the peridotite solidus in the presence of small amounts of H₂O and CO₂ (Olafsson & Eggler 1983) and typical oceanic and continental geotherms (after Bailey 1985, Fig. 1, p. 451).

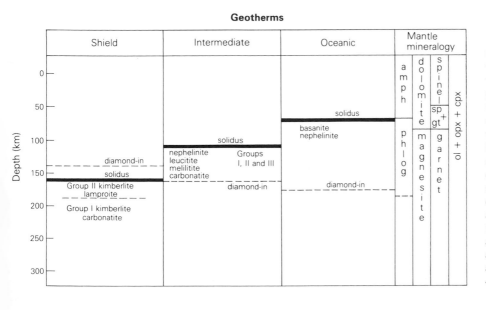

Figure 12.29 Schematic vertical sections through the mantle for the three geotherms shown in Figure 12.28, showing the changes in mineralogy of the mantle with depth, the beginning of melting (the solidus) and the field of diamond stability. Also shown are suggested depths of origin for Group I and II kimberlites, Group I, II and III ultrapotassic rocks and a range of silica-poor alkali basalts (modified after Bailey 1985, Fig. 2, p. 452).

higher than old continental shield geotherms due to mantle upwelling (Group II) and subduction-zone processes (Group III) respectively. Non-diamondiferous lamproites may be generated at shallower depths than diamond-bearing varieties, particularly the more SiO_2-rich examples. Ideally, we require diagrams such as Figure 12.28 and 12.29 for a range of metasomatized mantle source compositions before we should attempt to establish a petrogenetic grid for potassic intra-plate magmas. Unfortunately, there are insufficient experimental data available at present.

Further reading

Bergman, S.C. 1987. Lamproites and other potassium-rich igneous rocks: a review of their occurrence, mineralogy and geochemistry. In *Alkaline igneous rocks*, J.G. Fitton & B.G.J. Upton (eds), 103−89. Geol Soc. Spc. Publ. 30.

Foley, S.F., G. Venturelli, D.H. Green & L. Toscani 1987. The ultrapotassic rocks: characteristics, classification and constraints for petrogenetic models. *Earth-Sci. Rev.* 24, 81−134.

Hawkesworth, C.J., K.J. Fraser & N.W. Rogers 1985. Kimberlites and lamproites: extreme products of mantle enrichment processes. *Trans Geol Soc. S. Afr.* 88, 439−47.

Mitchell, R.H. 1986. *Kimberlites*. New York: Plenum Press, 442 pp.

Nelson, D.R., M.T. McCulloch & S.-S. Sun 1986. The origins of ultrapotassic rocks as inferred from Sr, Nd and Pb isotopes. *Geochim. Cosmochim. Acta* 50, 231−45.

Rock, N.M.S. 1987. The nature and origin of lamprophyres: an overview. In *Alkaline igneous rocks* In J.G. Fitton & B.G.J. Upton (eds), 191−226. Geol Soc. Spc. Publ. 30.

APPENDIX

Approximate range of D *values for the partitioning of trace elements between the common rock-forming minerals and liquids of basic−intermediate composition*

Quantitative modelling of differentiation in magmatic systems requires the precise knowledge of mineral/melt partition coefficients. Numerous measurements of such coefficients have been made on phenocryst−matrix pairs in volcanic rocks, and to a lesser extent in experimental systems. Figures A.1a−g clearly show that for some elements and some minerals the D values are poorly constrained by the data presently available. In some instances this places severe limitations on trace element modelling of the petrogenesis of suites of volcanic rocks.

In general, D values must vary as a function of temperature, pressure, fO_2 and bulk chemical composition of the melt and the mineral. Available data suggest that D values increase with increasing SiO_2 content of the magma (e.g. Henderson 1982, Lemarchand *et al.* 1987).

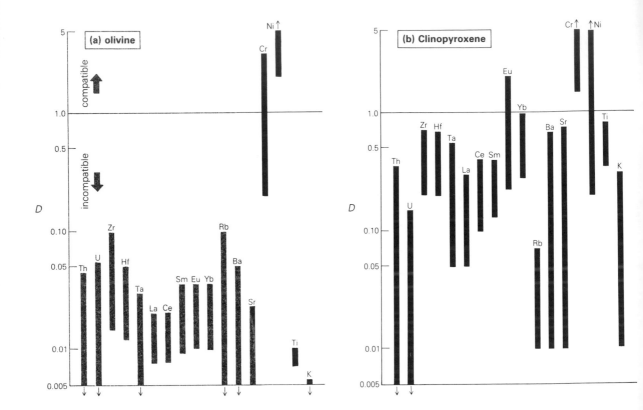

Figure A.1   Approximate ranges of D values. Data from Cox *et al.* (1979), Henderson (1982), Budahn & Schmitt (1985), Dunn (1987) and Le Marchand *et al.* (1987).

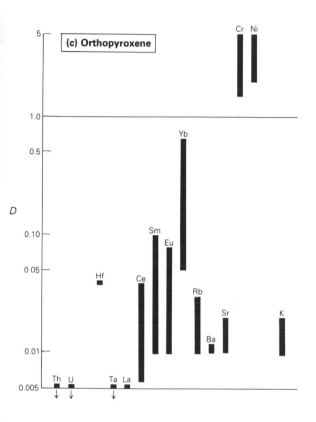

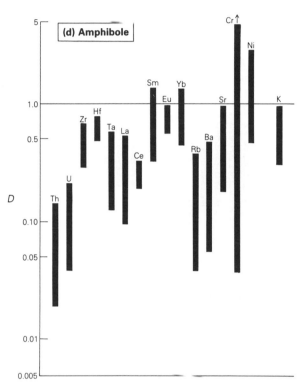

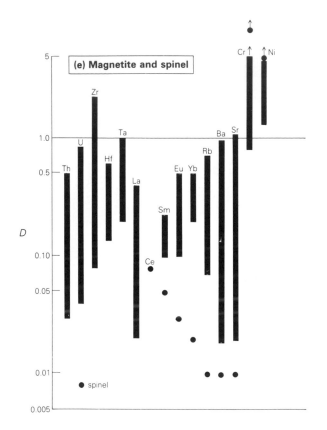

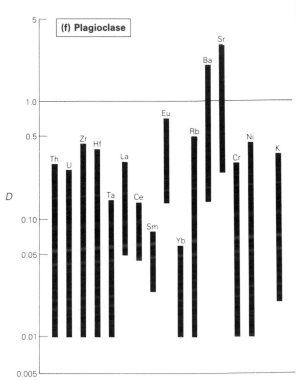

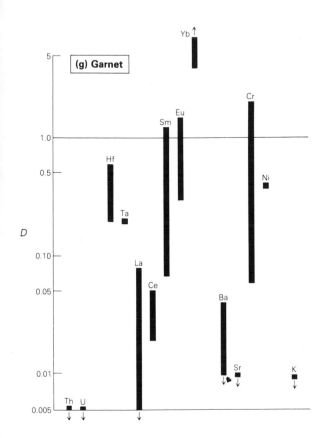

BIBLIOGRAPHY

Acharya, H. 1981. Volcanism and aseismic slip in subduction zones. *J. Geophys. Res.* **86**, 335–44.

Akella, J. & F. R. Boyd 1974. Petrogenetic grid for garnet peridotites. *Carnegie Inst. Wash. Yearbook.* **73**, 269–73.

Albarede, F. & A. Michard, 1986. Transfer of continental Mg., S, O, and U to the mantle through hydrothermal alteration of the oceanic crust. *Chem. Geol.* **57**, 1–15.

Aldritch, M. J. Jr, C.E. Chapin & A.W. Laughlin 1986. Stress history and tectonic development of the Rio Grande Rift, New Mexico. *J. Geophys. Res.* **91**, 6199–211.

Allègre, C. J. & M. Condomines 1982. Basalt genesis and mantle structure studied through Th-isotopic geochemistry. *Nature* **299**, 21–4.

Allègre, C. J., B. Dupré, B. Lambret & P. Richard 1981. The subcontinental versus the suboceanic debate I. Lead–neodymium–strontium isotopes in primary alkali basalts from a shield area: the Ahaggar volcanic suite. *Earth Planet. Sci. Lett.* **52**, 85–92.

Allègre, C. J., B. Dupré, P. Richard, D. Rousseau & C. Brooks 1982. Subcontinental versus suboceanic mantle, II. Nd–Sr–Pb isotopic comparison of continental tholeiites with mid-ocean ridge tholeiites and the structure of the continental lithosphere. *Earth Planet. Sci. Lett.* **57**, 25–34.

Allègre, C. J., B. Hamelin & B. Dupré 1984. Statistical analysis of isotopic ratios in MORB: the mantle blob cluster model and the convective regime of the mantle. *Earth Planet. Sci. Lett.* **71**, 71–84.

Allègre, C. J. & J. F. Minster, 1978. Quantitative models of trace element behavior in magmatic processes. *Earth Planet. Sci. Lett.* **38**, 1–25.

Allègre, C. J., T. Staudacher & P. Sarda 1986. Rare gas systematics: formation of the atmosphere, evolution and structure of the Earth's mantle. *Earth Planet. Sci. Lett.* **81**, 127–50.

Allègre, C. J., T. Staudacher, P. Sarda & M. Kurz 1983. Constraints on the evolution of the mantle from rare gas systematics. *Nature* **303**, 762.

Allègre C. J., M. Trevor, J. F. Minster, B. Minster & F. Albarede 1977. Systematic use of trace elements in igneous processes. *Contrib. Mineral. Petrol.* **60**, 55–75.

Allègre, C. J. & D. L. Turcotte 1985. Geodynamic mixing in the mesosphere boundary layer and the origin of oceanic islands. *Geophys. Res. Lett.* **12**, 207–10.

Allen, J. C. & A. Boettcher 1978. Amphiboles in andesite and basalt II. Stability as a function of $P-T-fH_2O-fO_2$. *Am. Mineral.* **63**, 1074–87.

Almond, D. C. 1986. Geological evolution of the Afro-Arabian dome. *Tectonophysics* **131**, 301–32.

Alvarez, L. W., W. Alvarez, F. Asaro & H. V. Michel 1980. Extraterrestrial cause for the Cretaceous–Tertiary extinction. *Science* **208**, 1095–108.

Anderson, R. N., S. E. Delong & W. M. Schwarz 1978. Thermal model for subduction with dehydration in the downgoing slab. *J. Geol.* **86**, 731–9.

Anderson, R. N., S. E. Delong, & W. M. Schwarz 1980. Dehydration, asthenospheric convection and seismicity in subduction zones. *J. Geol.* **88**, 445–51.

Ankeny, L. A., L. W. Braile, & K. H. Olsen 1986. Upper crustal structure beneath the Jemez Mountains Volcanic Field, New Mexico, determined by three-dimensional simultaneous inversion of seismic refraction and earthquake data. *J. Geophys. Res.* **91**, 6188–98.

Arculus, R. J. 1981. Island arc magmatism in relation to the evolution of the crust and mantle. *Tectonophysics* **75**, 113–33.

Arculus, R. J. 1987. The significance of source versus process in the tectonic controls of magma genesis. *J. Volcanol. Geotherm. Res.* **32**, 1–12.

Arculus, R. J. & J. W. Delano 1987. Oxidation state of the upper mantle: present conditions, evolution and controls. In *Mantle xenoliths*, P.H. Nixon (ed.), 589–98. Chichester: Wiley.

Arculus, R. J. & R. W. Johnson 1981. Island arc magma sources: a geochemical assessment of the roles of slab-derived components and crustal contamination. *Geochem. J.* **15**, 109–33.

Arculus, R. J. & R. Powell 1986. Source component mixing in the regions of arc magma generation. *J. Geophys. Res.* **91**, 5913–26.

Arima, M. & A. D. Edgar 1983a. High pressure experimental studies on a katungite and their bearing on the genesis of some potassium-rich magmas of the west branch of the African Rift. *J. Petrol.* **24**, 166–87.

Arima, M. & A. D. Edgar 1983b. A high pressure experimental study of a magnesian-rich leucite lamproite from the West Kimberley area, Australia: petrogenetic implications. *Contrib. Mineral. Petrol.* **84**, 228–34.

Armstrong, R. L. 1981. Radiogenic isotopes: The case for crustal recycling on a near-steady-state, no-continental-growth earth. *Phil Trans R. Soc. Lond.* A**310**, 443–72.

Atherton, M. P. & L. M. Sanderson 1985. The chemical variation and evolution of the super-units of the segmented coastal batholith. In Pitcher *et al.* (1985), op. cit. pp 177–202.

Bailey, D. K. 1974. Continental rifting and alkaline magmatism. In *The alkaline rocks*, Sorensen, H. (ed.),148–59. Chichester: Wiley.

Bailey, D. K. 1980a. Volcanism, Earth degassing and replenished lithosphere mantle. *Phil Trans R. Soc. Lond.* A**297**, 309–22.

Bailey, D. K. 1980b. Volatile flux, geotherms, and the generation of the kimberlite–carbonatite–alkaline magma spectrum. *Mineral Mag.* **43**, 695–9.

Bailey D. K. 1982. Mantle metasomatism – continuing chemical change within the Earth. *Nature* **296**, 525–30.

Bailey, D. K. 1983. The chemical and thermal evolution of rifts. *Tectonophysics* **94**, 585–97.

Bailey, D. K. 1985. Fluids, melts, flowage and styles of eruption in alkaline ultramafic magmatism. *Trans Geol Soc. S. Afr.* **88**, 449–57.

Bailey D. K. 1987. Mantle metasomatism – perspective and prospect. In *Alkaline igneous rocks*. J. G. Fitton & B. G. J. Upton (eds), 1–13. Geol Soc. Sp. Publ. 30.

Baker, B. H. 1987. Outline of the petrology of the Kenya rift alkaline province. In *Alkaline igneous rocks*, J. G. Fitton & B. G. J. Upton (eds), 293–311. Geol Soc. Sp. Publ. 30.

Baker B. H., G. G. Goles, W. P. Leeman, & M. M. Lindstrom 1977. Geochemistry and petrogenesis of a basalt–benmoreite–trachyte suite from the southern part of the Gregory Rift, Kenya. *Contrib. Mineral. Petrol.* **64**, 303–32.

Baker B. H. & A. R. McBirney 1985. Liquid fractionation. Part III: Geochemistry of zoned magmas and the compositional effects of liquid fractionation. *J. Volcanol. Geotherm. Res.* **24**, 55–81.

Baker, B. H., P. A. Mohr & L. A. J. Williams 1972. *Geology of the Eastern Rift System of Africa*. Geol Soc. Am., Sp. Publ. 186, 67 pp.

Baker, I. 1969. Petrology of the volcanic rocks of St. Helena island, South Atlantic. *Bull. Geol Soc. Am.* **80**, 1283–310.

Baker, P. E. 1973. Volcanism at destructive plate margins. *J. Earth Sci. Leeds* **8**, 183–95.

Baker, P. E. 1982. Evolution and classification of orogenic volcanic rocks. In Thorpe (1982), op. cit., pp. 11–23.

Baker, P.E., I.G. Gass, P. G. Harriss & R. W. Le Maitre 1984. The volcanologial report of the Royal Society Expedition to Tristan da Cunha, 1962. *Phil Trans R. Soc. Lond.* A256, 439–578.

Ballard, R. D., W. B. Bryan, J. R. Heirtzler, G. R. Keller, J. G. Moore & Tj. H. Van Andel 1975. Manned submersible observations in the FAMOUS area. *Science* 190, 103–8.

Ballard, R. D., J. Francheteau, Th. Juteau, C. Rangin & W. Normark 1981. East Pacific Rise at 21°N: the volcanic, tectonic and hydrothermal processes of the central axis. *Earth Planet. Sci. Lett.* 55, 1–10.

Ballard, R. D. & Tj. H. Van Andel 1977. Morphology and tectonics of the inner rift valley at Lat. 36°50′N on the Mid-Atlantic Ridge. *Bull. Geol Soc. Am.* 88, 507–30.

Barazangi, M. & B. L. Isacks 1979. Subduction of the Nazca plate beneath Peru: evidence from spatial distribution of earthquakes. *Geophys. J. R. Astron. Soc.* 57, 537–55.

Barberi, F., G. Ferrara, R. Santacroce, M. Treuil & J. Varet 1975. A transitional basalt–pantellerite sequence of fractional crystallisation, the Boina Centre, (Afar Rift, Ethiopia). *J. Petrol.* 16, 22–56.

Barberi, F., R. Santacroe & J. Varet 1982. Chemical aspects of rift magmatism. In *Continental and oceanic rifts*, G. Palmason (ed.), 223–58. Washington DC: American Geophysical Union.

Barker, D. S. 1983. *Igneous rocks*. Englewood Cliffs, N J: Prentice-Hall.

Barker, P. F. 1972. A spreading centre in the east Scotia Sea. *Earth Planet. Sci. Lett.* 15, 123–32.

Barreiro, B. 1983. Lead isotopic compositions of South Sandwich Island volcanic rocks and their bearing on magmagenesis in intra-oceanic island arcs. *Geochim. Cosmochim. Acta* 47, 817–22.

Bardsell, M., I. E. M. Smith & K. B. Sporli 1982. The origin of reversed geochemical zoning in the Northern New Hebrides volcanic arc. *Contrib. Mineral. Petrol.* 81, 148–55.

Barton, M. & D. L. Hamilton 1979. The melting relationships of a madupite from the Leucite Hills, Wyoming, to 30 kb. *Contrib. Mineral. Petrol.* 69, 133–42.

Barton, M. & D. L. Hamilton 1982. Water undersaturated melting experiments bearing upon the origin of potassium-rich magmas. *Mineral. Mag.* 45, 267–78.

Basaltic Volcanism Study Project 1981. *Basaltic volcanism on the terrestial planets*. New York. Pergamon Press, 1286 pp.

Batiza, R. 1982. Abundances, distribution and sizes of volcanoes in the Pacific Ocean and implications for the origin of non-hotspot volcanoes. *Earth Planet. Sci. Lett.* 60, 195–206.

Beane, J. E., C. A Turner, P. R. Hooper, K. V. Subbarao & J. N. Walsh 1986. Stratigraphy, composition and form of the Deccan basalts, Western Ghats, India. *Bull. Volcanol.* 48, 61–83.

Beckinsale, R. D., A. W. Sanchez-Fernandez, M. Brook, E. J. Cobbing, W. P. Taylor & N. D. Moore 1985. Rb–Sr whole rock isochron and K–Ar age determinations for the coastal batholith of Peru. In Pitcher (1985), op. cit., pp. 177–202.

Bellieni, G., P. Brotzu, P. Comin-Chiaramonti, M. Ernesto, A. Melfi, I. G. Pacca & E. M. Piccirillo 1984. Flood basalt to rhyolite suites in southern Parana Plateau (Brazil): palaeomagnetism, petrogenesis and geodynamic implications. *J. Petrol.* 25, 579–618.

Bellieni, G., P. Comin-Chiaramonti, L. S. Marques, A. J. Melfi, A. J. R. Nardy, C. Papatrechas, E. M. Piccirillo, A. Roisenberg & D. Stolfa 1986. Petrogenetic aspects of acid and basaltic lavas from the Parana Plateau (Brazil): geological, mineralogical and petrological relationships. *J. Petrol* 27, 915–44.

Bender, J. F., F. N. Hodges & A. E. Bence 1978. Petrogenesis of basalts from the project FAMOUS area: experimental study from 0 to 15 kbars. *Earth Planet. Sci. Lett.* 41, 277–302.

Bender, J. F., C. H. Langmuir & G. N. Hanson 1984. Petrogenesis of basalt glasses from the Tamayo region, East Pacific Rise. *J. Petrol.* 25, 213–54.

Bergman, S. C. 1987. Lamproites and other potassium-rich igneous rocks: a review of their occurrence, mineralogy and geochemistry. In *Alkaline igneous rocks*, J. G. Fitton & B. G. J. Upton (eds), 103−89. Geol. Soc. Sp. Publ. 30.

Best, M. G. 1982. *Igneous and metamorphic petrology*. New York: W. H. Freeman.

Betton, P. J. & L. Civetta 1984. Strontium and neodymium isotopic evidence for the heterogeneous nature and development of the mantle beneath Afar (Ethiopia). *Earth Planet.Sci. Lett.* 71, 59−70.

Boettcher, A. L., B. O. Mysen & P. J. Modreski 1975. Melting in the mantle: phase relationships in natural and synthetic peridotite−H_2O and peridotite−H_2O−CO_2 systems at high pressure. *Phys. Chem. Earth* 9, 855−65.

Boettcher, A. L. & J. R. O'Neil 1980. Stable isotope, chemical and petrographic studies of high-pressure amphiboles and micas: evidence for metasomatism in the mantle source regions of alkali basalts and kimberlites. *Am, J. Sci.* 280a, 594−621.

Bonatti, E. 1985. Punctiform initiation of seafloor spreading in the Red Sea during the transition from a continental to an oceanic rift. *Nature* 316, 33−7.

Bosworth, W. 1985. Geometry of propagating continental rifts. *Nature* 316, 625−7.

Bott, M. H. P. 1982. Origin of the lithospheric tension causing basin formation. *Phil Trans R. Soc. Lond.* A305, 319−24.

Brey, G., W. R. Brice, D. J. Ellis, D. H. Green, K. L. Harris & I. D. Ryabchikov 1983. Pyroxene−carbonate reactions in the upper mantle. *Earth Planet. Sci. Lett.* 62, 63−74.

Brey, G. P. & J. Huth 1984. The enstatite−diopside solvus to 60 kbar. *Proc. Third Int. Kimb. Conf* 2, 257−64.

Briqueu, L., H. Bougault & J. L. Joron 1984. Quantification of Nb, Ta, Ti and V anomalies in magmas associated with subduction zones: petrogenetic implications. *Earth Planet. Sci. Lett.* 68, 297−308.

Bristow, J. W. 1984. Picritic rocks of the north Lebombo and south-east Zimbabwe. In *Petrogenesis of the volcanic rocks of the Karoo Province*, A. J. Erlank (ed,), 105−23. Sp. Publ. Geol. Soc.

S. Afr. 13.

Brotzu, P., L. Morbidelli, E. M. Piccirillo & G. Traversa 1983. The basanite to peralkaline phonolite suite of the Plioquaternary Nyambeni multicentre volcanic range (East Kenya Plateau). *Neus. Jahrbuch Miner. Abh.* 147, 253−80.

Brown, C. & R. W. Girdyer 1980. Interpretation of African gravity and its implication for the breakup of the continents. *J. Geophys. Res.* 85, 6443−55.

Brown, G. C. & A. E. Mussett 1981. *The inaccessible Earth*. London: Allen and Unwin.

Brown, G. M., J. G. Holland, H. Sigurdsson, J. F. Tomblin & R. J. Arculus 1977. Geochemistry of the Lesser Antilles volcanic arc. *Geochim. Cosmochim. Acta* 41, 785−801.

Bryan, W. B. 1983. Systematics of modal phenocryst assemblages in submarine basalts: petrologic implications. *Contrib. Mineral. Petrol.* 83, 62−74.

Bryan, W. B., L. W. Finger & F. Chayes 1969. Estimating proportions in petrographic mixing equations by least squares approximation. *Science* 163, 926−7.

Bryan, W. B., G. Thompson, F. A. Frey & J. S. Dickey 1976. Inferred settings and differentiation in basalts from the Deep Sea Drilling Project. *J. Geophys. Res.* 81, 4285−304.

Buck, W. R. 1986. Small-scale convection induced by passive rifting: the cause for uplift of rift shoulders. *Earth Planet. Sci. Lett.* 77, 362−72.

Budahn, J. R. & R. A. Schmitt 1985. Petrogenetic modelling of Hawaiian tholeiitic basalts: a geochemical approach. *Geochim. Cosmochim. Acta* 49, 67−87.

Burke, K. & A. J. Whitemann 1973. Uplift, rifting and the break-up of Africa. In *Implications of continental drift to the Earth sciences*,vol. 2, D. H. Tarling & S. K. Runcorn, (eds) 735−55. New York: Academic Press.

Burke, K. C. & J. T. Wilson 1976. Hot spots on the Earth's surface. In *Volcanoes and the Earth's interior*, R. Decker & B. Decker, 31−42. New York: W. H. Freeman (1982).

Busse, F. H. 1983. Quadrupole convection in the lower mantle. *Geophys. Res. Lett.* 10, 285−8.

Bussell, M. A. 1976. Fracture control of high-level

plutonic contacts in the coastal Batholith of Peru. *Proc. Geol. Assoc.* **87**, 237−46

Bussell, M. A. & W. S. Pitcher 1985. The structural controls of batholith emplacement. In Pitcher *et al.* (1985), op. cit., pp. 167−76.

Byers, C. D., D. W. Muenow & M. O. Garcia 1983. Volatiles in basalts and andesites from the Galapagos spreading centre, 85° to 86°W. *Geochim. Cosmochim. Acta* **47**, 1551−8.

Cameron, W. E, E. G. Nisbet & V. J. Dietrich 1979. Boninites, komatiites and ophiolitic basalts. *Nature* **280**, 550−3.

Campbell, I. H. 1977. A study of macro-rythmic layering and cumulate process in the Jimberlana intrusion, Western Australia: I. The Upper Layered Series. *J. Petrol.* **18**, 183−215.

Campbell, I. H. 1978. Some problems with cumulus theory. *Lithos* **11**, 311−23.

Campbell, I. H. 1985. The difference between oceanic and continental tholeiites: a fluid dynamic explanation. *Contrib. Mineral. Petrol.* **91**, 37−43.

Campbell, I. H. & J. S. Turner 1986. The influence of viscosity on fountains in magma chambers. *J. Petrol.* **27**, 1−30

Cann, J. R. 1969. Spilites from the Carlsberg Ridge, Indian Ocean. *J.Petrol.* **10**, 1−19.

Cann, J. R. 1970. New model for the structure of the ocean crust. *Nature.* **226**, 928−30.

Cann, J. R. 1974. A model for oceanic crustal structure developed. *Geophys. J. R. Astron. Soc.* **39**, 169−87.

Carlson, R. W. 1984. Isotopic constraints on Columbia River flood basalt genesis and the nature of the subcontinental mantle. *Geochim. Cosmochim. Acta* **48**, 2357−72.

Carlson, R. W., G. W. Lugmair & J. D. Macdougall 1981. Columbia River volcanism: the question of mantle heterogeneity or crustal contamination. *Geochim. Cosmochim. Acta* **45**, 2483−99.

Carmichael, I. S. E. 1964. The petrology of Thingmuli, a Tertiary volcano in Eastern Iceland. *J. Petrol.* **5**, 435−60.

Carmichael, I. S. E., F. J. Tuner & J. Verhoogen 1974. *Igneous petrology.* New York: McGraw-Hill, 739 pp.

Carrigan, C. R. 1982. Multiple scale convection in the Earth's mantle: a three-dimensional study. *Science* **215**, 965−7

Carswell, D. A. 1980. Mantle derived lherzolite nodules associated with kimberlite, carbonatite and basaltic magmatism: a review. *Lithos* **13**, 121−38.

Carswell, D. A. & F. G. F. Gibb 1980. Geothermometry of garnet lherzolite nodules with special reference to those from the kimberlites of northern Lesotho. *Contrib. Mineral. Petrol.* **74**, 403−16.

Cas, R. A. F. & J. V. Wright 1987. *Volcanic successions.* London: Allen and Unwin.

Cawthorn, B. G. & T. S. McCarthy 1981. Bottom crystallization and diffusion control in layered complexes: evidence from Cr distribution in magnetite from the Bushveld Complex. *Trans Geol Soc. S. Afr.* **84**, 41−50.

Cawthorn, R. G. & M. J. O'Hara 1976. Amphibole fractionation in calcalkaline magma genesis. *Am. J. Sci.* **276**, 309−29.

Chase, C. G. 1981. Oceanic Island Pb: two-stage histories and mantle evolution. *Earth Planet. Sci. Lett.* **52**, 277−84.

Chayes, F. 1964. Variance−covariance relations in Harker diagrams of volcanic suites. *J. Petrol.* **5**, 219−37.

Chen, C. Y. 1987. Lead isotope constraints on the origin of Hawaiian basalts. *Nature* **327**, 49−52.

Chen, C. Y. & F. A. Frey 1983. Origin of Hawaiian tholeiite and alkalic basalt. *Nature* **302**, 785−9.

Chen, C. Y. & F. A. Frey 1985. Trace element and isotopic geochemistry of lavas from Haleakala volcano East Maui, Hawaii: implications for the origin of Hawaiian basalts. *J. Geophys. Res.* **90**, 8743−68.

Chen, J. H. & G. J. Wasserberg 1983. The least radiogenic Pb in iron meteorites. *Fourteenth Lunar and Planetary Science Conference, Abstracts*, Part 1, 103−4 Houston, Texas: Lunar and Planetary Institute.

Chen, W. P & P. Molnar 1983. Focal depths of intracontinental and intraplate earthquakes and their implications for the thermal and mechanical properties of the lithosphere. *J. Geophys. Res.*

88, 4183–214.

Christensen, R. L. 1984. Yellowstone magmatic evolution: its bearing on understanding large-volume explosive volcanism. In *Studies in geophysics. Explosive volcanism: inception, evolution and hazards*. Washington DC: National Academy Press, 176 pp.

Christiansen, N. I & M. H. Salisbury 1975. Structure and constitution of the lower oceanic crust. *Rev. Geophys. Space Phys.* 13, 57–86.

Christie, D. M. & J. M. Sinton 1981. Evolution of abyssal lavas along propagating segments of the Galapagos spreading centre. *Earth Planet. Sci. Lett.* 56, 321–33.

Church, S. G. 1985. Genetic interpretation of lead-isotopic data from the Columbia River Basalt Group, Oregon, Washington and Idaho. *Bull. Geol Soc. Am.* 96, 676–90.

Civetta, L., F. Innocenti, P. Manetti, A. Peccerillo & G. Poli 1981. Geochemical characteristics of potassic volcanics from Mt. Ernici (Southern Latium, Italy). *Contrib. Mineral. Petrol.* 78, 37–47.

Clague, D. A. 1987. Hawaiian alkaline volcanism. In *Alkaline igneous rocks*, J. G. Fitton & B. G. J. Upton (eds), 227–52. Geol Soc. Sp. Publ. 30.

Clague, D. A. & F. A. Frey 1982. Petrology and trace element geochemistry of the Honolulu volcanics, Oahu: implications for the oceanic mantle below Hawaii. *J. Petrol.* 23, 447–504.

Clague, D. A. & P. F. Straley 1977. Petrologic nature of the oceanic Moho. *Geology* 5, 133–6.

Clement, C. R., E. M. W. Skinner & B. H. Scott-Smith 1984. Kimberlite re-defined. *J. Geol.* 32, 223–8.

Cleverly, R. W., P. J. Betton, & J. W. Bristow 1984. Geochemistry and petrogenesis of the Lebombo rhyolites. *Sp. Publ. Geol Soc. S. Afr.* 13, 171–94.

Cobbing, E. J., W. S. Pitcher, J. Wilson, J. Baldock, W. P. Taylor, W. McCourt & N. J. Snelling 1981. *The geology of the western cordillera of Northern Peru*. Overseas Mem. Inst. Geol. Sci. Lond. 5, 143 pp.

Cohen, R. S. & R. K. O'Nions 1982a. The lead, neodymium and strontium isotopic structure of ocean ridge basalts. *J. Petrol.* 23, 299–324.

Cohen, R. S. & R. K. O'Nions 1982b. Identification of recycled continental material in the mantle from Sr, Nd and Pb isotope investigations. *Earth Planet. Sci. Lett.* 61, 73–84.

Cohen, R. S., R. K. O'Nions & J. B. Dawson 1984. Isotope geochemistry of xenoliths from East Africa: implications for development of mantle reservoirs and their interaction. *Earth Planet. Sci. Lett.* 68, 209–20.

Coleman, R. G. 1977. *Ophiolites*. Berlin: Springer-Verlag, 229 pp.

Condie, K. C. 1982. Plate tectonics and crustal evolution, 2nd edn. New York: Pergamon Press, 310 pp.

Cordell, L. 1978. Regional geophysical setting of the Rio Grande rift. *Bull. Geol Soc. Am.* 89, 1073–90.

Corliss, J. B., J. Dymond, L. I. Gordon, J. M. Edmond, R. P. Von Herzen, R. D. Ballard, K. Green, D. Williams, A. Bainbridge, K. Crane & Tj. H. Van Andel 1979. Submarine thermal springs on the Galapagos Rift. *Science* 203, 1073–83.

Corliss, J. B., M. Lyle, J. Dymond & K. Crane 1978. The chemistry of hydrothermal mounds near the Galapagos Rift. *Earth Planet. Sci. Lett.* 40, 12–24.

Couch, R., R. Whitsett, B. Huehn & L. Briceno-Guarupe 1981. Structures of the continental margin in Peru and Chile. In *Nazca plate: crustal formation and Andean convergence*, L. D. Kulm, D. Dymund, E. J. Dasch & D. M. Hussong (eds). *Mem. Geol Soc. Am.* 154, 703–26.

Courtillot, V., J. Besse, D. Vandamme, R. Montigny, J.-J. Jaeger & H. Cappetta 1986. Deccan flood basalts at the Cretaceous/Tertiary boundary? *Earth Planet. Sci. Lett.* 80, 361–74.

Cox, A. & R. B. Hart 1986. *Plate tectonics*. Palo Alto: Blackwell Scientific, 392 pp.

Cox, K. G. 1978. Flood basalts, subduction and the break-up of Gondwanaland. *Nature* 274, 47–9.

Cox, K. G. 1980. A model for flood basalt volcanism. *J. Petrol.* 21, 629–50.

Cox, K. G. 1983. The Karoo province of southern Africa: origin of trace element enrichment patterns. In *Continental basalts and mantle xenoliths*, C. J. Hawkesworth & M. J. Norry (eds), 139−57. Nantwich: Shiva.

Cox, K. G. 1987. Postulated restite fragments from Karoo picrite basalts: their bearing on magma segregation and mantle deformation. *J. Geol Soc. Lond.* 144, 275−80.

Cox, K. G., J. D. Bell & R. J. Pankhurst 1979. *The interpretation of igneous rocks*. London; Allen and Unwin, 450 pp.

Cox, K. G. & P. Clifford 1982. Correlation coefficient patterns and their interpretation in three basaltic suites. *Contrib. Mineral. Petrol.* 79, 268−78.

Cox, K. G., I. G. Gass & D. I. J. Mallick 1969. The evolution of the volcanoes of Aden and Little Aden, South Arabia. *Q. J. Geol Soc. Lond.* 124, 283−308.

Cox, K. G., C. J. Hawkesworth, R. K. O'Nions & J. D. Appleton 1976. Isotopic evidence for the derivation of some Roman region volcanics from anomalously enriched mantle. *Contrib. Mineral. Petrol.* 56, 173−80.

Cox, K. G. & C. J. Hawkesworth 1984. Relative contributions of crust and mantle to flood basalt magmatism, Mahabaleshwar area, Deccan Traps. *Phil Trans. R. Soc. Lond.* A310, 627−41.

Cox, K. G. & C. J. Hawkesworth 1985. Goechemical stratigraphy of the Deccan Traps, at Mahabaleshwar, Westerrn Ghats, India, with implications for open system magmatic processes. *J. Petrol.* 26, 355−77.

Cox, K. G. & B. G. Jamieson 1974. The olivine-rich lavas of Nuanetsi: a study of polybaric magmatic evolution. *J. Petrol.* 15, 269−301.

Cox, K. G., R. Macdonald & G. Hornung 1967. Geochemical and petrographic provinces in the Karoo basalts of southern Africa. *Am. Mineral.* 52, 1451−74.

Crane, K. & R. D. Ballard 1980. The Galapagos Rift at 86°W, 4: Structure and morphology of hydrothermal fields and their relationship to the volcanic and tectonic process of the rift valley. *J. Geophys. Res.* 85, 1443−54.

Crawford, A. J., L. Beccaluva & G. Serri 1981. Tectonomagmatic evolution of the West Philippine−Mariana region and the origin of boninites. *Earth Planet. Sci. Lett.* 54, 346−56.

Creager, K. C. & T. H. Jordan 1984. Slab penetration into the lower mantle. *J. Geophys. Res.* 89, 3031−49.

Cross, T. A. & R. H. Pilger Jr 1982. Controls of subduction geometry, location of magmatic arcs, and tectonics of arc and back-arc regions. *Bull. Geol Soc. Am.* 93, 545−62.

Cross, W., J. P. Iddings, L. V. Pirsson & H. S. Washington 1903. *Quantitative classification of igneous rocks*. University of Chicago Press.

Crough, S. T. 1981. Mesozoic hot spot epeirogeny in Eastern North America. *Geology* 9, 2−6.

Crough S. T. 1984. Seamounts as recorders of hot-spot epeirogeny. *Bull. Geol Soc. Am.* 95, 3−8.

Crough S. T., W. J. Morgan & R. B. Hargraves 1980. Kimberlites: their relation to mantle hot spots. *Earth Planet. Sci. Lett.* 50, 260−74.

Cundari, A. 1980. Role of subduction in the genesis of leucite-bearing rocks: facts or fashion? (Reply to A. D. Edgar's discussion paper.) *Contrib. Mineral. Petrol.* 73, 432−4.

CYAMEX 1978. First submersible study of the East Pacific Rise: R17A (Rivera−Tamayo) Project, 21°N. *EOS, Trans Am. Geophys. Union* 52, 1198.

CYAMEX 1981. First manned submersible dives on the East Pacific Rise, 21°N. general results. *Mar. Geophys. Res.* 4, 345−79.

Dalziel, I. W. D. (1986). Collision and cordilleran orogenesis: an Andean perspective. In *Collision tectonics*, M. P. Coward & A. C. Ries (eds), 389−404. Geol Soc. Sp. Publ. 19.

Davidson, J. P. 1984. *Petrogenesis of Lesser Antilles island arc magmas: isotopic and geochemical constraints*. Unpubl. PhD thesis, University of Leeds, UK.

Davidson, J. P. 1985. Mechanisms of contamination in Lesser Antilles island arc magmas from radiogenic and oxygen isotope relationships. *Earth Planet. Sci. Lett.* 72, 163−74.

Davidson, J. P. 1986. Isotopic and trace element constraints on the petrogenesis of subduction related lavas from Martinique, Lesser Antilles. *J. Geophys. Res.* 91, 5943−62

Davies, G. R. & F. E. Lloyd 1988. Pb−Sr−Nd isotope and trace element data bearing on the origin of the potassic subcontinental lithosphere beneath south west Uganda. In *Proc. 4th. Int. Kimberlite Conf.*, Perth, Western Australia. Oxford: Blackwell Scientific.

Davies, G. R. & R. Macdonald 1987. Crustal influences in the petrogenesis of the Naivasha basalt−rhyolite complex: combined trace element and Sr−Nd−Pb isotope constraints. *J. Petrol* 28, 1009−31.

Dawson, J. B 1980. *Kimberlites and their xenoliths*. New York: Springer-Verlag.

Dawson, J. B. 1987. The Kimberlite clan: relationship with olivine and leucite lamproites and inferences for upper-mantle metasomatism. In *Alkaline igneous rocks*, J. G. Fitton & B. G. J. Upton (eds), 95−101. Geol Soc. Sp. Publ. 30.

Delaney, J. M. & H. C. Helgeson 1978. Calculation of the thermodynamic consequences of dehydration in subducting oceanic crust to 100 kb and >800°C. *Am. J. Sci.* 278, 638−86.

DePaolo, D. J. 1979. Implications of correlated Nd and Sr isotopic variations for the chemical evolution of the crust and mantle. *Earth Planet. Sci. Lett.* 43, 201−11.

DePaolo, D. J. 1981. Trace element and isotopic effects of combined wallrock assimilation and fractional crystallisation. *Earth Planet. Sci. Lett.* 53, 189−202.

DePaolo, D. J. & R. W. Johnson 1979. Magma genesis in the New Britain island arc: constraints from Nd and Sr isotopes and trace element patterns. *Contrib. Mineral. Petrol.* 70, 367−79.

DePaolo, D. J. & G. J. Wasserburg 1976. Nd isotopic variations and petrogenetic models. *Geophys. Res. Lett.* 3, 249−52.

DePaolo, D. J. & G. J. Wasserburg 1977. The sources of island arcs as indicated by Nd and Sr isotopic studies. *Geophys. Res. Lett.* 4, 465−8.

DePaolo, D. J. & G. J. Wasserburg 1979. Petrogenetic mixing models and Nd−Sr isotopic patterns. *Geochim. Cosmochim. Acta* 43, 615−27.

Deruelle, B. 1978. Calc-alkaline and shoshonitic lavas from five Andean volcanoes (between latitudes 21°45 and 24°30'S) and the distribution of the Plio-Quaternary volcanism of the south-central and southern Andes. *J. Volcanol. Geotherm. Res.* 3, 281−98.

Deruelle, B., R. S. Harmon & S. Moorbath 1983. Combined Sr−O isotope relationships and petrogenesis of Andean volcanics of South America. *Nature* 302, 814−16.

Devey, C. W. & K. G. Cox 1987. Relationships between crustal contamination and crystallisation in continental flood basalt magmas with special reference to the Deccan Traps of the Western Ghats, India. *Earth Planet. Sci. Lett.* 84, 59−68.

Dewey, J. F. 1976. Ophiolite obduction. *Tectonics* 31, 93−120.

Dewey, J. F. & W. S. F. Kidd 1977. Geometry of plate accretion. *Bull Geol Soc. Am.* 88, 960−8.

Dickin, A. P. 1981. Isotope geochemistry of Tertiary igneous rocks from the Isle of Skye, N.W. Scotland. *J. Petrol.* 22, 155−89.

Dickin A. P., J. L. Brown, R. C. Thompson, A. N. Halliday & M. A. Morrison 1984. Crustal contamination and the granite problem in the British Tertiary Volcanic Province. *Phil Trans R. Soc. Lond.* A310, 755−80.

Dickinson, W. R. 1975. Potash-depth $(k−h)$ relations in continental margin and intra-ocean magmatic arcs. *Geology* 3, 53−6.

Dickinson, W. R. & T. Hatherton 1967. Andesitic volcanism and seismicity around the Pacific. *Science* 157, 801−3.

Doe, B. R., W. P. Leeman, R. L. Christiansen & C. E. Hedge 1982. Lead and strontium isotopes and related trace elelments as genetic tracers in the Upper Cenozoic rhyolite−basalt association of the Yellowstone Plateau volcanic field. *J. Geophys. Res.* 87, 4785−806.

Downes, H. 1984. Sr and Nd isotope geochemistry of coexisting alkaline series, Cantal, Massif Central, France. *Earth Planet. Sci. Lett.* 69, 321−34.

Duncan, A. R. 1987. The Karoo igneous province − a problem area for inferring tectonic setting from basalt geochemistry. *J. Volcanol. Geotherm. Res.* 32, 13−34.

Duncan, A. R. 1981. Hot spots in the southern oceans − an absolute frame of reference for the motion of Gondwana continents. *Tectonics* 74,

29–42.

Duncan, R. A. & D. H. Green 1987. The genesis of refractory melts in the formation of oceanic crust. *Contrib. Mineral. Petrol.* **96**, 326–42.

Dungan, M. A. & J. M. Rhodes 1978. Residual glasses and melt inclusions in basalts from D. S. D. P. Legs 45–46. Evidence for magma mixing. *Contrib. Mineral. Petrol.* **67**, 413–31.

Dungan, M. A., M. M. Lindstrom, N. J. McMilan, S. Moorbath, J. Hoefs & L. A. Haskin 1986. Open system magmatic evolution of the Taos plateau volcanic field, northern New Mexico I: The petrology and geochemistry of the Servilleta basalt. *J. Geophys. Res.* **91**, 5999–6028.

Dunham, A. C. & W. J. Wadsworth 1978. Cryptic variation in the Rhum layered intrusion. *Mineral. Mag.* **42**, 347–56.

Dunn, T. 1987. Partitioning of Hf, Lu, Ti and Mn between olivine, clinopyroxene and basaltic liquid. *Contrib. Mineral. Petrol.* **96**, 476–84.

Dupré, B. & C. J. Allègre 1980. Pb–Sr–Nd isotopic correlation and the chemistry of the North Atlantic mantle. *Nature* **286**, 17–22.

Dupré, B. & C. J. Allègre 1983. Pb–Sr isotopic variations in Indian Ocean basalts and mixing phenomena. *Nature* **303**, 142–6.

Dupuy, C. & J. Dostal 1984. Trace element geochemistry of some continental tholeiites. *Earth Planet. Sci. Lett.* **67**, 61–9.

Dupuy, C., J. Dostal, G. Marcelot, H. Bougault, J. L. Joron & M. Treuil 1982. Geochemistry of basalts from central and southern New Hebrides arc: implication for their source rock composition. *Earth Planet. Sci. Lett.* **60**, 207–25.

Dziewonski, A. M. & A. L. Anderson 1981. Preliminary reference Earth model. *Phys. Earth. Planet. Inter.* **25**, 297–356.

Dzurisin, D., R. Y. Koyanagi & T. T. English 1984. Magma supply and storage at Kilauea volcano, Hawaii, 1956–1983. *J. Volcanol. Geotherm. Res.* **21**, 177–206.

Eaton, J. P. & K. Murata 1960. How volcanoes grow. *Science* **132**, 925–38.

Edgar, A. D. 1980. Role of subduction in the genesis of leucite-bearing rocks: discussion. *Contrib. Mineral. Petrol.* **73**, 429–31.

Edgar, A. D. 1987. The genesis of alkaline magmas with emphasis on their source regions: inferences from experimental studies. In *Alkaline igneous rocks*, J. G. Fitton & B. G. J. Upton (eds), 29–52. Geol. Soc. Sp. Publ. 30.

Edgar, A. D., E. Condliffe, R. L. Barnett & R. J. Shirran 1980. An experimental study of an olivine ugandite magma and mechanisms for the formation of it's K-enriched derivatives. *J. Petrol.* **21**, 475–97.

Edgar, A. G., D. H. Green, & W. O. Hibberson 1976. Experimental petrology of a highly potassic magma. *J. Petrol.* **17**, 339–56.

Edmond, J. M. & K. Von Damm 1983. Hot springs on the ocean floor. *Scient. Am.* **284**, 70–85.

Eggler, D. H. & C. W. Burnham 1973. Crystallisation and fractionation trends in the system andesite–H_2O–CO_2–O_2 at pressures to 10 kb. *Bull. Geol Soc. Am.* **84**, 2517–32.

Eissler, H. K. & H. Kanamori 1986. Depth estimates of large earthquakes on the island of Hawaii since 1940. *J. Geophys. Res.* **91**, 2063–76.

Elliot, D. H. 1975. Tectonics in Antarctica: a review. *Am. J. Sci.* **275A**, 45–106.

Ellsworth, W. L. & R. Y. Koyanagi 1977. Three-dimensional crust and mantle structure of Kilauea Volcano, Hawaii. *J. Geophys. Res.* **82**, 5379–94.

Engel, A. E., C. G. Engel & R. G. Havens 1965. Chemical characteristics of oceanic basalts and the upper mantle. *Bull. Geol Soc. Am.* **76**, 719–34

England, P. 1983. Constraints on the extension of continental lithosphere. *J. Geophys. Res.* **88**, 1145–52.

England, P. & G. Houseman 1984. On the geodynamic setting of kimberlite genesis. *Earth Planet. Sci. Lett.* **67**, 109–22.

Erlank, A. J. (ed.) 1984. *Petrogenesis of the volcanic rocks of the Karoo Province.* Geol. Soc. S. Afr. Sp. Publ. 13, 395 pp.

Erlank, A. J., H. L. Allsopp, C. J. Hawkesworth & M. A. Menzies 1982. Chemical and isotopic characterisation of upper mantle metasomatism in peridotite nodules from the Bultfontein Kimberlite. *Terra Cognita* 2 (3), 262–3.

Ewart, A. 1982. The mineralology and petrology of Tertiary–Recent orogenic volcanic rocks: with special reference to the andesitic–basaltic compositional range. In *Andesites: orogenic andesites and related rocks*, R. S. Thorpe (ed.), 26–87. Chichester: Wiley.

Fairhead, J. D. 1979. The gravity link between the domally uplifted Cainozoic volcanic centres of North Africa and its similarity to the East African Rift System anomaly. *Earth Planet. Sci. Lett.* 42, 109–13.

Fairhead, J. D. & C. D. Reeves 1977. Teleseismic delay times, Bouger anomalies and inferred thickness of the African lithosphere. *Earth Planet. Sci. Lett.* 36, 63–76.

Faure, G. 1986. *Principles of isotope geology*. 2nd, ed. New York: Wiley.

Feigenson, M. D. 1984. Geochemistry of Kauai volcanics and a mixing model for the origin of Hawaiian alkali basalts. *Contrib. Mineral. Petrol.* 87, 109–19.

Feigenson, M. D., A. W. Hofmann & F. J. Spera 1983. Case studies on the origins of basalt II. The transition from tholeiitic to alkalic volcanism on Kohala volcano, Hawaii. *Contrib. Mineral. Petrol.* 84, 390–405.

Ferguson, A. K. & A. Cundari 1975. Petrological aspects and evolution of the leucite-bearing lavas from Bufumbira, South West Uganda. *Contrib. Mineral. Petrol.* 50, 25–46.

Ferrara, G., M. Preite-Martinez, H. P. Taylor Jr, S. Tonarini & B. Turi 1986. Evidence for crustal assimilation, mixing of magmas and a ^{87} Sr-rich upper mantle. *Contrib. Mineral. Petrol.* 92, 269–80.

Finnerty, A. A. & F. R. Boyd 1984. Evaluation of thermobarometers for garnet peridotites. *Geochim. Cosmochim. Acta* 48, 15–27.

Finnerty, A. A. & F. R. Boyd 1987. Thermobarometry for garnet peridotites: basis for the determination of thermal and composition structure of the upper mantle. In *Mantle xenoliths*, P. H. Nixon (ed.), 381–402. Chichester: Wiley.

Fisk, M. R. & A. E. Bence 1979. Experimental studies of spinel crystallisation in FAMOUS basalt 521–1–1 (Abstract). *EOS, Trans Am. Geophys. Union* 60, 420.

Fitton, J. G. & H. M. Dunlop 1985. The Cameroon Line, West Africa, and its bearing on the origin of oceanic and continental alkali basalt. *Earth Planet. Sci. Lett.* 72, 23–38.

Fleitout, L., C. Froidevaux & D. Yuen 1986. Active lithospheric thinning. *Tectonophysics* 132, 271–8.

Floyd, P. A & J. A. Winchester 1975. Magma type and tectonic setting discrimination using immobile elements. *Earth Planet. Sci. Lett.* 27, 211–18.

Foden, J. D. 1983. The petrology of the calkalkaline lavas of Rindjani volcano, East Sunda arc: a model for island arc petrogenesis. *J. Petrol.* 24, 98–130.

Fodor, R. V. 1987. Low- and high-TiO_2 flood basalts of southern Brazil: origin from picritic parentage and a common mantle source. *Earth Planet. Sci. Lett.* 84, 423–30.

Fodor, R. V., C. Corwin & A. Roisenberg 1985. Petrology of Serra Geral (Parana) continental flood basalts, southern Brazil: crustal contamination, source material and South Atlantic magmatism. *Contrib. Mineral. Petrol.* 91, 54–65.

Foley, S. F., W. R. Taylor & D. H. Green 1986. The role of fluorine and oxygen fugacity in the genesis of the ultrapotassic rocks. *Contrib. Mineral. Petrol.* 94, 183–92.

Foley, S. F., G. Venturelli, D. H. Green & L. Toscani 1987. The ultrapotassic rocks: characteristics, classification and constraints for petrogenetic models. *Earth Sci. Rev.* 24, 81–134.

Fowler, C. M. R. 1976. The crustal structure of the Mid-Atlantic Ridge crest at 37°N. *Geophys. J. R. Astron. Soc.* 47, 459–91.

Fowler, C. M. R. 1978. The Mid-Atlantic Ridge: structure at 45°N. *Geophys. J. R. Astron. Soc.* 54, 167–83.

Francheteau, J. & A. D. Ballard 1983. The East Pacific Rise near 21°N, 13°N and 20°S: inferences for along-strike variability of axial process of the Mid-Ocean Ridge. *Earth Planet. Sci. Lett.* 64, 93–116.

Fraser, K. J., C. J. Hawkesworth, A. J. Erlank, R. H. Mitchell & B.H. Scott-Smith 1985. Sr, Nd and Pb isotope and minor element geochemistry of lamproites and kimberlites. *Earth Planet. Sci. Lett.* 76, 57–70.

Frey, F. A. 1982. Rare earth element abundances in upper mantle rocks. In *Rare earth element geochemistry*, P. Henderson (ed.). Amsterdam: Elsevier.

Froidevaux, C. & H. C. Nataf 1981. Continental drift: what driving mechanism? *Geol. Rundsch.* 70, 166–76.

Furlong, K. P., D. S. Chapman & P. W. Alfeld 1982. Thermal modeling of the geometry of subduction with implications for the tectonics of the overriding plate. *J. Geophys. Res.* 87, 1786–802.

Galer, S. J. G. & R. K. O'Nions 1985. Residence time of thorium, uranium and lead in the mantle with implications for mantle convection. *Nature* 316, 778–82.

Galer, S. J. G. & R. K. O'Nions 1986. Magma genesis and the mapping of chemical and isotopic variations in the mantle. *Chem. Geol.* 56, 45–61.

Gasparik, T. 1984. Two-pyroxene thermobarometry with new experimental data in the system $CaO–MgO–Al_2O_3–SiO_2$. *Contrib. Mineral. Petrol.* 87, 87–97.

Gass, I. G., S. J. Lippard & A. W. Shelton 1984. *Ophiolites and oceanic lithosphere*. Oxford: Blackwell Scientific, 431 pp.

Gast, P. W., G. R. Tilton & C. Hedge 1964. Isotopic composition of lead and strontium from Ascension and Gough Islands. *Science* 145, 1181–85.

Ghiorso, M. S. 1985. Chemical mass transfer in magmatic processes: I. Thermodynamic relations and numerical algorithms. *Contrib. Mineral. Petrol.* 90, 107–20.

Ghiorso, M. S. & I. S. E. Carmichael 1985. Chemical mass transfer in magmatic process: II. Applications in equilibrium crystallisation, fractionation and assimilation. *Contrib. Mineral. Petrol.* 90, 121–41.

Giardini, D. & J. H. Woodhouse 1984. Deep seismicity and modes of deformation in Tonga subduction zone. *Nature* 307, 505–9.

Gibson, I. L. & A. D. Gibbs 1987. Accretionary volcanic processes and the crustal structure of Iceland. *Tectonophysics* 133, 57–64.

Gill, J. B. 1981. *Orogenic andesites and plate tectonics*. Berlin: Springer-Verlag, 358 pp.

Girdler, R. W. 1983. Processes of planetary rifting as seen in the rifting and breakup of Africa. *Tectonophysics* 94, 241–52.

Graham, C. M. & R. S. Harmon 1983. Stable isotope evidence on the nature of crust–mantle interactions. In *Continental basalts and mantle xenoliths*, C. J. Hawkesworth & M. J. Norry (eds), 20–45. Nantwich: Shiva.

Green, D. H. 1973. Experimental melting studies on a model upper mantle composition at high pressure under water-saturated and water-undersaturated conditions. *Earth Planet. Sci. Lett.* 19, 37–53.

Green, D. H. 1976. Experimental testing of equilibrium partial melting of peridotite under water-saturated, high pressure conditions. *Can. Mineral.* 14, 255–68.

Green, D. H., W. O. Hibberson & A. L. Jaques 1979. Petrogenesis of mid-ocean basalts. In *The Earth: its origin, structure and evolution*, W. H. McElhinny (ed.) 265–99. London: Academic Press.

Green, D. H. & A. E. Ringwood 1967. The genesis of basaltic magmas. *Contrib. Mineral. Petrol.* 15, 103–90.

Green, T. H. 1980. Island arc and continent-building magmatism: a review of petrogenetic models based on experimental petrology and geochemistry. *Tectonophysics* 63, 367–85.

Green, T. H. 1982. Anatexis of mafic crust and high pressure crystallisation of andesite. In *Andesites: orogenic andesites and related rocks*, R. S. Thorpe (ed.), 465–87. Chichester: Wiley.

Green, T. H. & A. E. Ringwood 1968. Genesis of the calkalkaline igneous rock suite. *Contrib. Mineral. Petrol.* 18, 163–74.

Greenbaum, D. 1972. Magmatic processes at ocean ridges, evidence from the Troodos Massif, Cyprus. *Nature* 238, 18–21.

Gurney, J. J. & B. Harte 1980. Chemical variations

in upper mantle nodules from South African kimberlites. *Phil Trans R. Soc. Lond.* **A297**, 273–93.

Haggerty, S. E. 1982. Kimberlites in western Liberia: An overview of the geological setting in a plate tectonic framework. *J. Geophys. Res.* **81**, 10811–26.

Haggerty, S. E. 1983. The mineral chemistry of new titanates from the Jagersfontein Kimberlite, South Africa: implications for metasomatism in the upper mantle. *Geochim. Cosmochim. Acta* **47**, 1833–54.

Haggerty, S. E. & L. A. Tompkins 1983. Redox state of the Earth's upper mantle from kimberlitic ilmenites. *Nature* **303**, 295–300.

Hamelin B. & C. J. Allègre 1985. Large-scale regional units in the depleted upper mantle revealed by an isotope study of the south-west Indian Ridge. *Nature* **315**, 196–8.

Hamelin, B., B. Dupré & C. J. Allègre 1984. Lead–strontium isotopic variations along the East Pacific Rise and the Mid-Atlantic Ridge: a comparative study. *Earth Planet. Sci. Lett.* **67**, 340–50.

Hamelin, B., B. Dupré & C. J. Allègre 1986. Pb–Sr–Nd isotopic data of Indian Ocean Ridges: new evidence of large-scale mapping of mantle heterogeneities. *Earth Planet. Sci. Lett.* **76**, 288–98.

Hanan, B. B., R. H. Kingsley & J.-G. Schilling 1986. Pb isotope evidence in the South Atlantic for migrating ridge-hotspot interactions. *Nature* **322**, 137–44.

Hanson, G. N. 1980. Rare earth elements in petrogenetic studies of igneous systems. *Ann. Rev. Earth Planet. Sci.* **8**, 371–406.

Harker, A. 1909. *The natural history of igneous rocks.* New York: Macmillan.

Harland, W. B., A. V. Cox, P. G. Llewellyn, C. A. Pickton, A. G. Smith & R. Walters 1982. *A geological time scale.* Cambridge: Cambridge University Press, pp. 131.

Harmon, R. S., B. A. Barreiro, S. Moorbath, J. Hoefs, P. W. Francis, R. S. Thorpe, B. Deruelle, J. McHugh & J. A. Viglino 1984. Regional O-, Sr- and Pb isotope relationships in late Cenozoic calk-alkaline lavas of the Andean Cordillera. *J. Geol Soc. Lond.* **141**, 803–22.

Harmon, R. S. & J. Hoefs 1984. O-isotope relationships in Cenozoic volcanic rocks: evidence for a heterogeneous mantle source and open-system magma genesis. In *Proc. ISEM Field Conf. Open Magmatic Systems*, M. A. Dungan, T. L. Grove & W. Hildreth (eds), 66–8.

Harmon, R. S., J. Hoefs & K. H. Wedepohl 1987. Stable isotope (O, H, S) relationships in Tertiary basalts and their mantle xenoliths from the Northern Hessian depression, W. Germany. *Contrib. Mineral. Petrol.* **95**, 350–69.

Harris, C. 1983. The petrology of lavas and associated plutonic inclusions of Ascension Island. *J. Petrol.* **24**, 424–70.

Harris, N. B. W., J. A. Pearce & A. G. Tindle 1986. Geochemical characteristics of collision-zone magmatism. In *Collision tectonics*, M. P. Coward & A. C. Ries (eds), 67–81. Geol. Soc. Sp. Publ. 19.

Harris, P. G. 1957. Zone refining and the origin of potassic basalts. *Geochim. Cosmochim. Acta* **12**, 195–208.

Harris, P. G. & E. A. K. Middlemost 1969. The evolution of kimberlites. *Lithos* **3**, 77–88.

Hart, R. K. 1985. Chemical and isotopic evidence for mixing between depleted and enriched mantle, northwestern U.S.A. *Geochim. Cosmochim. Acta* **49**, 131–44.

Hart, S. R. 1984. The DUPAL anomaly: a large-scale isotopic anomaly in the southern hemisphere. *Nature* **309**, 753–6.

Hart, S. R., J.-G. Schilling & J. L. Powell 1973. Basalts from Iceland and along the Reykjanes Ridge: Sr isotope geochemistry. *Nature* **246**, 104–7.

Hart, W. K. 1985. Chemical and isotopic evidence for mixing between depleted and enriched mantle, northwestern U.S.A. *Geochim. Cosmochim. Acta* **49**, 131–44.

Hart, W. K. & R. W. Carlson 1987. Tectonic controls on magma genesis in the northwestern United States. *J. Volcanol. Geotherm, Res.* **32**, 119–35.

Harte, B. 1983. Mantle peridotites and processes –

the kimberlite sample. In *Continental basalts and mantle xenoliths*, C. J. Hawkesworth & M. J. Norry, (eds), 46−91. Nantwich: Shiva.

Harte, B. 1987. Metasomatic events recorded in mantle xenoliths: an overview. In *Mantle xenoliths*, P. H. Nixon (ed.), 625−40. Chichester: Wiley.

Hasegawa, A., N. Umino & A. Takagi 1978. Double-planed structure of the deep seismic zone in the northeastern Japan Arc. *Tectonophysics* 47, 43−58.

Hawkesworth, C. J. 1982. Isotopic characteristics of magmas erupted along destructive plate margins. In R. S. Thorpe, (ed.), *Andesites: orogenic andesites and related rocks*, 549−71. Chichester: Wiley.

Hawkesworth, C. J., A. J. Erlank, J. S. Marsh, M. A. Menzies & P. Van Calsteren 1983. Evolution of the continental lithosphere: evidence from volcanics and xenoliths in Southern Africa. In *Continental basalts and mantle xenoliths*, C. J. Hawkesworth & M. J. Norry, (eds), 111−38. Nantwich: Shiva.

Hawkesworth, C. J., K. J. Fraser & N. W. Rogers 1985. Kimberlites and lamproites: extreme products of mantle enrichment processes. *Trans Geol Soc. S. Afr.* 88, 439−47.

Hawkesworth, C. J.,M. Hammill, A. R. Gledhill, P. Van Calsteren & G. Rogers 1982. Isotope and trace element evidence for late-stage intra-crustal melting in the high Andes. *Earth Planet. Sci. Lett.* 58, 240−54.

Hawkesworth, C. J., M. S. M. Mantovani, P. N. Taylor & Z. Palacz 1986. Evidence from the Paraná of south Brazil for a continental contribution to Dupal basalts. *Nature* 322, 356−9.

Hawkesworth, C. J., M. J. Norry, J. C. Roddick & P. E. Baker 1979. $^{143}Nd/^{144}Nd$, $^{87}Sr/^{86}Sr$ and incompatible element variations in calk-alkaline andesites and plateau lavas from South America. *Earth Planet. Sci. Lett.* 42, 45−57.

Hawkesworth, C. J., M. J. Norry, J. C. Roddick & R. Vollmer 1979. $^{143}/Nd^{144}Nd$ and $^{87}Sr/^{86}Sr$ ratios from the Azores and their significance in LIL-element enriched mantle. *Nature* 280, 28−31.

Hawkesworth C. J., R. K. O'Nions, R. J. Pank-hurst, P. J. Hamilton, & N. M. Evensen 1977. A geochemical study of island arc and back-arc tholeiites from the Scotia Sea. *Earth Planet. Sci. Lett.* 36, 253−62.

Hawkesworth, C. J. & M. Powell 1980. Magma genesis in the Lesser Antilles island arc. *Earth Planet. Sci. Lett.* 51, 297−308.

Hawkesworth, C. J., N. W. Rogers, P. W. C. Van Calsteren & M. A. Menzies 1984. Mantle enrichment processes. *Nature* 311, 331−5.

Hawkesworth, C. J. & R. Vollmer 1979. Crustal contamination versus enriched mantle: $^{143}Nd/^{144}Nd$ and $^{87}Sr/^{86}Sr$ evidence from Italian volcanics. *Contrib. Mineral. Petrol.* 69, 151−65.

Hawkins, J. W. 1974. Geology of the Lau Basin, a marginal basin behind the Tonga arc. In *The geology of continental margins*, C. A. Burke & C. L. Drake, (eds), 505−20. Berlin: Springer-Verlag.

Hekinian, R. 1982. *Petrology of the ocean floor*. Amsterdam: Elsevier, 393 pp.

Hekinian, R. & F. Aumento 1973. Rocks from the Gibbs Fracture Zone and the Minia Seamount near 53°N in the Atlantic Ocean. *Mar. Geol.* 14, 47−72.

Hekinian, R. & D. Walker 1987. Diversity and spatial zonation of volcanic rocks from the East Pacific Rise near 21°N. *Contrib. Mineral. Petrol.* 96, 265−80.

Henderson, P. 1982. *Inorganic geochemistry*. Oxford: Pergamon Press, 353 pp.

Hermance, J. F. 1982. Magnetotelluric and geomagnetic deep-sounding studies in rifts and adjacent areas: constraints on physical processes in the crust and upper mantle. In *Continental and oceanic rifts*, G. Palmason (ed.) 169−92. Washington, DC: Am. Geophys. Union.

Hess, H. H. 1962. History of ocean basins. In *Petrological studies: a volume in honor of A. F. Buddington*, A. E. J. Engel *et al.* (eds), 599−620. Boulder, Colorado: Geol Soc. Am.

Hickey, R. L. & F. A. Frey 1982. Geochemical characteristics of boninite series volcanics: implications for their source. *Geochim. Cosmochim. Acta* 46, 2099−115.

Hickey, R. L., F. A. Frey & D. C. Gerlach 1986. Multiple sources for basaltic arc rocks from the southern volcanic zone of the Andes (34−41°S): trace element and isotopic evidence for contributions from subducted oceanic crust, mantle and continental crust. *J. Geophys. Res.* 91, 5963−83.

Hiemstra, S. A. 1985. The distribution of some platinum-group elements in the UG-2 chromitite layer of the Bushveld Complex. *Econ. Geol.* 80, 944−57.

Hiertzler, J. R. & X. Le Pichon 1974. FAMOUS, a plate tectonic study of the genesis of the lithosphere. *Geology* 1, 273−4.

Higazy, R. A. 1954. Trace elements of volcanic ultrabasic potassic rocks of south western Uganda and the adjoining part of the Belgian Congo. *Bull. Geol Soc. Am.* 65, 39−70.

Hildreth, W. 1979. The Bishop Tuff: evidence for the origin of compositional zonation in silicic magma chambers. *Geol Soc. Am. Sp. Pap.* 180, 43−74.

Hill, D. P. 1972. Crustal and upper mantle structure of the Columbia Plateau from long range seismic-refraction measurements. *Bull. Geol Soc. Am.* 83, 1639−48.

Hill, R. & P. Roeder 1974. Stability of spinel in basaltic melts. *J. Geol.* 82, 709−29.

Hoffman, N. R. A. & D. P. McKenzie 1985. The destruction of geochemical heterogeneites by differential fluid motions during mantle convection. *Geophys. J. R. Astron. Soc.* 82, 163−206.

Hofmann, A. W 1986. Nb in Hawaiian magmas: constraints on source composition and evolution. *Chem. Geol.* 57, 17−30.

Hofmann, A. W., M. D. Feigenson & I. Raczek 1987. Kohala revisited. *Contrib. Mineral. Petrol.* 95, 114−22.

Hofmann, A. W. & W. M. White 1982. Mantle plumes from ancient oceanic crust. *Earth Planet. Sci. Lett.* 57, 421−36.

Hoffmann, N. R. A. & D. P. McKenzie 1985. The destruction of geochemical heterogeneities by differential fluid motions during mantle convection. *Geophys. J. R. Astron. Soc.* 82, 163−206.

Hole, M. J., A. D. Saunders, G. F. Marriner & J. Tarney 1984. Subduction of pelagic sediments: implications for the origin Ce-anomalous basalts from the Mariana Islands. *J. Geol Soc. Lond.* 141, 453−72.

Holm, P. E. 1982. Non-recognition of continental tholeiites using the Ti−Zr−Y diagram. *Contrib. Mineral. Petrol.* 79, 308−10.

Holm, P., S. Lou & A. Nielsen 1982. The geochemistry and petrogenesis of the lavas of the Vulsinian district, Roman province, Central Italy. *Contrib. Mineral. Petrol.* 80, 367−78.

Holm, P. & N. C. Munksgaard 1982. Evidence for mantle metasomatism: an oxygen and strontium study of the Vulsinian district, Central Italy. *Earth Planet. Sci. Lett.* 60, 376−88.

Hostetler, C. J. & M. J. Drake 1980. Predicting major element mineral/melt equilibria: a statistical approach. *J. Geophys. Res.* 85, 3789−96.

Houseman, G. 1983a. Large aspect ratio convection cells in the upper mantle. *Geophys. J. R. Astron. Soc.* 75, 309−34.

Houseman, G. 1983b. The deep structure of ocean ridges in a convecting mantle. *Earth Planet. Sci. Lett.* 64, 283−94.

Houseman. G. A., D. P. McKenzie & P. Molnar 1981. Convective instability of a thickened boundary layer and its relevance for thermal evolution of continental convergent belts. *J. Geophys. Res.* 86, 6115−32.

Houtz, R. & J. Ewing 1976. Upper crustal structure as a function of plate age. *J. Geophys. Res.* 81, 2490−8.

Huang, W. L. & P. J. Wyllie 1981. Phase relationships of S-type granite with H_2O to 35 kbar: muscovite granite from Harney Peak, South Dakota. *J. Geophys. Res.* 86, 10515−29.

Hughes, C. J. 1982. *Igneous petrology*. New York: Elsevier, 551 pp.

Humphris, S. E. & G. Thompson 1983. Geochemistry of rare earth elements in basalts from the Walvis Ridge: implications for its origin and evolution. *Earth Planet. Sci. Lett.* 66, 223−42.

Humphris, S. E., G. Thompson, J. G. Schilling & R. A. Kingsley 1985. Petrological and geochemical variations along the Mid-Atlantic Ridge between 46°S and 32°S: influence of the Tristan da Cunha mantle plume. *Geochim. Cosmochim. Acta.* 49, 1445−64.

Huppert, H. E. & R. S. J. Sparks 1980. The fluid dynamics of a basaltic magma chamber replenished by influx of hot, dense ultrabasic magma.

Contrib. Mineral. Petrol. 75, 279–89.

Huppert, H. E. & R. S. J. Sparks 1980. Restrictions on the compositions of mid-ocean ridge basalts: a fluid dynamic investigation. *Nature* 246, 46–8.

Huppert, H. E. & R. S. J. Sparks 1985. Cooling and contamination of mafic and ultramafic magmas during ascent through the continental crust. *Earth Planet. Sci. Lett.* 74, 371–86.

Huppert, H. E., R. S. J. Sparks & J. S. Turner 1982. The effects of volatiles on mixing in calkalkaline magma systems. *Nature* 297, 554–7.

Hutchison, C. S. 1983. *Economic deposits and their tectonic setting.* London: Macmillan, 365 pp.

Iddings, J. P. 1982. The origin of igeous rocks. *Bull. Phil Soc. Wash.* 12, 89–213.

Illies, J. H. 1981. Mechanism of graben formation. *Tectonophysics* 73, 249–66.

Imsland, P. 1983. Iceland and the ocean floor. Comparison of chemical characteristics of the magmatic rocks and some volcanic features. *Contrib. Mineral. Petrol.* 83, 31–7.

Irvine, T. N. 1980a. Infiltration metasomatism, accumulate growth and double-diffusive fractional crystallisation in the Muskox intrusion and other layered intrusions. In *Physics of magmatic processes*, R. B. Hargraves (ed.), 325–83. Princeton, N. J: Princeton University Press.

Irvine, T. N. 1980b. Magmatic density currents and cumulus processes. *Am. J. Sci.* 280A, 1–58.

Irvine, T. N. 1982. Terminology for layered intrusions. *J. Petrol.* 23, 127–62.

Irvine, T. N. & W. R. A. Baragar 1971. A guide to the chemical classification of the common rocks. *Can. J. Earth Sci.* 8, 523–48.

Ito, E., D. M. Harris & A. T. Anderson 1983. Alteration of oceanic crust and geologic cycling of chlorine and water. *Geochim. Cosmochim. Acta* 47, 1613–24.

Iyer, H. M. 1984. Geophysical evidence for the locations, shapes and sizes and internal structures of magma chambers beneath regions of Quaternary volcanism. *Phil. Trans. R. Soc.*

Lond. A310, 473–510.

Jackson, E. D. 1968. The character of the lower crust and upper mantle beneath the Hawaiian islands. *Int. Geol Congr.* 1, 135–50.

Jackson, H. R. & I. Reid 1983. Oceanic magnetic anomaly amplitudes: varation with sea-floor spreading rate and possible implications. *Earth Planet. Sci. Lett.* 63, 368 78.

Jakes, P. & J. B. Gill 1970. Rare earth elements and the island arc tholeiitic series. *Earth Planet. Sci. Lett.* 9, 17–28.

Jakes , P. & A. J. R. White 1972. Major and trace element abundances in volcanic rocks of orogenic areas. *Bull. Geol Soc. Am.* 83, 29–40.

James, D. E. 1981. The combined use of oxygen and radiogenic isotopes as indications of crustal contamination. *Ann. Rev. Earth Planet. Sci.* 9, 311–44.

James, D. E. 1982. A combined O, Sr, Nd and Pb isotopic and trace element study of crustal contamination in central Andean lavas. *Earth Planet. Sci. Lett.* 57, 47–62.

James, D. F. & L. A. Murcia 1984. Crustal contamination in northern Andean volcanics. *J. Geol Soc. Lond.* 141, 823–30.

Jaques, A. L. & D. H. Green 1980. Anhydrous melting of peridotite at 0–15 Kb pressure and the genesis of tholeiitic basalts. *Contrib. Mineral. Petrol.* 73, 287–310.

Jaques, A. L., J. D. Lewis, C. B. Smith, G. P. Gregory, J. Ferguson, B. W. Chappell & M. T. McCulloch 1984. The diamond-bearing ultrapotassic (lamproitic) rocks of the West Kimberley region, western Australia. In *Kimberlites, I. Kimberlites and related rocks*, J. Kornprobst, (ed.) 225–54 Amsterdam: Elsevier.

Jarrard, R. D. 1986. Relations among subduction parameters. *Rev. Geophys.* 24, 217–84.

Jarvis, G. T. & D. P. McKenzie 1980. Sedimentary basin formation with finite extension rates. *Earth Planet. Sci. Lett.* 48, 42–52.

Jeanloz, R. & S. Morris 1986. Temperature distribution in the crust and mantle. *Ann. Rev. Earth Planet. Sci.* 14, 377–415.

Jeanloz, R. & A. R. Thompson 1983. Phase transitions and mantle discontinuites. *Rev. Geo-*

phys. Space. Phys. **21**, 51–74.

Jenner, G. A. 1981. Geochemistry of high-Mg andesites from Cape Vogel, Papua New Guinea. *Chem. Geol.* **33**, 307–32.

Johnson, G. L. & P. R. Vogt 1973. The Mid-Atlantic Ridge from 47° to 51°N. *Bull. Geol Soc. Am.* **84**, 3443–62.

Jones, A. P., J. V. Smith & J. B. Dawson 1982. Mantle metasomatism in 14 veined peridotites from the Bultfontein Mine, South Africa. *J. Geol.* **90**, 435–53.

Jones, P. R. 1981. Crustal structures of the Peru continental margin and adjacent Nazca plate 9°S latitude. In *Nazca plate: crustal formation and Andean convergence*, L. D. Kulm., J. Dymond, E. J. Dasch & D. M. Hussong (eds). *Mem. Geol Soc. Am.* **154**, 423–43.

Jones, P. W. 1976. *Petrology and age of the Ethiopian Trap basalts*. Unpubl. Ph D thesis, Univ. Leeds, UK.

Jordan, T. E., B. L. Isacks, R. W. Allmendinger, J. A. Brewer, V. A. Ramos & C. J. Ando 1983. Andean tectonics related to the geometry of the subducted Nazca plate. *Bull. Geol Soc. Am.* **94**, 341–61.

Karig, D. E. 1971. Origin and development of marginal basins in the western Pacific. *J. Geophys. Res.* **76**, 2542–61.

Kay, R. W. 1980. Volcanic arc magmas: implications of a melting-mixing model for element recycling in the crust-upper mantle system. *J. Geol.* **88**, 497–522.

Kay, R. W. 1984. Elemental abundances relevant to identification of magma sources. *Phil Trans R. Soc. Lond.* A310, 535–47.

Keen, C. E. 1985. The dynamics of rifting: deformation of the lithosphere by active and passive driving forces. *Geophys. J. R. Astron. Soc.* **80**, 95–120.

Keller, J. 1983. Potassic lavas in the orogenic volcanism of the Mediterranean area. *J. Volcanol. Geotherm. Res.* **18**, 321–35.

Kellog, A. H & D. L. Turcotte 1986. Homogenization of the mantle by convective mixing and diffusion. *Earth Planet. Sci. Lett.* **81**, 371–8.

Kennett, J. P. 1982. *Marine geology*. Englewood Cliffs, NJ: Prentice-Hall 813 pp.

Kenyon, P. M. & D. L. Turcotte 1983. Convection in a two-layer mantle with a strongly temperature-dependent viscosity. *J. Geophys. Res.* **88**, 6403–14.

Kirkpatrick, R. J. 1979. The physical state of the oceanic crust: results of downhole geophysical logging in the Mid-Atlantic Ridge at 23°N. *J. Geophys. Res.* **84**, 178–88.

Klein, E. M. & C. H. Langmuir 1987. Global correlations of ocean ridge basalt chemistry with axial depth and crustal thickness. *J. Geophys. Res.* **92**, 8089–115.

Klein, F. W. 1982. Patterns of historical eruption at Hawaiian volcanoes. *J. Volcanol. Geotherm. Res.* **12**, 1–35.

Knittle, E., R. Jeanloz & G. L. Smith 1986. Thermal expansion of silicate perovskite and stratification of the Earth's mantle. *Nature* **319**, 214–15.

Kokelaar, B. P. & M. F. Howells (eds) 1984. *Marginal basin geology: volcanic and associated sedimentary and tectonic processes in modern and ancient marginal basins*. Geol. Soc. Sp. Publ. 16, 322 pp.

Koyanagi, R. Y. & E. T. Endo 1971. *Hawaiian seismic events during 1969*. US Geol Surv. Prof. Paper 750–C, C158–164.

Koyanagi, R. Y., E. T. Endo & J. S. Ebisu 1975. Reawakening of Mauna Loa volcano Hawaii: a preliminary evaluation of seismic evidence. *Geophys. Res. Lett.* **2**, 405–8.

Koyanagi, R. Y., J. D. Unger, E. T. Endo & A. T. Okamura 1976. Shallow earthquakes associated with inflation episodes at the summit of Kilauea volcano, Hawaii. *Bull. Volcanol.* **39**, 621–31.

Kramers, J. D. 1977. Lead and strontium isotopes in Cretaceous kimberlites and mantle derived xenoliths from southern Africa. *Earth Planet. Sci. Lett.* **34**, 419–43.

Kramers, J. D., J. C. M. Roddick & J. B. Dawson 1983. Trace element and isotope studies on veined metasomatic and MARID xenoliths from Bultfontein, South Africa. *Earth Planet. Sci. Lett.* **65**, 90–106.

Kramers, J. D., C. D. Smith, N. Lock, R. Harmon & F. R. Boyd 1981. Can kimberlite be generated from ordinary mantle? *Nature* 291, 53−6.

Krishnamurthy, P. & K. G. Cox 1977. Picrite basalts and related lavas from the Deccan Traps of western India. *Contrib. Mineral. Petrol.* 62, 53−75.

Krishnamurthy, P. & K. G. Cox 1980. A potassium-rich alkalic suite from the Deccan Traps, Rajpipla, India. *Contrib. Mineral. Petrol.* 73, 179−89.

Kuno, H. 1959. Origin of Cenozoic petrographic provinces of Japan and surrounding areas. *Bull. Volcanol.* 20, 37−76.

Kuno, H. 1969. Plateau basalts. In *The Earth's crust and upper mantle*, P. Hart, (ed.) 495−501. Am Geophys. Union Geophys. Monogr. 13.

Kurz, M. D., W. J. Jenkins, J.-G. Schilling & S. R. Hart 1982. Helium isotopic systematics of ocean islands and mantle heterogeneity. *Nature* 297, 43−6.

Kushiro, I. 1972. Effects of water on the composition of magmas formed at high pressures. *J. Petrol.* 13, 311−34.

Kuznir, N. J. 1980. Thermal evolution of the oceanic crust: its dependence on spreading rate and effect on crustal structure. *Geophys. J. R. Astron. Soc.* 61, 167−81.

Kyser, T. K., J. J. O'Neil & I. S. E. Carmichael 1982. Genetic relations among basic lavas and ultramafic nodules: evidence from oxygen isotope compositions. *Contrib. Mineral. Petrol.* 81, 88−102.

Langmuir, G. H. & G. N. Hanson 1981. Calculating mineral-melt equilibria with stoichiometry, mass balance and single-component distribution coefficients. In *Thermodynamics of Minerals and melts*, R. C. Newton, A. Navrotsky & B. J. Wood, (eds), 247−72. Berlin: Springer-Verlag.

Lanphere, M. A. & F. A. Frey 1987. Geochemical evolution of Kohala volcano, Hawaii. *Contrib. Mineral. Petrol.* 95, 100−13.

Le Bas, M. J. 1971. Peralkaline volcanism, crustal swelling and rifting. *Nature* 230, 85−7.

Le Bas, M. J. 1987. Nephelinites and carbonatites. In *Alkaline igneous rocks*, J. G. Fitton & B. G. J. Upton (eds), 53−83. Geol. Soc. Sp. Publ. 30.

Le Bas, M. J., R. W. Le Maitre, A. Streckeisen & B. Zanettin 1986. A chemical classification of volcanic rocks based on the total alkali-silica diagram. *J. Petrol.* 27, 745−50.

Le Douaran, S. E. & J. Francheteau 1981. Axial depth anomalies from 10 to 50° north along the Mid-Atlantic Ridge: correlation with other mantle properties. *Earth Planet. Sci. Lett.* 54, 29−47.

Le Maitre, R. W. 1980. A generalised petrological mixing model program. *Comput. Geosci.* 7, 229−47.

Le Marchand, F., B. Villemant & G. Calas 1987. Trace element distribution coefficients in alkaline series. *Geochim. Cosmochim. Acta.* 51, 1071−81.

Le Pichon, X. & P. Huchon 1984. Geoid, Pangea and convection. *Earth Planet. Sci. Lett.* 67, 123−35.

Le Roex, A. P. 1985. Geochemistry, mineralogy and magmatic evolution of the basaltic and trachytic lavas from Gough Island, South Atlantic. *J. Petrol.* 26, 149−86.

Le Roex, A. P. 1986. Geochemical correlation between southern African kimberlites and south Atlantic hotspots. *Nature* 324, 243−5.

Le Roex, A. P. 1987. Source regions of mid-ocean ridge basalts: evidence for enrichment processes. In *Mantle metasomatism*, M. A. Menzies & C. J. Hawkesworth, (eds), 389−422. London: Academic Press.

Le Roex, A. P., H. J. B. Dick, A. J. Erlank, A. M. Reid, F. A. Frey & S. R. Hart 1983. Geochemistry, mineralogy and petrogenesis of lavas erupted along the southwest Indian Ridge between the Bouvet triple junction and 11 degrees East. *J. Petrol.* 24, 267−318.

Le Roex, A. P., H. J. B. Dick, A. M. Reid, F. A. Frey, A. J. Erlank & S. R. Hart 1985. Petrology and geochemistry of basalts from the American−Antarctic Ridge, Southern Ocean: implications for the westward influence of the Bouvet mantle plume. *Contrib. Mineral. Petrol.* 90, 367−80.

Leeds, A. 1975. Lithospheric thickness in the western Pacific. *Phys. Earth Planet. Inter.* 11, 61−4.

Leeman, W. P. 1983. The influences of crustal

structure on compositions of subduction-related magmas. *J. Volcanol. Geotherm. Res.* **18**, 561–88.

Leeman, W. A., J. R. Budahn, D. C. Gerlach, D. R. Smith & B. N. Powell 1980. Origin of Hawaiian tholeiites: trace element constraints. *Am. J. Sci.* **280A**, 794–819.

Leeman, W. P. & C. J. Hawkesworth 1986. Open magma systems: trace element and isotopic constraints. *J. Geophys. Res.* **91**, 5901–12.

Lewis, B. T. R. & J. D. Garmany 1982. Constraints on the structure of the East Pacific Rise from seismic refraction data. *J. Geophys. Res.* **87**, 8417–25.

Lightfoot, P. C., C. J. Hawkesworth & S. F. Sethna 1987. Petrogenesis of rhyolites and trachytes from the Deccan Traps: Sr, Nd and Pb isotope and trace element evidence. *Contrib. Mineral. Petrol.* **95**, 44–54.

Lindsley, D. H., J. E. Grover & P. M. Davidson 1981. The thermodynamics of the $Mg_2Si_2O_6$–$CaMgSi_2O_6$ join: a review and an improved model. In *Advances in physical geochemistry*, vol. 1, 149–75. New York: Springer-Verlag.

Liotard, J. M., H. G. Barsczus, C. Dupuy & J. Dostal 1986. Geochemistry and origin of basaltic lavas from Marquesas Archipelago, French Polynesia. *Contrib. Mineral. Petrol.* **92**, 260–8.

Lipman, P. W. 1980. Cenozoic volcanism in the western United States: implications for continental tectonics. In *Studies in geophysics*, 161–74. Washington, DC: National Academy of Sciences.

Lippard, S. J. 1973. The petrology of phonolites from the Kenyan rift. *Lithos* **6**, 217–34.

Lloyd, F. E. 1981. Upper mantle metasomatism beneath a continental rift: clinopyroxenes in alkalic mafic lava and nodules from south west Uganda. *Mineral. Mag.* **44**, 315–23.

Lloyd, F. E., M. Arima & A. D. Edgar 1985. Partial melting of a phlogopite–clinopyroxenite nodule from south-west Uganda: An experimental study bearing on the origin of highly potassic continental rift volcanics. *Contrib. Mineral. Petrol.* **91**, 321–9.

Lloyd, F. E. & D. K. Bailey 1975. Light element

metasomatism of the continental mantle: the evidence and the consequences. *Phys. Chem. Earth.* **9**, 389–416.

Loper, D. E. 1985. A simple model of whole-mantle convection. *J. Geophys. Res.* **90**, 1809–36.

Luff, I. W. 1982. *Petrogenesis of the island arc tholeiite series of the South Sandwich Islands.* Unpubl. PhD thesis, Univ. Leeds, UK.

Maaløe, S. 1981. Magma accumulation in the ascending mantle. *J. Geol. Soc. Lond.* **138**, 223–36.

Maaløe, S. 1982. Geochemical aspects of permeability controlled partial melting and fractional crystallisation. *Geochim. Cosmochim. Acta.* **46**, 43–57.

Maaløe, S. 1985. *Principles of igneous petrology.* Berlin: Springer-Verlag, 371 pp.

Maaløe, S. & K. Aoki 1977. The major element composition of the upper mantle estimated from the composition of lherzolites. *Contrib. Mineral. Petrol.* **63**, 161–73.

Maaløe, S. & A. Scheie 1982. The permeability controlled accumulation of primary magma. *Contrib. Mineral. Petrol.* **81**, 350–7.

Macdonald, G. A. 1968. Composition and origin of Hawaiian lavas. *Mem. Geol Soc. Am.* **116**, 477–522.

Macdonald, G. A. & T. Katsura 1964. Chemical composition of Hawaiian lavas. *J. Petrol.* **5**, 82–133.

Macdonald, R. 1987. Quaternary peralkaline silicic rocks and caldera volcanoes of Kenya. In *Alkaline igneous rocks*, J. G. Fitton & B. G. J. Upton, (eds), 313–33. Geol Soc. Sp. Publ. 30.

Macdonald, R., G. R. Davies, C. M. Bliss, P. T. Leat, D. K. Bailey & R. L. Smith 1987. Geochemistry of high-silica peralkaline rhyolites, Naivasha, Kenya Rift Valley. *J. Petrol.* **28**, 979–1008.

Mahoney, J., J. D. Macdougall, G. W. Lugmair, M. Sankar Das, A. V. Murali & K. Gopalan 1982. Origin of the Deccan Trap flows at Mahabaleshwar inferred from Nd and Sr isotopic and chemical evidence. *Earth Planet. Sci. Lett.* **60**, 47–60.

Mahoney, J. J., J. D. Macdougall, G. W. Lugmair, K. Gopalan & P. Krishnamurthy 1985. Origin of contemporaneous tholeiitic and K-rich alkalic lavas: a case study from northern Deccan Plateau, India. *Earth Planet. Sci. Lett.* **72**, 39–53.

Mantovani, M. S. M., L. S. Marques, M. A. De Sousa, L. Civetta, T. Atalla & F. Innocenti 1985. Trace element and strontium isotope constraints on the origin and evolution of Parana continental flood basalts of Santa Catarina State (Southern Brazil). *J. Petrol.* **26**, 187–209.

Margaritz, M., D. J. Whitford & D. E. James 1978. Oxygen isotopes and the origin of high $^{87}Sr/^{86}Sr$ andesites. *Earth Planet. Sci. Lett.* **40**, 220–30.

Marriner, G. F. & D. Millward 1984. The petrology and geochemistry of Cretaceous to Recent volcanism in Colombia: the magmatic history of an accretionary plate margin. *J. Geol Soc. Lond.* **141**, 473–86.

Marsh, B. D. 1982. The Aleutians. In *Andesites: Orogenic andesites and related rocks*, R. S. Thorpe (ed.) 99–114. Chichester: Wiley.

Marsh, B. D. & I. S. E. Carmichael 1974. Benioff zone magmatism. *J. Geophys. Res.* **79**, 1196–206.

Marsh, J. S. 1987. Basalt geochemistry and tectonic discrimination within continental flood basalt provinces. *J. Volcanol. Geotherm. Res.* **32**, 35–49.

Martin, D., R. W. Griffiths & I. H. Campbell 1987. Compositional and thermal convection in magma chambers. *Contrib. Mineral. Petrol.* **96**, 465–75.

Mason, G. H. 1985. The mineralogy and textures of the Coastal Batholith, Peru. In *Magmatism at a plate edge*, W. S. Pitcher, M. P. Atherton, E. J. Cobbing, R. D. Beckinsale (eds), 156–66. London: Blackie.

Mason, R. G. & A. D. Raff 1961. A magnetic survey off the west coast of North America, 32°N to 42°N. *Bull. Geol Soc. Am.* **72**, 1259–65.

McBirney, A. R. 1979. Effects of assimilation. In *The evolution of igneous rocks: fiftieth anniversary perspectives*, H. S. Yoder, (ed.) 307–38. Princeton NJ: Princeton University Press.

McBirney, A. R. 1984. *Igneous petrology*. San Francisco: Freeman, Cooper, 504 pp.

McBirney, A. R., B. H. Baker & R. H. Nilson 1985. Liquid fractionation, Part 1: Basic principles and experimental simulations. *J. Volcanol. Geotherm. Res.* **24**, 1–24.

McBirney, A. R. & R. M. Noyes 1979. Crystallisation and layering of the Skaergaard intrusion. *J. Petrol.* **20**, 487–54.

McClain, J. S. & B. T. R. Lewis 1980. A seismic experiment at the axis of the East Pacific Rise. *Mar. Geol.* **35**, 147–69.

McClain, J. S., J. A. Orcutt & M. Burnett 1985. The East Pacific Rise in cross-section: A seismic model. *J. Geophys. Res.* **90**, 8627–39.

McCulloch, M. T., A. L. Jaques, D. R. Nelson & J. D. Lewis 1983. Nd and Sr isotopes in kimberlites and lamproites from Western Australia: enriched mantle origin. *Nature* **302**, 400–3.

McCulloch, M. T. & M. R. Perfit 1981. $^{143}Nd/^{144}Nd$, $^{87}Sr/^{86}Sr$ and trace element constraints on the petrogenesis of Aleutian island arc magmas. *Earth Planet. Sci. Lett.* **56**, 167–79.

McDonough, W. F., M. T. McCulloch & S. S. Sun 1985. Isotopic and geochemical systematics in Tertiary–Recent basalts from southeastern Australia and implications for the evolution of the sub-continental lithosphere. *Geochim. Cosmochim. Acta.* **49**, 2051–67.

McKenzie, D. P. (1978a). Active tectonics of the Alpine–Himalayan belt: the Aegean Sea and surrounding regions. *Geophys. J. R. Astron. Soc.* **55**, 217–54.

McKenzie, D. P. (1978b). Some remarks on the development of sedimentary basins. *Earth Planet. Sci. Lett.* **40**, 25–32.

McKenzie, D. P. (1984a). The generation and compaction of partially molten rock. *J. Petrol.* **25**, 713–65.

McKenzie, D. P. (1984b). A possible mechanism for epeirogenic uplift. *Nature* **307**, 616–18.

McKenzie, D. 1985. The extraction of magma from the crust and mantle. *Earth Planet. Sci. Lett.* **74**, 81–91.

McKenzie, D & R. K. O'Nions 1983. Mantle reservoirs and ocean island basalts. *Nature* **301**,

229–31.

McMillan, N. J. & M. A. Dungan 1986. Magma mixing as a petrogenetic process in the development of the Taos Plateau Volcanic Field, New Mexico. *J. Geophys. Res.* **91**, 6029–45.

Meijer, A. 1976. Pb and Sr isotopic data bearing on the origin of volcanic rocks from the Mariana island-arc system. *Bull. Geol Soc. Am.* **87**, 1358–69.

Meijer, A. 1980. Primitive arc volcanism and a boninite series: Examples from western Pacific island arcs. In *tectonics and geological evolution of Southeast Asia seas and islands*, D. F. Hayes, (ed.), 269–82. Am. Geophys. Union Monogr. 23.

Meijer, A. & M. Reagan 1981. Petrology and geochemistry of the island of Sarigan in the Mariana Arc; calc-alkaline volcanism in an oceanic setting. *Contrib. Mineral. Petrol.* **77**, 337–54.

Melson, W. G., T. L. Vallier, T. L. Wright, G. Byerly & J. Nelen 1976. Chemical diversity of abyssal volcanic glass erupted along Pacific, Atlantic and Indian Ocean sea-floor spreading centers. In *The geophysics of the Pacific Ocean basin and its margin*, 351–67. Washington, DC: Am. Geophys. Union.

Melson, W. G. & T. H. Van Andel 1966. Metamorphism in the Mid-Atlantic Ridge, 22° latitude. *Mar. Geol.* **4**, 165–86.

Menzies, M. A. 1983. Mantle ultramafic xenoliths in alkaline magmas: evidence for mantle heterogeneity modified by magmatic activity. In *Continental basalts and mantle xenoliths*, C. J. Hawkesworth & M. J. Norry, (eds), 92–110. Nantwich: Shiva.

Menzies. M. 1987. Alkaline rocks and their inclusions: a window on the Earth's interior. In *Alkaline igneous rocks*, J. G. Fitton & B. G. J. Upton, (eds), 15–27. Geol Soc. Sp. Publ. 30.

Menzies, M. A. & C. J. Hawkesworth 1987. Upper mantle processes and composition. In *Mantle xenoliths*, P. H. Nixon, (ed.), 725–38. Chichester: Wiley.

Menzies, M. A., W. R. Leeman & C. J. Hawkesworth 1983. Isotope geochemistry of Cenozoic volcanic rocks reveals mantle heterogeneity below western U.S.A. *Nature* **303**, 205–9.

Menzies, M. A., W. P. Leeman & C. J. Hawkesworth 1984. Geochemical and isotopic evidence for the origin of continental flood basalts with particular reference to the Snake River Plain, Idaho, U.S.A. *Phil Trans R. Soc. Lond.* **A310**, 643–60.

Menzies, M. & V. R. Murthy 1980a. Enriched mantle: Nd and Sr isotopes in diopsides from kimberlite nodules,. *Nature* **283**, 634–6.

Menzies, M. & V. R. Murthy 1980b. Mantle metasomatism as precursor to the genesis of alkaline magmas – isotopic evidence. *Am. J. Sci.* **280A**, 622–38.

Menzies, M. A. & S. Y. Wass 1983. CO_2 rich mantle below eastern Australia: REE, Sr and Nd isotopic study of Cenozoic alkaline magmas and apatite-rich xenoliths, Southern Highlands province, New South Wales, Australia. *Earth Planet. Sci. Lett.* **65**, 287–302.

Meschede, M. 1986. A method of discriminating between different types of mid-ocean ridge basalts and continental tholeiites with the Nb–Zr–Y diagram. *Chem. Geol.* **56**, 207–18.

Michael, P. J. & R. L. Chase 1987. The influence of primary magma composition, H_2O and pressure on mid-ocean ridge basalt differentiation. *Contrib. Mineral. Petrol.* **96**, 245–63.

Michaelson, C. A. & C. S. Weaver 1986. Upper mantle structure from teleseismic P wave arrivals in Washington and Northern Oregon. *J. Geophys. Res.* **91**, 2077–94.

Middlemost, E. A. K. 1975. The basalt clan. *Earth Sci. Rev.* **11**, 337–64.

Middlemost, E. A. K. 1980. A contribution to the nomenclature and classification of volcanic rocks. *Geol Mag.* **117**, 51–7.

Mitchell, A. H. & H. G. Reading 1969. Continental margins, geosynclines and ocean floor spreading. *J. Geol.* **77**, 629–46.

Mitchell, R. H. 1985. A review of the mineralogy of lamproites. *Trans Geol Soc. S. Afr.* **88**, 411–37.

Mitchell, R. H. 1986. *Kimberlites: mineralogy, geochemistry and petrology*. New York: Plenum Press, 442 pp.

Mitchell, R. H. & K. Bell 1976. Rare earth element

geochemistry of potassic lavas from the Birunga and Toro−Ankole regions of Uganda, Africa. *Contrib. Mineral. Petrol.* **58**, 293−303.

Miyashiro, A. 1974. Volcanic rock series in island arcs and active continental margins. *Am. J. Sci.* **274**, 321−55.

Miyashiro, A. 1978. Nature of alkalic volcanic rock series. *Contrib. Mineral. Petrol.* **66**, 91−104.

Miyashiro, A., F. Shido & M. Ewing 1971. Metamorphism in the Mid-Atlantic Ridge near 24°N and 30°N. *Phil Trans R. Soc. Lond.* **A268**, 589−603.

Moberly, R. & J. F. Campbell 1984. Hawaiian hotspot volcanism mainly during geomagnetic normal intervals. *Geology* **12**, 459−63.

Mohr, P. 1982. Musings on continental rifts. In *Continental and oceanic rifts*, G. Palmason (ed.), 293−309. Washington DC: Am. Geophys. Union.

Molnar, P. & T. Atwater 1978. Interarc spreading and Cordilleran tectonics as alternates related to the age of subducted oceanic lithosphere. *Earth Planet. Sci. Lett.* **41**, 330−40.

Molnar, P. & J. Stock 1987. Relative motions of hotspots in the Pacific, Atlantic and Indian Oceans since late Cretaceous time. *Nature* **327**, 587−91.

Moorbath, S. & R. N. Thompson (eds) 1984. The relative contributions of mantle, oceanic crust and continental crust to magma genesis. *Phil Trans R. Soc. Lond.* **A310**, 437−780.

Moore, J. G. 1965. Petrology of deep-sea basalt near Hawaii. *Am. J. Sci.* **263**, 40−52.

Moore, J. G., H. S. Fleming & J. D. Phillips 1974. Preliminary model for extrusion and rifting at the axis of the Mid-Atlantic Ridge 36°48′ North. *Geology* **2**, 437−40.

Morgan, P. 1983. Constraints on rift thermal processes from heat flow and uplift. *Tectonophysics* **94**, 277−98.

Morgan, W. J. 1971. Convection plumes in the lower mantle. *Nature* **230**, 42−3.

Morgan, W. J. 1972a. *Plate motions and deep mantle convection.* Geol Soc. Am. Mem. 7−22.

Morgan, W. J. 1972b. Deep mantle convection plumes and plate motions. *AAPG Bull.* **56**, 203−13.

Morgan, W. J. 1983. Hotspot tracks and the early rifting of the Atlantic. *Tectonophysics* **94**, 123−39.

Morrison, G. W. 1980. Characteristics and tectonic setting of the shoshonite rock association. *Lithos* **13**, 97−108.

Mukasa, S. B. 1986. Common Pb isotopic compositions of the Lima, Arequipa and Toquepala segments in the coastal batholith, Peru: implications for magmagenesis. *Geochim. Cosmochim. Acta.* **50**, 771−82.

Mullen, E. D. 1983. $MnO/TiO_2/P_2O_5$: a minor element discriminant for basaltic rocks of oceanic environments and its implications for petrogenesis. *Earth Planet. Sci. Lett.* **62**, 53−62.

Mysen, B. O. 1982. The role of mantle anatexis. In *Andesites: orogenic andesites and related rocks*, R. S. Thorpe (ed.), 489−522. Chichester: Wiley.

Mysen, B. O. & A.L. Boettcher 1975. Melting of a hydrous mantle: II. Geochemistry of crystals and liquids formed by anatexis of mantle peridotite at high pressures and high temperatures as a function of controlled activities of water, hydrogen and carbon dioxide. *J. Petrol.* **16**, 549−93.

Mysen, B. O. & I. Kushiro 1977. Compositional variations of coexisting phases with degree of melting of peridotite in the upper mantle. *Am. Mineral.* **62**, 843−65.

Nakamura, N. 1974. Determination of REE, Ba, Fe, Mg, Na and K in carbonaceous and ordinary chondrites. *Geochim. Cosmochim. Acta.* **38**, 757−73.

Nathan, H. D. & C. K. Van Kirk 1978. A model of magmatic crystallisation. *J. Petrol.* **19**, 66−94.

Natland, J. H. 1978. Crystal morphologies in basalts from DSDP site 395, 23°N, 46°W, Mid-Atlantic Ridge. In *Initial Reports of the Deep Sea Drilling Project* **45**, 423−46. Washington, DC: US Goverment Printing Office.

Navon, D. & E. Stolper 1987. Geochemical consequences of melt percolation: the upper mantle as a chromatographic column. *J. Geol.* **95**, 285−307.

Nelson, D. R., M. T. McCulloch & S.-S. Sun 1986. The origins of ultrapotassic rocks as inferred from Sr, Nd and Pb isotopes. *Geochim. Cosmochim. Acta.* 50, 231–45.

Nielson, R. L. & M. A. Dungan 1983. Low pressure mineral-melt equilibria. *Contrib. Mineral. Petrol.* 84, 310–26.

Nicholls, I. A. & D. J. Whitford 1983. Potassium rich volcanic rocks of the Muriah Complex, Java, Indonesia: products of multiple magma sources? *J. Volcanol. Geotherm. Res.* 18, 337–59.

Nicholls, I. A., D. J. Whitford, K. L. Harris & S. R. Taylor 1980. Variation in the geochemistry of mantle sources for tholeiitic and calk-alkaline mafic magmas, western Sunda volcanic arc, Indonesia. *Chem. Geol.* 30, 177–99.

Nicholas, A. 1985. Novel type of crust produced during continental rifting. *Nature* 315, 112–15.

Nicholas, A. 1986. A melt extraction model based on structural studies in mantle peridotites. *J. Petrol.* 27, 999–1022.

Nielsen, R. L. & M. A. Dungan 1983. Low pressure mineral-melt equilibria in natural anhydrous systems. *Contrib. Mineral. Petrol.* 84, 310–26.

Nilson, R. H., A. R. McBirney & B. H. Baker 1985. Liquid fractionation, Part II: Fluid dynamics and quantitative implications for magmatic systems. *J. Volcanol. Geotherm. Res.* 24, 25–54.

Nisbet, E. G. & C. M. R. Fowler 1978. The Mid-Atlantic Ridge at 37° and 45°N: some geophysical and petrological constraints. *Geophys. J. R. Astron. Soc.* 54, 631–60.

Nixon, P. H. (1987). *Mantle xenoliths*. Chichester: Wiley, 844pp.

Nixon, P. H. & G. R. Davies 1987. Mantle xenolith perspectives. In *Mantle xenoliths*, P. H. Nixon (ed.), 741–56. Chichester: Wiley.

Nixon, P. H., N. W. Rogers, J. L. Gibson & A. Grey 1981. Depleted and fertile mantle xenoliths from South African kimberlites. *Ann. Rev. Earth Planet. Sci.* 9, 285–309.

Nixon, P. H., M. F. Thirlwall, F. Buckley & C. J. Davies 1984. Spanish and Western Australian lamproites: aspects of whole rock geochemistry. In *Kimberlites I: kimberlites and related rocks*, J. Kornprobst, (ed.), 285–96. Amsterdam: Elsevier.

Nolet, G. & S. Mueller 1982. A model for the deep structure of the East African rift system from simultaneous inversion of teleseismic data. *Tectonophysics* 84, 151–78.

Norry, M. J. & J. G. Fitton 1983. Compositional differences between oceanic and continental basic lavas and their significance. In *Continental basalts and mantle xenoliths*, C. J. Hawkesworth & M. J. Norry (eds), 5–19. Nantwich: Shiva.

Norry, M. J., P. H. Truckle, S. J. Lippard, C. J. Hawkesworth, S. D. Weaver & G. F. Marriner 1980. Isotopic and trace element evidence from lavas bearing on mantle heterogeneity beneath Kenya. *Phil Trans R. Soc. Lond.* A297, 259–71.

Norton, I. O. & J. G. Sclater 1979. A model for the evolution of the Indian Ocean and the break-up of Gondwanaland. *J. Geophys. Res.* 84, 6803–30.

Nur, A. & Z. Ben-Avraham 1982. Oceanic plateaus, the fragmentation of continents and mountain building. *J. Geophys. Res.* 87, 3644–61.

Nur, A. & Z. Ben-Avraham 1983. Displaced terranes and mountain building. In *Mountain building processes*, K. Hsu (ed.), 73–83. London: Academic Press.

Ocala, L. C. & R. P. Meyer (1972). Crustal low velocity zone under the Peru–Bolivia Altiplano. *Geophys. J. R. Astron. Soc.* 30, 199–209.

Officer, C. B. & C. L. Drake 1985. Terminal Cretaceous environmental events. *Science* 227, 1161–7.

O'Hara, M. J. 1968a. The bearing of phase equilibria studies in synthetic and natural systems and the origin and evolution of basic and ultrabasic rocks. *Earth Planet. Sci. Lett.* 4, 69–133.

O'Hara, M. J. 1968b. Are ocean floor basalts primary magmas? *Nature* 220, 683–6.

O'Hara, M. J. 1973. Non-primary magmas and dubious mantle plume beneath Iceland. *Nature* 243, 507–8.

O'Hara, M. J. 1977. Geochemical evolution during fractional crystallisation of a periodically refilled magma chamber. *Nature* 266, 503–7.

O'Hara, M. J. 1982. MORB – a Mohole misbegotten. *EOS, Trans Am. Geophys. Union.* **63**, 537–9.

O'Hara, M. J. 1985. Importance of the 'shape' of the melting regime during partial melting of the mantle. *Nature* **314**, 58–61.

O'Hara, M. J. & R. E. Mathews 1981. Geochemical evolution in an advancing, periodically replenished, periodically tapped, continuously fractionated magma chamber. *J. Geol Soc. Lond.* **138**, 237–77.

Olafsson, M. & D. H. Eggler 1983. Phase relations of amphibole–carbonate and phlogopite–carbonate peridotite: petrologic constraints on the asthenosphere. *Earth Planet. Sci. Lett.* **64**, 305–15.

Olsen, K. H., L. W. Braile, J. N. Stewart, C. R. Daudt, G. R. Keller, L. A. Ankeny & J. J. Wolff 1986. Jemez Mountains Volcanic field, New Mexico: time term interpretation of the CARDEX seismic experiment and comparison with Bouger gravity. *J. Geophys. Res.* **91**, 6175–87.

Olsen, K. H., G. R. Keller & J. N. Stewart 1979. Crustal structure along the Rio Grande Rift from seismic refraction profiles. In *Rio Grande Rift: tectonics and magmatism*, R. H. Riecker, (ed.), 127–43. Washington DC: Am. Geophys. Union.

Olson, P. 1987. Drifting mantle hotspots. *Nature* **327**, 559–60.

O'Nions, R. K. 1987. Relationships between chemical and convective layering in the Earth. *J. Geol Soc. Lond.* **144**, 259–74.

O'Nions, R. K., P. J. Hamilton & N. M. Evenson 1977. Variations in $^{143}Nd/^{144}Nd$ and $^{87}Sr/^{86}Sr$ ratios in oceanic basalts. *Earth Planet. Sci. Lett.* **39**, 13–22.

Orcutt, J. B., B. Kennett, L. Dorman & W. A. Protherow 1975. Evidence for a low velocity zone underlying a fast spreading rise crest. *Nature* **256**, 475–6.

Osborn, E. F. 1962. Reaction series for subalkaline igneous rocks based on different oxygen pressure conditions. *Am. Mineral.* **47**, 211–26.

Osborn, E. F. & D. B. Tait 1952. The system diopside–forsterite–anorthite. *Am. J. Sci.* (Bowen Volume), 413–33.

Oxburgh, E. R. & E. M. Parmentier 1977. Compositional and density stratification in oceanic lithosphere – causes and consequences. *J. Geol. Soc. Lond.* **133**, 343–55.

Palacz, Z. A. 1985. Sr–Nd–Pb isotopic evidence for crustal contamination in the Rhum intrusion. *Earth Planet. Sci. Lett.* **74**, 35–44.

Palmason, G. 1982. *Continental and oceanic rifts.* Washington, DC: Am. Geophys. Union.

Parmentier, E. M., D. L. Turcotte & K. E. Torrance 1975. Numerical experiments on the structure of mantle plumes. *J. Geophys. Res.* **80**, 4417–24.

Parsons, B. & J. G. Sclater 1977. An analysis of the variation of ocean floor bathymetry and heat flow with age. *J. Geophys. Res.* **82**, 803–27.

Parsons, I. 1978. Feldspars and fluids in cooling plutons. *Mineral. Mag.* **42**, 1–17.

Patchett, P. J. 1980. Thermal effects of basalt on continental crust and crustal contamination of magmas. *Nature* **283**, 559–61.

Pearce, J. A. (1976) Statistical analysis of major element patterns in basalt. *J. Petrol.* **17**, 15–43.

Pearce, J. A. 1982. Trace element characteristics of lavas from destructive plate boundaries. In *Andesites: orogenic andesites and related rocks*, R. S. Thorpe (ed.), 525–48. Chichester: Wiley.

Pearce, J. A. 1983. The role of sub-continental lithosphere in magma genesis at destructive plate margins. In *Continental basalts and mantle xenoliths.* C. J. Hawkesworth & M. J. Norry (eds), 230–49. Nantwich: Shiva.

Pearce, J. A. 1987. An expert system for the tectonic characterization of ancient volcanic rocks. *J. Volcanol. Geotherm. Res.* **32**, 51–65.

Pearce, J. A. & J. R. Cann 1973. Tectonic setting of basic volcanic rocks determined using trace element analysis. *Earth Planet. Sci. Lett.* **19**, 290–300.

Pearce, T. H., B. E. Gorman & T. C. Birkett 1975. The TiO_2–K_2O–P_2O_5 diagram: a method of discriminating between oceanic and non-oceanic basalts. *Earth Planet. Sci. Lett.* **24**, 419–26.

Pearce, T. H., B. E. Gorman & T. C. Birkett 1977. The relationship between major element chemistry and tectonic environment of basic and intermediate volcanic rocks. *Earth Planet. Sci. Lett.* 36, 121–32.

Peccerillo, A. 1985. Roman comagmatic province (Central Italy): Evidence for subduction related magma genesis. *Geology* 13, 103–6.

Peccerillo, A. & P. Manetti 1985. The potassium alkaline volcanism of central-southern Italy: a review of the data relevant to petrogenesis and geodynamic significance. *Trans Geol Soc. S. Afr.* 88, 379–94.

Peccerillo, A., G. Poli & L. Tolomeo 1984. Genesis, evolution and tectonic significance of K-rich volcanics from the Alban Hills (Roman comagmatic region) as inferred from trace element geochemistry. *Contrib. Mineral. Petrol.* 86, 230–40.

Peccerillo, A. & S. R. Taylor 1976. Geochemistry of Eocene calc-alkaline volcanic rocks from the Kastamonu area, northern Turkey. *Contrib. Mineral. Petrol.* 58, 63–81.

Perfit, M. R., D. A. Gust, A. E. Bence, R. J. Arculus & S. R. Taylor 1980. Chemical characteristics of island arc basalts: implications for mantle sources. *Chem. Geol.* 30, 277–56.

Perry, F. V., W. S. Baldridge & D. S. DePaolo 1987. Role of asthenosphere and lithosphere in the genesis of Late Cenozoic basaltic rocks from the Rio Grande Rift and adjacent regions of the south western United States. *J. Geophys. Res.* 92, 9193–213.

Petrini, R., L. Civetta, E. M. Piccirillo, G. Bellieni, P. Comin-Chiaramonti, L. S. Marques & A. J. Melfi 1987. Mantle heterogeneity and crustal contamination in the genesis of low-Ti continental flood basalts from the Parana Plateau (Brazil): Sr−Nd isotope and geochemical evidence. *J. Petrol.* 28, 701–26.

Philpotts, A. R. 1979. Silicate liquid immiscibility in tholeiitic basalts. *J. Petrol.* 20, 99–118.

Pilger, R. H. Jr 1981. Plate reconstructions, aseismic ridges and low angle subduction beneath the Andes. *Bull. Geol Soc. Am.* 92, 448–56.

Pilger, R. H. Jr (1984). Cenozoic plate kinematics, subduction and magmatism: South American Andes. *J. Geol Soc. Lond.* 141, 793–802.

Pineau, F., M. Javoy, J. W. Hawkins & H. Craig 1976. Oxygen isotope variations in marginal basin and ocean-ridge basalts. *Earth Planet. Sci. Lett.* 28, 299–307.

Pinet, C. & C. Jaupart 1987. A thermal model for the distribution in space and time of the Himalayan granites. *Earth Planet. Sci. Lett.* 84, 87–99.

Pitcher, W. S. 1979. The nature, ascent and emplacement of granitic magmas. *J. Geol Soc. Lond* 136, 627–62.

Pitcher, W. S., M. P. Atherton, E. J. Cobbing & R. D. Beckinsale (eds) 1985. *Magmatism at a plate edge − the Peruvian Andes.* London: Blackie, 328 pp.

Pitcher, W. S. & E. J. Cobbing 1985. Phanerozoic plutonism in the Peruvian Andes. In *Magmatism at a plate edge*, W. S. Pitcher, M. P. Atherton, E. J. Cobbing & R. D. Beckinsale (eds), 19–25. London: Blackie.

Pollack, H. N. (1986). Cratonization and thermal evolution of the mantle. *Earth Planet. Sci. Lett.* 80, 175–82.

Pollack, H. N. & D. S. Chapman 1977. On the regional variation of heat flow, geotherms and lithospheric thickness. *Tectonophysics* 38, 279–96.

Potts, P. J. 1987. *A handbook of silicate rock analysis.* London: Blackie, 622 pp.

Powell, M. 1978. Crystallisation conditions of low-pressure cumulate nodules from the Lesser Antilles island arc. *Earth Planet. Sci. Lett.* 39, 162–72.

Powell, R. 1984. Inversion of the assimilation and fractional crystallisation (AFC) equations: Characterization of contaminants from isotope and trace element relationships in volcanic suites. *J. Geol Soc. Lond.* 141, 447–52.

Presnall, D. C., J. R. Dixon, T. H. O'Donnell & S. A. Dixon 1979. Generation of mid-ocean ridge tholeiite. *J. Petrol.* 20, 3–35.

Press, F. & R. Siever 1982. *Earth* 3rd edn. New York: W. H. Freeman, 613 pp.

Prestvik, T. & G. G. Goles 1985. Comments on the petrogenesis and the tectonic setting of Columbia River Basalts. *Earth Planet. Sci. Lett.* 72, 65–73.

Price, R. C., R. W. Johnson, C. M. Gray & F. A.

Frey 1985. Geochemistry of phonolites and trachytes from the summit region of Mt. Kenya. *Contrib. Mineral. Petrol.* **89**, 394−409.

Rabinowicz, M., A. Nicholas & J. L. Vigneresse 1984. A rolling mill effect in asthenosphere beneath oceanic spreading centers. *Earth Planet. Sci. Lett.* **67**, 97−108.

Ramberg, I. B., F. A. Cook & S. B. Smithson 1978. Structure of the Rio Grande Rift in southern New Mexico and west Texas based on gravity interpretation. *Geol Soc. Am. Bull.* **89**, 107−23.

Rea, D. K. & T. L. Vallier 1983. Two Cretaceous volcanic episodes in the western Pacific Ocean. *Bull. Geol Soc. Am.* **94**, 1430−7.

Reeves, C. V., F. M. Karanja & I. N. Macleod (1987). Geophysical evidence for a failed Jurassic rift and triple junction in Kenya. *Earth Planet. Sci. Lett.* **81**, 299−311.

Reidel, S. P. 1983. Stratigraphy and petrogenesis of the Grande Ronde Basalt from the deep canyon country of Washington, Oregon and Idaho. *Bull. Geol Soc. Am.* **94**, 519−42.

Renard, V., R. Hekinian, J. Francheteau, R. D. Ballard & H. Backer 1985. Submersible observations at the axis of the ultra-fast spreading East Pacific Rise (17°30′ to 21°30′S). *Earth Planet. Sci. Lett.* **75**, 339−53.

Ribe, N. M. 1985. The generation and compaction of partial melts in the earth's mantle. *Earth Planet. Sci. Lett.* **73**, 361−76.

Richardson, S. H., A. J. Erlank, A. R. Duncan & D. L. Reid 1982. Correlated Nd, Sr and Pb isotope variation in Walvis Ridge basalts and implications for the evolution of their mantle source. *Earth Planet. Sci. Lett.* **59**, 327−42.

Richardson, S. H., A. J. Erlank & S. R. Hart 1985. Kimberlite-born garnet peridotite xenoliths from old enriched sub-continental lithosphere. *Earth Planet. Sci. Lett.* **75**, 116−28.

Richardson, S. H., J. J. Gurney, A. J. Erlank & J. W. Harris (1984). Origin of diamonds in old enriched mantle. *Nature* **310**, 198−202.

Richter, F. M. (1986). Simple model for trace element fractionation during melt segregation. *Earth Planet. Sci. Lett.* **79**, 333−44.

Richter, F. M. & D. P. McKenzie 1981. On some consequences and possible causes of layered mantle convection. *J. Geophys. Res.* **86**, 6133−42.

Richter, F. M. & D. McKenzie 1984. Dynamic models for melt segregation from deformed matrix. *J. Geol.* **92**, 729−40.

Ringwood, A. E. 1975. *Composition and petrology of the Earth's mantle.* New York: McGraw-Hill, 618 pp.

Rhodes, J. M. & M. A. Dungan 1979. The evolution of ocean floor basaltic magmas. In *Deep drilling results in the Atlantic Ocean: ocean crust,* C. M. Talwani, C. G. Harrison & D. E. Hayes (eds), 262−72. Washington DC: Am. Geophys. Union.

Robson, D. & J. R. Cann 1982. A geochemical model of mid-ocean ridge magma chambers. *Earth Planet. Sci. Lett.* **60**, 93−104.

Rock, N. M. S. 1986. The nature and origin of ultramafic lamprophyres: alnoites and allied rocks. *J. Petrol.* **27**, 155−96.

Rock, N. M. S. 1987. The nature and origin of lamprophyres: an overview. In *Alkaline igneous rocks,* J. G. Fitton & B. G. J. Upton (eds), 191−226. Geol Soc. Sp. Publ. 30.

Roden, M. F., F. A. Frey & D. A. Clague 1984. Geochemistry of tholeiitic and alkalic lavas from the Koolau Range, Oahu, Hawaii: Implications for Hawaiian volcanism. *Earth Planet. Sci. Lett.* **69**, 141−58.

Roedder, E. 1979. Silicate liquid immiscibility in magmas. In *The evolution of igneous rocks: fiftieth anniversary perspectives,* H. S. Yoder, Jr (ed.) 483−520. Princeton, NJ: Princeton University Press.

Roeder, P. L. & R. F. Emslie 1970. Olivine−liquid equilibrium. *Contrib. Mineral. Petrol.* **29**, 275−89.

Rogers, N. W., C. J. Hawkesworth, R. J. Parker & J. S. Marsh 1985. The geochemistry of potassic lavas from Vulsini, central Italy and implications for mantle enrichment processes beneath the Roman region. *Contrib. Mineral. Petrol.* **90**, 244−57.

Rosendahl, B. R. 1976. Evolution of oceanic crust, 2: constraints, implications and inferences. *J.*

Geophys. Res. 81, 5305—14.

Rosendahl, B. R. 1987. Architecture of continental rifts with special reference to East Africa. *Ann. Rev. Earth Planet. Sci.* 15, 445—503.

Rosendahl, B. R., R. W. Raitt, L. M. Dormon, L. D. Bibee, D. M. Hussong & G. H. Sutton 1976. Evolution of oceanic crust 1: a physical model of the East Pacific Rise crest derived from seismic refraction data. *J. Geophys. Res.* 81, 5294—304.

Rowell, W. F. & A. D. Edgar 1983. Cenozoic potassium rich mafic volcanism in the western U.S.A.: its relationship to deep subduction. *J. Geol.* 91, 338—41.

Rubie, D. C. & W. D. Gunter 1983. The role of speciation in alkaline igneous fluids during fenite metasomatism. *Contrib. Mineral. Petrol.* 82, 165—75.

Ryan, M. P., R. Y. Koyanagi & R. S. Fiske 1981. Modeling the three-dimensional structure of macroscopic magma transport systems: application to Kilauea volcano, Hawaii. *J. Geophys. Res.* 86, 7111—29.

Sacks, I. S. 1983. The subduction of young lithosphere. *J. Geophys. Res.* 88, 3355—66.

Sahama, Th. G. 1973. Evolution of the Nyiragongo magma. *J. Petrol.* 14, 33—48.

Sahama, T. G. 1974. Potassium-rich alkaline rocks. In *The alkaline rocks*, H. Sorenson (ed.) 96—109. New York: Wiley.

Sandford, A. R. & P. Einarsson 1982. Magma chambers in rifts. In *Continental and oceanic rifts*, G. Palmason (ed.). 147—68. Washington, DC: Am. Geophys. Union.

Saunders, A. D. & J. Tarney 1979. The geochemistry of basalts from a back-arc spreading centre in the East Scotia Sea. *Geochim. Cosmochim. Acta.* 43, 555—72.

Saunders, A. D. & J. Tarney 1984. Geochemical characteristics of basaltic volcanism within back-arc basins. In *Marginal basin geology: volcanic and associated sedimentary and tectonic processes in modern and ancient marginal basins*, B. P. Kokelaar & M. F. Howells, (eds), 59—76. Geol Soc. Lond. Sp. Publ. 16.

Saunders, A. D., J. Tarney, N. G. Marsh & D. A.

Wood 1980. Ophiolites as ocean crust or marginal basin crust: a geochemical approach. In *Proc. Int. Ophiolite Conf.* Nicosia, Cyprus, A. Panayiotou (ed.), 193—204.

Saunders, A. D., J. Tarney, C. Stern & I. W. D. Dalziel 1979. Geochemistry of Mesozoic marginal basin floor igneous rocks from southern Chile. *Bull. Geol Soc. Am.* 90, 237—58.

Savage, J. E. E. & R. E. Long 1985. Lithospheric structure beneath the Kenya dome. *Geophys. J. R. Astron. Soc.* 82, 461—77.

Schilling, J.-G. 1973. Iceland mantle plume: geochemical evidence along the Reykjanes Ridge. *Nature* 242, 565—71.

Schilling, J.-G. 1975. Azores mantle blob: rare earth evidence. *Earth Planet. Sci. Lett.* 25, 103—15.

Schilling, J.-G. 1985. Upper mantle heterogeneities and dynamics. *Nature* 314, 62—7.

Schilling, J.-G., G. Thompson, R. Kingsley & S. Humphris 1985. Hotspot — migrating ridge interaction in the South Atlantic. *Nature*, 313, 187—91.

Schilling, J.-G., M. Zajac, R. Evans, T. Johnston, W. White, J. D. Devine & R. Kingsley 1983. Petrologic and geochemical variations along the Mid-Atlantic Ridge from 27°N to 73°N. *Am. J. Sci.* 283, 510—86.

Schneider, M. E. & D. H. Eggler 1986. Fluids in equilibrium with peridotite minerals: implications for mantle metasomatism. *Geochim. Cosmochim. Acta.* 50, 711—24.

Scholl, D. W., R. Van Huene, T. L. Vallier & D. G. Howell 1980. Sedimentary masses and concepts about tectonic processes at underthrust ocean margins. *Geology* 8, 564—8.

Schouten, H. & K. D. Klitgord 1982. The memory of the accreting plate boundary and the continuity of fracture zones. *Earth Planet. Sci. Lett.* 59, 255—66.

Schouten, H., K. D. Klitgord & J. A. Whitehead 1985. Segmentation of mid-ocean ridges. *Nature* 317, 225—9.

Sclater, J. G., C. Bowin, R. Hey, H. Haskins, J. Peirce, J. Phillips & C. Tapscott 1976. The Bouvet triple junction. *J. Geophys. Res.* 81, 1857—69.

Sclater, J. G., J. W. Hawkins, J. Mammerickz & C. G. Chase 1972. Crustal extension between the Tonga and Lau ridges: petrologic and geophysical evidence. *Bull. Geol Soc. Am.* **83**, 505–18.

Sclater, J. G., C. Jaupart & D. Galson 1980. The heat flow through oceanic and continental crust and the heat loss of the earth. *Rev. Geophys. Space Phys.* **18**, 269–311.

Sclater, J. G., L. A. Lawver & B. Parsons 1975. Comparison of long wavelength residual elevation and free-air gravity anomalies in the North Atlantic and possible implications for the thickness of the lithospheric plate. *J. Geophys. Res.* **80**, 1031–52.

Scott, D. R. & D. J. Stevenson 1986. Magma ascent by porous flow. *J. Geophys. Res.* **91**, 9283–96.

Scott-Smith, B. H. & E. M. W. Skinner 1984a. Diamondiferous lamproites. *J. Geol.* **92**, 433–8.

Scott-Smith, B. H. & E. M. W. Skinner 1984b. A new look at Prairie Creek, Arkansas. In *Kimberlites I: kimberlites and related rocks*, J. Kornprobst (ed.), 255–83. Amsterdam: Elsevier.

Searle, R. C. 1970. Evidence from gravity anomalities for thinning of the lithosphere beneath the rift valley in Kenya. *Geophys. J. R. Astron. Soc.* **21**, 13–31.

Sekine, T., T. Irifune, A. E. Ringwood & W. O. Hibberson 1986. High-pressure transformation of eclogite to garnetite in subducted oceanic crust. *Nature* **319**, 584–6

Sekine, T. & P. J. Wyllie 1982a. Phase relationships in the system $KAlSiO_4–Mg_2SiO_4–SiO_2–H_2O$ as a model for hybridization between hydrous siliceous melts and peridotite. *Contrib. Mineral. Petrol.* **79**, 368–74.

Sekine, T. & P. J. Wyllie 1982b. The system granite–peridotite–H_2O at 30 kbar, with applications to hybridization in subduction zone magmatism. *Contrib. Mineral. Petrol.* **81**, 190–202.

Sengor, A. H. C. & K. Burke 1978. Relative timing of rifting and volcanism in earth and its tectonic implications. *Geophys. Res. Lett.* **5**, 419–21.

Shaw, H. R. 1965. Comments on viscosity, crystal settling and convection in granitic magmas. *Am. J. Sci.* **263**, 120–52.

Shaw, H. R., E. D. Jackson & K. E. Bargar 1980. Volcanic periodicity along the Hawaiian–Emperor chain. *Am. J. Sci.* **280A**, 667–707.

Shepherd, G. L. & R. Moberly 1981. Coastal structure of the continental margin of northwest Ecuador. In *Nazca plate: crustal formation and Andean convergence*. L. D. Kulm, J. Dymond, E. J. Dasch & D. M. Hussong (eds). *Mem. Geol Soc. Am.* **154**, 351–91.

Sheppard, S. M. F. & C. Harris 1985. Hydrogen and oxygen isotope geochemistry of Ascension Island lavas and granites: variation with crystal fractionation and interaction with seawater. *Contrib. Mineral. Petrol.* **91**, 74–81.

Shervais, J. W. 1982. Ti–V plots and the petrogenesis of modern and ophiolitic lavas. *Earth Planet. Sci. Lett.* **57**, 101–18.

Shudofsky, G. N. 1985. Source mechanisms and focal depths of East African earthquakes using Rayleigh-wave inversion and body wave modelling. *Geophys. J. R. Astron. Soc.* **83**, 563–614.

Siders, M. A. & D. H. Elliot 1985. Major and trace element geochemistry of the Kirkpatrick Basalt, Mesa Range, Antarctica. *Earth Planet. Sci. Lett.* **72**, 54–64.

Sinno, Y. A., P. H. Daggett, G. R. Keller, P. Morgan & S. H. Harder 1986. Crustal structure of the southern Rio Grande Rift determined from seismic refraction profiling. *J. Geophys. Res.* **91**, 6143–56.

Sinton, J. M., D. S. Wilson, D. M. Christie, R. N. Hey & J. R. Delaney 1983. Petrologic consequences of rift propagation on oceanic spreading ridges. *Earth Planet. Sci. Lett.* **62**, 193–207.

Sleep, N. H. 1975. Formation of oceanic crust: some thermal constraints. *J. Geophys. Res.* **80**, 4037–42.

Smith, A. L., M. J. Roobol & B. M. Gunn 1980. The Lesser Antilles – a discussion of the island arc magmatism. *Bull. Volcanol.* **43**, 287–302.

Smith, C. B. 1983. Pb, Sr and Nd isotopic evidence for sources of southern African Cretaceous kimberlites. *Nature* **304**, 51–4.

Smith, C. B., J. J. Gurney, E. M. W. Skinner, C. R. Clement & N. Ebrahim 1985. Geochemical character of southern Africa kimberlites: a new approach based on isotopic constraints. *Trans Geol Soc. S. Afr.* **88**, 267–80.

Smith, R. B. & L. W. Braile 1984. Crustal structure and evolution of an explosive silicic volcanic system at Yellowstone National Park. In *Studies in geophysics, explosive volcanism: inception, evolution and hazards.* 96–109. New York: National Academy Press.

Smith, R. L. 1979. Ash-flow magmatism. *Geol Soc. Am. Sp. Pap.* **180**, 5–27.

Sorensen, H. 1974. *The alkaline rocks.* Chichester: Wiley, 622 pp.

Sparks, R. S. J. & H. E. Huppert 1984. Density changes during fractional crystallisation of basaltic magmas: fluid dynamic implications. *Contrib. Mineral. Petrol.* **85**, 300–9.

Sparks, R. S. J., H. E. Huppert & J. S. Turner 1984. The fluid dynamics of evolving magma chambers. *Phil Trans R. Soc. Lond.* **A310**, 511–34.

Spence, D. A. & D. L. Turcotte 1985. Magma-driven propagation of cracks. *J. Geophys. Res.* **90**, 575–80.

Spera, F. J. 1980. Aspects of magma transport. In *physics of magmatic processes* R. B. Hargraves (ed.), 265–323. Princeton, NJ: Princeton University Press.

Spera, F. J. 1984. Carbon dioxide in petrogenesis III: Role of volatiles in the ascent of alkaline magma with special reference to xenolith-bearing mafic lavas. *Contrib. Mineral. Petrol.* **88**, 217–32.

Spiess, F. N., K. C. Macdonald, T. Atwater, R. Ballard, D. Carrenzo, D. Cordoba, V. Cox, V. M. Diazgarcia, J. Francheteau, J. Guerrero, J. Hawkins, R. Hayman, R. Hessler, T. Juteau, M. Kastner, R. Larson, B. Luyendyk, I. D. Macdougall, S. Miller, W. Normaric, J. Orcutt & C. Ranger 1980. East Pacific Rise: hot springs and geophysical experiments. *Science* **207**, 1421–33.

Spohn, T. & G. Schubert 1982. Convective thinning of the lithosphere: a mechanism for the initiation of continental rifting. *J. Geophys. Res.* **87**, 4669–81.

Staudigel, H., S. R. Hart & S. H. Richardson 1981. Alteration of the oceanic crust: processes and timing. *Earth Planet. Sci. Lett.* **52**, 311–27.

Staudigel, H., A. Zindler, S. R. Hart, T. Leslie, C. Y. Chen & D. Clague 1984. The isotope systematics of a juvenile intra-plate volcano: Pb, Nd and Sr isotope ratios of basalts from Loihi Seamount, Hawaii. *Earth Planet. Sci. Lett.* **69**, 13–29.

Stern, C. R. 1980. Geochemistry of Chilean ophiolites: evidence for the compositional evolution of the mantle source of back-arc basin basalts. *J. Geophys. Res.* **85**, 955–66.

Stern, R. J. 1982. Strontium isotopes from circum-Pacific intra-oceanic island arcs and marginal basins: regional variations and implications for magma genesis. *Bull. Geol Soc. Am.* **93**, 477–86.

Stille, P., D. M. Unruh & M. Tatsumato 1983. Pb, Sr, Nd and Hf isotopic evidence of multiple sources for Oahu, Hawaiian basalts. *Nature* **304**, 25–9.

Stolper, E. 1980. A phase diagram for mid-ocean ridge basalts: preliminary results and implications for petrogenesis. *Contrib. Mineral. Petrol.* **74**, 13–28.

Stolper, E., D. Walker, B. H. Hager & J. F. Hays 1981. Melt segregation from partially molten source regions: the importance of melt density and source region size. *J. Geophys. Res.* **86**, 6261–71.

Stosch, H. G., R. W. Carlson & G. W. Lugmuir 1980. Episodic mantle differentiation: Nd and Sr isotopic evidence. *Earth Planet. Sci. Lett.* **47**, 263–71.

Streckeisen, A. 1976. To each plutonic rock its proper name. *Earth Sci. Rev.* **12**, 1–33.

Suen, C. J., F. A. Frey & J. Malpas 1979. Bay of Islands ophiolite suite, Newfoundland: petrologic and geochemical characteristics with emphasis on rare earth element geochemistry. *Earth Planet. Sci. Lett.* **45**, 333–48.

Sugimura, A. 1973. Multiple correlation between composition of volcanic rocks and depth of earthquake foci. In *The Western Pacific; island arcs, marginal seas and geochemistry*, P. Coleman (ed.), 471–82. Perth: Western Australia University Press.

Sun, S.-S. 1980. Lead isotopic study of young

volcanic rocks from mid-ocean ridges, ocean islands and island arcs. *Phil Trans R. Soc. Lond.* A297, 409–45.

Sun, S.-S. & R. W. Nesbitt 1977. Chemical heterogeneity of the Archean mantle: composition of the earth and mantle evolution. *Earth Planet. Sci. Lett.* 35, 429–48.

Sun, S.-S., R. W. Nesbitt & A. Ya. Sharaskin 1979. Geochemical characteristics of mid-ocean ridge basalts. *Earth Planet. Sci. Lett.* 44, 119–38.

Swanson, D. A., W. A. Duffield & R. S. Fiske 1976. *Displacement of the south flank of Kilauea Volcano: the result of forceful intrusion of magma into the rift zones.* US Geol Surv. Prof. Paper 963, 39 pp.

Swanson, D. A., T. L. Wright, P. R. Hooper & R. D. Bentley 1979. Revisions in stratigraphic nomenclature of the Columbia River Basalt group. *US Geol Surv. Bull.* 1457–G, G1–G59.

Takahashi, E. & I. Kushiro 1983. Melting of a dry peridotite at high pressures and basalt magma genesis. *Am. Mineral.* 68, 859–79.

Talwani, M., C. G. Harrison & D. E. Hayes (eds) 1979. *Deep drilling results in the Atlantic Ocean crust.* Washington DC: Am. Geophys. Union.

Talwani, M., C. C. Windish & M. G. Langseth Jr. 1971. Reykjanes Ridge crest: a detailed geophysical study. *J. Geophys. Res.* 76, 473–517.

Tarney, J., A. D. Saunders & S. D. Weaver 1977. Geochemistry of volcanic rocks from the island arcs and marginal basins of the Scotia Arc region. In *Island arcs, deep sea trenches and back arc basins*, M. Talwani & W. C. Pitman III (eds), 367–77. Washington DC: Am. Geophys. Union.

Tarney, J., D. A. Wood, A. D. Saunders, J. R. Cann & J. Varet 1980. Nature of mantle heterogeneity in the North Atlantic: evidence from deep sea drilling. *Phil Trans R. Soc. Lond.* 297A, 179–202

Tatsumi, Y. 1981. Melting experiments on a high-magnesian andesite. *Earth Planet. Sci. Lett.* 54, 357–65.

Tatsumoto, M. 1978. Isotopic composition of lead in oceanic basalt and its implication to mantle evolution. *Earth Planet. Sci. Lett.* 38, 63–87.

Tatsumoto, M., R. J. Knight & C. J. Allègre 1973. Time differences in the formation of meteorites as determined by the ratio of lead-207 to lead-206. *Science* 180, 1279.

Taylor, B. & G. D. Karner 1983. On the evolution of marginal basins. *Rev. Geophy.* 21, 1727–41.

Taylor, H. P., B. Turi & A. Cundari 1984. $^{18}O/^{16}O$ and chemical relationships in K-rich rocks from Australia, East Africa, Antarctica and San Venanzo–Cupaello, Italy. *Earth Planet. Sci. Lett.* 69, 263–76.

Taylor, L. A. 1984. Kimberlite magmatism in the Eastern United States: relationships to Mid-Atlantic tectonism. In *Kimberlites I: kimberlites and related rocks*, J. Kornprobst (ed.), 417–24. New York: Elsevier.

Taylor, S. R. & S. M. McLennan 1985. *The continental crust: its composition and evolution.* Oxford: Blackwell Scientific.

Tera, F., L. Brown, J. Morris, I. S. Sacks, J. Klein & R. Middleton 1986. Sediment incorporation in island-arc magmas: inferences from ^{10}Be. *Geochim. Cosmochim. Acta.* 50, 535–50.

Thirlwall, M. F. & A. M. Graham 1984. Evolution of high-Ca, high-Sr C-series basalt from Grenada, Lesser Antilles: The effects of intra-crustal contamination. *J. Geol Soc. Lond.* 141, 427–45.

Thirlwall, M. F. & N. W. Jones 1983. Isotope geochemistry and contamination mechanics of Tertiary lavas from Skye, northwest Scotland. In *Continental basalts and mantle xenoliths*, C. J. Hawkesworth & M. J. Norry (eds), 186–208. Nantwich: Shiva.

Thompson, R. N. 1972. Melting behaviour of two Snake River lavas at pressures up 35 kb. *Carnegie Inst. Wash. Geophys. Lab. Yearbook* 71, 406–10.

Thompson, R. N. 1977. Primary basalts and magma genesis III: Alban Hills, Roman co-magmatic province, Central Italy. *Contrib. Mineral. Petrol.* 50, 91–108.

Thompson, R. N. 1982. Magmatism of the British Tertiary Volcanic Province. *Scott. J. Geol.* 18, 49–107.

Thompson, R. N. 1985. Asthenospheric source of Ugandan ultrapotassic magma. *J. Geol.* 93,

603–8.

Thompson, R. N. & M. B. Fowler 1986. Subduction-related shoshonitic and ultrapotassic magmatism: a study of Siluro-Ordovician syenites from the Scottish Caledonides. *Contrib. Mineral. Petrol.* 94, 507–22.

Thompson, R. N., M. A. Morrison, A. P. Dickin, I. L. Gibson & R. S. Harmon 1986. Two contrasting styles of interaction between basic magmas and continental crust in the British Tertiary Volcanic Province. *J. Geophys. Res.* 91, 5985–97.

Thompson, R. N., M. A. Morrison, A. P. Dickin & G. L. Hendry 1983. Continental flood basalts... arachnids rule OK? In *Continental basalts and mantle xenoliths*, C. J. Hawkesworth & M. J. Norry (eds), 158–85. Nantwich: Shiva.

Thompson, R. N., M. A. Morrison, G. L. Hendry & S. J. Parry 1984. An assessment of the relative roles of a crust and mantle in magma genesis: an elemental approach. *Phil Trans R. Soc. Lond.* A310, 549–90.

Thorpe, R. S. (ed.) 1982. *Andesites: orogenic andesites and related rocks*. Chichester: Wiley, 697 pp.

Thorpe, R. S., P. W. Francis, M. Hamill & M. C. W. Baker 1982. The Andes. In *Andesites: orogenic andesites and related rocks*, R. S. Thorpe (ed.), 187–205. Chichester: Wiley.

Thorpe, R. S., P. W. Francis & R. S. Harmon 1981. Andean andesites and continental growth. *Phil Trans R. Soc. Lond.* A301, 305–20.

Thorpe, R. S., P. W. Francis & L. O'Callaghan 1984. Relative roles of source composition, fractional crystallisation and crustal contamination in the petrogenesis of Andean volcanic rocks. *Phil Trans R. Soc. Lond.* A310, 675–92.

Thy, P. 1983. Spinel minerals in transitional and alkali basaltic glasses from Iceland. *Contrib. Mineral. Petrol.* 83, 141–9.

Toksöz, M. N. (1975). The subduction of the lithosphere. In *Volcanoes and the Earth's interior*, R. W. Decker & B. Decker (eds), 6–16. Scientific American Inc. (1982).

Toksöz, M. N. & P. Bird 1977. Formation and evolution of marginal basins and continental plateaus. In *Island arcs, deep sea trenches and back-arc basins*, M. Talwani & W. C. Pitman III (eds), 379–93. Washington DC: Am. Geophys. Union.

Toksöz, M. N. & A. T. Hsui 1978. Numerical studies on back-arc convection and the formation of marginal basins. *Tectonophysics* 50, 177–96.

Turcotte, D. L. & L. H. Kellogg 1986. Isotope modeling of the evolution of the mantle crust. *Rev. Geophys* 24, 311–28.

Turcotte, D. L. & E. R. Oxburgh 1972. Mantle convection and the new global tectonics. *Ann. Rev. Fluid Mech.* 4, 33–68.

Turcotte, D. L. & E. R. Oxburgh 1978. Intra-plate volcanism. *Phil Trans R. Soc. Lond.* A288, 561–79.

Turi, B. & H. P. Taylor 1976. Oxygen isotope studies of potassic volcanic rocks of the Roman Province, Central Italy. *Contrib. Mineral. Petrol.* 55, 1–31.

Turner, J. S. & I. H. Campbell 1986. Convection and mixing in magma chambers. *Earth Sci. Rev.* 23, 255–352.

Turner, J. S., H. E. Huppert & R. S. J. Sparks 1983. Experimental investigations of volatile exsolution in evolving magma chambers. *J. Volcanol. Geotherm. Res.* 16, 263–77.

Twyman, J. D. & J. Gittens 1987. Alkalic carbonatite magmas: parental or derivative? In *Alkaline igneous rocks*, J. G. Fitton & B. G. J. Upton (eds), 85–94. Geol. Soc. Sp. Publ. 30.

Uyeda, S. (1982). Subduction zones: an introduction to comparative subductology. *Tectonophysics* 81, 133–59.

Van Andel, Tj. H. & R. D. Ballard 1979. The Galapagos Rift at 86°W, 2; volcanism, structure and evolution of the rift valley. *J. Geophys. Res.* 84, 5390–406.

Varne, R. 1985. Ancient subcontinental mantle: a source for K-rich orogenic volcanics. *Geology* 13, 405–8.

Venturelli, G., S. Capedri, G. Di Battistini, A. Crawford., L. N. Kogarko & S. Celestini 1984. The ultrapotassic rocks from southeastern Spain. *Lithos* 17, 37–54.

Vidal, Ph., C. Chauvel & R. Brousse 1984. Large mantle heterogeneity beneath French Polynesia. *Nature* 307, 536−8.

Vine, F. J. & D. H. Matthews 1963. Magnetic anomalies over oceanic ridges. *Nature* 199, 947−9.

Vollmer, R. 1976. Rb−Sr and U−Th−Pb systematics of alkaline rocks: the alkaline rocks from Italy. *Geochim. Cosmochim. Acta.* 40, 283−95.

Vollmer, R. 1977. Isotopic evidence for genetic relations between acidic and alkaline rocks in Italy. *Contrib. Mineral. Petrol.* 60, 109−18.

Vollmer, R., K. Johnston, M. R. Ghiara, L. Lirer & R. Munno 1981. Sr isotope geochemistry of megacrysts from continental rift and converging plate margin alkaline volcanism in south Italy. *J. Volcanol. Geotherm. Res.* 11, 317−27.

Vollmer, R. & M. J. Norry 1983a. Possible origin of K-rich volcanic rocks from Virunga, East Africa, by metasomatism of continental crustal material: Pb, Nd and Sr isotopic evidence. *Earth Planet. Sci. Lett.* 64, 374−86.

Vollmer, R. & M. J. Norry 1983b. Unusual isotopic variations in Nyiragongo nephelinites. *Nature* 301, 141−3.

Vollmer, R., P. Ogden, J.-G. Schilling, R. H. Kingsley & D. G. Waggoner 1984. Nd and Sr isotopes in ultrapotassic volcanic rocks from the Leucite Hills, Wyoming. *Contrib. Mineral. Petrol.* 87, 359−68.

Wager, L. R. & G. M. Brown 1968. *Layered igneous rocks*. Edinburgh: Oliver and Boyd, 588 pp.

Walker, D., T. Shibata & S. E. Delong 1979. Abyssal tholeiites from the Oceanographer Fracture Zone II. Phase equilibria and mixing. *Contrib. Mineral. Petrol.* 70, 111−25.

Wass, S. Y. 1980. Geochemistry and origin of xenolith-bearing and related alkali basaltic rocks from the southern Highlands, New South Wales, Australia. *Am. J. Sci.* 280A, 639−66.

Wass, S. Y. & N. W. Rogers 1980. Mantle metasomatism: precursor to continental alkaline volcanism. *Geochim. Cosmochim. Acta.* 44, 1811−23.

Wasserburg, G. J., S. B. Jacobsen, D. J. DePaolo, M. T. McCulloch & T. Wen 1981. Precise determination of Sm and Nd isotopic abundances in standard solutions. *Geochim. Cosmochim. Acta.* 45, 2311−23.

Watson, E. B. 1982. Basalt contamination by continental crust: some experiments and results. *Contrib. Mineral. Petrol.* 80, 73−87.

Watts, A. B., U. S. ten Brink, P. Buhl & T. M. Brocher 1985. A multichannel seismic study of lithospheric flexure across the Hawaiian−Emperor seamount chain. *Nature* 315, 105−11.

Weaver, B. L. & J. Tarney 1981. Lewisian geochemistry and Archean crustal development models. *Earth Planet. Sci. Lett.* 55, 171−80.

Weaver, S. D., J. S. C. Sceal & I. L. Gibson 1972. Trace element data relevant to the origin of trachytic and pantelleritic lavas in the East African Rift system. *Contrib. Mineral. Petrol.* 36, 181−94.

Weis, D., D. Demaiffe, S. Cauet & M. Javoy 1987. Sr, Nd, O and H isotopic ratios in Ascension Island lavas and plutonic inclusions; cogenetic origin. *Earth Planet. Sci. Lett.* 82, 255−68.

Wendlandt, R. F. 1984. An experimental and theoretical analysis of partial melting in the system $KAlSiO_4−CaO−MgO−SiO_2−CO_2$ and applications to the genesis of potassic magmas, carbonatites and kimberlites. In *Kimberlites I: kimberlites and related rocks*, J. Kornprobst (ed.), 359−70. New York: Elsevier.

Wendlandt, R. F. & D. H. Eggler 1980a. The origin of potassic magmas 1. Melting relations in the system $KAlSiO_4−Mg_2SiO_4−SiO_2$ and $KAlSiO_4−MgO−SiO_2−CO_2$ to 30 kilobars. *Am. J. Sci.* 280, 385−420.

Wendlandt, R. F. & D. H. Eggler 1980b. The origin of potassic magmas 2. Stability of phlogopite in natural spinel lherzolite and in the system $KAlSiO_4−MgO−SiO_2−H_2O−CO_2$ at high pressures and high temperatures. *Am. J. Sci.* 280, 421−58.

Wendlandt, R. F. & B. O. Mysen 1980. Melting phase relations of natural peridotite + CO_2 as a function of melting at 15 and 30 kbar. *Am. Mineral.* 65, 37−44.

Wheller, G. E., R. Varne, J. D. Foden & M. J. Abbott 1987. Geochemistry of Quaternary volcanism in the Sunda—Banda arc, Indonesia, and the three-component genesis of island-arc basaltic magmas. *J. Volcanol. Geotherm. Res.* 32, 137—60.

White, R. S. 1984. Atlantic oceanic crust: seismic structure of a slow spreading ridge. In *Ophiolites and oceanic lithosophere*, I. G. Gass, S. J. Lippard & A. W. Shelton (eds), 101—11. Oxford: Blackwell Scientific.

White, W. M. 1985. Sources of oceanic basalts: radiogenic isotope evidence. *Geology* 13, 115—18.

White, W. M. & B. Dupré 1986. Sediment subduction and magma genesis in the Lesser Antilles: isotopic and trace element constraints. *J. Geophys. Res.* 91, 5927—41.

White, W. M., B. Dupré & P. Vidal 1985. Isotope and trace element geochemistry of sediments from the Barbados Ridge — Demerara Plain region, Atlantic Ocean. *Geochim. Cosmochim. Acta.* 49, 1857—86.

White, W. M. & A. W. Hofmann 1982. Sr and Nd isotope geochemistry of mantle evolution,. *Nature* 296, 821—5.

White, W. M., A. W. Hofmann & H. Puchelt 1987. Isotope geochemistry of Pacific mid-ocean ridge basalt. *J. Geophys. Res.* 92, 4881—93.

White, W. M. & J. Patchett 1984. Hf—Nd—Sr isotopes and incompatible element abundances in island arcs: implications for magma origins and crust—mantle evolution. *Earth Planet. Sci. Lett.* 67, 167—85.

White, W. M. & J.-G. Schilling 1978. The nature and origin of geochemical variation in Mid-Atlantic Ridge basalts from the central north Atlantic. *Geochim. Cosmochim. Acta.* 42, 1501—16.

White, W. M., M. D. M. Tapia & J.-G. Schilling 1979. The petrology and geochemistry of the Azores islands. *Contrib. Mineral. Petrol.* 69, 201—13.

Whitford, D. J. & P. A. J zek 1982. Isotopic constraints on the role of subducted sialic material in Indonesian island-arc magmatism. *Bull. Geol Soc. Am.* 93, 504—13.

Wilkinson, J. F. G. 1982. The genesis of Mid-Ocean Ridge Basalt. *Earth Sci. Rev* 18, 1—57.

Wilkinson J. F. G. (1986). Classification and average chemical compositions of common basalts and andesites. *J. Petrol.* 27, 31—62.

Wilkinson, J. F. G. & R. A. Binns 1977. Relatively iron-rich lherzolite xenoliths of the Cr-diopside suite: a guide to the primary nature of anorogenic tholeiitic andesite magmas. *Contrib. Mineral. Petrol.* 65, 199—212.

Wilkinson, J. F. G. & R. W. Le Maitre 1987. Upper mantle amphiboles and micas and TiO_2, K_2O and P_2O_5 abundances and $100Mg/(Mg + Fe^{2+})$ ratios of common basalts and andesites: implications for modal mantle metasomatism and undepleted mantle compositions. *J. Petrol.* 28, 37—73.

Williams, L. A. J. 1982. Physical aspects of magmatism in continental rifts. In *Continental and oceanic rifts*, G. Palmason (ed.), 193—222. Washington DC: Am. Geophys. Union.

Williams, S. & V. R. Murthy 1979. Sources and genetic relationships of volcanic rocks from the northern Rio Grande Rift: Rb—Sr and Sm—Nd evidence. *EOS, Trans Am. Geophys. Union* 60, 407.

Wills, J. K. A. 1974. *The geological history of southern Dominica and plutonic nodules from the Lesser Antilles.* Unpubl. PhD thesis, Univ. Durham, UK.

Wilson, J. T. 1963. A possible origin of the Hawaiian islands. *Can. J. Phys.* 41, 863—70.

Wilson, J. T. 1973. Mantle plumes and plate motion. *Tectonophysics* 19, 149—64.

Wilson, M. & J. P. Davidson 1984. The relative roles of crust and upper mantle in the generation of oceanic island arc magmas. *Phil Trans R. Soc. lond.* A310, 661—74.

Wood, B. J. & S. Banno 1973. Garnet—orthopyroxene and orthopyroxene—clinopyroxene relationships in simple and complex systems. *Contrib. Mineral. Petrol.* 42, 109—24.

Wood, B. J. & D. G. Fraser 1976. *Elementary thermodynamics for geologists.* Oxford: Oxford University Press.

Wood, D. A., J. L. Joron & M. Treuil 1979a. A re-appraisal of the use of trace elements to classify and discriminate between magma series erupted in different tectonic settings. *Earth Planet. Sci. Lett.* 45, 326–36.

Wood, D. A., J. L. Joron, M. Treuil, M. Norry & J. Tarney 1979b. Elemental and Sr isotope variations in basic lavas from Iceland and the surrounding sea floor. *Contrib. Mineral. Petrol.* 70, 319–39.

Wood, D. A., J. Tarney, A. D. Saunders, H. Bougault, J. L. Joron, M. Treuil & J. R. Cann 1979. Geochemistry of basalts drilled in the north Atlantic by IPOD Leg 49:implications for mantle heterogeneity. *Earth Planet Sci. Lett.* 42, 77–97.

Woodhead, J. D. & D. E. Fraser 1985. Pb, Sr and [10]Be isotopic studies of volcanic rocks from the Northern Mariana Islands: implications for magma genesis and crustal recycling in the Western Pacific. *Geochim. Cosmochim. Acta.* 49, 1925–30.

Woolley, A. R. 1982. A discussion of carbonatite evolution and nomenclature, and the generation of sodic and potassic fenites. *Mineral. Mag.* 46, 13–17.

Worner, G., A. Zindler, H. Staudigel & H. U. Schmincke 1986. Sr, Nd and Pb isotope geochemistry of Tertiary and Quatenary alkaline volcanics from West Germany. *Earth Planet. Sci. Lett.* 79, 107–19.

Wörner, G., R. S. Harmon & J. Hoefs 1987. Stable isotope relations in an open magma system, Laacher See, Eifel (FRG). *Contrib. Mineral. Petrol.* 95, 343–9.

Wortel, M. J. R. 1984. Spatial and temporal variations in the Andean subduction zone. *J. Geol Soc. Lond.* 141, 783–91.

Wright, E. & White, W. M. 1986. The origin of Samoa: new evidence from Sr, Nd and Pb isotopes. *Earth Planet. Sci. Lett.* 81, 151–62.

Wright, T. L. 1971. *Chemistry of Kilauea and Mauna Loa lava in space and time.* U.S. Geol Surv. Prof. Pap. 735, 40 pp.

Wright, T. L. & P. C. Doherty 1970. A linear programming and least squares computer method for solving petrological mixing problems. *Bull. Geol Soc. Am.* 81, 1995–2008.

Wright, T. L., D. A. Swanson & W. A. Duffield 1975. Chemical composition of Kilauea East-Rift lava 1968–1971. *J. Petrol.* 16, 110–33.

Wyllie, P. J. 1975. The Earth's mantle. In *Volcanoes and the Earth's Interior* R. Decker & B. Decker (eds), 86–97. Scientific American Inc. (1982).

Wyllie, P. J. 1979. Magmas and volatile components. *Am. Mineral.* 64, 469–500.

Wyllie, P. J. 1980. The origin of kimberlites. *J. Geophys. Res.* 85, 6902–10.

Wyllie, P. J. 1981. Plate tectonics and magma genesis. *Geol. Rundsch.* 70, 128–53.

Wyllie, P. J. 1982. Subduction products according to experimental prediction. *Bull. Geol Soc. Am.* 93, 468–76.

Wyllie, P. J. 1984. Constraints imposed by experimental petrology on possible and impossible magma sources and products. *Phil Trans R. Soc. Lond.* A310. 439–56.

Wyllie, P. J. 1987. Discussion of recent papers on carbonated peridotite, bearing on mantle metasomatism. *Earth Planet. Sci. Lett.* 82, 391–7.

Wyllie, P. J. & W. L. Huang 1976. Carbonation and melting reactions in the system $CaO-MgO-SiO_2-CO_2$ at mantle pressures with geophysical and petrological applications. *Contrib. Mineral. Petrol.* 54, 79–107.

Wyllie, P. J. & T. Sekine 1982. The formation of mantle phlogopite in subduction zone hybridisation. *Contrib. Mineral. Petrol.* 79, 375–80.

Yamada, H. & E. Takahashi 1984. Subsolidus phase relations between coexisting garnet and two pyroxenes at 50 to 100 kbar in the system $CaO-MgO-Al_2O_3-SiO_2$. *Proc. Third Int. Kimb. Conf.* 2, 247–55.

Yoder, H. S. Jr 1976. *Generation of basaltic magma.* Washington DC: National Academy of Sciences.

Yuen, D. A. & L. Fleitout 1985. Thinning of the lithosphere by small-scale convective destabilization. *Nature* 313, 125–8.

Zartman, R. E. & B. R. Doe 1981. Plumbotectonics – the model. *Tectonophysics* 75, 135–62.

Zindler, A. & S. Hart 1986. Chemical geodynamics. *Ann. Rev. Earth Planet. Sci.* 14, 493–571.

Zindler, A., E. Jagoutz & S. Goldstein 1982. Nd, Sr and Pb isotopic systematics in a three component mantle: a new perspective. *Nature* **298**, 519–23.

Zindler, A., H. Staudigel & R. Batiza 1984. Isotope and trace element geochemistry of young Pacific seamounts: implications for the scale of upper mantle heterogeneity. *Earth Planet. Sci. Lett.* **70**, 175–95.

Zindler, A. & S. Hart 1986. Chemical geodynamics. *Ann. Rev. Earth Planet. Sci.* **14**, 493–571.

Zorin, Yu, A. 1981. The Baikal Rift: an example of the intrusion of asthenospheric material into the lithosphere as a cause of disruption of lithospheric plates. *Tectonophysics* **73**, 91–104.

INDEX